Oil and Gas: Processing and Production Technologies

Oil and Gas: Processing and Production Technologies

Editor: Oliver Haghi

www.callistoreference.com

Callisto Reference,
118-35 Queens Blvd., Suite 400,
Forest Hills, NY 11375, USA

Visit us on the World Wide Web at:
www.callistoreference.com

ISBN: 978-1-63239-864-2 (Hardback)

The publisher's policy is to use permanent paper from mills that operate a sustainable forestry policy. Furthermore, the publisher ensures that the text paper and cover boards used have met acceptable environmental accreditation standards.

Printed in China.

Cataloging-in-publication Data

Oil and gas : processing and production technologies / edited by Oliver Haghi.
p. cm.
Includes bibliographical references and index.
ISBN 978-1-63239-864-2
1. Petroleum--Prospecting. 2. Gas engineering. 3. Petroleum engineering. 4. Petroleum industry and trade--Technological innovations. I. Haghi, Oliver.

TN271.P4 O35 2017
622.1828--dc23

Table of Contents

Preface

This book elucidates the concepts and innovative models around prospective developments with respect to oil and gas processing. It presents researches and studies performed by experts across the globe. Oil and gas processing and production falls under the field of petroleum geology. The main aim of this field is to remove, separate and transform hydrocarbons from the mixed hydrocarbons which contain various components like nitrogen, possibly hydrogen sulfide (H2S), carbon dioxide (CO2), dirt, scale, etc. The different sub-fields of this subject along with technological progress that have future implications are glanced at in this text. From theories to research to practical applications, case studies related to all contemporary topics of relevance to this field have been included in it. Researchers and students in this area will be assisted by this text. A number of latest researches have been included to keep the readers up-to-date with the global concepts in this area of study.

The information shared in this book is based on empirical researches made by veterans in this field of study. The elaborative information provided in this book will help the readers further their scope of knowledge leading to advancements in this field.

Finally, I would like to thank my fellow researchers who gave constructive feedback and my family members who supported me at every step of my research.

Editor

1

Towards the Development of Bitumen Carbonates: An Integrated Analysis of Grosmont Steam Pilots

C.C. Ezeuko*, J. Wang, M.S. Kallos and I.D. Gates

Department of Chemical and Petroleum Engineering, Schulich School of Engineering, University of Calgary - Canada
e-mail: cezeuko@ucalgary.ca - jwang@ucalgary.ca - mskallos@ucalgary.ca - idgates@ucalgary.ca

* Corresponding author

Résumé — Vers le développement des carbonates bitumineux : une analyse intégrée des pilotes vapeur de Grosmont — La Formation de Grosmont de l'Alberta, au Canada, est un réservoir de carbonates riches en bitume, hautement fracturé, karstifié et vacuolaire, situé à l'ouest et en-dessous du gisement de sables bitumineux de l'Athabasca. La plate-forme de carbonates bitumineux s'étend sur environ 500 km en longueur et jusqu'à 150 km en largeur et contient un volume de pétrole estimé à 64,5 milliards de m^3 (406,5 milliards de barils). La Formation de Grosmont est plus vaste que le total cumulé de tous les autres dépôts de carbonates bitumineux connus au monde. Dans le cas présent, nous analysons les premiers pilotes vapeur de Grosmont afin d'améliorer la conception des futurs pilotes ainsi que le développement commercial de cet imposant dépôt de bitume. En accord avec les conclusions des analyses précédentes de ces pilotes de Grosmont, ceux-ci étaient raisonnablement satisfaisants compte tenu de la nature hétérogène de la Formation de Grosmont. Des facteurs opérationnels tels qu'une qualité médiocre de la vapeur, des pressions d'injection élevées non optimisées et des problèmes de réalisation semblent avoir lourdement affecté les performances de récupération. À l'évidence, les opérations de récupération à base de vapeur détiennent un potentiel intéressant pour Grosmont, en particulier étant donné qu'il s'agit d'une technologie commerciale maîtrisée. Suivant une analyse intégrée des premiers pilotes de Grosmont, nous pouvons affirmer que la stimulation cyclique à la vapeur (*Cyclic Steam Stimulation*, CSS) au moyen de puits horizontaux présente un potentiel supérieur pour le développement des carbonates de Grosmont par rapport à la technologie de drainage gravitaire assisté par injection de vapeur (*Steam-Assisted Gravity Drainage*, SAGD).

Abstract* — *Towards the Development of Bitumen Carbonates: An Integrated Analysis of Grosmont Steam Pilots — *The Grosmont Formation in Alberta, Canada is a highly fractured, karstified and vuggy bitumen-rich carbonate reservoir located west of and below the Athabasca oil sands deposit. The bitumen carbonate platform extends about 500 km in length and up to 150 km in width and contains an estimated 64.5 billion m^3 (406.5 billion barrels) of oil. The Grosmont Formation is larger than the combined total of all other known carbonate bitumen deposits in the world. Here, we analyze early Grosmont steam pilots to improve the design of future pilots and commercial development of this massive bitumen deposit. In agreement with the conclusions of earlier analysis of these Grosmont pilots, they were reasonably successful considering the heterogeneous nature of the Grosmont Formation. Operational factors such as poor steam quality, non-optimized high injection pressures and completion issues appear to have heavily impacted recovery performances. Clearly, steam-based recovery operations have good potential for Grosmont, especially considering that it* *is mature commercial technology. Following an integrated analysis of early Grosmont pilots, we posit that Cyclic Steam Stimulation (CSS) using horizontal wells exhibits greater potential for the development of Grosmont carbonate, compared with Steam-Assisted Gravity Drainage technology (SAGD).*

INTRODUCTION

The Grosmont Formation is a highly fractured, karstified and vuggy bitumen-rich carbonate located West of the Athabasca oil sands deposit in Alberta, Canada. The bitumen carbonate platform in Northern Alberta is a massive resource which extends about 500 km in length and up to 150 km in width and is often referred to as the carbonate triangle. With an original bitumen in-place estimated to be as high as 64.5 billion m^3 (406.5 billion barrels) (ERCB, 2011), the Grosmont Formation is larger than the combined total of all other known carbonate bitumen deposits in the world and represents an immense economic potential for the province of Alberta and Canada in general.

Aggressive development of Grosmont is expected within the next decade, championed by *Laricina Energy Ltd.* However, two other energy giants, *Royal Dutch Shell plc* and *Husky Energy*, are also primed to develop their large mineral rights within Grosmont. Primary conventional oil recovery is driven by natural (*in situ*) energy, typically a combination of oil expansion, gas ex-solution and expansion, formation compaction and aquifer expansion. Unfortunately, heavy oil and bitumen have minimal dissolved gas and large viscosity contrast with formation water resulting in typically poor or essentially zero primary recovery. An example is the primary recovery of less than 1% of the original oil-in-place reported for the Ikiztepe heavy oil reservoir in Turkey (Nakamura *et al.*, 1995). A major additional challenge associated with bitumen recovery from carbonates is the often extreme complexity of carbonate geology. A common tendency is for heavy oil carbonate reservoirs to be oil-wetted, making traditional secondary recovery such as water injection equally inefficient due to excessive channeling and early breakthrough. Grosmont bitumen fluid characterization indicates that viscosity in reservoir conditions is in excess of 1 million cP (°API 5-9) and a reasonable lateral and vertical viscosity heterogeneity/variation exists (Zhao and Machel, 2012). Steam-based thermal processes which have successfully been used to exploit Athabasca bitumen offer a viable option for decreasing *in situ* viscosity, thereby mobilizing Grosmont bitumen.

Steam processes that have been field-tested in carbonates include Thermally-Assisted Gas-Oil Gravity Drainage (TA-GOGD) (Macaulay *et al.*, 1995); Cyclic Steam Stimulation (CSS) (Jiang *et al.*, 2010; Samir, 2010; Yuan *et al.*, 2010; Novak *et al.*, 2007; Zhou *et al.*, 1998; Ahmad and Milhem, 1989, Milhem and Ahmad, 1987) and steamflooding (Courderc *et al.*, 1990; Sahuquet *et al.*, 1990; Sahuquet and Ferrier, 1982). Also, heavy oil carbonate pilots involving Steam-Assisted Gravity Drainage (SAGD) and solvent variants of SAGD are underway at Grosmont. For a steam-based process to be successful in heavy oil carbonate, it must increase the *in situ* oil mobility while simultaneously providing or exploring sufficient drive energy for moving mobile oil to production wells. A number of processes have been identified as responsible (to varying degrees) for several successful applications of steam-based recovery from heavy oil carbonates. These include; free and forced imbibition (enhanced by thermal wettability reversal from oil to water-wet) of steam condensate into the matrix and attendant displacement of oil from the matrix into fracture networks, viscosity reduction of heavy oil both due to the thermal effect and dissolution of CO_2 into oil, gravity drainage of oil, an internal depletion drive associated with steam flashing, thermal oil expansion and forced oil displacement (Mollaei and Maini, 2010; Babadagli and Al-Bemani, 2007; Briggs *et al.*, 1992; Hjelmeland and Larrondo, 1986; Dreher *et al.*, 1986).

In view of the complex geology of Grosmont, a key challenge is to maximize the transfer of injected heat to the bitumen by minimizing *in situ* heat losses associated with an insufficiently characterized network of vugs, fractures and caverns. Even when the bitumen viscosity has been reduced to a reasonable value (< 200 cP), drive energy is critical to deliver heated bitumen to production wells. Regardless of the steam-based method used in previous carbonate pilots TA-GOGD (Yates field, Qarn Alarn) (Snell and Close, 1999; Macaulay *et al.*, 1995), CSS (Grosmont, Issaran, Ikiztepe, Oude-Shiranish, Wafra) (Jiang *et al.*, 2010; Samir, 2010; Yuan *et al.*, 2010; Novak *et al.*, 2007; Zhou *et al.*, 1998; Ahmad and Milhem, 1989, Milhem and Ahmad, 1987), steamflooding (Emeraude, Grosmont, Lacq Supérieur, Wafra, Ikiztepe) (Brown *et al.*, 2011; Courderc *et al.*, 1990; Sahuquet *et al.*, 1990; Sahuquet and Ferrier, 1982; Union Oil Company of Canada Limited, 1975), SAGD and SC-SAGD (Grosmont) – each process exploits one or two predominant drive mechanisms (may vary with stages of application) to move heated oil to production wells. In TA-GOGD, the expanding gas cap and gravity drainage are the key mechanisms for moving oil to production wells. Early stages of CSS are usually dominated by formation re-compaction and the solution-gas drive, whereas later cycles are dominated by gravity drainage. The primary driving force in SAGD and SC-SAGD is gravity drainage of heated oil into a production well usually placed a few meters below an injection well. Steamflooding, however, exploits the external pressure-drive for this purpose. The applicability and efficiency of each method depends on reservoir configuration, properties and initial conditions. As an example, although heat transmission into the reservoir during CSS in oil sands is facilitated by steam fracturing in early cycles, the implications of fractures, vugs and caverns for a similar process in the uniquely complex Grosmont geology is less obvious. Nevertheless, high injectivity recorded during early Grosmont pilots is ascribed to the network of fractures. A critical analysis of Grosmont pilot data will reveal where a window of opportunity exists for exploiting certain recovery technologies which may have been absent during these early pilots.

Somewhat traditional pessimism associated with Enhanced Oil Recovery (EOR) from carbonates (typically fluid injection), the low market price of bitumen and the mixed results from earlier pilots have hampered commercial development of Grosmont. To date, no commercial projects have been operated in Grosmont and as a result, reserve estimates are yet to be made by the regulatory authority (Energy Resources Conservation Board, ERCB) in Alberta. Although a price differential still exists for heavy crude compared with lighter crude, a high 5-year average value of about US$70 per barrel for the Canadian benchmark heavy crude (West Canadian Select) makes prospects for Grosmont commercialization attractive. In addition, an excellent opportunity lies in an in-depth analysis of previous Grosmont pilots conducted by Alberta Oil Sands Technology Research Authority (AOSTRA), Union Oil Company of Canada Limited (UNOCAL) and Canadian Superior, in the light of technological advancement in bitumen recovery. This paper presents a condensed analysis based on a comprehensive integration of early pilots (between 1975 and 1987) which serves to ascertain to what degree factors such as the reservoir, fluid properties and operational variables affected recovery performances in comparable bitumen carbonates. Such insight is then integrated into currently available technology to inform viable ideas for commercial development of the Grosmont bitumen carbonate. Bitumen production from a SAGD pilot which became operational in 2010 offers additional opportunities for future evaluation of bitumen carbonate development and in particular, Grosmont, based on field experiences.

1 GEOLOGY AND DIAGENESIS

The Grosmont is an Upper Devonian shallow marine to peritidal platform with coarsening-up carbonate sequences. Efforts to characterize carbonate-bitumen deposits in Alberta have been ongoing since the 1980s (Harrison and McIntyre, 1981; Theriault and Hutcheon, 1987; Theriault, 1988). A SW-NE stratigraphic section highlighting Grosmont Formation in relation to other key successions in northern Alberta is shown in Figure 1. The Grosmont carbonate strata dip gently ($\approx$ 0.3 to 1 degree) westward and are underlain by the shaley limestones and shales of the Lower Ireton Formation (Walker, 1986; Luo and Machel, 1994). It is conformably overlain by the Upper Ireton Formation and the Nisku Formation, while the McMurray Formation overlays it along the sub-Cretaceous unconformity. The Grosmont Formation is divided into four units; A, B, C and D (formerly described as LG, UG1, UG2 and UG3, respectively), where A is the lowermost unit and D is the uppermost. Each of the four units is separated from the other by marl (*i.e.* a calcium carbonate which contains variable amounts of clays and aragonite) layers. The C-D marl (*i.e.* marl separating the C and D units) occurs at 346 m (Edmunds *et al.*, 2009). The most prolific hydrocarbon reservoirs (based on porosity and thickness) are the C and D units, which contain 80% of total Grosmont bitumen in-place. The geologic sequences represent a series of diagenetic events on sediment deposits, culminating in dolomitization, uplift, erosion and karstification to produce a highly heterogeneous reservoir containing

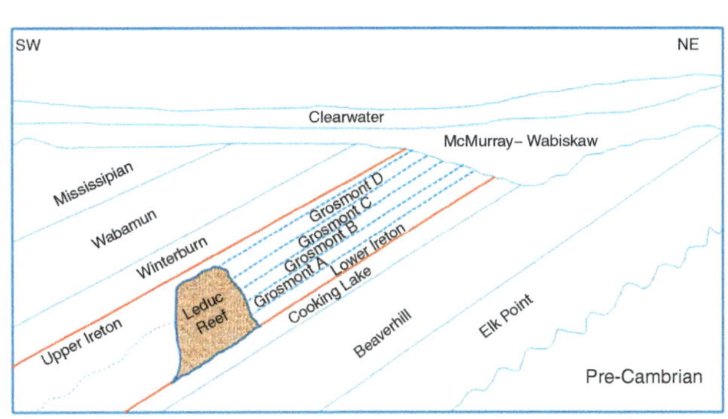

Figure 1

A SW-NE stratigraphic section highlighting Grosmont Formation.

heavily degraded oil (bitumen). In summary, the Grosmont carbonate reservoirs are highly heterogeneous.

Dolomitization and karstification processes which occurred during the diagenesis of Grosmont formation resulted in a near complete dissolution of calcite within the C and D units, leaving nearly pure dolomite. Some of the dolomites have also been dissolved. Breccia (rubble) zones also formed following continued dissolution of rock by meteoric water and eventual collapse as the porosity of the zones reached 40% (Barrett *et al.*, 2008). It is also interesting to note that features such as the breccia and mega-porosity zones are well correlated for distances of the order of 100 km. The Grosmont also has very dense concentration of small (10-cm scale) subvertical fractures. These fractures are thought to have originated from stresses induced by local dissolution and/or differential collapse of lower zones. Vugs appear to be connected by these fractures and dissolution often led to further widening of the fractures. Several fractures are commonly observable in a single cross-section of a core. The average porosity of the C and D units is about 20%, while a porosity range of 5 to 40% has been reported for individual units. Tomography studies suggest that more than half of the total porosity is greater than 0.5 mm (pore diameter or fracture width) and about 100 mD matrix permeabilities are typical in Grosmont.

The diagenesis of the Grosmont Formation has been presented in the literature and involves a significant number of events in a paragenetic sequence (Luo and Machel, 1995; Buschkuehle *et al.*, 2007; Machel and Hawlader, 1990). The paragenetic sequence that describes Grosmont's diagenetic processes can be presented in a chronological order of the first appearance of an event. However, several events occurred more than once and it is therefore significantly more insightful to classify the paragenetic sequence based on the burial history of the Grosmont area. We therefore adopt the approach presented in Buschkuehle *et al.* (2007) and describe the main diagenetic features at Grosmont, synchronous with the burial history of the area, within a group of three (3) time periods, as illustrated in Figure 2.

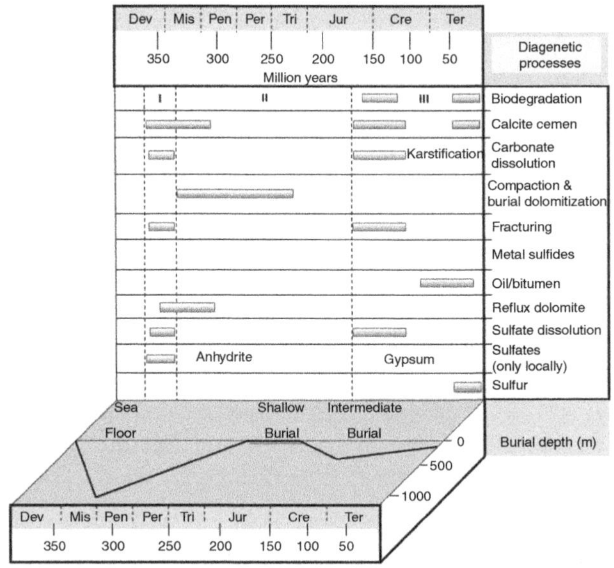

Figure 2

Diagenetic sequence in Grosmont platform categorized into three main stages: (I) early stage, late Devonian (Dev) to early Mississippian (Mis); (II) middle stage, early Mississippian (Mis) to late Jurassic (Jur); (III) late stage, late Jurassic to Tertiary (Ter). Other abbreviations include: Pen (Pennsylvanian), Per (Permian), Tri (Triassic), Cre (Cretaceous). Revised from Buschkuehle *et al.* (2007).

1.1 Late Devonian to Early Mississippian

Shallow water marine deposits were formed followed by calcite cementation and an early dolomitization episode. An early fracture event related to desiccation and dewatering also occurred. These were then followed by sediment compaction resulting in an increase in burial depth to about 700 m. Dolomitization, sulfate dissolution and local precipitation of anhydrite also occurred simultaneously. Dissolution of evaporite minerals produced moldic and vuggy porosity (Luo and Machel, 1994; Buschkuehle *et al.*, 2007).

1.2 Early Mississippian to Late Jurassic

This stage of diagenesis was characterized by continuing precipitation, compaction, dolomitization and dissolution of anhydrite. Uplift of the sediment to intermediate depths also occurred during this period (Buschkuehle *et al.*, 2007).

1.3 Late Jurassic to Tertiary

Several crucial events occurred during this stage of diagenesis including uplifting, tilting, erosion, karstification, oil migration and biodegradation. Uplifting continued during this stage, resulting in sediments being uplifted to shallow burial depth. This eventually led to erosion and karstification. Karstification played a dominant role in the evolution of reservoir properties and this is indicated by the peak amount of secondary/tertiary dissolution porosity and oil saturation near the sub-Cretaceous unconformity (Luo *et al.*, 1993). While karstification contributed significantly to producing high

TABLE 1
Average reservoir properties of the Grosmont (ERCB, 2011)

Grosmont zones	Average pay thickness (m)	Water saturation (%)	Porosity (%)	Initial bitumen in-place (10^6 m^3)
D (UG3)	21.0	19	23	32 860
C (UG2)	13.6	22	17	18 755
B (UG1)	4.9	24	15	4 450
A (LG)	6.5	28	14	8 472

porosity, it compromises seal integrity, which is vital for selecting *in situ* recovery schemes. Oil migration then occurred, followed by degradation and enrichment with sulfur. Degradation of oil to bitumen was mainly by biodegradation (Brooks *et al.*, 1988). Diagenesis influenced fundamental Grosmont reservoir properties, a summary of which is listed in Table 1.

2 RESEARCH SCOPE AND METHODS

2.1 Research Scope

In the late 1970s and early 1980s, thermal EOR pilots were conducted by AOSTRA, UNOCAL and Canadian Superior in the Grosmont Formation, mostly in Grosmont Zone C (formally named UG2), to evaluate the productivity of CSS, steam drive or steamflooding and *In Situ* Combustion (ISC). Vertical wells were used for all the pilots performed during this period and documented in various reports (Union Oil Company of Canada Limited, 1975, 1977, 1980, 1982-83; Chevron Canada Resources Limited, 1983; Ziegler and Bay, 1984; Ziegler, 1985). Due to poor results from the steamflood pilots, most of the pilots were tested on CSS. Results from these Grosmont pilots have generally been described as 'mixed'. Some of these CSS pilots gave very encouraging bitumen production rates and cumulative Steam-Oil Ratio (cSOR) – comparable with CSS in Clearwater sandstone reservoirs – while very poor bitumen production and cSOR were recorded in some other Grosmont CSS pilots. The first aim of this research is to understand the performances of early Grosmont pilots and the finer details of how reservoir, operational and available technology influenced performances, Figure 3. A second goal is to explore how current commercial technology can be adapted and recommend EOR technology and modifications with more potential to facilitate commercial development of the Grosmont carbonate reservoirs. Bearing in mind that these early pilots mostly utilized steam and the mature commercial

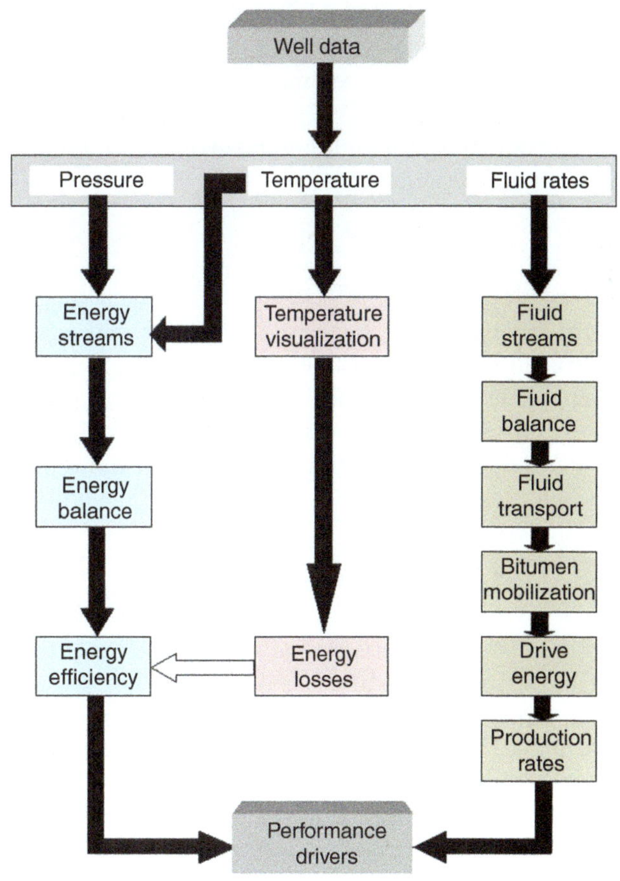

Figure 3

Flowchart describing process workflow applied for pilot history analysis.

development of steam technology for heavy oil/bitumen recovery, we focus our discussion on steam-based EOR.

To achieve the defined research goals, a three-part work flow was designed that describes the research scope and consists of:

- fluid analysis (including the impact of the aquifer),
- energy flow analysis,
- spatial temperature time-lapse analysis.

The implications of each phase of the three-part work flow are systematically evaluated individually and in an integrative manner. Indices for pilot performance include the bitumen production rate, steam-oil ratio and degree of steam conformance.

2.2 Methods

About 203 digital datafiles were analyzed during this research. A consistency quality check was performed to reconcile information in different reports with raw data. Details of the scheme employed for data analysis and energy balance calculations are summarized below:

1. the steam injection volume Q_w for each cycle, Cold Water Equivalent (CWE), was calculated from daily injection raw data (q_w) cumulated over injection periods (t) as described by:

$$Q_w^{cycle} = \sum_{t_{start}}^{t_{end}} q_w \times t \tag{1}$$

2. the initial reservoir temperature for each well was calculated from the arithmetic average of all thermocouple temperature readings for the well;
3. the injected energy for each cycle was calculated by summing daily energy injection for the duration of each cycle. Equation (2) describes the calculation of injected energy:

$$\dot{Q} = \sum \dot{m}[c_l \Delta T + f(h_v - h_l)] \tag{2}$$

   where $\dot{m}$ is the mass rate of steam injection, f is the injected steam quality, c_l is the specific heat capacity of liquid, ΔT is the temperature difference, h_v is the specific vapor enthalpy and h_l is the specific liquid enthalpy;
4. the net injected energy for each cycle was calculated as the difference between injected energy and sensible heat of water at the initial reservoir temperature – specific heat capacity of liquid water was assumed to be 4.1805 kJ/kgK;
5. the sensible heat of produced bitumen in each well was calculated from the average production temperature, the mass of bitumen produced in the cycle (bitumen density approximated to 1 037 kg/m^3) and the specific heat of bitumen (derived from correlation) at the production temperature (Gates, 2011). A similar approach was used to calculate the sensible heat of produced water, assuming water density of 1 000 kg/m^3;
6. to obtain the net produced energy in water (for each cycle), the sensible heat of produced water was calculated by using the heat capacity of water at the initial reservoir temperature and subtracted from injected energy. This was also calculated for the bitumen phase;
7. the net energy balance was calculated as the difference between net injected energy (steam) and produced energy (water + bitumen). This represents the amount of energy retained in the reservoir and/or lost to the surroundings. Visualization of 3D temperature profiles revealed the retention or loss of heat energy beyond the conformance zone;
8. the fraction of energy produced was calculated as the ratio of net produced energy to the net injected energy. Similarly, the fraction of energy lost was calculated as the ratio of net lost energy (injected energy – net produced energy) to the injected energy;
9. it is important to state that the energy balance calculations presented here did not incorporate the sensible heat of produced sediment. The implication is that calculated energy lost may be slightly higher than actual energy lost. The reference state chosen for the energy calculations was liquid at the initial reservoir temperature;
10. the steam-to-oil ratio was calculated as the ratio of CWE volume of injected steam to the volume of produced bitumen for each cycle (cyclic SOR, cySOR) and for all the cycles (cumulative SOR, cSOR).

The majority of key pilot tests were conducted at the Buffalo Creek and McLean sites. Regardless, a brief analysis of the Chipewyan River pilot is presented in the next section based on insights from available data. This is then followed by a more concentrated analysis of the Buffalo Creek and McLean pilots.

3 CHIPEWYAN RIVER PILOT (SECTION 21, TOWNSHIP 89, RANGE 21, W4M)

Pilot tests in Chipewyan River were conducted in two phases, jointly by UNOCAL, AOSTRA and Canadian Superior Resources (CSR). The first phase was in 1975 and comprised of air injection, followed by a single-cycle CSS in two wells, 14B-21-89-21 W4M (tested zone C of the Grosmont) and 11E-21 (tested zone D of the Grosmont), summarized in Table 2. Following 23 days of steam injection into Well 14B-21, reservoir pressure measured at the well increased from an initial value of 1 255 kPa to 6 550 kPa. Also, more than 4 000 kPa increase in reservoir pressure was recorded after 16 days of steam injection into Well 11E-21. Despite the large pressure increase, no evidence of hydraulic fracturing was reported. The pilot exhibited good injectivity, with the water injection rate in each well averaging more than 18.7 m^3/d (118 bbl/d). During the production cycle, bitumen flow rates dropped to about 0.16 m^3/day (≈ 1 bbl/d)

TABLE 2
Summary of Chipewyan River Pilot data. cySOR is the cyclic steam-to-oil ratio (steam expressed as CWE volume)

Well	Phase 1 (1975)		Phase 2 (1977)
	14B-21 (Zone C)	11E-21 (Zone D)	14C-21 (Zone D)
Variables			
Formation pressure (kPa)	1 255	1 103	1 103
Air injection tests			
Injection rate (cfm)	350	350	-
Pressure (kPa)	2 344	1 586	-
Steam injection tests			
Steam injection (days)	≈23	≈16	≈61
Pressure (kPag)	6 550	5 171	7 000-11 032
Volume (m^3)	1 970	1 893	6 551
Steam quality (%)	-	-	55
Soak period (days)	≈6	≈5	≈3
Production period (days)	≈49	≈33	≈13
Oil (m^3)	10.3 (20 → 3 BPD in 4 days)	2.7 (5 → 1 BPD in 4 days)	27 (12 → 3 BPD in 3 days)
Water (m^3)	3 037	2 292	245
cySOR (m^3/m^3)	191	701	243

in less than 5 days and cumulative water production far exceeded the injected steam volume in both wells. The cSOR in Wells 14B-21 and 11E-21 were 191 and 701 m^3/m^3, respectively. Three scenarios have been identified as likely drivers of pilot performance:

- a combination of poor injection steam quality and heat loss to formation water contrived to produce very poor thermal viscosity reduction of bitumen;
- high permeability water zones were encountered which, in addition to being heat sinks, also delivered large volumes of water during the production cycle;
- in the absence of hydraulic fracturing during steam injection, the large externally induced pressure gradient drove injected fluids deep into the formation which in turn displaced oil from the near-wellbore regions of the reservoir.

The excessive production of water, while being undesirable, gave an important insight into pilot performance. There was certainly a driving energy to move reservoir fluid into the wellbore. Therefore, regardless of the exact individual or combination of scenarios that was in place, integration of available data strongly implicates poor bitumen mobilization as the key CSS pilot performance inhibitor.

A second phase single-cycle CSS test was conducted in 1977 in Well 14C-21, to evaluate the response of the Upper Ireton and Grosmont C reservoirs to steam injection, the effects of clay and to evaluate well completion techniques. Poor injectivity was evident during steam injection as indicated by very high injection pressure up to 11 000 kPa. This was attributed to three cement squeezes performed during well completion. Poor quality of injected steam – averaging between 50 and 60% – was also reported. Recovery performance was poor, with the bitumen production rate declining to 0.47 m^3/d (3 bbl/d) within three days and cSOR averaging 243 m^3/m^3. It is noteworthy that despite the high cSOR, unlike the first pilot phase, most of the injected water was not produced. Loss of injected steam *via* a network of fractures, vugs and sinkholes is a very likely reason for this. Ultimately, a balance must be met to ensure injection pressures are high enough to overcome injectivity issues and distribute heat in the formation, but not excessive to induce steam losses and consequent poor bitumen mobilization. Optimized injection pressure promotes sufficient production to develop a sustainable vapor chamber required for the gravity drainage phase of CSS. In addition, effective thermal heating associated with steam zone conformance should stimulate other important thermal recovery processes such as thermal oil expansion, wettability reversal and CO_2 production, which combine to improve bitumen release from the matrix into the wellbore, either directly or through interconnected fractures.

4 BUFFALO CREEK PILOT (SECTION 5, TOWNSHIP 88, RANGE 19, W4M)

The Buffalo Creek pilot was conducted by UNOCAL between 1980 and 1986 to test CSS in Well 10A completed on Zone C at a depth of 286 m. A total of 12 steam cycles were conducted and spatial evolution of heated zones in the reservoir was monitored by four surrounding observation wells. Figure 4 shows the well configuration of Buffalo Creek pilot and indicates approximate distances between the injector/producer well from the observation wells. Compared with the Chipewyan River pilot, a significantly improved average injection steam quality of 80% was reported for Buffalo Creek.

4.1 Fluid Stream Analysis

Average injected steam slug volume per cycle for Well 10A was about 7 900 m^3 CWE. A graph of time duration of injection, soak, production periods and bottomhole pressure, displayed in Figure 5, reveals that the ratio of production to injection times generally increased with an increase in the number of cycles. Impressive oil production rates as high as 70 m^3/d were achieved. The volume of injected steam was doubled in Cycles 9 and 10, resulting in relatively long injection times and equally long production periods. An important deduction from the bottomhole pressure plot is the good steam injectivity, which increased with the number of cycles. In Figure 5, the bottomhole pressure history corroborates the trend in duration of phases for steam cycles and points towards good steam conformance. Injection pressure dropped from a peak value of over 4 MPa in the first cycle to about 1.8 MPa in Cycle 10. This is indicative of good steam conformance, which precludes the displacement of oil deep into the reservoir and away from the wellbore. In addition, good CSS steam conformance promotes gravity drainage of bitumen and offers additional energy for driving heated oil towards the wellbore, thereby improving late cycle recovery performance.

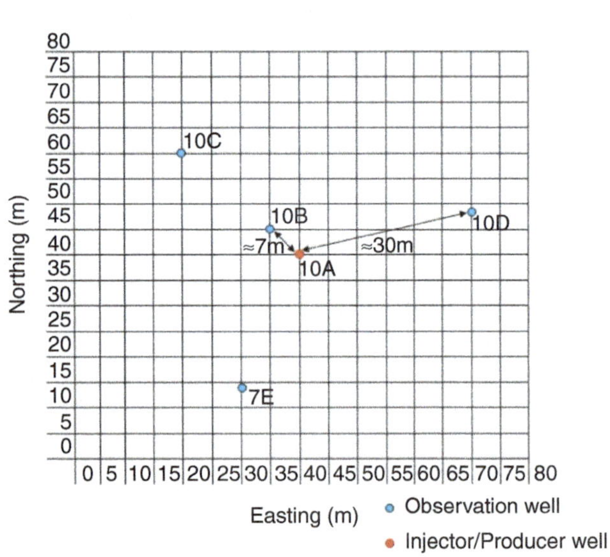

Figure 4

Buffalo Creek pilot well configuration.

Figure 6 shows the cyclic volumes of injected steam (CWE) and produced liquid (bitumen + water). Effectively, cumulative injected volume surpassed produced volume throughout the CSS test periods, and this is indicative of fluid loss in the reservoir. Nevertheless, the cSOR declined from about 14 m^3/m^3 in the first cycle to a respectable value of about 5.8 m^3/m^3. Also noteworthy is the fact that except for Cycle 1 and Cycle 6, cyclic SOR for all cycles was less than 7 m^3/m^3, as shown in Figure 7. While a large cyclic SOR may be expected for the first cycle due to a poorly developed steam zone, increased cyclic SOR in Cycle 6 is attributed to the injection of a large volume (3 000 m^3) of hot water prior to the start of steam injection in this cycle. This was done to evaluate the possibility of improving bitumen production *via* a hot water pre-flood. The results, however, showed that hot water pre-flood neither improved oil production nor overall thermal efficiency. The poor performance of hot water pre-flood is intuitive considering that unlike steam, hot water only contains sensible heat which implies less thermal energy per unit mass of injected fluid. As a result, there was insufficient thermal mobilization and production of bitumen which eventually led to poorer growth of the steam conformance zone and the associated decline of gravity drainage.

Figure 8 shows net water injection or production for each cycle and cumulative for all cycles. Clearly, there was a net water injection for each cycle and overall. The net cumulative water injection at the end of Cycle 10 was 33 000 m^3, which translates to 60% recovery of injected water. Compared with the Chipewyan pilot, it is evident that heat loss to formation water or aquifer zones was minimal during the Buffalo Creek CSS pilot tests. At this point, we postulate that the CSS wells in the Buffalo Creek pilot traversed a favorable network of fractures such that steam injection was facilitated;

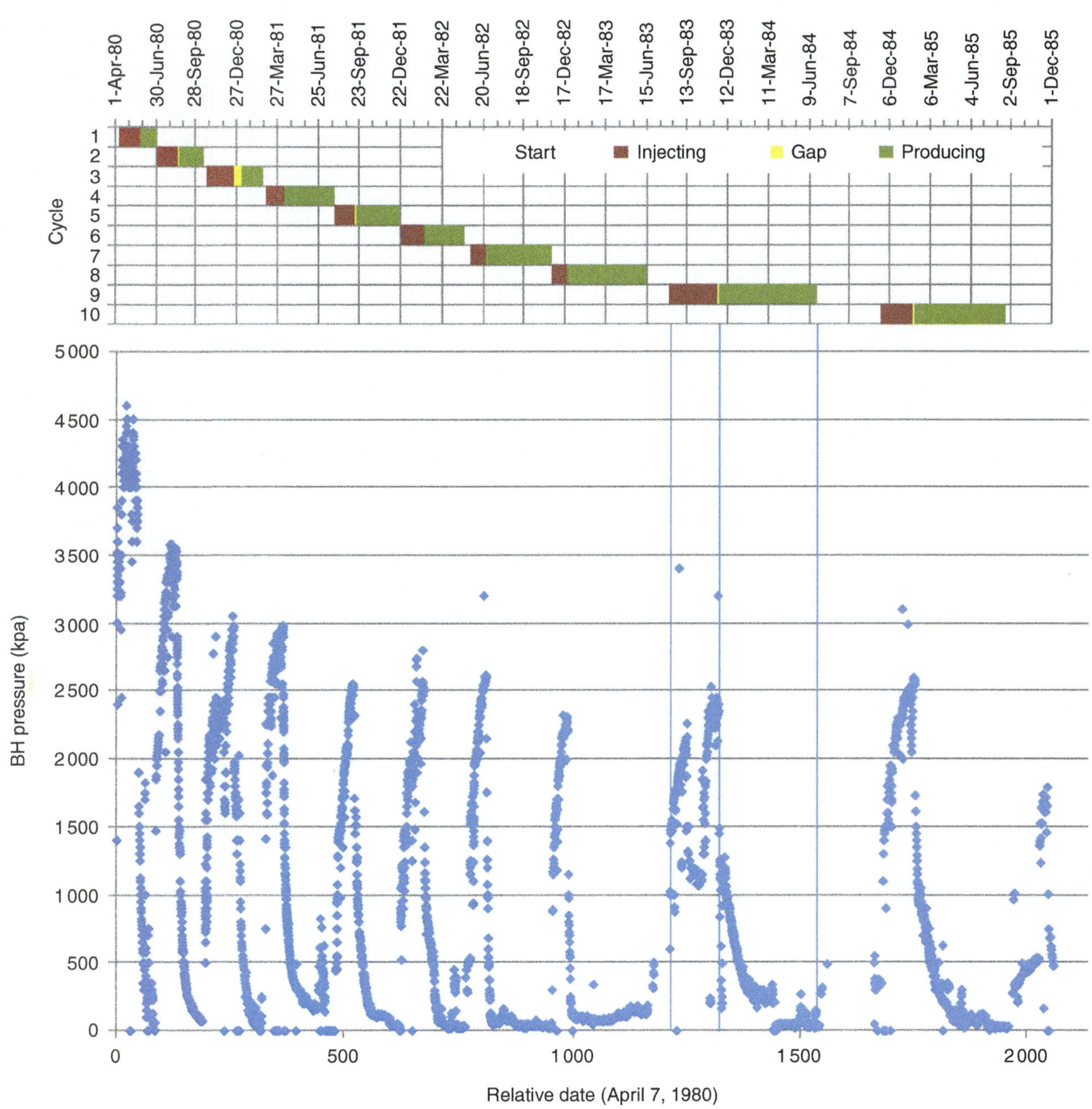

Figure 5

Graphical representation of time duration for steam injection, soak (shown as GAP) and production phases; the bottom plot shows bottomhole pressure for each cycle of Buffalo Creek CSS pilots.

similar to induced fractures in sandstone CSS projects. This ensured that injection pressure was low enough to prevent displacement of heated bitumen too far away from the wellbore, while also providing an effective steam conduit for convective and conductive heating of a large reservoir volume. The subsequent energy balance analysis will also test the accuracy of this postulation.

4.2 Energy Balance Analysis

We analyzed the energy balance during each CSS cycle to ascertain how much of the injected energy (sensible and latent heat) was produced or retained in the reservoir using an initial reservoir temperature of 12°C. This information was combined with the spatial temperature

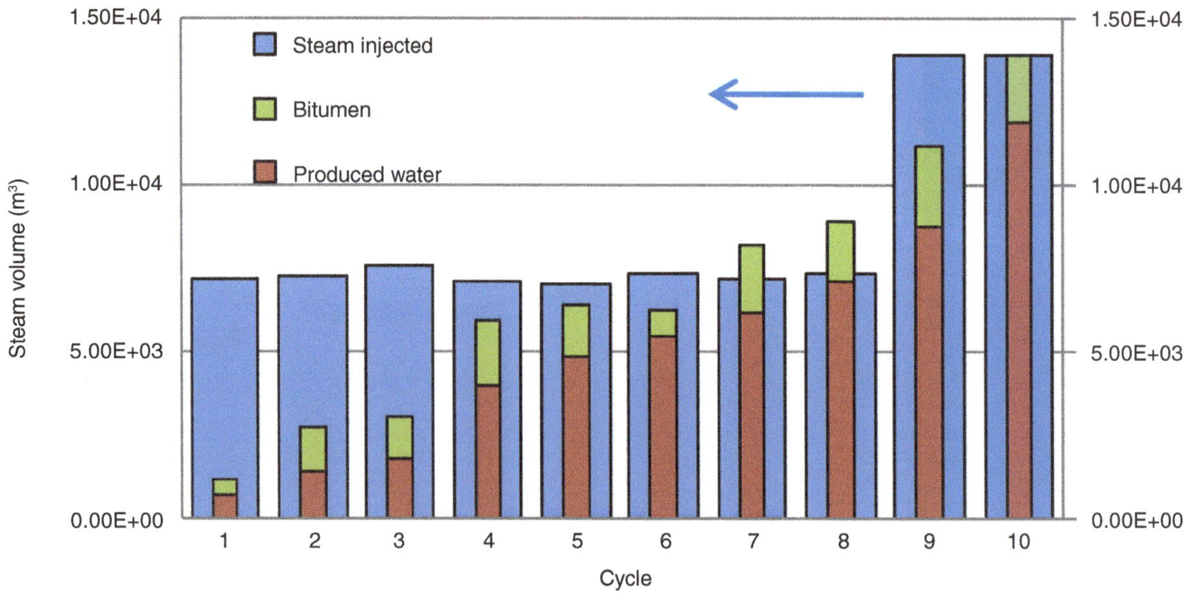

Figure 6

Cyclic volumes of injected steam and produced fluid during Buffalo Creek CSS pilots.

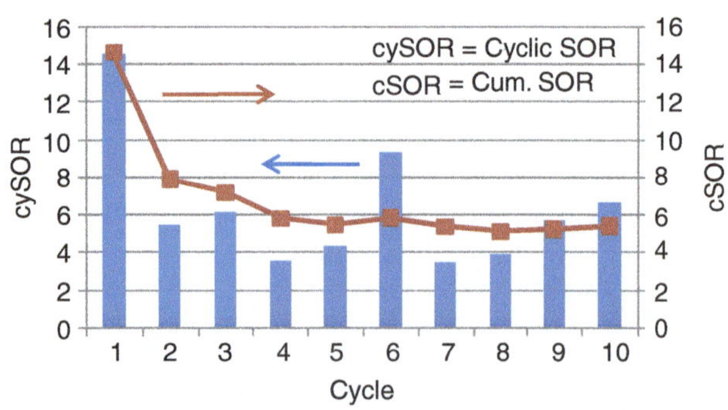

Figure 7

Cyclic and cumulative steam-to-oil ratio for Buffalo Creek CSS pilots.

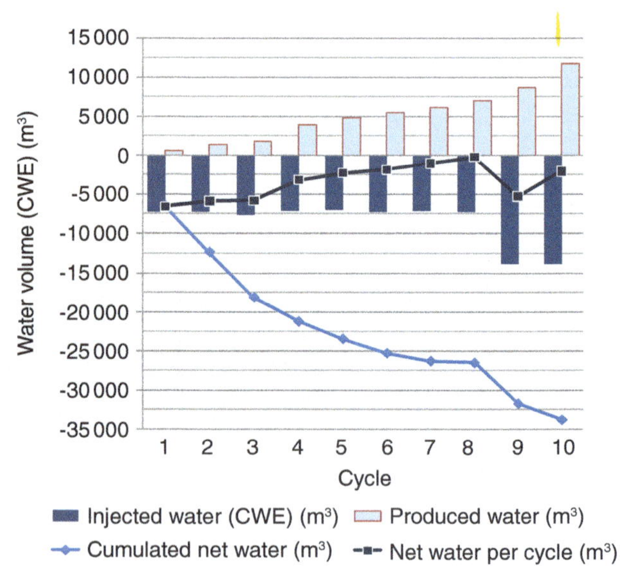

Figure 8

Water balance for Buffalo Creek CSS pilot showing a net water injection of 33 000 m^3 corresponding to 60% recovery of injected water.

profile to diagnose whether this energy was within the steam conformance zone or lost to sinkholes and vugs. Figure 9 shows the result of energy calculations for injected energy (sensible + latent heat) and produced energy (sensible heat). It is evident that as a result of consistent injection volumes of fairly high-quality steam (≈80%), the corresponding amounts of injected energy for the first 8 cycles are comparable. Although less than 2% of injected energy was produced in Cycle 1, it increased to more than 14% by Cycle 8. This was a combined effect of the reservoir getting warmer and an increased production volume. A doubling of injected steam volume in Cycles 9 and 10 had a reverse effect on produced energy. This may be due to the displacement of injected fluid too far from the wellbore by an excessively large injected slug. It therefore appears that the more appropriate size of injection slug for the Buffalo Creek CSS pilot was in the range of 7 000 m^3. Further

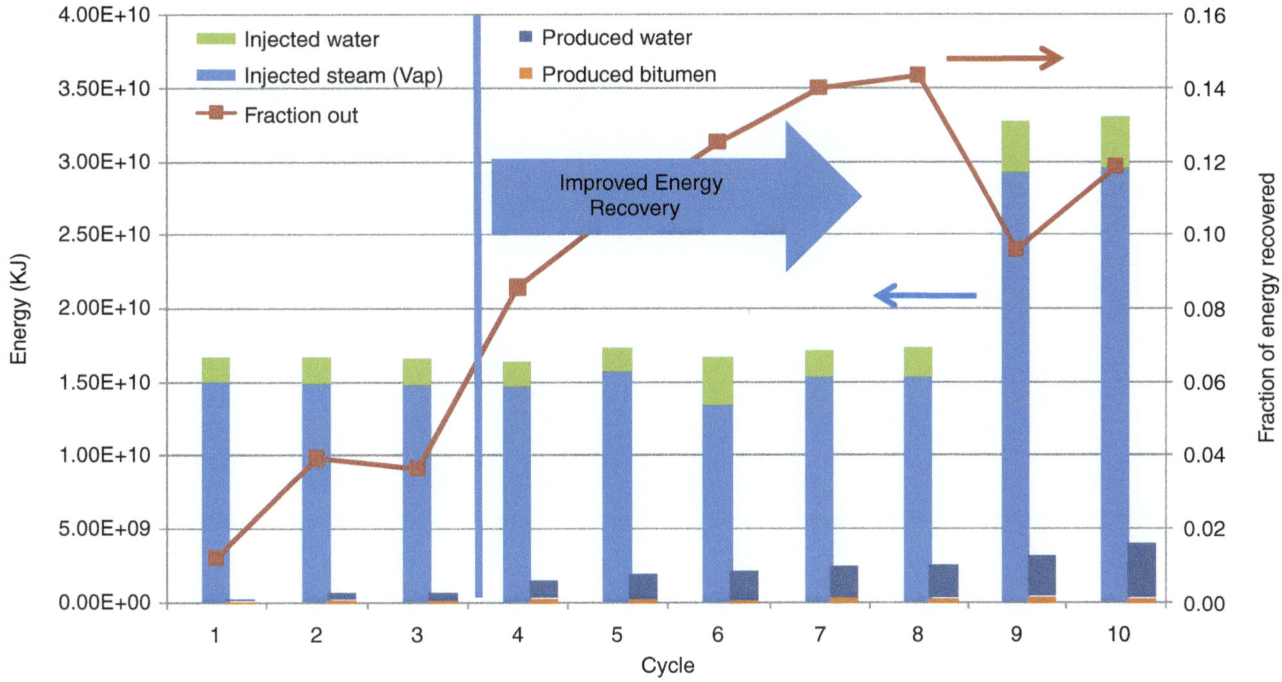

Figure 9

A plot of energy distribution within injected and produced fluids during Buffalo Creek CSS pilots.

analysis of the overall energy balance for all cycles showed that only about 9% of injected energy was produced and that produced bitumen contained about 1% while the rest was in produced water. Considering that 90% of injected energy was retained in the reservoir, it is important to understand how this energy was distributed to conceptualize and engineer improved CSS design. Figure 10 shows that a large portion of the retained energy was outside the steam conformance zone. The magnitude of energy lost outside the steam conformance zone in Cycles 9 and 10 corroborates our earlier assertion that the injected slug size in those cycles (≈14 000 m^3) was excessive for this pilot and detrimental to heat retention in the steam conformance zone. The correlation between a number of performance indices and bottomhole injection pressure was evaluated and is presented in Figure 11. It emerged that a strong correlation (r^2 = 0.78) exists between the fraction of energy produced and steam injection pressure. A small fraction of injected energy was recovered at high injection pressures (early cycles ≈4 000 kPa) while significantly larger fractional energy recovery was calculated at lower (late ≈2 000 kPa) injection pressures. This is consistent with observations from Figure 9. Cyclic SOR also shows a reasonable correlation with injection pressure and

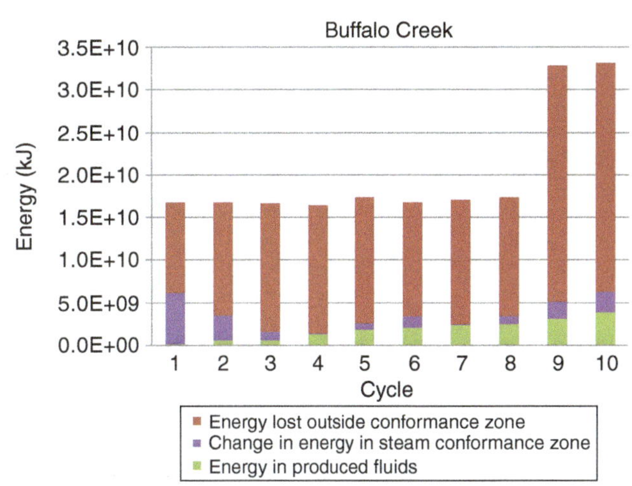

Figure 10

Overall energy balance for Buffalo Creek Pilot comparing energy retained in the conformance zone with that lost outside the conformance zone in each cycle. Total height of bars represents injected energy.

indicates an increase in cyclic SOR with increasing injection pressure. In contrast, the bitumen production rate and volumes are poorly correlated with injection

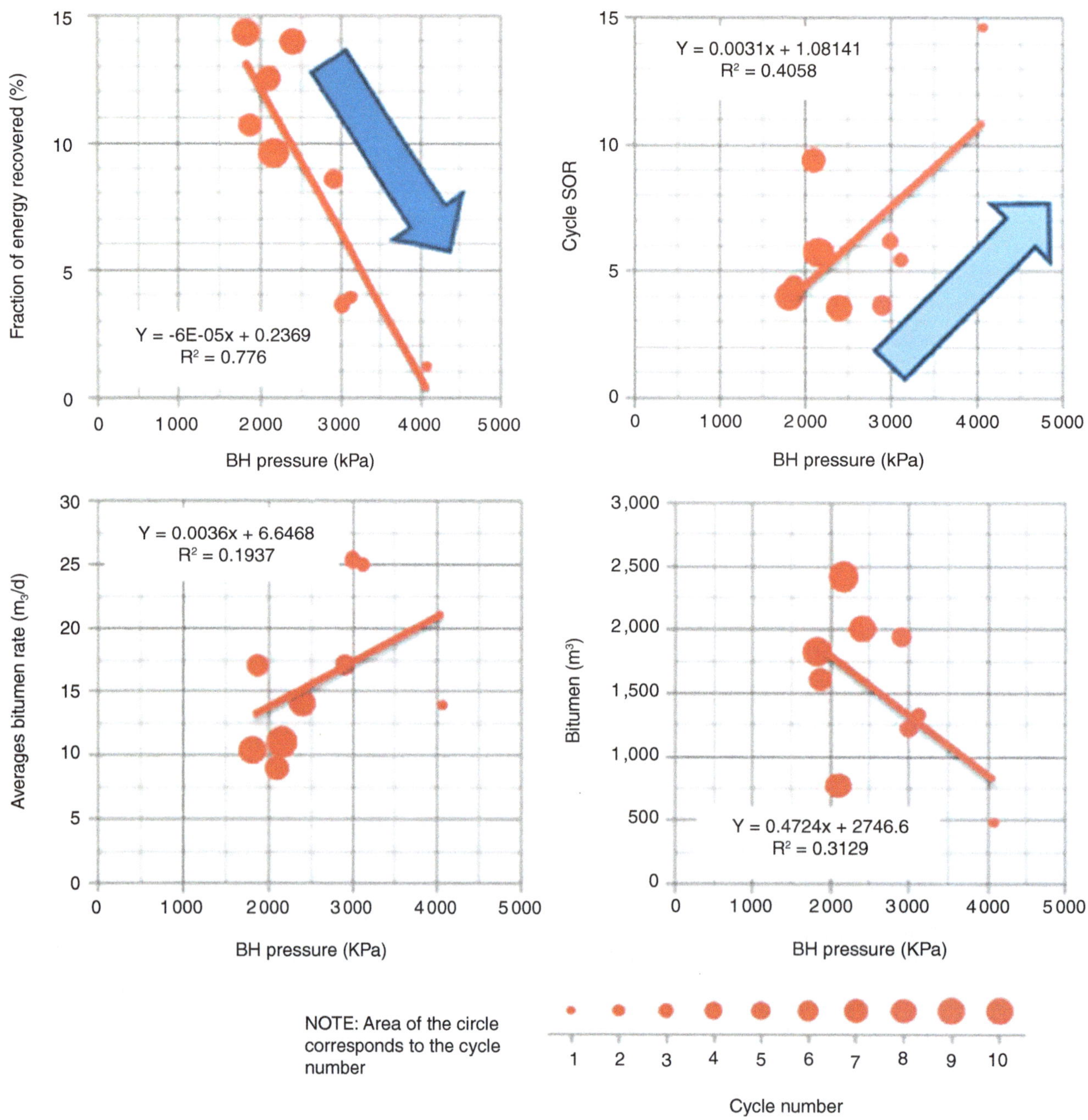

Figure 11

Effect of bottomhole injection pressure on selected performance indices during Buffalo Creek CSS pilots.

pressure when all the cycles are considered. However, the correlation factor for the bitumen rate *versus* injection pressure improves to around 0.63 when Cycle 1 is ignored. Considering that CSS is at a very early stage during Cycle 1, effective correlation may be more representative when Cycle 1 is ignored. Regardless, the produced bitumen volume remains poorly correlated to injection even without Cycle 1 data.

4.3 Temperature Visualization

Although temperature data for Well 10A were sparse, an extensive profile of temperature data was available for the observation wells. A rigorous quality check was conducted and an algorithm developed to transform this data into very informative 3D depth, time and space data, presented in Figure 12. At day 360, the CSS pilot

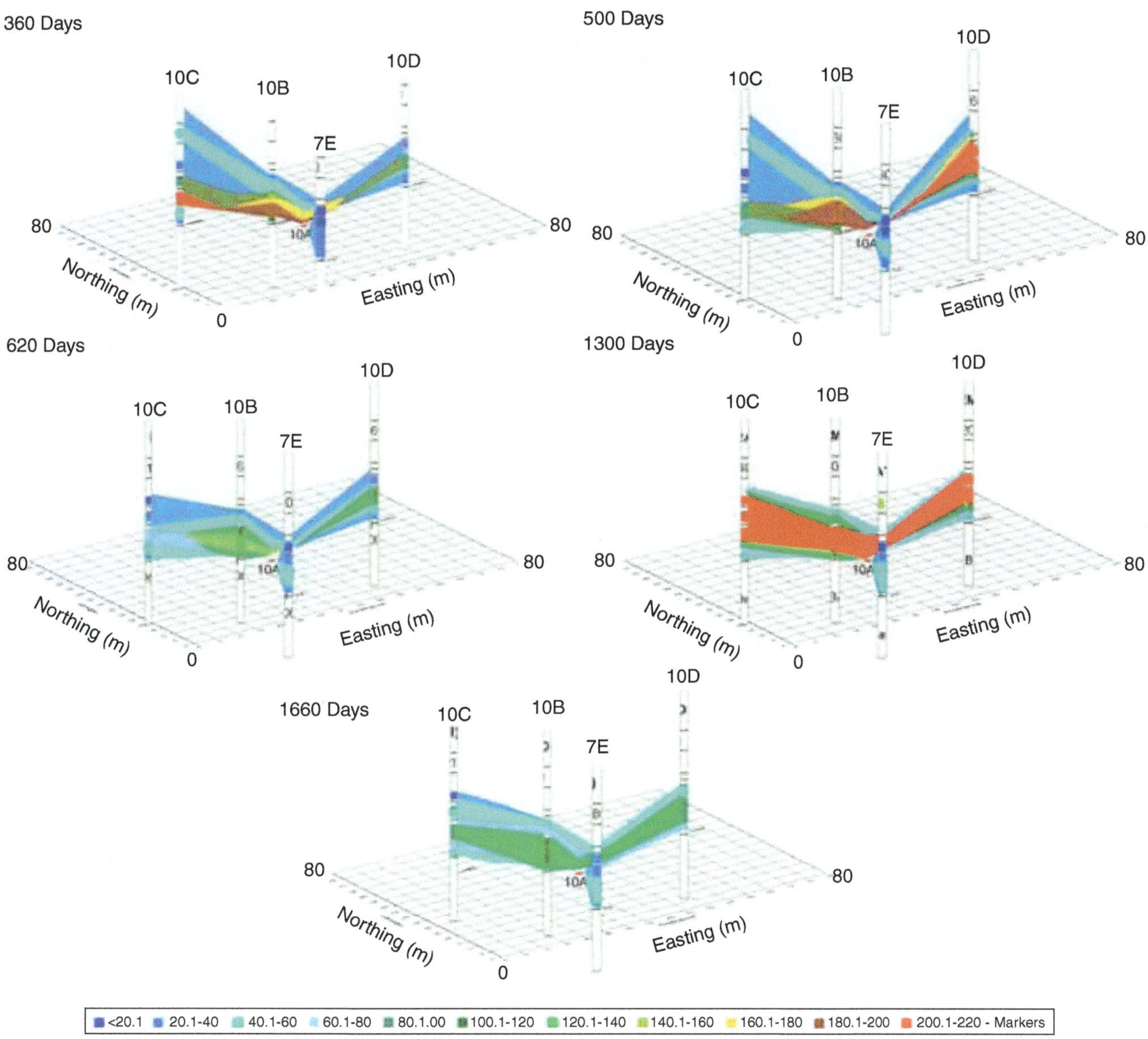

Figure 12

Graphical display of interwell spatial temperature at different time intervals representing injection or production during Buffalo Creek CSS pilot.

was on a production cycle and the very high temperatures (>180°C) observed at Wells 10B and 10C clearly indicate hydraulic communication in the NW direction. While this may facilitate heat distribution, it also highlights the likelihood of energy loss when excessively large injection slugs were implemented. Day 500 represents a Cycle 5 injection period while Day 620 is near the end of the Cycle 5 production interval. Day 500 shows that besides the NW direction, there is also hydraulic communication in the NE direction where temperatures exceeded 180°C at Well 10D. In contrast, just over 40°C was recorded in Well 7E at Day 500 and this is attributed to conductive heating. The reservoir temperature in high temperature areas, however, dropped to less than 100°C close to the end of Cycle 5 production shown at Day 620. Similar trends in heat distribution were also evident at Days 1 300 and 1 660, which represent the Cycle 9 injection and end of production periods, respectively. Again, temperatures at Well 7E remained below 60°C during Cycle 9 steam injection. The drastic drop in temperature from above 200 down to 100°C during the production period of Cycle 9 (at Day 1 660) highlights inefficiency in the CSS process. This implies that during subsequent steam injection, a large heat loss occurred within the conformance zone and compromised heat that should have been delivered to bitumen at the conformance zone perimeter. Poor temperature communication at Well 7E demonstrates that owing to Grosmont reservoir heterogeneity, fluid displacement was directional rather than radial. Directional hydraulic communication can be associated with either a network of fractures, vugs and karsts or high permeability layering. In view of Grosmont stratigraphic information (*Fig. 1*), layering may play a role in promoting NE directional flow. However, in the absence of additional information, directional flow in the NW direction can only be attributed to a network of fractures, vugs and karsts.

5 MCLEAN PILOT (SECTION 28, TOWNSHIP 87, RANGE 19, W4M)

The generally successful Buffalo Creek CSS pilot was followed by the McLean Pilot about 3 km south of Buffalo Creek. The McLean pilot was conducted by UNOCAL with an initial inverted 10 117-m^2 (2.5 acres) spaced five-spot pattern designed to test a steam drive process, as shown in Figure 13. Grosmont Zone C was the target reservoir at a depth of about 273 m. Steam injection (≈80% quality) commenced in 1982 and terminated in 1987. In addition, three observation wells were drilled within the pattern, with an additional observation well south of the pattern. The initial steamflood test was designed to inject steam through Well CA13 and produce through the four corner wells. Unfortunately, just after about two days of steam injection in Well CA13, steam loss to the upper reservoir (Zone D) was observed through very high temperatures recorded at the observation well EA-13. This suggests that the shale layer separating Zones C and D (termed CD-Marl) was compromised. Further field investigation revealed a karst zone in place of the CD-Marl at a location near Well CA13 and this was deemed responsible for upward steam channeling to Zone D. To avoid process inefficiency associated with steam loss to the upper reservoir, Well CA13 was shut in and the corner wells (*i.e.* BD13, DB13, DA13 and DD12) were converted to CSS wells. Prior to a fourth CSS cycle, the McLean pilot was expanded in 1984 with the addition of two wells (C12 and C13) to create a larger 5-spot pattern. The measured bottomhole temperature of the new wells suggested that they were not affected by previous CSS cycles. However, three wells (BD13, DB13 and DD12) were part of both the old and expanded pattern, making it more difficult to interpret test results. cSOR varied from just above 100 m^3/m^3 to less than 5 m^3/m^3. There were several issues related to casing damage and production reported during this pilot. It is also noteworthy that the McLean pilot produced a significant volume of sediments. Nevertheless, some of the wells performed very well and achieved a peak bitumen production rate of 100 m^3/d (628 bbl/d).

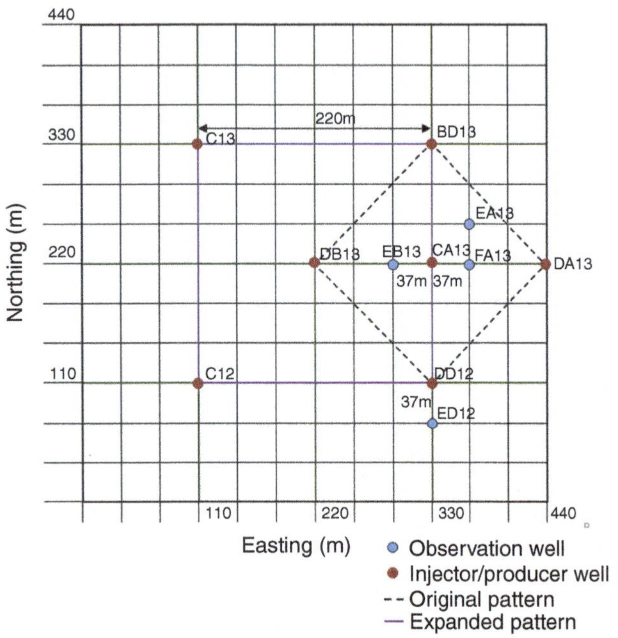

Figure 13

McLean pilot original (1982) and expanded (1984) patterns.

5.1 Fluid Stream Analysis

It can be seen from Figure 14 that prior to pattern expansion, injection volumes per cycle (*i.e.* steam slug size) in all wells within the original pattern peaked at 7 000 m^3. However, while wells DB13 and DA13 achieved this peak steam slug size in Cycle 1, significantly lower volume was injected in wells BD13 and DD12 during this cycle. Considering that peak injection volumes were achieved by Cycle 3 in wells BD13 and DD12, the ramping up of injected steam in both wells was most likely due to operational issues related to reported casing problems rather than a reservoir effect. Throughout the operation of the original well pattern, the water produced from wells DD12 and DB13 was only a fraction of injected steam (CWE). In contrast, wells BD13 and DA13 produced more water than injected volume during Cycles 3 and 1, respectively. During Cycle 2 of the expanded pattern, a large net steam injection was recorded in Well DB13 and corresponded to a large net water production in Well BD13. This suggests a hydraulic communication between

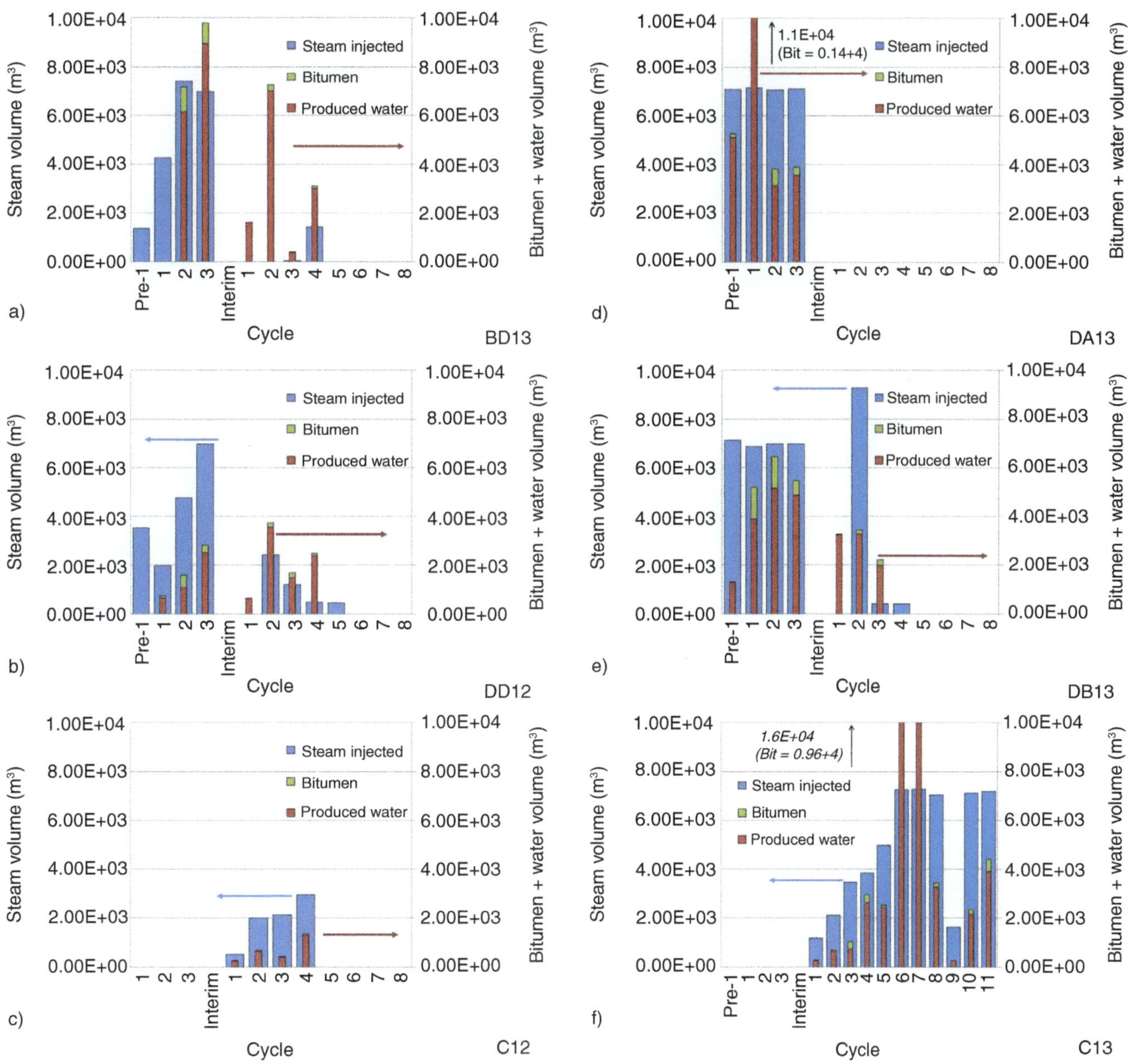

Figure 14

Cyclic injection and production volumes for different McLean pilot CSS wells. Interim indicates McLean pattern expansion and Well C13 was operated as a single CSS pilot from Cycle 5.

the wells and is consistent with the NE directional communication noted in the Buffalo Creek pilot. Well C12 produced an insignificant volume of bitumen, whereas Well C13 showed considerable cyclic variation between the volume of injected water and produced bitumen.

A foam (*Chevron* Chaser SD1000) pilot test was conducted on Well C13 during Cycle 9 to evaluate its ability to improve steam conformance. The foam test was deemed successful based on increased bitumen production volume in Cycles 10 and 11, as displayed in Figure 14. This suggests that CSS bitumen recovery from Grosmont can be improved by mitigating excessive steam channeling. Figure 15 shows that excluding wells BD13 and C12, the cumulative

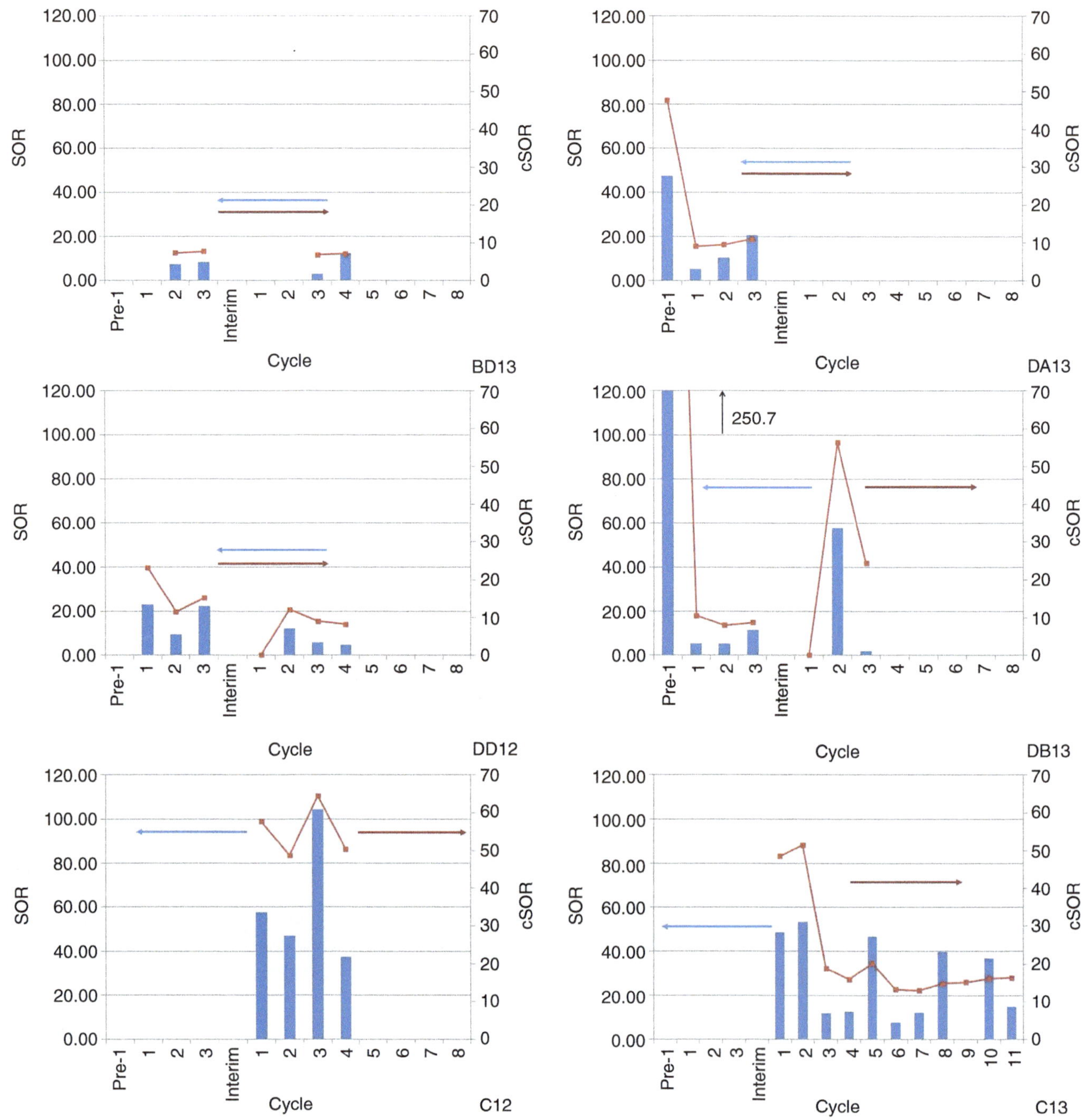

Figure 15

Cyclic and cumulative Steam Oil Ratio (SOR) for different McLean pilot CSS wells. Interim indicates McLean pattern expansion and Well C13 was operated as a single CSS pilot from Cycle 5.

SOR generally decreased with increase in the number of cycles. During operation of the original pattern, Well DB13 showed a rapid improvement in recovery performance; cSOR declined from about 250 m^3/m^3 to less than 20 m^3/m^3. Nevertheless, Well DB13 performed poorly during operation of the expanded

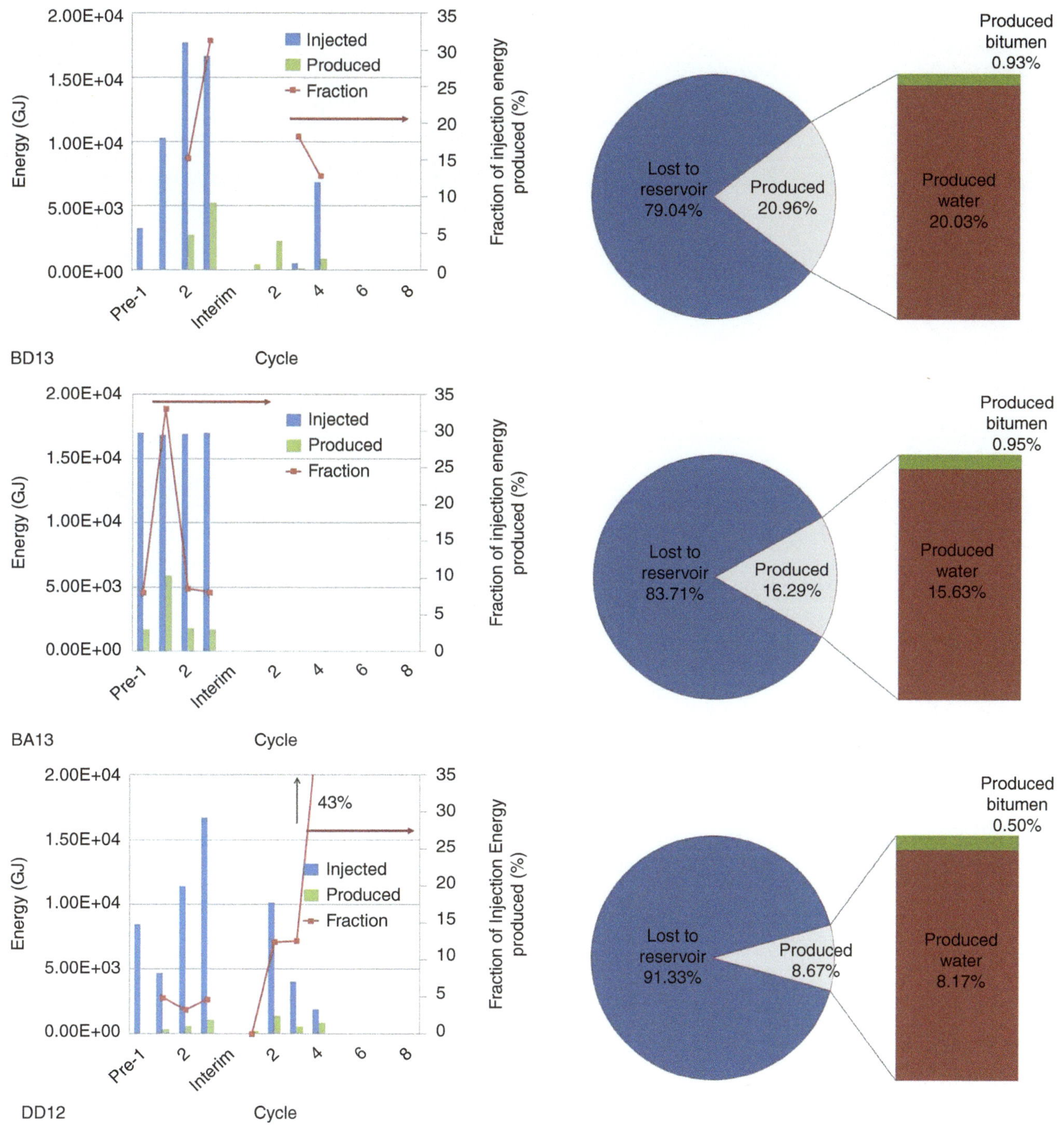

Figure 16

Calculated energy balance for different McLean pilot CSS wells. Interim indicates McLean pattern expansion and Well C13 was operated as a single CSS pilot from Cycle 5.

pattern, whereas Well BD13 showed consistently reasonable performance (cSOR < 10 m^3/m^3). It is therefore suspected that steam injected into Well DB13 mobilized bitumen in a network of vugs/karsts/fractures interconnected to Well BD13 and preferentially advanced towards Well BD13. The rapid improvement in recovery

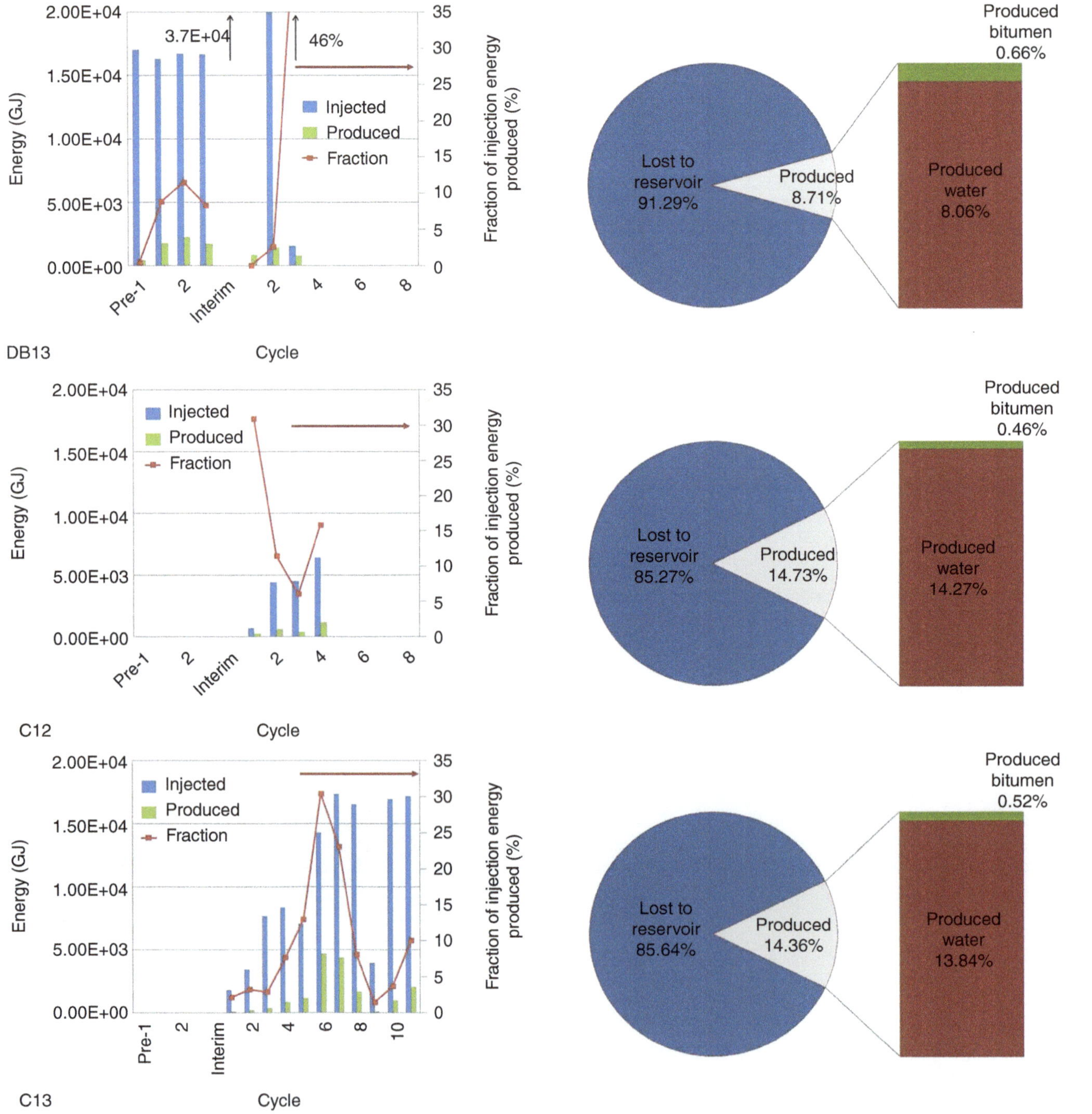

Figure 16
(continued)

performance of Well DB13 would have occurred prior to steam breakthrough, followed by a decline in performance as injected steam channeled into Well BD13. Analysis of the water balance for McLean indicates that excluding Well BD13, all other wells had a net water injection leading to overall 78% recovery of injected water. Therefore, interference from the aquifer zone during the pilot was unlikely.

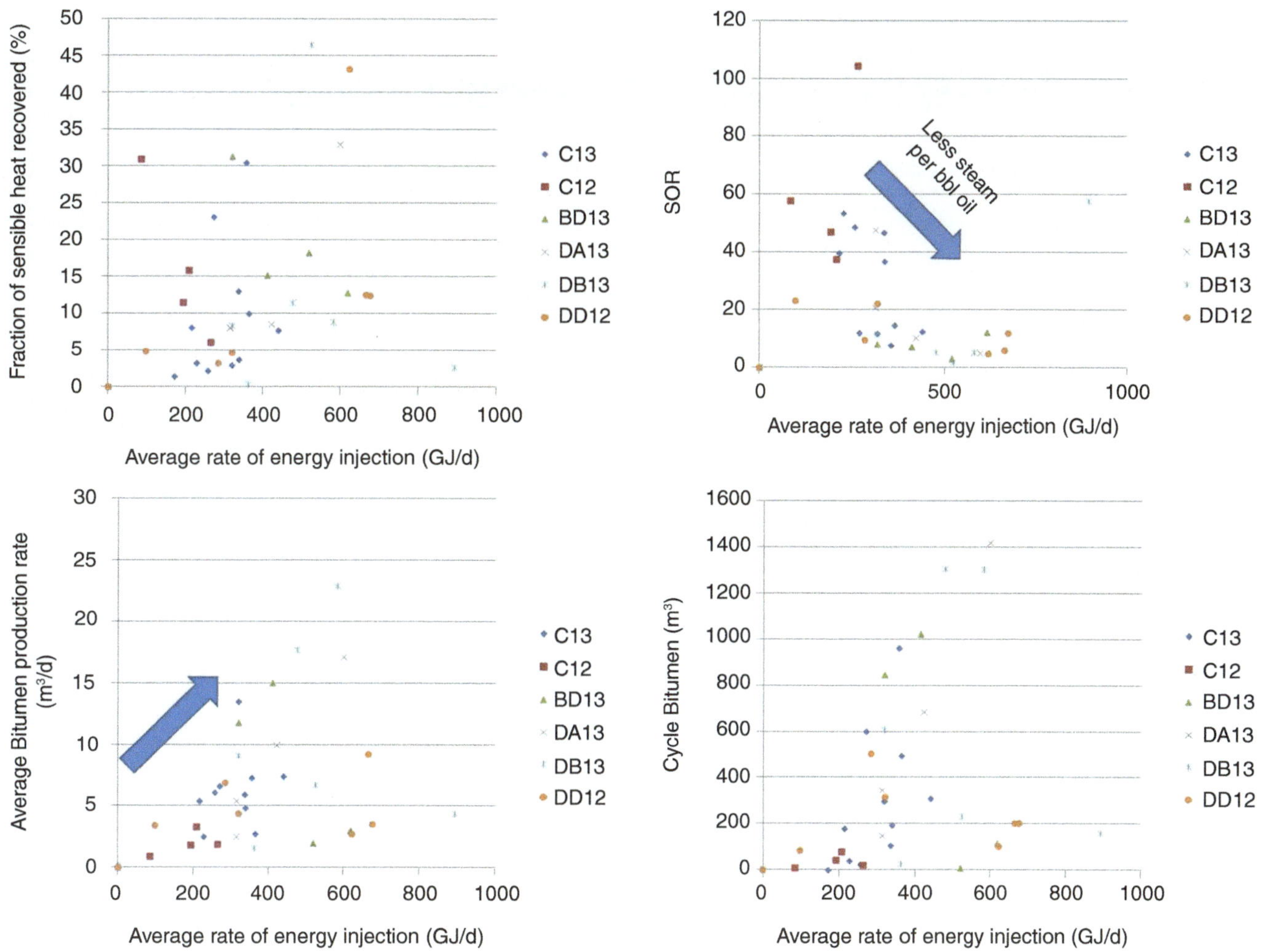

Figure 17

Effect of steam injection rate on performance indices for different McLean pilot CSS wells.

5.2 Energy Balance Analysis

Consistent with the approach described in the research scope and methods section above, energy balance calculations were performed for the McLean pilot. Figure 16 shows the results including cyclic injected and produced energy, fraction of injected energy recovered and distribution of energy in separate phases. Similar to earlier Grosmont pilots, the majority of injected energy (>79%) was retained in the reservoir. In addition, only a small fraction (<1%) of produced energy was in the mobilized bitumen phase. Unlike the Buffalo Creek pilot, where the fraction of injected energy recovered increased with an increase in the number of steam cycles, the result of the McLean pilot does not show any trend. In addition to interference of steam conformance zones, the variability in calculated energy results was most likely influenced by operational issues associated with casing problems in some of the wells. The increased bitumen production after the Cycle 9 foam pilot test led to an increased fraction of recovered energy during this period, as illustrated in Figure 16. The effects of the average energy injection rate on the fraction of sensible heat recovered, SOR, bitumen production rate and cyclic produced bitumen volume are shown in Figure 17. The rate of energy injection is somewhat related to injection pressure in the sense that the energy injection rate was generally lower at higher injection pressures (*i.e.* lower injectivity). The SOR decreased with increase in the energy injection rate. This is related to high injectivity at lower pressures precluding displacement of bitumen too far away from the production well. Figure 17 also shows a positive correlation between the energy injection rate and bitumen production rate. In contrast, a large

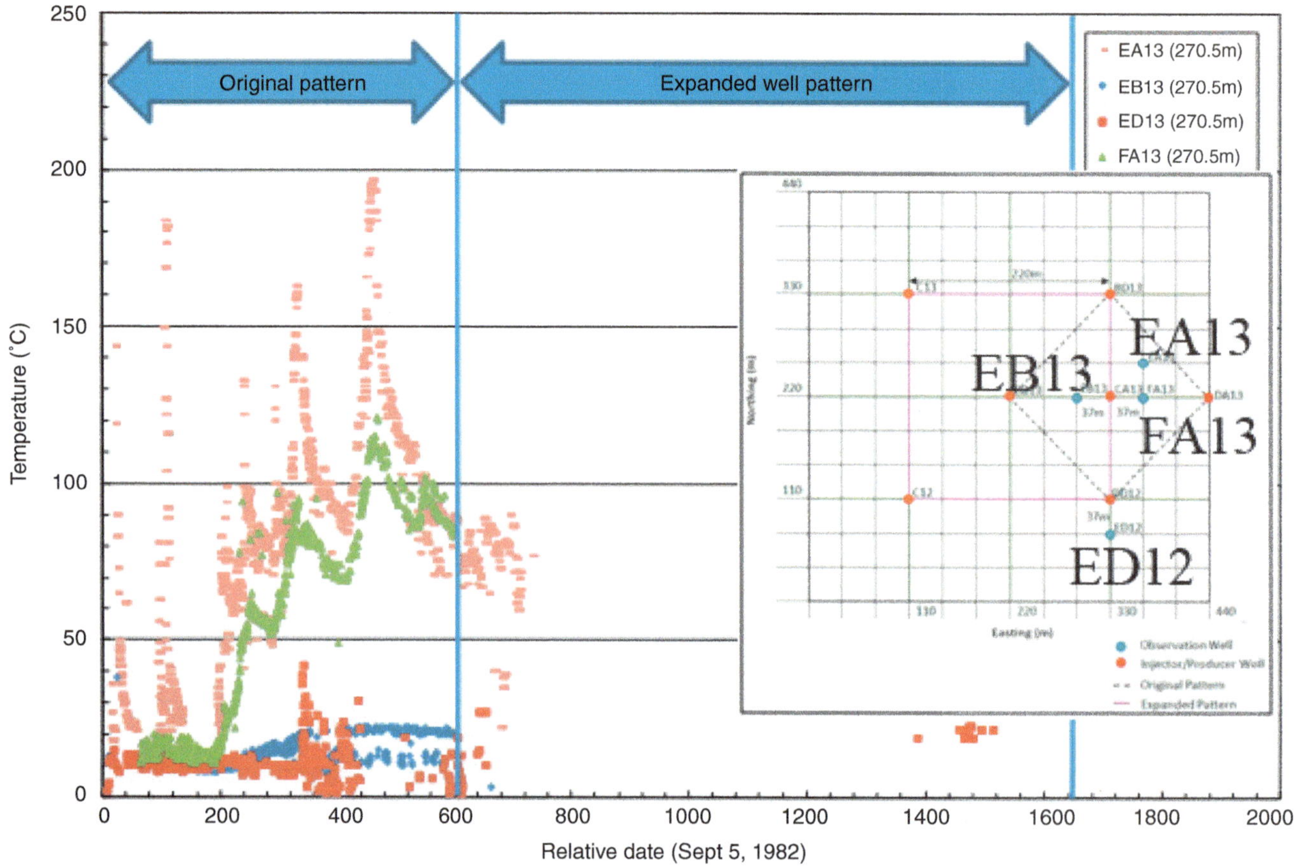

Figure 18

Temperature profile recorded in the observation wells (blue colored in inset pattern map) at the top of Grosmont Zone C (270 m) during McLean CSS pilot tests.

amount of scatter existed for the produced bitumen volume and fraction of energy recovered, highlighting their poor correlation with the rate of energy injection. This indicates that although the bitumen production rate may increase with the rate of energy injection, this may not necessarily translate to a larger volume of produced bitumen, due primarily to rapid decline in the bitumen rate during the production cycle.

5.3 Temperature Visualization

One major design flaw of the McLean pilot was the absence of observation wells in strategic locations. Only four observation wells were available. Worse still, a change in operational method (from steamflood to CSS) occasioned by steam losses from Well CA-13 to the overlying reservoir (Zone D) made these observation wells less strategic. This unfortunately devalued the McLean pilot in terms of improved understanding of *in situ* distribution and directional flow of heat energy. Temperature data recorded at observation wells plotted against time are shown in Figure 18. It is possible that upward steam channeling still continued even after shut in of Well CA13. Along-well temperature profiles showed evidence of convective heating at the top of Zone C in wells EA13 and FA13 (temperatures exceeding 120°C), while wells EB13 and ED12 did not show any hydraulic communication with injected steam (temperatures averaging 25°C). Nevertheless, the heterogeneity of Grosmont carbonate even on a scale of less than 40 m is evident. The difficulties associated with interpreting the temperature data of the McLean pilot make it imperative that a carefully designed steam strategy must be associated with future pilots employing multiple CSS wells.

6 DISCUSSION

An extended analysis of early pilots in the Grosmont points towards an attractive potential for steam-based commercial development of this reservoir by currently available technology. In agreement with previous authors, these early pilots were generally successful. Previous analysis of these early Grosmont pilots (Yuan *et al.*, 2010; Novak *et al.*, 2007) identified a number of technology advancements as particularly relevant to the commercial development of Grosmont carbonate. These include; horizontal wells and advancement in completions, SAGD, solvent-assisted SAGD, 3D/4D seismic, VAPor EXtraction (VAPEX) and improved CSS. Following our integrated analysis, however, we identified specific adaptations of some of these reported technologies to the Grosmont. In particular, we propose the alignment of future horizontal wells along the NE direction (parallel to the direction of observed preferential steam channeling) to promote transverse flow of steam into lower permeability matrix objects in the Grosmont. Although SAGD has proved significantly successful in sandstone reservoirs of Athabasca and Cold Lake, we suspect, based on this pilot analysis, that continuous steam injection (associated with SAGD) will most likely result in excessive heat loss to 'steam sinks', leading to a poor SAGD performance in the Grosmont reservoir. In addition, although both cases of horizontal well CSS and SAGD benefit from gravity drainage, it is quite likely that mobilized bitumen at the edges of the SAGD steam zone may be trapped in low permeability paths and not reach the production well. A key attraction of the CSS process is that unlike SAGD, heated bitumen is produced *via* the same path that injected steam had traveled. In addition, the potential to expel matrix oil *via* steam flashing during CSS offers a realistic chance of improved recovery over SAGD. Of course, lower pressure CSS is desirable to mitigate instability of the steam zone. For these reasons, we therefore suggest the use of horizontal well CSS (instead of SAGD) as a more technically appropriate protocol for the Grosmont and similar bitumen carbonate reservoirs. This hypothesis is contrary to the assertion in favor of SAGD by previous authors (Yuan *et al.*, 2010). However, the ongoing SAGD pilot at Grosmont (Saleski) will offer definitive field data on the performance of SAGD in the Grosmont while confirming the consistency or otherwise of our suspected challenges with the application of SAGD at Grosmont.

SUMMARY AND CONCLUSIONS

In summary, the integrated analysis of early Grosmont steam pilots presented here has revealed many positive insights that should help to improve the design of future pilots and commercial development of this massive bitumen deposit. In agreement with earlier discussions (Jiang *et al.*, 2010; Yuan *et al.*, 2010; Novak *et al.*, 2007), these Grosmont steam pilots were considerably successful based on reservoir performance factors such as high injectivity, peak bitumen delivery rate of about 70 m^3/d and a low cyclic steam-to-oil ratio of about 3.6 vol/vol (steam expressed as CWE). Considering the heterogeneous nature of the Grosmont Formation, it was not surprising that significant performance variation was exhibited by shortly spaced wells.

Operational factors such as poor steam quality, non-optimized high injection pressures and completion issues appear to have heavily impacted recovery performances. As an example, the very low injection steam quality for the Chipewyan River pilot for highly viscous bitumen prevents effective assessment of reservoir effects.

Single-well CSS tests (Buffalo Creek Well 10A and McLean Well C13) performed much better than multiple-well tests. This is a tremendously positive driver as it points to the viability of development with sparsely spaced wells, thereby making Grosmont development economically attractive.

Spatial analysis showed that directional steam flow was predominantly towards the NE direction. For future CSS processes in the Grosmont Formation, we propose utilizing horizontal wells and aligning the horizontal length along the NE direction. This may encourage steam flow in the transverse direction, thereby reducing steam channeling and improving matrix recovery. Positive results obtained from the Grosmont foam pilot support the viability of this proposed improvement in process development.

ACKNOWLEDGMENTS

The authors acknowledge funding and data support from *Laricina Energy Ltd.* and funding from the Natural Sciences and Engineering Research Council (NSERC) (MSK, IDG).

REFERENCES

Ahmed K.N., Milhem M.M. (1989) Operation of a second steam stimulation pilot project in Kuwait, *Middle East Oil Show*, Bahrain, 11-14 March, *SPE Paper* 17987.

Babadagli T., Al-Bemani A. (2007) Investigation on matrix recovery during steam injection into heavy-oil containing carbonate rocks, *J. Petrol. Sci. Eng.* **58**, 259-274.

Barrett K., Hopkins J., Wilde K.N., Connelly M.E. (2008) The origin of matrix and fracture mega-porosity in a carbonate bitumen reservoir, Grosmont Formation, Saleski, Alberta, *2nd World Heavy Oil Congress*, Edmonton, Canada, 10-12 March, Paper 2008-344.

Briggs P.J., Beck D.L., Black C.J.J., Bissell R. (1992) Heavy oil from fractured carbonate reservoirs, *SPE Reserv. Eng. J.* **7**, 2, 173-179.

Brooks P.W., Fowler M.G., MacQueen R.W. (1988) Biomarker geochemistry of Cretaceous oilsands, heavy oils and Paleozoic carbonate trend bitumens, Western Canada basin, In: Meyer R.F., Wiggins E.J. (eds), *4th International UNITAR/UNDP Conference on Heavy Crude and Tar Sands, Vol. 2, geology, chemistry*: International Conference on Heavy Crude and Tar Sands; pp. 594-606.

Brown J., Deemer A., Al-Dhafeeri F., Lekia S., Hoadley S., Al-Mutairi G., Al-Yami F., Barge D. (2011) Early results from a carbonate steamflood pilot in 1st Eocene reservoir, Wafra field, PZ, *SPE Heavy Oil Conference and Exhibition*, Kuwait City, Kuwait, 12-14 Dec, *SPE Paper* 150605.

Buschkuehle B.E., Hein F.J., Grobe M. (2007) An overview of the geology of the Upper Devonian Grosmont Carbonate Bitumen Deposit, Northern Alberta, Canada, *Nat. Resour. Res.* **16**, 1, 3-15.

Chevron Canada Resources Limited (1983) Algar Thermal Recovery Project, Final Report.

Couderc B.M., Verpeaux J.F., Monfrin D., Quettler U.H. (1990) Emeraude vapeur: a steam pilot in an offshore environment, *SPE Reservoir Eng. J.* **5**, 4, 508-516.

Dreher K.D., Kenyon D.E., Iwere F.O. (1986) Heat flow during steam injection into a fractured carbonate reservoir, *Proceeding SPE Enhanced Oil Recovery Symposium*, Tulsa, USA, *SPE Paper* 14902.

Edmunds N., Barret K., Solanki S., Cimolai M., Wong A. (2009) Prospects for commercial bitumen recovery from the Grosmont carbonate, Alberta, *J. Can. Pet. Technol.* **48**, 9, 26-32.

ERCB (2011) Energy Resources Conservation Board, Alberta's energy reserves 2010 and supply/demand outlook 2011-2020, ST98-2011, pp. 3-6.

Gates I.D. (2011) *Basic Reservoir Engineering*, 1st edn., Kendall-Hunt, Dubuque.

Harrison R.S., McIntyre B.G. (1981) The geologic setting of the Grosmont thermal recovery project, Northeastern Albert, Proceeding Alberta Oil Sands and Technology Research Authority (AOSTRA), *Seminar on Advance in Petroleum Recovery and Upgrading Technology*.

Hjelmeland O.S., Larrondo L.E. (1986) Experimental investigation of the effects of temperature, pressure, and crude oil composition on interfacial properties, *SPE Reserv. Eng.* **1**, 4, 321-328.

Jiang Q., Yuan J., Russel-Houston J., Thornton B., Squires A. (2010) Evaluation of recovery technologies for the Grosmont carbonate reservoirs, *J. Can. Pet. Technol.* **49**, 5, 56-64.

Luo P., Machel H.G. (1994) Petrophysical properties of matrix blocks of a heterogeneous dolostone reservoir – the upper Devonian Grosmont Formation, Alberta, Canada, *Bull. Can. Petrol. Geol.* **42**, 465-481.

Luo P., Machel H.G. (1995) Pore size and pore throat types in a heterogeneous reservoir, Devonian Grosmont formation, Western Canada Sedimentary Basin, *AAPG Bull.* **79**, 11, 1698-1720.

Luo P., Dembicki E.A., Huebscher H., Machel H.G. (1993) Diagenesis and reservoir characteristics of the heavy-oil carbonate trend in Western Canada – refined evaluation of reservoir characteristics of the Grosmont Formation, AOSTRA Report **7**, p. 170.

Macaulay R.C., Krafft J.M., Hartemink M., Escovedo B. (1995) Design of a steam pilot in a fractured carbonate reservoir – Qarn Alam field, Oman, *Internatioal Heavy Oil Symposium*, Calgary, Canada, 19-21 June, SPE Paper 30300.

Machel H.G., Hawlader H. (1990) Diagenesis and reservoir characteristics of the heavy-oil carbonate trend in western Canada, preliminary investigations of facies, diagenesis, porosity, and bitumen saturation of the Grosmont Formation: AOSTRA, reports 2, p.169.

Milhem M.M., Ahmed K.N. (1987) Performance of a pilot cyclic steam stimulation project in Kuwait, *Middle East Oil Show*, Bahrain, 7-10 March, SPE Paper 15733.

Mollaei A., Maini B. (2010) Steam flooding of naturally fractured reservoirs: basic concepts and recovery mechanisms, *J. Can. Pet. Technol.* **49**, 1, 65-70.

Nakamura S., Sarma H.K., Umucu T., Issever K., Kanemitsu M. (1995) A critical evaluation of a steamflood pilot in a deep heavy carbonate reservoir in Ikiztepe field, Turkey, *SPE Annual Technical Conference and Exhibition*, Dallas, USA, 22-25 Oct., SPE paper 30727.

Novak J., Edmunds N., Cimolai M. (2007) A history match of CSS recovery in the Grosmont, *8th Canadian International Petroleum Conference*, Calgary, Canada, 12-14 June, SPE Paper 2007-154.

Sahuquet B.C., Ferrier J.J. (1982) Steam-drive pilot in a fractured carbonated reservoir: Lacq Supérieur field, *J. Petrol. Technol.* **34**, 4, 873-880.

Sahuquet B.C., Spreux A.M., Corre B., Guittard M.P. (1990) Steam injection in a low-permeability reservoir through a horizontal well in Lacq Supérieur field, *SPE Annual Technical Conference and Exhibition*, Louisiana, USA, 23-26 Sept, SPE Paper 20526.

Samir M. (2010) Rejuvenation Issaran field – success story, *North Africa Technical Conference and Exhibition*, Cairo, Egypt, 14-17 Feb., SPE Paper 127847.

Snell J.S., Close A.D. (1999) Yates field steam pilot applies latest seismic and logging monitoring techniques *SPE Annual Technical Conference and Exhibition*, Texas, USA, 3-6 Oct., SPE Paper 56791.

Theriault F. (1988) Lithofacies, diagenesis, and related reservoir properties of the Upper Devonian Grosmont Formation, Northern Alberta, *Bull. Can. Petrol. Geol.* **36**, 1, 52-69.

Theriault F., Hutcheon I. (1987) Dolomitization and calcitization of the Devonian Grosmont Formation, Northern Alberta, *J. Sediment. Res.* **57**, 6, 955-966.

Union Oil Company of Canada Limited (1975) UNOCAL, Progress Report Injectivity and Sample Recovery Test, Grosmont Formation – Chipewyan River Test Site.

Union Oil Company of Canada Limited (1977) UNOCAL, Progress Report Steam Stimulation and Communication Heating Test, Chipewyan River Test Site – Grosmont and Ireton Formations.

Union Oil Company of Canada Limited (1980) UNOCAL, Progress Report Steam Stimulation Buffalo Creek Test Site, ERCB Approval No. 2367C.

Union Oil Company of Canada Limited (1982-1983) UNOCAL, Progress Report McLean Heavy Oil Pilot, ERCB Approval No. 3603.

Walker D. (1986) Regional Stratigraphy of the Upper Devonian Grosmont Formation, Northern Alberta, AOSTRA/ARC Oil Sands Geology Agreement 158B, Open File Report-2, June.

Yuan J.-Y., Jiang Q., Russel-Houston J., Thornton B., Putman P. (2010) Evolving recovery technologies directed towards commercial development of the Grosmont carbonate reservoirs, *Canadian Unconventional Resources and International Petroleum Conference*, Calgary, Canada, 19-21 Oct., SPE Paper 137941.

Zhao Y., Machel H. (2012) Viscosity and other rheological properties of bitumen from the upper devonian Grosmont reservoir, Alberta, Canada, *AAPG Bull.* **96**, 1, 133-153.

Zhou Y., Lin Y., Liu L., Chen G. (1998) Steam pilot in the Cao-20 fracture limestone with extra heavy oil, *7th UNITAR Conference on Heavy Crude and Tar Sands*, Beijing, China, 27-30 Oct., *SPE Paper* 1998-131.

Ziegler V.M. (1985) Chevron Algar 10, Simulation of Steam Injection at the House River Field, Alberta, Canada, Project 48,817, January.

Ziegler V.M., Bay A.R. (1984) Chevron Algar 9, Special Core Tests Well Chevron CCl – House 10–36-81-17W4 M, House River Field, Alberta, Canada, Project 48,991, January.

2

Crack Features and Shear-Wave Splitting Associated with Fracture Extension during Hydraulic Stimulation of the Geothermal Reservoir in Soultz-sous-Forêts

M. Adelinet[1]*, C. Dorbath[2], M. Calò[2]**, L. Dorbath[2] and M. Le Ravalec[1]

[1] IFP Energies nouvelles, 1-4 avenue de Bois-Préau, 92852 Rueil-Malmaison Cedex - France
[2] École et Observatoire des Sciences de la Terre, EOST, Université de Strasbourg, Strasbourg Cedex - France
**Now at Instituto de Geofísica, UNAM, Circuito de la investigación Científica s/n Ciudad Universitaria Delegación Coyoacán C.P. 04510, México D.F
e-mail: mathilde.adelinet@ifpen.fr

* Corresponding author

Abstract — *The recent tomography results obtained within the scope of the Enhanced Geothermal System (EGS) European Soultz project led us to revisit the meso-fracturing properties of Soultz test site. In this paper, we develop a novel approach coupling effective medium modeling and shear-wave splitting to characterize the evolution of crack properties throughout the hydraulic stimulation process. The stimulation experiment performed in 2000 consisted of 3 successive injection steps spanning over 6 days. An accurate 4-D tomographic image was first carried out based upon the travel-times measured for the induced seismicity [Calò M., Dorbath C., Cornet F.H., Cuenot N. (2011) Large-scale aseismic motion identified through 4-D P-wave tomography,* Geophys. J. Int. ***186****, 1295-1314]. The current study shows how to take advantage of the resulting compressional wave (Calò et al., 2011) and shear-wave velocity models. These are given as input data to an anisotropic effective medium model and converted into crack properties. In short, the effective medium model aims to estimate the impact of cracks on velocities. It refers to a crack-free matrix and 2 families of penny-shaped cracks with orientations in agreement with the main observed geological features: North-South strike and dip of 65°East and 65°West [Genter A., Traineau H. (1996) Analysis of macroscopic fractures in granite in the HDR geothermal well EPS-1, Soultz-sous-Forêts, France,* J. Vol. Geoth. Res. ***72****, 121-141], respectively. The resulting output data are the spatial distributions of crack features (lengths and apertures) within the 3-D geological formation. We point out that a flow rate increase results in a crack shortening in the area imaged by both compressional and shear waves, especially in the upper part of the reservoir. Conversely, the crack length, estimated during continuous injection rate phases, is higher than during the increasing injection rate phases. A possible explanation for this is that cracks remain large because the system has time to relax. We also calculate the extension and opening rates during all hydraulic stimulation sets. While the opening rate is unchanged, the extension rate varies depending on the stimulation phase. It is also shown to be higher around and above the open-hole section than below. This can indicate a potential upward path that makes fluid percolation easier within the granite formation, this path being induced by the temperature gradient. We also compare the evolution of crack extension during injection with shear-wave splitting. Split shear waves were recorded at 2 stations during hydraulic stimulation and processed in terms of splitting parameters. The fast shear-wave polarization remains constant and parallel to the maximum horizontal stress orientation while the amplitude of splitting varies with time. We observe a good agreement between travel-time differences and crack extension rates during the first 4 days of the stimulation experiment. Afterwards, these two* *parameters depart from each other. This study emphasizes the added value of the coupling between effective medium modeling and shear-wave splitting to monitor meso-scale cracks in reservoirs submitted to hydraulic stimulation.*

Résumé — Caractérisation et évolution des fissures au cours de la simulation hydraulique du réservoir géothermique de Soultz-sous-Forêts — Les récents résultats obtenus en tomographie sismique dans le cadre du projet européen EGS (*Enhanced Geothermal System*) offrent un nouvel éclairage sur les propriétés de fissure du site expérimental de Soultz-sous-Forêts. Dans cet article, une nouvelle approche associant modélisation par milieu effectif et bi-réfringence des ondes S a été développée afin de caractériser les propriétés des fissures au cours d'une stimulation hydraulique. Le test hydraulique mené en 2000 comporte 3 phases d'injection réparties sur 6 jours. Une première tomographie 4D précise a été réalisée à partir des temps de trajet mesurés sur la sismicité induite [Calò M., Dorbath C., Cornet F.H., Cuenot N. (2011) Large-scale aseismic motion identified through 4-D P-wave tomography, *Geophys. J. Int.* **186**, 1295-1314]. La présente étude s'attache à utiliser les modèles de vitesse obtenus en ondes P (Calò *et al.*, 2011) mais également en ondes S. Ces vitesses sont les données d'entrée d'un modèle anisotrope de milieu effectif que l'on construit pour traduire les vitesses en propriété de fissures. Le modèle est fait d'une matrice saine et de 2 familles de fissures modélisées sous forme de pièces de monnaie. Leurs orientations géométriques sont établies à partir des données géologiques : orientation Nord-Sud et pendages respectifs de 65° Est et 65° Ouest [Genter A., Traineau H. (1996) Analysis of macroscopic fractures in granite in the HDR geothermal well EPS-1, Soultz-sous-Forêts, France, *J. Vol. Geoth. Res.* **72**, 121-141]. Les résultats en sortie du modèle de milieu effectif sont les longueurs et les ouvertures de fissures dans le réservoir géologique. Nous mettons en évidence qu'une augmentation du débit d'injection entraine un rétrécissement en longueur des fissures dans la zone tomographique couverte en vitesses P et S, notamment dans la partie supérieure du réservoir. Inversement, les longueurs des fissures calculées au cours des phases d'injection continue sont supérieures à celles calculées au cours des phases d'augmentation brutale du débit. Une explication possible réside dans le fait que les fissures restent ouvertes sur une longueur importante au cours des phases de plateau car le système a le temps de se relâcher. Nous avons également calculé les taux d'ouverture et d'allongement des fissures au cours des différentes phases de l'injection. Alors que le taux d'ouverture reste globalement le même, le taux d'allongement est très variable. Il est plus important autour du puits ouvert et au-dessus qu'en dessous. Ceci pourrait être le signe d'un chemin préférentiel de drainage des fluides vers le haut au sein de la formation granitique, peut-être induit par un gradient de température. Nous avons également comparé les évolutions des taux d'allongement durant les phases d'injection avec les résultats de bi-réfringence des ondes S. Deux types d'ondes S ont été enregistrées à deux stations au cours de l'injection et ont donc pu être interprétées en terme de paramètres de bi-réfringence. L'orientation du plan de polarisation de l'onde S rapide reste constante et parallèle à l'orientation de la contrainte horizontale maximale alors que l'amplitude de la réfringence varie au cours du temps. On observe une bonne cohérence entre les différences de temps de trajet entre les ondes S lentes et rapides et les taux d'allongement des fissures au cours des 4 premiers jours de la stimulation hydraulique. Au-delà, ces deux paramètres divergent. Finalement, cette étude met en évidence l'importance de coupler différentes méthodes issues de la géophysique et de la géomécanique, ici la modélisation en milieu effectif et la bi-réfringence des ondes S, dans le but de mieux caractériser la fracturation dans des réservoirs géologiques.

INTRODUCTION

The European Enhanced Geothermal System (EGS) project at Soultz-sous-Forêts (France) started in 1987. Its main objective was the extraction of energy from hot fractured granites. The test site is located in France on the western edge of the Rhine graben, some 50 km north of Strasbourg close to the German border. Large-scale faults are observed with distinct major orientations that can be characterized from outcrop measurements. The main orientation is known as the Rhenish orientation: it corresponds to normal faults striking N10°-N20°. These faults were created due to the development of the graben during the Oligocene. The analysis of outcrop observations is consistent with the imagery logs collected for different wells.

The concept of a geothermal system consists in extracting the heat from the rock mass by circulating fluids between injectors and producers. With the natural permeability of granite being too low, it was enhanced through hydraulic stimulation. The objective was to create a network of connected and permeable fractures between the wells. Two wells, named GPK1 and GPK2, were first drilled respectively in 1987 and 1995, down to depths of about 3 600 and 3 900 m. The depths reported in this paper are the True Vertical Depths (TVD) measured from the drilling platform. At these depths, the bottom hole temperature was 160°C. In 1999, the GPK2 well was deepened to about 5 000 m, increasing the bottom hole temperature to 200°C. Then, in 2003 and 2004, two deviated wells reaching depths of 5 000 m each, GPK3 and GPK4, were drilled from the same platform as GPK2. The GPK2 well was subjected to a stimulation program in 2000 to enhance the rock mass permeability. Water was injected with an increasing stepwise rate over a period of 6 days and the well was then shut in (Fig. 1). The well head pressure was monitored all along the stimulation process plus a few days after shut-in.

Many geophysical studies in reservoir characterization focus on the variations in the elastic properties of rocks. They commonly involve seismic data, which are processed in terms of seismic attributes: amplitudes, time-shifts, velocities, impedances, etc. These processed data still have to be related to the physical properties of the rock mass and the fluids saturating the pore space. This is essential to understand the impact of stimulation on the rock system. Such need motivated the development of research projects referring to the Effective Medium Theory (EMT).

The EMT makes it possible to evaluate the macroscopic properties of a medium within a representative elementary volume provided that its microstructural properties are known. In particular, the EMT has focussed on fractured media (Kachanov, 1993; Sayers and Kachanov, 1991; Schoenberg and Sayers, 1995), where fracturing is a major factor affecting the velocities of the seismic waves propagating in the rock mass. In short, velocity variations can be explained in terms of variations in crack properties: crack density, aspect ratio or orientation. A very special characteristic of the Soultz site is the intense alteration of the granitic basement due to fluid flows. This has led to the identification of various fracture networks depending on their contributions to fluid flow within the reservoir (Sausse and Genter, 2005; Sausse *et al.*, 2006, 2010; Dezayes *et al.*, 2005, 2010). At the first order, the flow behavior of the fracture networks in granite is driven by the connectivity between fractures. This parameter is very sensitive to the fracture density and fracture geometry.

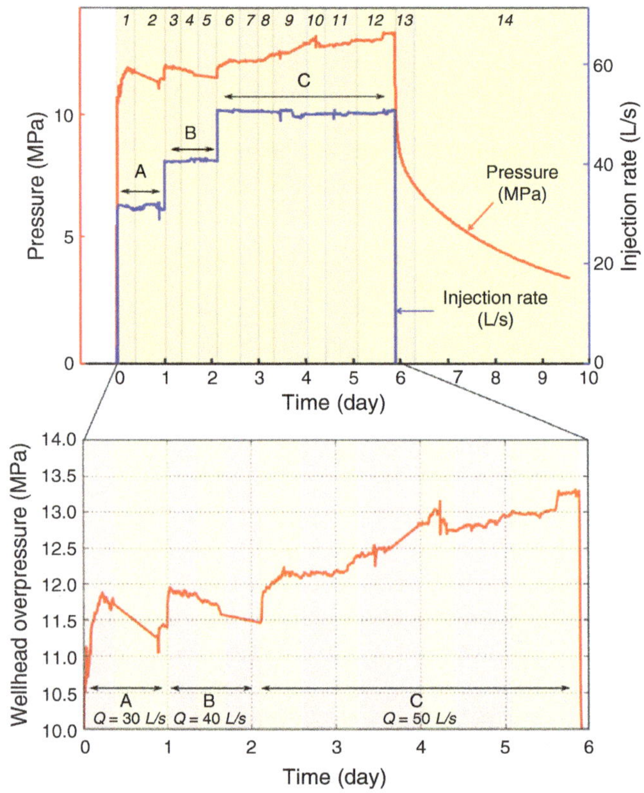

Figure 1

Wellhead injection rate (blue) and overpressure (red) in GPK2 during the 2000 stimulation experiment. The three main steps are tagged by A, B and C. The 14 steps used for the 4-D tomography, according to Calò *et al.* (2011), are annotated above the two curves with black italic numbers.

The inverse problem consists in deriving information about the microstructure from the available seismic attributes. This can be achieved by integrating the forward effective model into an optimization process. Such idea was developed in active geological areas by Zhao and Mizuno (1999) and Mishra and Zhao (2003) who characterized the hypocenter locations of the Kobe (1995) and Buhj (2001) earthquakes in terms of crack density and saturation rate. On the other hand, Adelinet *et al.* (2011) focused on passive microseismicity data: they estimated the spatial distributions of the crack density and the fluid bulk modulus beneath the Reykjanes Peninsula thanks to accurate tomography data.

In this paper, we refer to EMT to analyze how hydraulic stimulations influence fracturing in the Soultz granitic reservoir. The leading idea is to take advantage of the available velocities to characterize the evolution of meso-crack features during stimulation. In other words, we develop an anistropic effective medium model that provides numerical responses for seismic velocities. These responses are then compared to the velocities obtained from tomography. The overall procedure consists in adjusting the input parameters of the EMT model, which are crack densities and crack apertures, to minimize the velocity mismatch. We then address the following questions: can seismic information be used to infer crack features (length and aperture)? Do cracks remain the same during water injection?

The first section of this paper recaps how EMT modeling is used to infer crack features from the data collected during the hydraulic stimulation of GPK2. The second section focuses on the analysis of the shear-wave splitting evidenced at 2 stations close to GPK2 in terms of anisotropy (amplitude and orientation). The original contribution of this paper is the subject of the third section where the EMT results are compared to the information derived from the shear-wave splitting analysis. Finally, it makes it possible to establish a physically-based model to explain the reservoir behavior during hydraulic stimulation.

1 INPUT DATA

1.1 Hydraulic Parameters and 4-D Tomography Data

The stimulation test performed in 2000 at well GPK2 aimed to enhance the connectivity between the open section of GPK2 and the surrounding fractured medium. Well GPK2 was open between 4 403 and 4 955 m over a section of about 500 m length. The seismic monitoring performed during this test includes two distinct periods: monitoring during injection over 6 days and monitoring during shut over 4 days after. The first period of time encompasses 3 consecutive steps during which the injection rate was first set to 30 L/s for 24 h (indicated as injection A in Fig. 1), then to 40 L/s for 27 h (injection B) and last to 50 L/s for 90 h (injection C). The total volume of fluid injected was about 23 000 m^3. The injection rates and pressures collected at well GPK2 during stimulation are reported in Figure 1. At the very beginning, the overpressure increases very quickly to 11 MPa (bottom plot of Fig. 1). Then, it goes on increasing, but moderately, to about 12 MPa, before declining. The pressure behavior observed during the second injection step is very similar: a fast rise to 12 MPa followed by a gradual drop. The third step led to a different pressure response. The pressure again rises up at once to 12 MPa and then, it keeps on increasing smoothly till the well shut in, followed by a slow pressure decrease. Dorbath *et al.* (2009) concluded that the rock mass around well GPK2 could be still considered at this stage as quasi-closed. In other words, the stimulation test did not succeed in connecting the well to a major fracture network.

Intense microseismic activity was monitored during and after the stimulation test from a dense seismic network consisting of 3 down-hole array and 14 surface stations (Fig. 2) (Cuenot *et al.*, 2008). This makes it possible to perform a 4-D tomography study from about 7 000 well-located events (Fig. 2) which had duration magnitude from −0.9 to 2.5. Calò *et al.* (2011) reported the results for the P-wave tomography. The shear-wave velocities used in the present study are presented in Appendix A. The total set of events was split into 14 subsets based upon the time injection scheme, according to Calò *et al.* (2011). These subsets are indicated by the black italic numbers 1 to 14 in Figure 1. The corresponding periods of time have distinct durations (Tab. 1). This choice was preferred to better capture the influence of

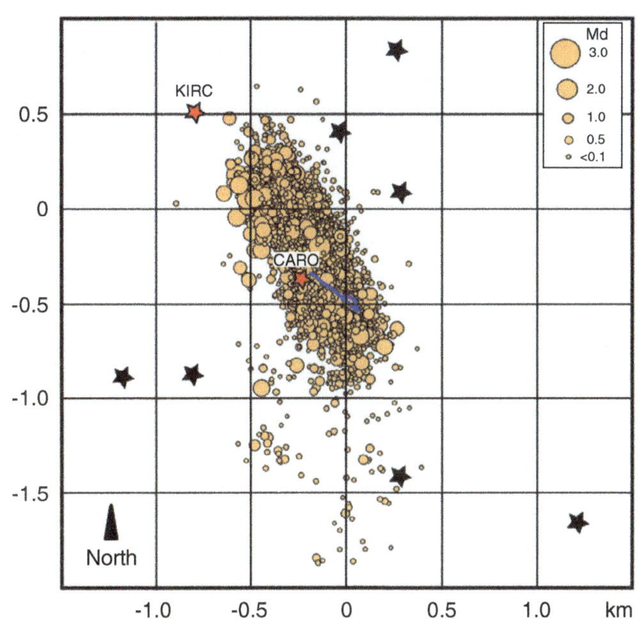

Figure 2

The 2000 stimulation of GPK2: horizontal projection of the cloud of induced seismic events. The radius of each circle is proportional to the magnitude. The thick blue line indicates the open-hole section of the injection well. Surface seismic stations are identified by stars, the red ones are used for the shear-wave splitting study.

TABLE 1

Concordance table between sets of tomography data and the corresponding time of the middle of the sets expressed in days

Sets	1	2	3	4	5	6	7	8	9	10	11	12	13	14
Days	0.3	0.95	1.6	2.1	2.6	3.25	3.85	4.25	4.95	5.75	6.4	7.4	8.3	10.8

the successive stimulation phases on the variations in seismic velocities. A double-difference tomography code (Zhang and Thurber, 2003) has been used to solve event locations and velocity structure simultaneously by taking into account both absolute and differential traveltime data. The initial 1-D reference velocity models used in the tomography workflow are presented in Table 2. The tomography results include velocity maps at depths between 4 000 and 5 000 m and centered on well GPK1 for every subsets. The initial horizontal nodes spacing for the tomography is 250 m in the *X* and *Y* directions near the center of the grid (Calò *et al.*, 2011). After coupling a tomographic method (tomoDD) and an averaging method (WAM, described in Calò, 2009), the final grid has a resolution of 100 m in the three space directions (Calò *et al.*, 2011). Whatever the subset under consideration, the maximal seismic velocity standard deviation is less than 0.03 km/s over the resolved grid. The resolved area depends on the seismic velocity models and subset studied: it is larger for compressional-wave than for shear-wave velocities (Fig. 3). We focus hereafter on the area jointly resolved in compressional and shear-wave velocities. Calò *et al.* (2011) observed a low compressional-wave velocity anomaly located around the GPK2 open-hole section with a magnitude of about 10% of the mean velocity. The anomaly is detected every time the injection rate is increased. Otherwise, during the periods of constant injection rates, the anomaly almost vanishes.

TABLE 2

1-D reference velocity model used in the tomography workflow

Depth (m)	Compression-wave velocity (km/s)	Shear-wave velocity (km/s)
4 100	5.860	3.349
4 600	5.870	3.354
5 100	5.885	3.363

1.2 Log Data Available at Well GPK2

The GPK2 well was drilled in 1995 down to 3 883 m. Two preliminary hydraulic stimulation tests were carried out in 1995 and 1996. Then, GPK2 was deepened in 1999 to 4 955 m before performing the 2000 massive hydraulic stimulation. The casing of the well was installed to a depth of 4 403 m, meaning that the open-hole section was between 4 403 and 4 955 m. Sonic and Gamma-Ray logs were recorded between 1 500 and 3 720 m depth after the first drilling. Deeper, the well crosses large cavities. They have been explained as the intersection with a major kilometric fault (Sausse *et al.*, 2008). There is no sonic log below 3 900 m due to a casing collapse close to a significant cavity.

An initial set of Ultrasonic Borehole Imager (UBI) images collected at the end of the first drilling phase provides a description of the fracture pattern along the well for depths between 1 422 and 3 807 m. During the deepening phase, a second UBI set was acquired to image the borehole

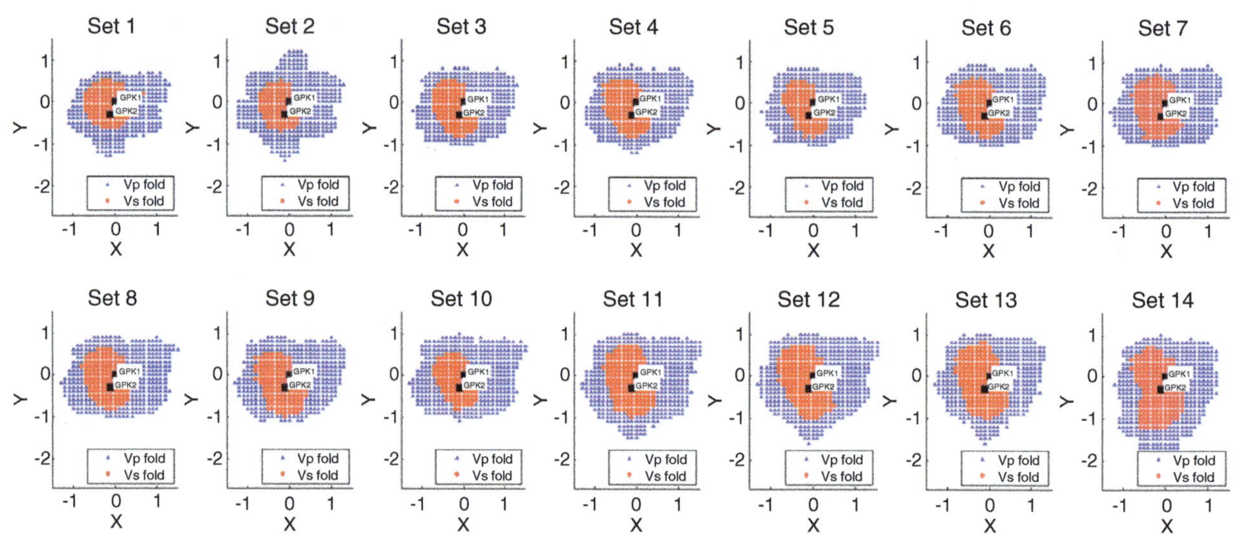

Figure 3

Resolved area for velocities from tomography investigation at 4 600 m depth for both compressional (blue) and shear (red) waves. The area resolved for compressional waves is the same than the one presented in Calò *et al.* (2011). GPK1 wellhead is the geographical origin for the gridpoints. Axes *X* and *Y* correspond to geographical distances expressed in km. North direction is vertical upward. Black squares: GPK1 and GPK2 locations.

between 3 479 and 3 866 m. The analysis of these UBI data sets reveals two crack families: natural fractures and Drilling Induced Tensile (DIT) fractures (Bérard and Cornet, 2003). As a result, the crack density estimated from UBI data collected in GPK2 cannot be directly used to quantify the natural fracturing within the reservoir. UBI data were also collected in deeper wells GPK3 and GPK4.

2 EFFECTIVE MEDIUM MODELING AND CALIBRATION

2.1 Theoretical Background

A challenging issue in many geophysical studies is to express the variations in the measured responses (*e.g.*, seismic travel time or waveform amplitude) in terms of variations in the physical properties of rocks. In the case studied in this paper, this consists in explaining the evolution of seismic velocities with time against the evolution of fracture properties with time. The link is ensured referring to effective medium modeling. The EMT relies on the Representative Elementary Volume (REV), *i.e.* the smallest volume over which a property is considered as a value representative for the whole. The effective medium model used hereafter is a homogenization approach based on Eshelby's inclusion theory (Eshelby, 1957). It involves an ellipsoidal inclusion embedded within an infinite medium. The final objective is to evaluate the disruption in the strain field in the medium due to this inclusion. Following Eshelby (1957), the strain tensor ε^i in inclusion i can be expressed as:

$$\varepsilon^i = \left(I + P : \left(C^i - C^0\right)\right)^{-1} : E^{\text{inf}} \tag{1}$$

C^i and C^0 are the elastic tensors for the inclusion and matrix, respectively. I is the four-order identity tensor, P the polarization Hill's tensor, and E^{inf} the stress field far away from the inclusion. If the inclusion is flattened, as in cracks for instance, parts of the four-order tensor $(I + P : (C^i - C^0)$ approach infinity. In this case, the relevant tensor to describe strain is (Barthélémy, 2005):

$$\lim_{\xi \to 0} \left(I + P : \left(C^i - C^0\right)\right)^{-1} \tag{2}$$

where ξ is the crack aspect ratio. In this study, we consider "penny-shaped" inclusions to approximate cracks and assume that these cracks are embedded into an isotropic matrix. Cracks are filled with water whose bulk modulus is 2.2 GPa. In addition, as there are many cracks within the matrix, we refer to homogenization to derive the macroscopic elasticity at the REV scale. A number of techniques have been studied in the literature as the diluted scheme, the self-consistent scheme (Budiansky and O'Connell, 1976) or the differential effective modeling (Henyey and Pomphrey, 1982; Norris, 1985). As granite has a very low porosity resulting from cracks only (Schubnel *et al.*, 2003; Guéguen and Schubnel, 2003), we chose to use the Mori-Tanaka approach (Mori and Tanaka, 1973) in which we consider cracks embedded into a matrix with the elastic properties of an inclusion-free medium. Such an approach is appropriate for rocks characterized by low porosities (< 10%) and clearly identified matrices. Thus, this approach is well-suited for estimating the elastic properties of the Soultz granite.

Within this framework, cracks are described by two parameters: volumetric density ρ and aspect ratio ξ. The crack density is defined in a volumetric way following Walsh (1965) and Budiansky and O'Connell (1976):

$$\rho = \frac{1}{V}\sum_{i=1}^{n} {c_i}^3$$

V is the total volume of the REV and c_i the radius of the i-th penny-shaped crack. This volumetric density is related to the P_{32} fracture intensity (ratio between area of fractures and volume of rock mass) and the mean radius of cracks, noted $\overline{c}$, by:

$$\rho = P_{32}\frac{\overline{c}}{\pi} \tag{3}$$

The elastic moduli being known at the REV scale, the compressional and shear-wave velocities can be derived from Christoffel's equation:

$$||\boldsymbol{\Gamma}_{ik} - \boldsymbol{\mu}(\mathbf{V}^{\text{model}})^2\delta_{ik}|| = 0 \tag{4}$$

with $\boldsymbol{\Gamma}_{ik}$ the symetrical Christoffel's matrix, δ_{ik} the Kronecker symbol, $\boldsymbol{\mu}$ the matrix density and $\mathbf{V}^{\text{model}}$ the velocity vector. The subscripts i and k correspond to the three main directions of space (ranging from 1 to 3). Solving this equation provides three real and positive eigenvalues, which are the three compressional-wave, slow shear-wave and fast shear-wave velocities. They obviously depend on the direction of propagation. In our model, we account for the wave paths derived from the tomography study and consider a propagation cone with a maximum deviation angle of 40° with respect to the vertical axis.

The elastic behavior of a medium with randomly oriented cracks has been exhaustively studied in the literature (Barthélémy, 2009; Budiansky and O'Connell, 1976; Henyey and Pomphrey, 1982; Hudson, 1980; Kachanov, 1993; Kachanov *et al.*, 1994; Kemeny and Cook, 1986; Sayers and Kachanov, 1991; Schoenberg and Sayers, 1995; Walsh, 1965). However, natural fractures are not randomly distributed within the Soultz EGS site (Valley, 2007).

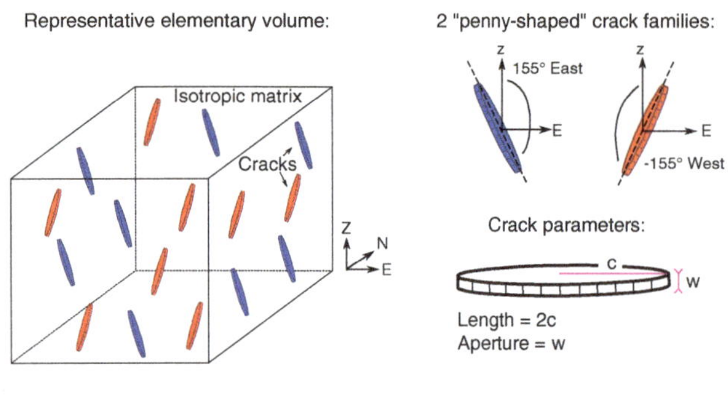

Figure 4

Effective medium modeling: parameters of the fractured model.

TABLE 3

Crack orientation for the effective medium model

Crack family	Azimuth angle	Dip angle	Geological features
Family 1	180°	155° E	Fracture set striking N-S with dip direction to the East
Family 2	180°	−155° W	Fracture set striking N-S with dip direction to the West

Their orientations have been driven by the complex geological history of the upper Rhine graben. The effective medium model is built assuming a preferential penny-shaped crack orientation in line with the Rhenish direction. In addition, the Soultz zone is within a horst area, which is bounded by subvertical N-S trending normal faults (Genter and Traineau, 1996). As a result, we develop an anisotropic effective medium model (Fig. 4) with two sets of penny-shaped cracks with orientations defined from two angles in the spherical coordinate system: the θ dip angle measured from the vertical axis and the φ azimuth angle measured from the East axis, following the mathematical convention. Referring to both outcrop observations and borehole imagery, we consider the following crack orientations (Tab. 3): two North-South crack families with 155°dip, respectively eastward and westward. Two parameters characterize crack shape: length (two times the radius c_i) and aperture (denoted w in Fig. 4).

2.2 Inverse Modeling

As shown above, the simulated velocities depend on microstructural properties or parameters (crack density, crack radius and aperture) and matrix elastic parameters (bulk and shear moduli, denoted K_0 and G_0, respectively). We now aim to solve the inverse problem. In other words, we want to estimate the microstructural parameters from the known compressional- and shear-wave velocities. From previous studies, we assume that water injection influences only the shape of the cracks, not their number. Indeed, Horálek *et al.* (2010) studied a pool of 45 micro-earthquakes with magnitudes between 1.4 and 2.9, which occurred during and after the 2003 massive fluid injection in the GPK3 borehole. They pointed out a massive dominance of double-couple events indicating that shear-slip is the main mechanism of the investigated micro-earthquakes. Therefore, microseismicity is mainly due to sliding on preexisting failure planes and tensile fractures, assumed to be created during injection phases, are rare.

Inverse problems can be solved using various optimization techniques (Tarantola, 2005). Within this framework, the leading idea consists in minimizing a given objective or cost function in order to find the unknown parameters. In the case under consideration, the objective function (J) is defined as:

$$J(x) = \frac{1}{2}\left(1 - \frac{V_P^{\text{model}}(\mathbf{x})^2}{V_P^{\text{tomo}^2}}\right)^2 + \frac{1}{2}\left(1 - \frac{V_S^{\text{model}}(\mathbf{x})^2}{V_S^{\text{tomo}^2}}\right)^2 \quad (5)$$

This two-term function quantifies the data mismatch in a least-square sense. The first term measures the misfit between the squared compressional-wave velocity obtained from tomography data (V_P^{tomo}) and the corresponding numerical responses ($V_P^{\text{model}}(x)$). The second term is very similar, but focuses on shear waves. Superscripts "tomo" and "model" refer to the tomography data and the corresponding numerical modeling results, respectively. Vector $\mathbf{x}$ encompasses the unknown parameters. It includes four components: the mean crack radius, the mean crack aperture and the matrix bulk and shear moduli. The minimization of the J objective function provides a $\mathbf{x}$ vector so that the difference between the simulated responses and data is as small as possible. To alleviate some of the drawbacks related to ill-posed problems, the unknown parameters are submitted to bound constraints (Tab. 4). The ones for matrix bulk and shear

TABLE 4

Bound constraints and initial values used during optimization process

Optimization parameter	Bound constraints	Initial value for optimization process
Crack length	[0-3]	1
Crack aperture	$[1\times10^{-5}\text{-}1\times10^{-3}]$	1×10^{-4}
Matrix bulk modulus (GPa)	[47-53]	50
Matrix shear modulus (GPa)	[25-35]	30

moduli are selected on the basis of the 1-D reference velocity model used in the tomography workflow (Tab. 2) as explained hereafter. Crack aperture is initialized to 10^{-4} (corresponding to an aspect ratio of 10^{-4} with a length of 1). It can vary between 10^{-5} and 10^{-3}, which is a reasonable range for crack aperture in granites (Schubnel *et al.*, 2003, 2006). Crack length boundaries are chosen regarding on the relationship with volumetric crack density expressed in Equation (3). Indeed a unit density can be more or less considered as a threshold for mechanical failure (Guéguen *et al.*, 1997). In order to be far from this threshold, the upper crack length boundary is fixed to 3 in order to reach a maximal crack density of 0.5 according Equation (3). The minimization problem is then solved using the (Bound Optimization BY Quadratic Approximation BOBYQA) local algorithm. This derivative-free algorithm solves bound constrained optimization problems using a trust region methods that forms quadratic models for the objective function by interpolation. The interested reader can refer to Powell (2009) for more details. This method has been chosen principally because of its speed of convergence.

At this stage, an essential issue is related to scale. It is commonplace to apply the EMT at the rock sample scale, less at the field scale. At the sample scale, the EMT relates microcracks and ultrasonic velocities. At the field scale, it links meso- or macro- fracturing to log and/or seismic data. In both cases, the scaling relationship between interrogating wavelength and REV size are similar. Furthermore, EMT methods allow to derive unrelaxed elastic moduli. These models are scale independent as long as the wavelength is larger than the REV scale. The tomography analysis used as a basis in this study resulted in passive seismic data in the frequency range 1-50 Hz with a prominent frequency around 10 Hz. Assuming a mean P-wave velocity of 6 km/s (for the whole ray paths, from the surface to the foci zone), and a mean frequency of 10 Hz corresponds to a wavelength of 600 m. Further, the horizontal dimension of the tomography mesh is 100 × 100 m. As a result, we consider that the REV of our EMT model is the same size as the tomography grid. This hypothesis is relevant because the REV is smaller than the wavelength. Thus, we claim that the fracturing scale of interest in this study is of the order of meters. We aim to characterize the meso-scale fracturing, which is the same scale than UBI observations (Valley, 2007).

2.3 Calibration of the Initial Crack Density and Matrix Parameters

Surface crack density is assumed to be constant within the REV. Its value could be derived from borehole data. Unfortunately, there is no UBI data available in well GPK2 below 3 860 m depth. Instead, we refer to the UBI data collected in deeper wells GPK3 and GPK4. Dezayes *et al.* (2000) estimated the cumulative fracture density in these two wells from 1 500 to 5 200 m depth. In the resolved tomography area, we inferred a 1-D crack density, often named P_{10}, which is the number of fractures divided by length of scan-line. Thus, a constant P_{10} crack density equal to 0.5 fractures.m^{-1}. P_{10} is independent of the fracture length. In our modeling we need the surface crack density P_{32} which corresponds to the area of fracture divided by the volume of rock mass. Even if we know that it is not obvious to link 3D crack density and 1D parameter like P10, especially in such anisotropic medium, we make the assumption to have a linear relationship between P_{10} and P_{32} and thus we put into the EMT model a constant P_{32} of 1 m^{-1}.

Another issue is the estimation of the crack-free matrix bulk and shear moduli, K_0 and G_0. These two parameters are considered as unknown and are adjusted when minimizing the objective function. However, the inverse problem being ill-posed, the variations in these parameters have to be bounded to ensure regularization. To do so, we refer to the 1-D reference P-wave velocity model inferred from log data and calibration explosions within the framework of the tomography survey (Tab. 2). Then, the Vp/Vs ratio is assumed to be constant and set to 1.75. This provides the shear-wave velocities. The derivation of the bulk and shear moduli from these compressional- and shear-wave velocities makes it possible to estimate reasonable upper and lower bounds for these parameters. Thus, K_0 is considered to fall in the range of 47 to 53 GPa and G_0 of 25 to 35 GPa.

The primary calibration phase being completed, we move to the inversion process itself. The inversion is independently performed for every grid blocks populating the resolved sub-domain within the tomography grid. It entails the minimization of the J objective function by adjusting the unknown parameters: the matrix elastic moduli and above all, the crack lengths and apertures. Inversion results were not smoothed from one mesh to others. A sensitivity study performed to assess the reliability of our optimization results is presented in Appendix B.

3 SHEAR-WAVE SPLITTING METHODOLOGY

One of the main objectives of this paper is the comparison of the previous effective medium-based results with geophysical data such as the shear wave splitting.

Cracked isotropic media may exhibit anisotropic properties with cylindrical symmetry (Nur, 1971; Anderson *et al.*, 1974). A lot of attention has been paid to such media, which are of great interest in seismic exploration and reservoir production (Al-Harrasi *et al.*, 2011). The interested reader can refer to Crampin (1984) for more details.

Shear wave splitting is typically used as a tool for detecting anisotropy, in particular at reservoir scale

(Verdon *et al.*, 2009; Wuestefeld *et al.*, 2011). When a polarized shear wave enters an anisotropic homogeneous medium, it splits into two orthogonally polarized shear waves with distinct velocities. There are two splitting parameters of interest: δt, the travel-time difference between the fast and slow waves and the orientation of the fast shear-wave polarization that is defined by the ψ angle in the plane perpendicular to the ray path. These two parameters make it possible to estimate the orientation and intensity of the anisotropy in the medium. Splitting has been attributed to the effects of extensive-dilatancy anisotropy, the distributions of stress-aligned fluid filled microcracks and preferentially oriented pore-space (Crampin and Booth, 1989). The pre-existing vertical cracks parallel to the direction of the maximum compressive horizontal stress (Sh_{max}) can open during injection while the others cracks remain closed. The polarization of the vertically fast shear-wave particle motion direction is parallel to the direction of the preferred orientation of the open cracks. In other words, the fast wave polarization is parallel to the direction of the maximum horizontal compression.

High-frequency shear waves split completely into two separate shear waves, which ease the determination of the splitting parameters. They are derived from two linear orthogonal components of the particle motion. However, the splitting parameters are even more accurately estimated from an elliptical particle motion due to the interference of fast and slow shear-waves. Parameter identification was performed using a version of the SPLIT shear code (Vecsey *et al.*, 2008) that has been modified to handle the Soultz reservoir scale.

Shear-wave splitting is routinely observed in Soultz at the 6 surface 3-component stations. However, the records for only 2 of them, CARO and KIRK (Fig. 2), could be used to analyze the full particle motion ellipse as they were not affected by any disturbance due to converted arrivals.

The data processing was manually performed following in two steps (Fig. 5):

- signal pre-processing: we selected events with magnitude greater than 1 and high signal to noise ratio. Then, we chose a 0.5 s length window containing shear-wave and re-sampled the signal to 1 000 samples/s using the Wiggins (1976) interpolation method;
- splitting evaluation: we input station and event coordinates to rotate the signal to ray coordinate system (Sileny and Plomerova, 1996), set interactively the window length from the beginning of the S signal until the particle motion was properly constructed. We applied an Eigenvalue method as recommended by Vecsey *et al.* (2008) for direct shear-waves, and calculated error estimates using a bootstrap method.

Finally, we only considered the results with δt and ψ errors less than 0.15×10^{-2} s and 10°, respectively.

4 RESULTS

4.1 Crack Features Maps

Crack properties evolve with time because of hydraulic stimulation. Figures 6 and 7 display the results obtained for inversion of crack features (aperture and length). Figure 6 corresponds to North-South vertical cross-sections including well GPK2. Both crack lengths and crack apertures are presented in Figure 6. Besides, Figure 7 shows horizontal slices with only the evolution of it normalized, the reference state being 1. Therefore, values larger than 1 indicate crack extension while values lower than 1 are associated with crack shortening. While it is easy to interpret crack extension from a physical point of view, it could be difficult to understand the meaning of crack shortening. In this case, we have to consider that cracks are not stimulated on all their lengths. We may envision that a pore pressure drop within the crack network can contribute to close part of pre-existing cracks.

Figure 6 shows that crack aperture is not an influential parameter during stimulation test: it remains more or less the same as the initial value of 10^{-4} as shown by large green areas for all sets. This result is in agreement with theoretical studies (Simpson *et al.*, 2001; de Dreuzy *et al.*, 2002). On the contrary, crack length exhibits significant variations during hydraulic stimulation. Right after the beginning of injection, for the time periods identified by sets 1 and 2, large blue areas are detected on the crack length maps: they correspond to high crack length values. When the flow rate is set to 40 L/s, the crack length first decreases, especially close to the open-hole section of GPK2 (set 3), before increasing again but in a moderate and diffuse way (sets 4 and 5). Again, the following flow rate increase to 50 L/s contributes to reduce crack length in the vicinity of GPK2 open-hole section (white cloud visible on set 6), similar to what is observed for set 3. For the following sets of the C phase, (sets 7 to 12), we observe higher values in the vicinity of the open-hole section (below 4.6 km depth). This general picture does not change within this period of time whatever the depth is. When hydraulic stimulation is over (sets 13 and 14), crack length tends to reach the reference state, even if there is still extension in the vicinity of the GPK2 well.

The horizontal cross-sections reported in Figure 7 evidence lateral variations in crack lengths at 3 depths: 4.2 (close-hole section), 4.6 (open-hole section) and 5.0 (below GPK2 borehole). It shows that a disturbed area, *i.e.* an area with larger cracks, is centered close to the injection well. The injection phases lead to an enlargement of this area. After shut-in, the area reduces once again around GPK2 (sets 13 and 14).

The relevance of our inversion results can be strengthened by the analysis of the *J* objective function evolution (Tab. 5). Whatever the target set, the final normalized objective

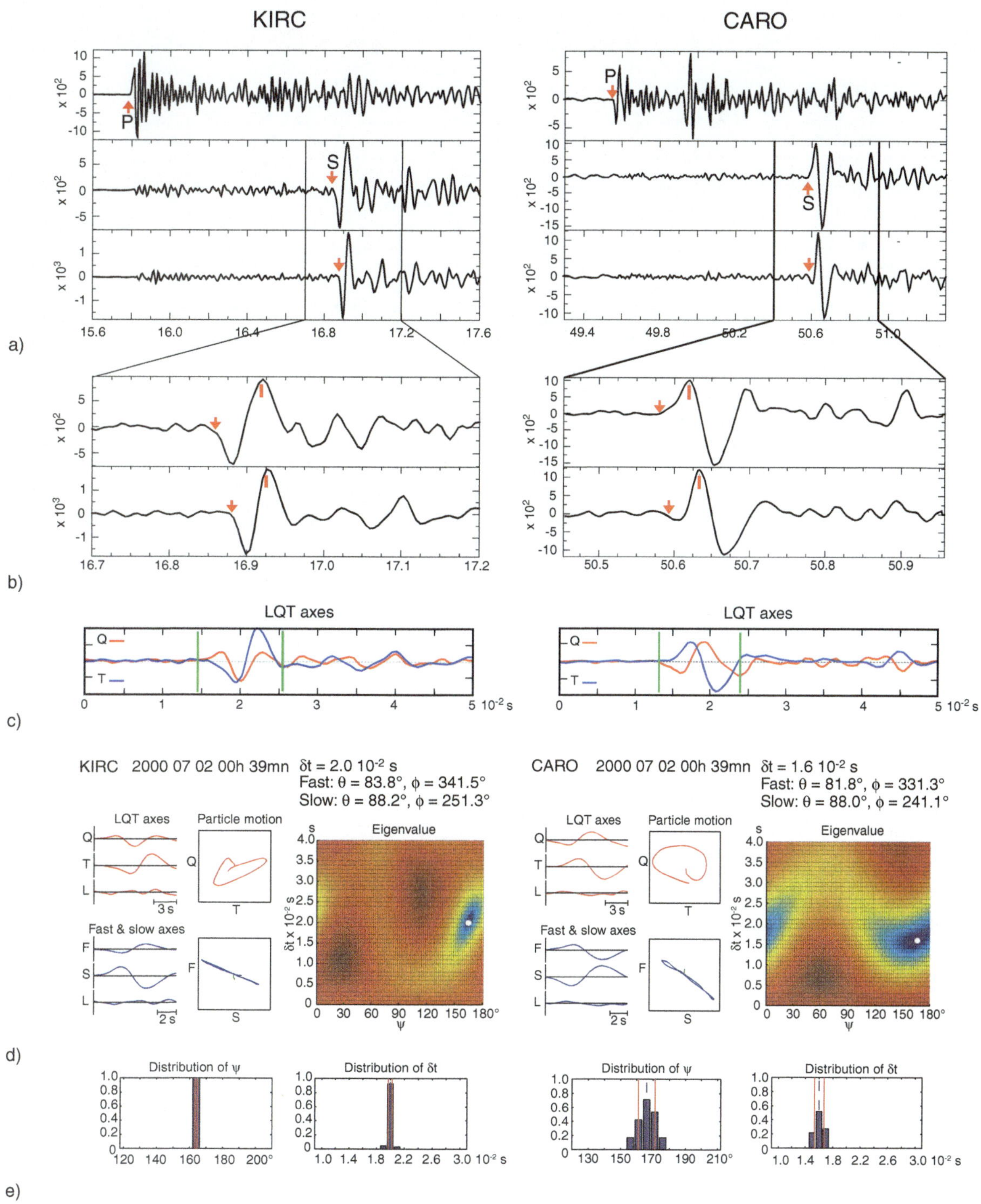

Figure 5

Steps for the shear-wave splitting processing at stations KIRC and CARO. The example shown is an event that occurred on July 2, 2000, at 00h 39min UTC. a) the three components of the seismic signal (Vertical, North-South, East-West), b) magnification of the shear-waves on the horizontal components, c) rotation of the seismic signal to the ray coordinate LQT system (L – longitudinal, Q – radial, normal to [L-T (transverse)] plane) and selection of the time window for splitting evaluation, d) left: construction of the elliptically polarized particle motion in the (L, Q, T) coordinates and rotation to the F (fast) and S (slow) axes, where the particle motion should be linearly polarized; right: diagram of the misfit function obtained by Eigenvalue method. The white circle in the blue minimum domain shows the splitting parameters: time delay δt of the slow split shear-wave and rotation angle Ψ of the fast wave polarization in the (Q, T) plane. The fast polarization direction Ψ is defined by two Euler angles, azimuth ϕ and inclination θ, e) histograms for splitting parameters Ψ and δt. See Vecsey *et al.* (2008), for a complete description.

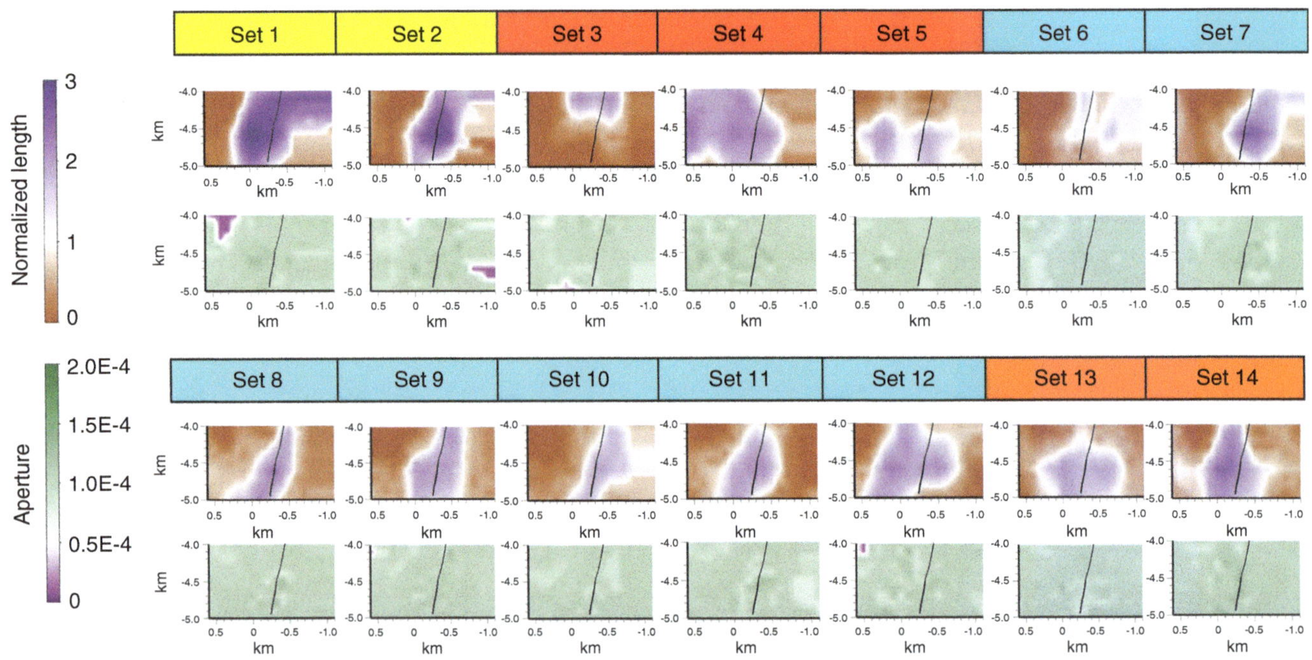

Figure 6

North-South vertical cross sections with crack length (orange – purple color scale) and crack aperture (purple – green color scale) variations with time. The dark curve in the middle of the maps corresponds to well GPK2 trajectory, the thicker parts of it being the open-hole section.

function is very small (mean value around 10^{-5}). In addition, we check that the objective function was decreasing monotonically and we observed a single valley with a single minimum, which makes the minimization process easier.

4.2 Extension and Opening Rates

In this study, the two crack parameters are the length and the aperture. We now calculate extension and opening rates. The ε_{ext} extension rate and the ε_{op} opening rate are defined as:

$$\varepsilon_{ext} = \frac{L_i - L_{ref}}{L_{ref}} \text{ and } \varepsilon_{op} = \frac{\omega_i - \omega_{ref}}{\omega_{ref}} \quad (7)$$

L_i is the optimized length, L_{ref} the reference crack length (equal to 0.5 and corresponding to the smallest mean value obtained for set 3 between 4 and 5 km depth), ω_i the optimized aperture and ω_{ref} the reference crack aperture (equal to 5×10^{-5} and corresponding to closed cracks). An extension rate of 100% corresponds to a crack length of 1 and reciprocally an opening rate of 100% corresponds to a crack aperture of 10^{-4}. The reason why we focus on global extension and opening rates is that we want to compare a tomography set to another one.

We first determine the rates for all depths investigated by tomography (Fig. 8a). Second, we focus on given levels in an attempt to capture the influence of the well (Fig. 8b, c): at the level of the open-hole section (4.5-4.9 km depth), above the open-hole section (4.0-4.4 km depth) and below (5.0 km depth).

As already suggested by Figure 6, no significant variation in the opening rate is evidenced when considering the whole depth range. However, if studying more specifically the area above close-hole, it turns out that sets 1 and 2 are associated with fracture opening (Fig. 8b). Then, after set 3, all the curves converge and the rate gets close to 100%. This result may be puzzling if considering only crack aperture. However, as suggested by lab experiments, elastic waves are very sensitive to variations in crack aspect ratio, not crack aperture only. Precisely, the extension rate shows even more variations with time. The same general trend is observed for the three depth sections with some differences for sets 2, 5 and 7 (Fig. 8c). Except the close-hole section, the extension rate is minimal for set 3. Again, this was already emphasized from the analysis of the crack length maps (Fig. 6, 7). A second minimal value is achieved for set 6. Remember that sets 3 and 6 correspond to the time period right after a sharp flow rate increase. Other maximal values are obtained with sets 4 and 9 (Fig. 8a). These sets actually correspond to the middle of the B and C injection phases. The extension rate is higher in the area around the open-hole section of GPK2 (violet/purple curve in Fig. 8c) than in other areas, only except

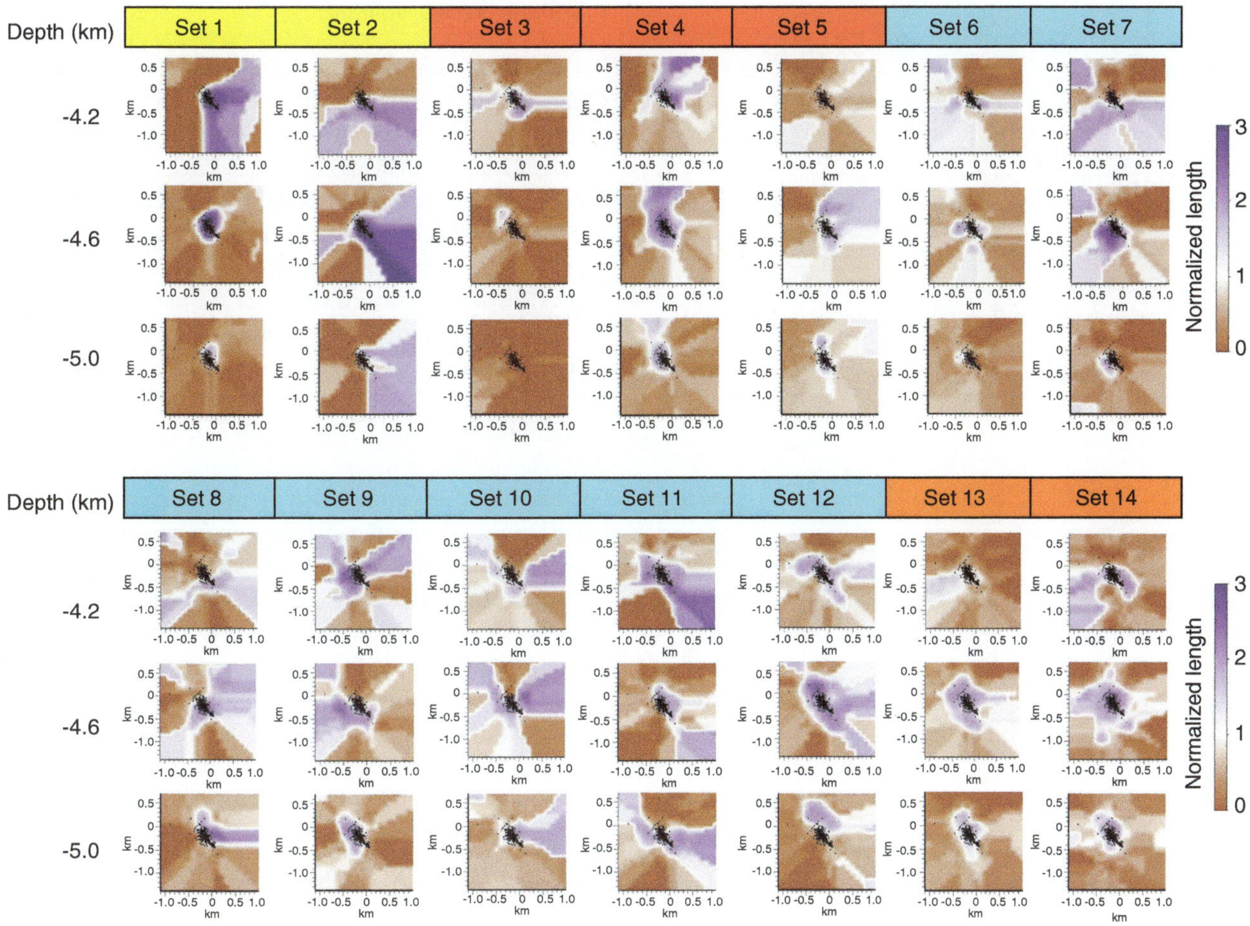

Figure 7

Horizontal cross sections at selected depths with the inverted crack lengths. Black dots are the projections of events used to obtain the velocity models.

TABLE 5

Statistical data for objective function during all the sets. Std-dev stands for 'standard-deviation'

Sets	1	2	3	4	5	6	7
Min value	0.00E+00	0.00E+00	0.00E+00	0.00E+00	0.00E+00	0.00E+00	0.00E+00
Max value	**8.95E-03**	**7.82E-04**	**6.86E-04**	**1.20E-03**	**2.19E-24**	**4.57E-07**	**4.10E-04**
Mean	**1.49E-04**	**3.57E-06**	**1.74E-06**	**1.89E-06**	**2.03E-26**	**3.32E-10**	**3.44E-07**
Std-dev	7.13E-04	3.53E-05	2.68E-05	3.85E-05	9.56E-26	1.23E-08	1.04E-05
Sets	8	9	10	11	12	13	14
Min value	0.00E+00	0.00E+00	0.00E+00	0.00E+00	0.00E+00	0.00E+00	0.00E+00
Max value	**1.66E-24**	**1.44E-04**	**1.54E-04**	**1.19E-03**	**1.82E-04**	**2.01E-24**	**6.41E-07**
Mean	**2.37E-26**	**1.78E-07**	**3.14E-07**	**2.74E-06**	**3.79E-07**	**2.01E-26**	**1.52E-10**
Std-dev	9.65E-26	4.93E-06	5.84E-06	3.73E-05	6.73E-06	8.00E-26	3.34E-08

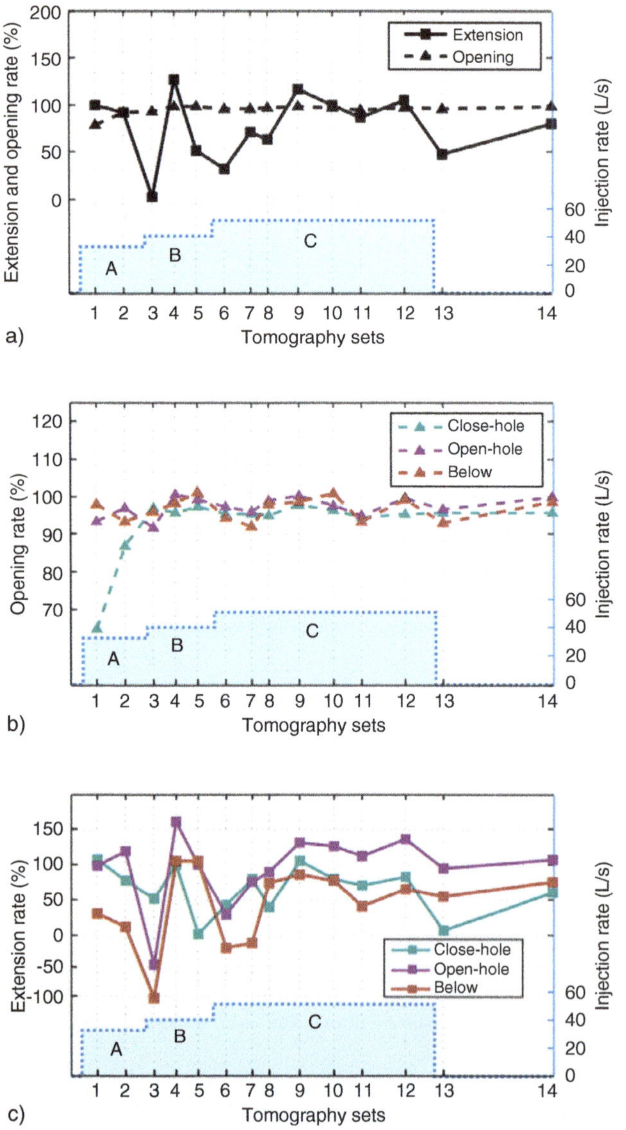

Figure 8

Time evolution of crack feature rates in the geothermal Soultz reservoir during GPK2 stimulation. Injection flow rates and injection phases are also mentioned: A = 30 L/s, B = 40 L/s and C = 50 L/s. a) Rates calculated for all tomography depths, *i.e.* between 4 000 and 5 000 m depth, b) opening rates divided for the open-hole section (4 413-4 955 m depth), the casing section (4 000-4 413 m depth) and the part below GPK2 well (below 4 955 m depth), c) extension rates divided for the same sections than opening rate.

for set 3.This is in agreement with the results shown on the vertical cross-sections (Fig. 6). Maximum values for extension rates are reached for set 4. Second maximum values are reached for close-hole section during set 9 and for the open-hole section during set 12.

4.3 Shear-Wave Splitting

We ended up with about 15 events with high quality splitting evaluation for each tomography subset. We calculated ψ_{mean} and an index of anisotropy $I\alpha = (\Sigma\delta t_i / t_{si}) / N$. δt_i and t_{si} are the significant shear wave splitting value and the shear wave splitting time delay along the ray of the fast shear-wave for the ith event. N is the number of events. We also checked that the centroids of the hypocenters of the selected events are very close (less than 100 m difference) for all subsets. Figure 5 explains the shear-wave processing used in this study.

The anisotropy evidenced on the surface stations can result from various sources like the sedimentary layers above the granitic basement or the stimulation process performed a few years earlier in the granite. As our analysis is based only upon the relative variations in the splitting parameters at the two stations during the 2 000 stimulation, our results are independent of the aforementioned sources and they can be used to quantify the influence of stimulation on crack features. First, we observe two main features:

- the direction of the fast shear-wave polarization does not significantly change during stimulation and,
- its mean value is 155° (Fig. 9).

This direction is roughly parallel to the maximum horizontal stress orientation estimated to be 158° by Dorbath *et al.* (2010) from the inversion of well-constrained focal mechanisms.

The two $I\alpha$ curves (Fig. 10) present similar behaviors: anisotropy decreases when the injection rate increases. This is particularly clear for set 3, when the water rate is increased from 30 to 40 L/s. The same behavior is also observed for set 6, but not as clearly, when the rate is set to 50 L/s. For these two sets, the seismic velocities came back to their initial values (Calò *et al.*, 2011). Another common feature of the two curves is the continuous increase in $I\alpha$ from set 10 till the end of seismic monitoring. This result could be related to the spreading of seismicity in an area getting wider with time at the end of stimulation. The increase is even reinforced after shut-in (Cuenot *et al.*, 2008).

From a mechanical point of view, two trends can be distinguished from $I\alpha$ curves: (1) between set 1 and sets 5/6, the main trend is a slight decrease with many variations and (2) beyond sets 5/6, there is a constant increase. Inspiring from results of Guéguen and Sarout (2011) concerning EMT in the case of vertical cracks, one could suggest that trend (1) corresponds to a re-saturation regime, the injected water replacing the fluid in place, perturbing the saturation state of fractures and also probably the crack features but not permanently; and trend (2) corresponds more to a "damage" regime, with permanent changes of crack features, implying a constant $I\alpha$ increase.

Last, the differences in the absolute $I\alpha$ values (Tab. 6) are easily explained by the locations of the seismic stations

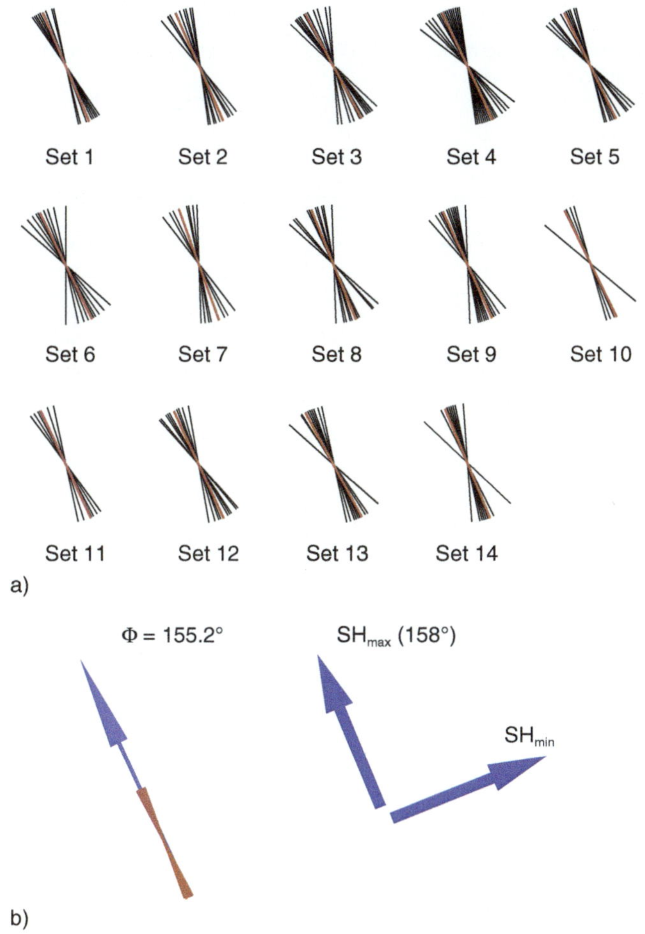

Figure 9

Fast shear-wave polarization angle Ψ at station KIRC; a) comparison of Ψ for the 14 sets: black lines represent individual results, red line the mean value for each set; b) comparison of the mean Ψ value for the 14 sets (left, in blue) to the direction of the maximum horizontal stress Sh_{max} deduced from inversion of focal mechanisms.

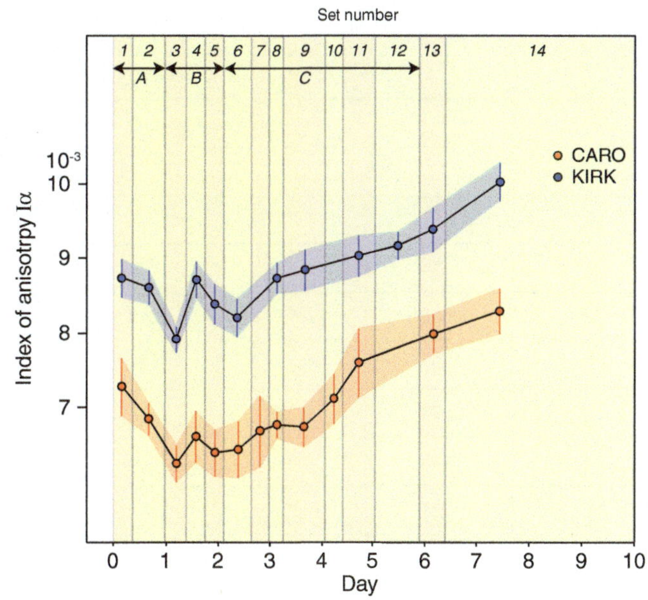

Figure 10

Evolution of the index of anisotropy of the S waves for two seismic stations during GPK2 stimulation.

relatively to the reservoir. CARO is located just above the open hole section of GPK2, while KIRC is located to the northwest. Therefore, the rays travelling to KIRC cross a larger stimulated area.

5 DISCUSSION

If we focus on set 3 in Figures 5 and 6, an interesting local trend is pointed out. Crack lengths are larger around the close-hole section of GPK2 than in the vicinity of the open-hole section. However, microseismic events do not preferentially migrate upwards (Calò *et al.*, 2011), usually meaning that there is no preferential flow direction. The EMT results could be explained by considering what is going on right after a sharp increase in the flow rate. During a very short period of time, a transient regime takes place. As described in Weidler *et al.* (2002), the reservoir is split into two sub-domains with distinct flow behaviors: a shallow part with a well-connected system where cracks are open on all their entire lengths and a deep part behaving like a closed system. However, effective medium modeling assumes a pure elastic behavior for the material, which is not suitable for describing a transient regime in a closed system. This is why our inversion process entails crack shortening in the small area close to the open-hole section (Fig. 8c).

Shear-wave splitting is an useful method for characterizing anisotropy from travel-time differences. This makes it possible to estimate anisotropy indices. In this section, we compare these indices with the global extension rates introduced previously (Fig. 8a, 10, Tab. 6). These extension rates are the ones computed at all depths sampled by the velocity models, *i.e.*, between 4.0 and 5.0 km depth.

The agreement is good for the first 9 sets, which correspond to about the first 4 days of the injection test. The lowest $I\alpha$ values (set 3, Fig. 10) correspond to negative extension rate values (Fig. 8, crack shortening in fact). The maximum extension rate calculated for set 4 (Fig. 8a) is associated with a relatively high $I\alpha$ value (Fig. 10). However, the comparison

TABLE 6

Shear-wave anisotropy parameters and crack extension and opening rate values. Errors correspond respectively to the standard deviation and to the 90% confidence interval for anisotropy and crack rates

	Seismological stations				Crack rates (%)			
Sets	KIRK		CARO		Extension		Opening	
No.	$\delta t \times 10^{-3}$	+/−	$\delta t \times 10^{-3}$	+/−	Rate	+/−	Rate	+/−
1	8.74	0.26	7.28	0.39	98.74	4.94	78.20	3.91
2	8.62	0.23	6.85	0.22	91.91	4.60	91.56	4.58
3	7.92	0.18	6.26	0.25	1.68	0.08	92.77	4.64
4	8.73	0.25	6.62	0.35	126.34	6.32	97.91	4.90
5	8.38	0.27	6.40	0.32	51.07	2.55	98.33	4.92
6	8.21	0.26	6.44	0.38	32.37	1.62	96.05	4.80
7			6.68	0.48	71.26	3.56	95.26	4.76
8	8.73	0.21	6.78	0.18	63.40	3.17	96.86	4.84
9	8.85	0.28	6.74	0.27	115.63	5.78	98.69	4.93
10			7.11	0.34	99.14	4.96	97.09	4.85
11	9.03	0.28	7.61	0.47	86.57	4.33	94.57	4.73
12	9.17	0.19			103.92	5.20	97.34	4.87
13	9.38	0.30	8.00	0.27	47.06	2.35	95.88	4.79
14	10.44	0.26	8.29	0.31	80.17	4.01	97.62	4.88

is poor for the final data sets. The index of anisotropy is then shown to depart from the crack extension rate. In short, it tends to increase for sets 10 to 14 while the extension rate ratio remains more or less the same or even slightly decreases. Shear wave splitting is mostly due to vertical cracks parallel to the direction of the maximum compressive horizontal stress. An increasing index of anisotropy (Fig. 10) is associated with a stronger splitting, hence to the extension of preferential cracks. Wuestefeld *et al.* (2011) also noted a temporal increase in splitting delays for steep arrivals during the stimulation of a gas reservoir. They explain this result as a "halo" of increased fracture density around the stimulated zone. Crack density is assumed to be constant within the EMT model, but it is clear that an increase in the number or the length of favorably oriented cracks will lead to the same anisotropy effect.

Two hypotheses of our EMT model are:

- cracks are oriented according to the regional geological direction;
- the variation in the extension rate is a macro parameter derived at the REV scale, *i.e.* an average for all crack populations embedded within the effective medium.

The influence of two factors could be strengthen due to hydraulic stimulation. During the first stimulation stages, the crack network is subjected to a huge fluid pressure change; injection flow rates impact crack properties, hence shear wave splitting. In such conditions, the system is said under hydraulic control. During the last stages of the stimulation process, the flow rate remains constant. This means that we potentially move from a hydraulic control to a geological control through regional stress orientation. In other words, the horizontal field stress could vary locally due to the hydraulic stimulation: this would explain the discrepancy observed between the extension rate and the time-shift between slow and fast shear arrivals.

An important issue to be discussed at this point is the assumption according to which surface crack density P_{32} is constant during the whole injection experiment. The underlying question is about the capability of hydraulic stimulation to generate new cracks. Cornet *et al.* (2007) showed that the Soultz granite was in a critically stressed state from a mechanical point of view. Pore pressure variations of only 2 to 3 MPa induce microseismic activity and values hardly larger produce macroscopic failure. However, there are cracks in the granite both at the micro and meso-scales. As stated above, our EMT-based inversion method strictly focuses on the meso-scale. Thus, we only pay attention to macroscopic failure localization. The processes inducing microcracking such as thermal cooling are beyond the scope of this paper. This approach is consistent with previous studies (Zhao and Mizuno, 1999; Mishra and Zhao, 2003; Adelinet *et al.*, 2011). The next step consists in identifying the mechanisms that can explain our results keeping in mind that crack density is constant.

During sets 2 and 4, wellhead pressure slightly decreases after the sharp increases related to flow rate jumps (Fig. 1). This suggests an increase in the available storage volume. The injection flow rate may be not significant enough to fill up this additional space. During the same periods of time, our modeling shows higher crack extension rates (Fig. 6-8). The induced dilation of cracks due to their extension could explain such a pressure reduction at the wellhead. This kind of process has been already observed during a hydraulic stimulation test in the Cornubian granite in England (Batchelor, 1982; Crampin and Booth, 1989). Finally the effective medium model built in this study seems to suit well to the very specific problem of geothermal stimulation of a fractured reservoir.

CONCLUSION

This paper introduces a novel approach to explain the behavior of the Soultz reservoir, which was submitted to a massive hydraulic stimulation. We solved an inverse problem using an effective medium model that makes the link between the available tomography data and the unknown parameters describing the state of fractures. The results obtained stressed that a strong increase in the injection rate leads to a decrease in crack length without influence on crack aperture. This provides evidence in favor of crack shortening. The equilibrium is achieved after some time with injection performed at the same flow rate. In such conditions, crack length keeps from fluctuating.

Extension rates deduced from the effective medium based inversion process correlate with time shifts recorded between the slow and fast shear-wave arrivals. The good agreement between the two independent data sets suggest that the processes occurring at the meso-scale (extension/shortening of cracks) are quite similar to those occurring at larger scales even though the corresponding crack networks are different. In addition, in this paper, we distinguished clearly the effect of crack extension from the effect of crack closure. This helps better understand the effect of stimulation.

Finally, effective medium modeling is a very useful tool to explain reservoir and geophysical data in terms of microstructural evolutions. This is all the more essential than the maximization of productivity depends on reservoir structure, whatever the nature of the target resource (thermal energy or fossil resources). The Soultz reservoir is an ideal study case because of its homogeneous petrophysical properties and the large number of data collected at different scales. To date, an important issue is to extend the workflow proposed in this paper for the Soultz reservoir to more complicated field cases.

ACKNOWLEDGMENTS

We would particularly thank Jean-François Barthélémy for the numerical development in the effective medium. We also thank GEIE for providing log data and François Cornet for fruitful discussions. The work of Marco Calo' was founded by the European project GEISER (FP7:241321).

REFERENCES

Adelinet M., Dorbath C., Le Ravalec M., Fortin J., Guéguen Y. (2011) Deriving microstructure and fluid state within the Icelandic crust from the inversion of tomography data, *Geophys. Res. Lett.* **38**, L03305.

Al-Harrasi O.H., Al-Anboori A., Wüstefeld A., Kendall J.-M. (2011) Seismic anisotropy in a hydrocarbon field estimated from microseismic data, *Geophysical Prospecting* **59**, 227-243.

Anderson D.L., Minster B., Cole D. (1974) The effect of oriented cracks on seismic velocities, *J. Geophys. Res.* **79**, 4011-4015.

Barthélémy J.F. (2005) Approche micromécanique de la rupture et de la fissuration dans les geomatériaux, *PhD Thesis*, École nationale des ponts et chaussées.

Barthélémy J.F. (2009) Compliance and Hill polarization tensor of a crack in an anisotropic matrix, *Int. J. Solids Str.* **46**, 22-23, 4064-4072.

Batchelor A.S. (1982) The stimulation of a Hot Dry Rock geothermal reservoir in the Cornubian granite, England, *Proceedings 8th workshop on Geothermal Reservoir Engineering*, Stanford, California, USA, pp. 237-247.

Bérard T., Cornet F.H. (2003) Evidence of thermally induced borehole elongation: a case study at Soultz, France, *Int. J. Rock Mech. Min.* **40**, 7-8, 1121-1140.

Budiansky B., O'Connell R.J. (1976) Elastic moduli of a cracked solid, *Int. J. Solids Structures* **12**, 81-97.

Calò M. (2009) Tomography of subduction zones using regional earthquakes: methodological developments and application to the Ionian slab, *PhD Thesis*, EOST, University of Strasbourg.

Calò M., Dorbath C., Cornet F.H., Cuenot N. (2011) Large-scale aseismic motion identified through 4-D P-wave tomography, *Geophys. J. Int.* **186**, 1295-1314.

Cornet F.H., Bérard T., Bourouis S. (2007) How close to failure is a granite rock mass at a 5 km depth? *Int. J. Rock Mech. Min.* **44**, 1, 47-66.

Crampin S. (1984) Effective anisotropic elastic constants for wave propagation through cracked solids, *Geophys. J. R. Astr. Soc.* **76**, 1, 135-145.

Crampin S., Booth D.C. (1989) Shear-wave splitting showing hydraulic dilation and pre-existing joints in granite, *Scientific Drilling* **1**, 21-26.

Cuenot N., Dorbath C., Dorbath L. (2008) Analysis of the microseismicity induced by fluid injection in the Hot Dry Rock site of Soultz-sous-Forêts (Alsace, France): implications for the characterization of the geothermal reservoir properties, *Pure Appl. Geophys.* **165**, 797-828.

Dezayes C., Valley B., Maqua E., Syren G., Genter A. (2000) Natural fracture system of the Soultz granite based on UBI data in the GPK3 and GPK4 wells, BRGM, Synthetic final report. contrat ENK5-2000-00301.

Dezayes C., Genter A., Hooijkaas G.R. (2005) Deep-seated geology and fracture system of the EGS Soultz Reservoir (France) based on recent 5 km depth boreholes, *Proceedings World Geothermal Congress 2005*.

Dezayes C., Genter A., Valley B. (2010) Structure of the low permeable naturally fractured geothermal reservoir at Soultz, *C. R. Geoscience* **342**, 7-8, 517-530.

Dorbath L., Cuenot N., Genter A., Frogneux M. (2009) Seismic response of the fractured and faulted granite of Soultz-sous-Forêts (France) to 5 km deep massive water injections, *Geophys. J. Int.* **177**, 653-675.

Dorbath L., Evans K., Cuenot N., Valley B., Charléty J., Frogneux M. (2010) The stress field at Soultz-sous-Forêts from focal mechanisms of induced seismic events: cases of the wells GPK2 and GPK3, *C. R. Geoscience* **342**, 7, 600-606.

de Dreuzy J.R., Davy P., Bour O. (2002) Hydraulic properties of two-dimensional random fracture networks following power law distributions of length and aperture, *Water Resources Research* **38**, 12, 12-1.

Eshelby J.D. (1957) The determination of the elastic field of an ellipsoidal inclusion and related problems, *Proceedings of the Royal Society of London A* **241**, 376-396.

Genter A., Traineau H. (1996) Analysis of macroscopic fractures in granite in the HDR geothermal well EPS-1, Soultz-sous-Forêts, France, *J. Vol. Geoth. Res.* **72**, 121-141.

Guéguen Y., Chelidze T., Le Ravalec M. (1997) Microstructures, percolation thresholds, and rock physical properties, *Tectonophysics* **279**, 1, 23-35.

Guéguen Y., Schubnel A. (2003) Elastic wave velocities and permeability of cracked rocks, *Tectonophysics* **370**, 1-4, 163-176.

Guéguen Y., Sarout J. (2011) Characteristics of anisotropy and dispersion in cracked medium, *Tectonophysics* **503**, 1, 165-172.

Henyey F., Pomphrey N. (1982) Self-consistent elastic moduli of a cracked solid, *Geophys. Res. Lett.* **9**, 903-906.

Horálek J., Jechumtálová Z., Dorbath L., Šílený J. (2010) Source mechanisms of micro-earthquakes induced in a fluid injection experiment at the HDR site Soultz-sous-Forêts (Alsace) in 2003 and their temporal and spatial variations, *Geophys. J. Int.* **181**, 1547-1565.

Hudson J.A. (1980) Overall properties of a cracked solid, *Mathematical Proceedings of the Cambridge Philosophical Society* **88**, 371-384.

Kachanov M. (1993) Elastic solids with many cracks and related problems, *Adv. Appl. Mech.* **30**, 259-445.

Kachanov M., Tsukrov I., Shafiro B. (1994) Effective moduli of solids with cavities of various shapes, *Appl. Mech. Rev.* **47**, 151-174.

Kemeny J., Cook N.G.W. (1986) Effective moduli, non-linear deformation and strength of a cracked elastic solid, *Int. J. Rock Mech. Min.* **23**, 2, 107-118.

Mori T., Tanaka K. (1973) Average stress in matrix and average elastic energy of materials with misfitting inclusions, *Acta Metallurgica* **21**, 5, 571-574.

Mishra O., Zhao D. (2003) Crack density, saturation rate and porosity at the 2001 Bhuj, India, earthquake hypocenter: A fluid-driven earthquake? *Earth Planet. Sci. Lett.* **212**, 393-405.

Norris A.N. (1985) A differential scheme for the effective moduli of composites, *Mechanics of Materials* **4**, 1-16.

Nur A. (1971) Effects of stress on velocity anisotropy in rocks with cracks, *J. Geophys. Res.* **76**, 2022-2034.

Powell M.J.D. (2009) The BOBYQA algorithm for bound constrained optimization without derivatives, *Department of Applied Mathematics and Theoretical Physics*, Cambridge England, technical report NA2009/06.

Sausse J., Genter A. (2005) Types of permeable fractures in granite, *Geological Society, London, Special Publications* **240**, 1-14.

Sausse J., Fourar M., Genter A. (2006) Permeability and alteration within the Soultz granite inferred from geophysical and flow log analysis, *Geothermics* **35**, 5-6, 544-560.

Sausse J., Dezayes C., Genter A., Bisset A. (2008) Characterization of fracture connectivity and fluid flow pathways derived from geological interpretation and 3D modelling of the deep seated EGS reservoir of Soultz (France), *Proceedings, thirty-third workshop on Geothermal Reservoir Engineering*, Stanford, California.

Sausse J., Dezayes C., Dorbath L., Genter A. (2010) 3D model of fracture zones at Soultz-sous-Forêts based on geological data, image logs, induced microseismicity and vertical seismic profiles, *C. R. Geoscience* **342**, 7-8, 531-545.

Sayers C.M., Kachanov M. (1991) A simple technique for finding effective elastic constants of cracked solids for arbitrary crack orientation statistics, *Int. J. Solids Structures* **27**, 6, 671-680.

Schoenberg M., Sayers C.M. (1995) Seismic anisotropy of fractured rock, *Geophysics* **60**, 204-211.

Schubnel A., Nishizawa O., Masuda K., Lei X.J., Xue Z., Guéguen Y. (2003) Velocity Measurements and crack density determination during wet triaxial experiments on Oshima and Toki granites, *Pure Appl. Geophys.* **160**, 869-887.

Schubnel A., Benson P., Thompson B., Hazzard J., Young R. (2006) Quantifying Damage, Saturation and Anisotropy in Cracked Rocks by Inverting Elastic Wave Velocities, *Pure Appl. Geophys.* **163**, 947-973.

Sileny J., Plomerova J. (1996) Inversion of shear-wave splitting parameters to retrieve three-dimensional orientation of anisotropy in continental lithosphere, *Phys. Earth Planet. Inter.* **95**, 277-292.

Simpson G., Guéguen Y., Schneider F. (2001) Permeability enhancement due to microcrack dilatancy in the damage regime, *Journal of Geophysical Research Solid Earth* **106**, B3, 3999-4016.

Tarantola A. (2005) Inverse problem theory and methods for model parameter estimation, *Society for Industrial and Applied Mathematics*.

Valley B.C. (2007) The relation between natural fracturing and stress heterogeneities in deep-seated crystalline rocks at Soultz-sous-Forêts (France), *PhD Thesis*, Swiss Federal Institute of Technology Zurich.

Vecsey L., Plomerova J., Babuska V. (2008) Shear-wave splitting measurements – Problems and solutions, *Tectonophys.* **462**, 178-196.

Verdon J.P., Kendall J.-M., Wuestefeld A. (2009) Imaging fractures and sedimentary fabrics using shear wave splitting measurements made on passive seismic data, *Geophys. J. Int.* **179**, 2, 1245-1254.

Walsh J. (1965) The effect of cracks on the compressibility of rock, *J. Geophys. Res.* **70**, 381-389.

Weidler R., Gérard A., Baria R., Baumgärtner J., Jung R. (2002) Hydraulic and micro-seismic results of massive stimulation test at 5 km depth at the European Hot-Dry-Rock test site Soultz, France, *Proceedings 27th workshop on Geothermal Reservoir Engineering*, Stanford, California, USA, pp. 95-100.

Wiggins R.A. (1976) Interpolation of digitized curves, *Bul. Seism. Soc. Am.* **66**, 2077-2081.

Wuestefeld A., Verdon J.P., Kendall J.M., Rutledge J., Clarke H., Wookey J. (2011) Inferring rock fracture evolution during reservoir stimulation from seismic anisotropy, *Geophysics* **76**, 6, WC159-WC168.

Zhang H., Thurber C.H. (2003) Double-difference tomography: The method and its application to the Hayward fault, California, *Bulletin of the Seismological Society of America* **93**, 5, 1875-1889.

Zhao D., Mizuno T. (1999) Crack density and saturation rate in the 1995 Kobe Earthquake Region, *Geophys. Res. Lett.* **26**, 3213-3216.

APPENDIX A

Evolution of the shear-wave seismic velocity at 4.6 km depth (Fig. A1) and corresponding vertical cross-sections (Fig. A2) during the 2000 stimulation in GPK2. Figures are chronologically ordered from set 1 to 14. The same results for compressional wave velocity were published in Calò *et al.* (2011).

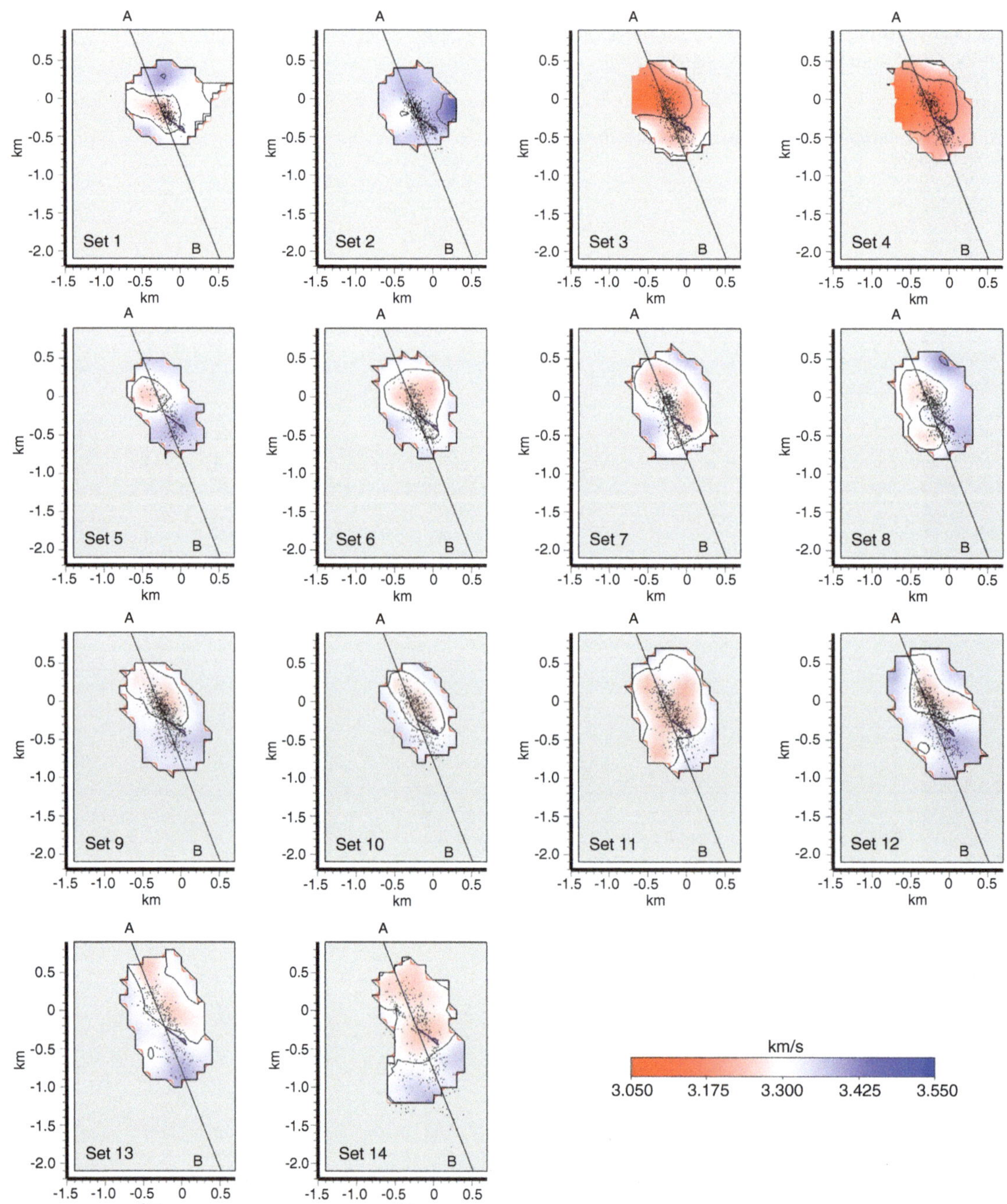

Figure A1

Evolution of the shear-wave seismic velocity at 4.6 km depth during the 2000 stimulation test. Images are in chronological order from set 1 to set 14. The black dots are the projections of the events used to obtain the shear-wave velocity models.

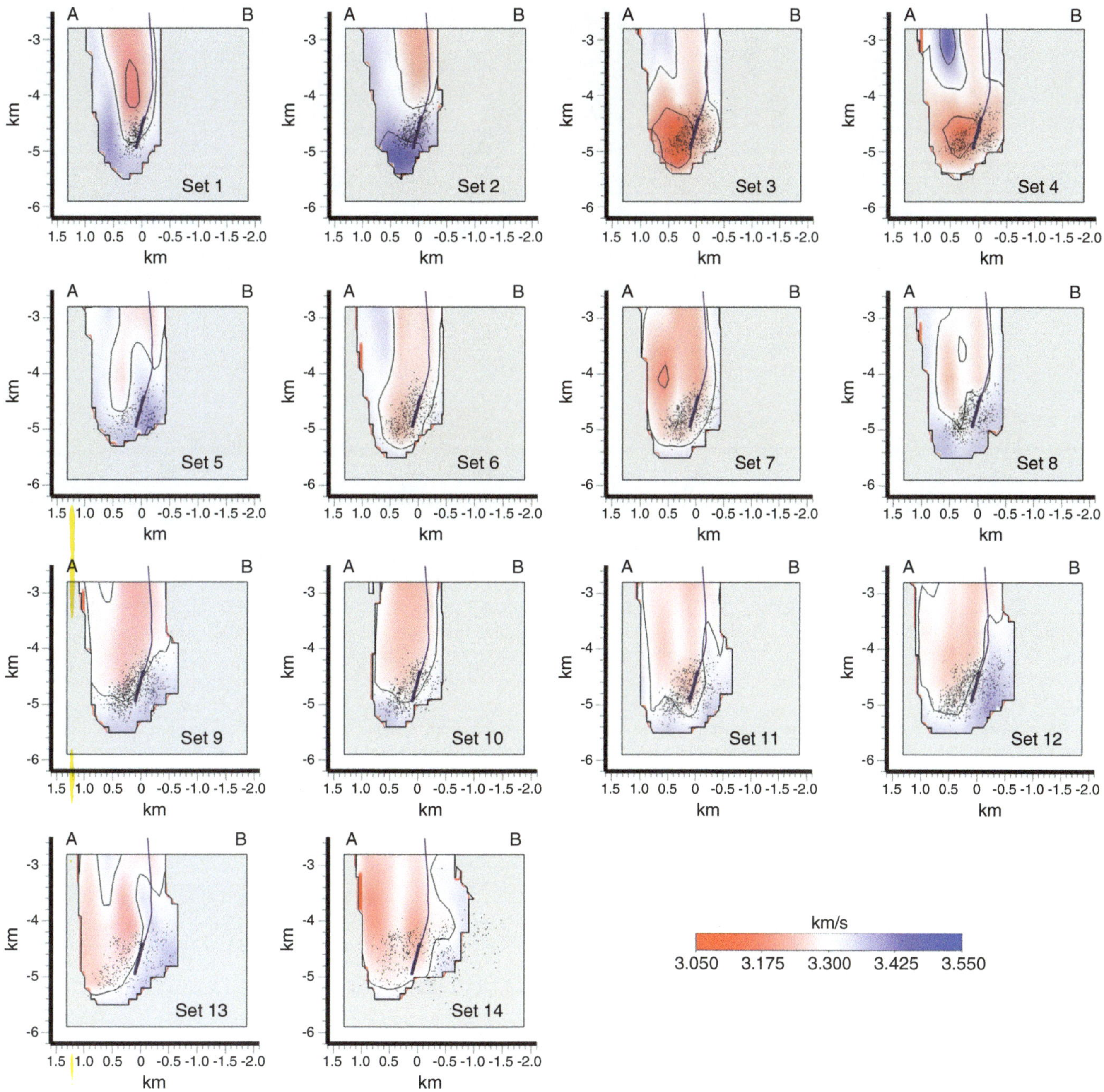

Figure A2

Vertical sections of the shear-wave velocity models along traces A–B reported in Figure A1. The black dots are the projections of the events used to obtain the shear-wave velocity models. Blue lines are the projections of GPK2. The thick part of the borehole trajectory corresponds to the open-hole section of GPK2 well.

APPENDIX B

We repeated our inversion process several times to assess the reliability of our results and to evaluate the sensitivity of the numerical responses obtained with respect to the effective medium approach. The data used for this purpose correspond to the 5th set and the 4.6 km depth.

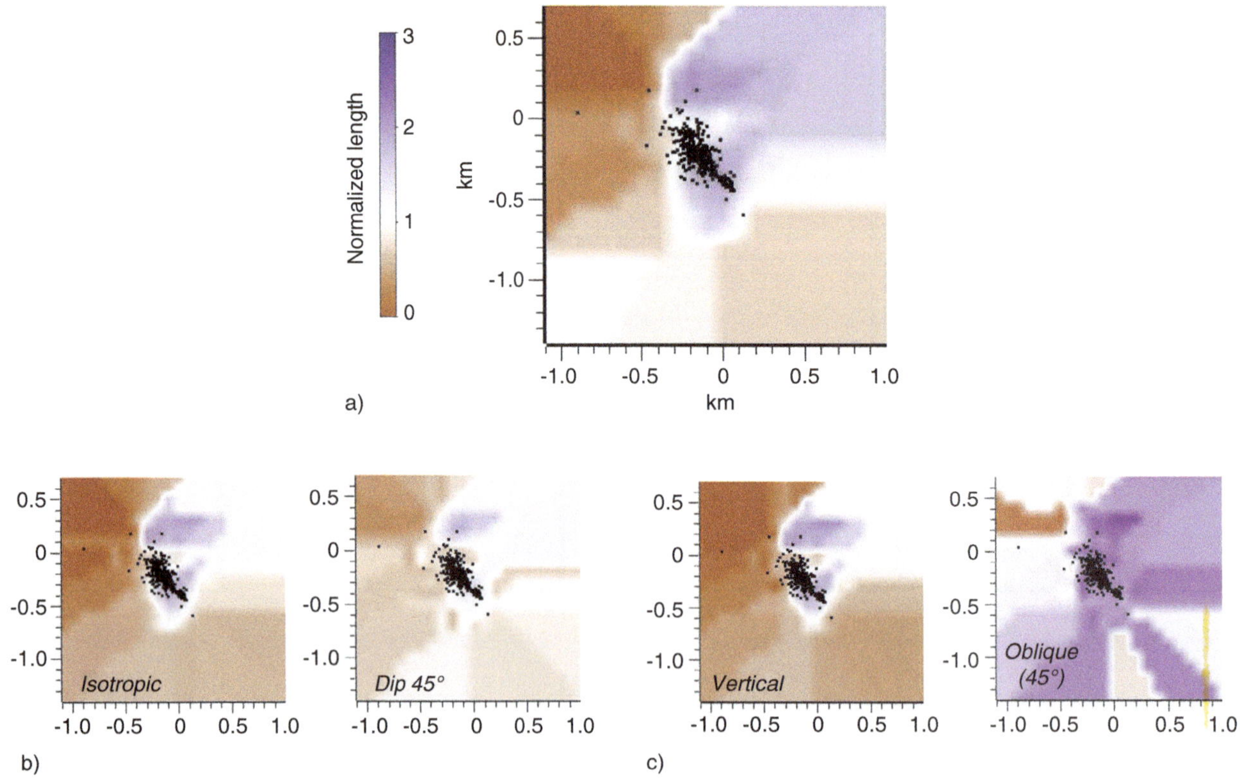

Figure B1

Sensitivity study on direct parameters using data extracted from set 5 at 4.6 km depth. Comparison is made on the inverted crack length. Black dots are the projections of events used to obtain velocity models. a) True model: geological distribution of cracks and direction of propagation in agreement with tomography input data. b) Effect of the crack orientation on the inverted length, isotropic distribution (left picture) and cracks with dip 45° (right picture). c) Effect of the direction of the wave propagation: purely vertical (left picture) and oblique propagation (right picture).

In a first step, we investigated the influence of crack orientation and direction of wave propagation on the inversion of crack length (Fig. B1). We considered as the true model the one obtained with parameters described in this paper (Fig. B1a): two NS crack families with dip 65° and direction of wave propagation in agreement with tomography data (maximum deviation angle of 40° with respect to the vertical axis). Then, two crack distributions are considered (Fig. B1b): an isotropic one and another one for NS cracks with a dip 45°. The inverted crack length inferred from the isotropic distribution is quite close from the true model. Conversely, the single family of cracks with 45° dip yields a crack length distribution quite different. Finally, two wave propagation directions are tested (Fig. B1c): a vertical and an oblique one (45°). The inverted crack length obtained when considering a vertical propagation is quite similar to the true model although the oblique propagation leads to very different results. This sensitivity study emphasizes the fact that it is essential to properly characterize the true geological medium in order to input realistic parameters in the effective medium modeling.

In a second step, we studied the sensitivity and the robustness of the optimization process (Fig. B2). We considered as the true model the compressional wave velocities derived from the tomography study (Calò *et al.*, 2011) displayed in Figure B2a. Then, we carried out various tests to evaluate the influence of the number of optimization parameters: optimization with one parameter (crack radius) in Figure B2b, 2 parameters (crack radius and aperture) in Figure B2c and 4 parameters (crack radius, crack aperture, bulk and shear matrix moduli) in Figure B2d. The results obtained showed that the number of optimization parameters is not a crucial parameter for the achievement of the optimization process. Whatever this number, the general trend is preserved, even if extreme compressional wave velocity values are not reached.

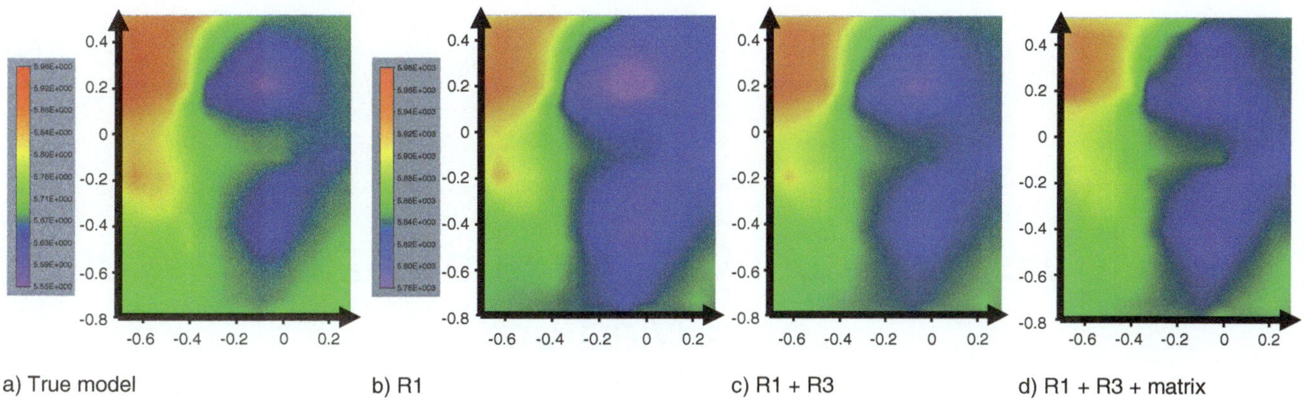

Figure B2

Sensitivity study on inversion parameters using data extracted from set 5 at 4.6 km depth. Comparison is made on the compression-wave velocity (expressed in km/s). Different inversion parameters are tested: crack main radius (R1), crack aperture (R3) and bulk and shear moduli of the matrix (matrix). a) The true model is the compression-wave seismic velocity model calculated from the tomography study (Calò *et al.*, 2011); b) compression-wave velocity model inverted with only R1 as optimisation parameter; c) P-wave velocity model inverted with R1 and R3 as optimisation parameters; d) compression-wave velocity model inverted with R1, R3 and matrix moduli as optimisation parameters.

3

Recyclable PhotoFuel Cell for Use of Acidic Water as a Medium

Yuta Ogura, Mao Yoshiba and Yasuo Izumi*

Department of Chemistry, Graduate School of Science, Chiba University, Yayoi 1-33, Inage-ku, Chiba 263-8522 - Japan
e-mail: tyojuryoku@hotmail.co.jp - yubari.com@ae.auone-net.jp - yizumi@faculty.chiba-u.jp

* Corresponding author

Abstract — *We reported a new PhotoFuel Cell (PFC) comprising two photocatalysts for use of acidic water as a recyclable medium. Nitrogen and oxygen flow was required in the photoanode and photocathode, respectively. In this study, we developed a gas-circulating PFC that needs no gas supply from outside. In the gas-circulating PFC, the reverse reaction of water oxidation at the anode was prevented by the gas flow of photogenerated* O_2 *from the anode to the cathode inside the PFC. The gas-circulating PFC accommodated an organic solvent layer over the aqueous electrolyte for the anode, and also a vent hole in the upper part of the Proton-Conducting Polymer (PCP) film.* O_2 *transferred from the anode electrolyte to the organic solvent due to the solubility difference between the HCl solution and organic solvent.* O_2 *transfer from the gas phase in the anode to that in the cathode was achieved by the vent hole in the PCP film due to the pressure difference due to the progress of the reaction. By the addition of a hexane layer to the anode of the PFC, it was demonstrated to achieve a photocurrent value of 69.7 μA per 1.3* cm^2 *of photocatalysts. However, in the stability tests for more than 7 h, the small amount of remaining* O_2 *in the electrolyte (2.85 μmol* L^{-1}*) exhibited serious effects on the PFC performance. The* I_{SC}*,* V_{OC} *and* P_{Max} *values of the gas-circulating PFC were 29.2 μA, 1.18 V and 6.10 μW, that were 40%, 74% and 44%, respectively, of those for a* N_2 *and* O_2 *flow-type PFC. Apparently, photocurrents were dramatically suppressed by the reverse reaction at the photoanode in the extended tests for the gas-circulating PFC.*

Résumé — Cellule photocombustible recyclable pour l'utilisation d'eau acide en tant que milieu — Nous décrivons une nouvelle cellule photocombustible (*PhotoFuel Cell*, PFC) comprenant deux photocatalyseurs utilisant de l'eau acide recyclée. Un flux d'azote et d'oxygène a été nécessaire dans respectivement la photoanode et la photocathode. Dans cette étude, nous avons développé une PFC à circulation de gaz qui n'a pas besoin d'être alimentée en gaz de l'extérieur. Dans la PFC à circulation de gaz, la réaction inverse d'oxydation de l'eau à l'anode a été évitée par le flux de gaz d'O_2 photo-généré de l'anode à la cathode à l'intérieur de la PFC. La PFC à circulation de gaz présentait une couche de solvant organique au-dessus de l'électrolyte aqueux pour l'anode, et également un orifice d'aération dans la partie supérieure du film polymère conducteur de proton (*Proton-Conducting Polymer*, PCP). L'O_2 était transféré de l'électrolyte de l'anode au solvant organique, à cause de la différence de solubilité entre la solution d'HCl et le solvant organique.
Le transfert d'O_2 de la phase gazeuse à l'anode vers celle dans la cathode a été réalisé par l'orifice d'aération dans le film PCP, du fait de la différence de pression due à la progression de la réaction. En ajoutant une couche d'hexane à l'anode de la PFC, il a été démontré que l'on atteignait une valeur de courant photo-électrique de 69,7 μA par 1,3 cm^2 de photocatalyseurs. Toutefois, dans les tests de stabilité, pendant plus de 7 h, la faible quantité d'O_2 restant dans l'électrolyte (2,85 μmol L^{-1}) a eu des

effets importants sur les performances de la PFC. Les valeurs I_{sc}, V_{oc} et P_{Max} de la PFC à circulation de gaz ont atteint 29,2 μA, 1,18 V, et 6,10 μW, soit 40%, 74% et 44%, respectivement de celles d'une PFC de type à flux de N_2 et d'O_2. Apparemment, les courants photo-électriques ont été considérablement supprimés par la réaction inverse à la photoanode dans les tests prolongés pour la PFC à circulation de gaz.

INTRODUCTION

Fossil fuels have been utilized as the essential energy source for industrialization. Industrial CO_2 emissions have led to an increase in the level of atmospheric CO_2 concentration (400 ppm), and the effects of this increase on global warming cannot be underestimated. The development of renewable energy as a replacement for fossil fuels has been slow [1, 2].

Among the renewable energies, solar energy has the greatest potential. Although the Silicon Solar Cell (Si SC) has been commercialized, the technology that can convert solar energy to electricity often needs subsidies to spread more widely [3]. Other types of SC, *e.g.* the dye-sensitized SC and compound SC [4], and Fuel Cells (FC) that use hydrogen/methanol fuel [5-8], potentially obtained using solar energy [9], have been extensively investigated. However, all the requisites (sustainability, durability and an electromotive force of 1-3 V per cell) have not been fully satisfied.

Recently, a new device was demonstrated: a PFC utilizing two photocatalysts of TiO_2 and silver(0/I)-doped TiO_2 [10] on an electrode film, both immersed in acidic solutions separated by a PCP film [11]. Platinum is not used [5-8, 12-16], and only inexpensive materials were used in the PFC for sustainability. The PFC is similar to photoelectrochemical cells (PEC) consisting of a photoanode and a photocathode to form O_2 and H_2, respectively, from water [17, 18] and also PEC consisting of a photoanode to oxidize water [19] or organic dye fuel [20] and a photocathode to reduce protons. The PFC is different from these PEC in that the medium, water, is recycled in the cell and no fuel is needed except for natural light [21].

Although PFC enable higher electromotive force (1.8 V) [11, 21] and use only acidic water as a medium and inexpensive catalysts, PFC need a gas supply to promote surface reactions at both the anode and cathode. At the anode, O_2 gas needs to be evacuated by N_2 flow to prevent a reverse surface reaction:

$$O_2 + 4H^+ + 4e^- \rightarrow 2H_2O \quad (1)$$

On the other hand, at the cathode reactant O_2 should be supplied for the reaction (1) to proceed. Therefore, PFC need gas cylinders to supply these gases for effective power generation. This situation limits the design of compact enough PFC that can be accommodated in portable electronic devices.

Herein, we developed a PFC in which gas circulates internally. By circulating formed O_2 gas from the anode to the cathode, the reverse reaction at the anode could be prevented. The concept of gas-circulating PFC is shown in Figure 1a, Test 3. The PFC has two photocatalysts (TiO_2 and Ag-TiO_2) immersed in HCl solution. The electrolyte at the anode and that at the cathode (Fig. 1b2, 3) are separated by a PCP (Fig. 1b4, 5). As the mechanism of gas circulation, the PFC designed in this study comprises an organic solvent phase on an anode electrolyte (Fig. 1b1) and a vent hole in the upper part of the PCP film (Fig. 1b4, 5). The organic solvent can dissolve a greater amount of O_2 gas than water can.

O_2 gas circulated in the PFC as follows. O_2 generated by the anode reaction transferred preferentially into the organic solvent due to the difference in O_2 solubility between the HCl solution and the organic solvent. Then, the dissolved O_2 in the organic solvent moved to the upper gas phase due to the increased O_2 concentration in the solvent in the compartment of the anode, and then to the upper gas phase in the compartment of the cathode through the vent hole in the PCP film due to the pressure difference between the O_2 formed at the anode and O_2 consumed at the cathode (Fig. 1a, 1b1). Finally, the gaseous O_2 that moved to the cathode side reached the photocatalyst on the cathode and was photoreduced.

1 EXPERIMENTAL SECTION

1.1 Sample Preparation of the Photoelectrodes

TiO_2 powder [1.00 g; P25, *Degussa*; anatase/rutile phases = 7/3; Brunauer-Emmett-Teller (BET) surface area (S_{BET}) = 60 m^2 g^{-1}] was suspended in 3.0 mL of deionized water (<0.06 μS cm^{-1}) and then stirred well. The obtained suspension was dried at 373 K for 24 h and heated in air at 673 K for 2 h. The resultant powder was suspended in 75 μL of deionized water and placed on an Indium Tin Oxide (ITO, thickness 1.2-1.6 μm)-coated Pyrex glass plate. The TiO_2/ITO/Pyrex was dried at 373 K for 18 h and heated in air at 573 K for 30 min. The amount of TiO_2 deposited on the ITO-coated glass plate was 5.0 mg and covered an area of 1.3 cm^2.

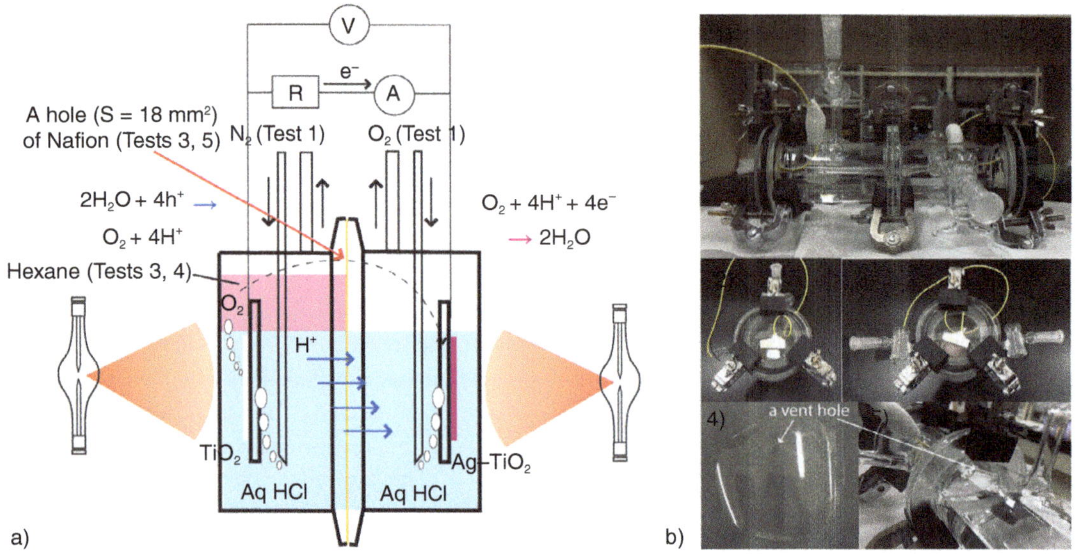

Figure 1

a) Components of the PFC comprising TiO_2 and Ag–TiO_2 photocatalysts used for Tests 1-5. Oxidation by photogenerated holes at the anode and reduction by photogenerated electrons take place. Remaining electrons at the anode flow to the cathode and combine with the remaining holes. b) Photographs of the PFC for Test 3 (1). Photoanode side (2), photocathode side (3), and the PCP with a vent hole (4, 5).

Silver nitrate (159 mg; 99.8%, *Wako Pure Chemicals*) was dissolved in 10 mL of deionized water. The solution was mixed with 3.33 g of untreated TiO_2 (P25). The mixture was magnetically stirred at a rate of 850 rotations per minute (rpm) and the water was distilled at 353 K. The obtained powder was dried at 373 K for 24 h and heated in air at 673 K for 2 h. The resultant Ag-TiO_2 powders contained 3.0 wt% of Ag. They were suspended in a minimum amount of water and placed on ITO-coated glass (Ag-TiO_2/ITO/Pyrex) in a manner similar to that for the TiO_2/ITO/Pyrex.

1.2 Photocurrent Tests

1.2.1 Test 1: Gas-Flow Conditions

TiO_2/ITO/Pyrex and Ag-TiO_2/ITO/Pyrex electrodes were immersed in HCl solutions (40 mL in each compartment; initial pH 4.0-2.0). The two compartments were separated by a 50-µm-thick PCP film (Nafion, *DuPont*; acid capacity > 9.2×10^{-4} equivalent g^{-1}). N_2 (purity > 99.999%) and O_2 gases (purity > 99.6%) were bubbled 30 mm away from each photoelectrode at a flow rate of 100 mL min^{-1} (Fig. 1a). The PFC was equipped with quartz windows (ϕ = 80 mm) on both sides. Both the TiO_2 and Ag-TiO_2 photocatalysts were irradiated with UV–visible light through the quartz windows using a two-way branched quartz fiber light guide (Model 5ϕ-2B-1000L, *San-ei Electric Co.*) from a 500-W xenon arc lamp (Model SX-UID502XAM, *Ushio*). The distance between the light exit (ϕ = 5 mm) and the TiO_2 or Ag-TiO_2 film was 46 mm. The light intensity was 8 mW cm^{-2} at the center of the photocatalyst film on each electrode.

1.2.2 Test 2: Gas-Closed Conditions

PFC comprising TiO_2/ITO/Pyrex and Ag-TiO_2/ITO/Pyrex electrodes were prepared and immersed in HCl solutions (40 mL in each compartment; initial pH 2.0) in a similar procedure to that in Test 1. The two compartments were separated by a 50-µm-thick PCP film. N_2 and O_2 gases purged the compartments of TiO_2/ITO/Pyrex and Ag-TiO_2/ITO/Pyrex, but the gas supply was stopped before the start of Test 2. Both the TiO_2 and Ag-TiO_2 photocatalysts were irradiated with UV–visible light in the closed compartments filled with N_2 and O_2, respectively.

1.2.3 Test 3: Gas-Circulating Conditions inside the PFC

An Ag-TiO_2/ITO/Pyrex electrode was immersed in 50 mL of HCl solution (initial pH 2.0), whilst TiO_2/ITO/Pyrex was immersed in 50 mL of the HCl solution phase (initial pH 2.0) and 50 mL of hexane phase was added above the HCl solution phase (Fig. 1). All the area of the TiO_2 film was immersed in the HCl solution phase and not in contact with the hexane phase. N_2 and O_2 gases purged the compartments of TiO_2/ITO/Pyrex and Ag-TiO_2/ITO/Pyrex,

TABLE 1

The conditions for current tests of the PFC

Test	N_2/O_2 supply	Hexane@anode	Vent hole@PCP film
1	○	×	×
2	×	×	×
3	×	○	○
4	×	○	×
5	×	×	○

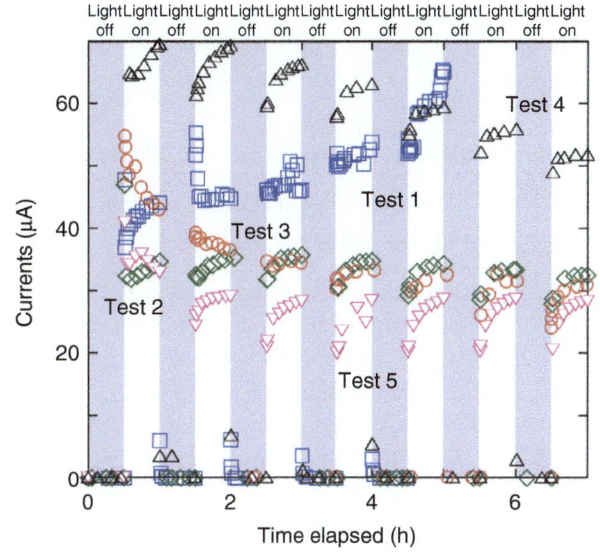

Figure 2

The time course of photocurrents of the PFC comprising TiO_2 and Ag-TiO_2 photocatalysts in Tests 1 (gas flow), 2 (gas closed), 3 (gas circulating inside the PFC), 4 (gas-closed conditions plus hexane@anode), and 5 (gas-circulating conditions inside the PFC minus hexane@anode). The pH values of the electrolytes in both the anode and cathode were 2.0.

respectively, but the gas supply was stopped before the start of Test 3. Both the TiO_2 and Ag-TiO_2 photocatalysts were irradiated with UV–visible light, set in the closed compartments filled with N_2 and O_2, respectively. The two compartments were separated by a 50-μm-thick PCP film; however, in Test 3, gas circulation inside the PFC was enabled through a vent hole (6.0 mm × 3.0 mm) in the upper part of the PCP film (Fig. 1a, b4, b5) from the anode compartment in which O_2 was formed to the cathode compartment in which O_2 was consumed. The conditions of UV–visible light irradiation were similar to those for Tests 1 and 2 (Tab. 1).

1.2.4 Test 4: Gas-Closed Conditions Plus Hexane@Anode

An Ag-TiO_2/ITO/Pyrex electrode was immersed in 55 mL of HCl solution (initial pH 2.0), whilst TiO_2/ITO/Pyrex was immersed in 55 mL of the HCl solution phase (initial pH 2.0) and 55 mL of the hexane phase above the HCl solution phase (Fig. 1). All the area of the TiO_2 film was immersed in the HCl solution phase and not in contact with the hexane phase. N_2 and O_2 gases purged the compartments of TiO_2/ITO/Pyrex and Ag-TiO_2/ITO/Pyrex, respectively, but the gas supply was stopped before the start of Test 4. Both the TiO_2 and Ag-TiO_2 photocatalysts were irradiated with UV–visible light, set in the closed compartments filled with N_2 and O_2, respectively. The two compartments were separated by a 50-μm-thick PCP film. The film was complete and did not comprise any vent holes as in Test 3. The conditions of UV–visible light irradiation were similar to those for Tests 1-3 (Tab. 1).

1.2.5 Test 5: Gas-Circulating Conditions inside the PFC Minus Hexane@Anode

An Ag-TiO_2/ITO/Pyrex electrode was immersed in 40 mL of HCl solution (initial pH 2.0), whilst TiO_2/ITO/Pyrex was immersed in 40 mL of the HCl solution phase (initial pH 2.0). N_2 and O_2 gases purged the compartments of TiO_2/ITO/Pyrex and Ag-TiO_2/ITO/Pyrex, respectively, but the gas supply was stopped before the start of Test 5. Both the TiO_2 and Ag-TiO_2 photocatalysts were irradiated with UV–visible light, set in the closed compartments filled with N_2 and O_2, respectively. The two compartments were separated by a 50-μm-thick PCP film. The conditions of UV–visible light irradiation were similar to those for Tests 1-4. This PFC had a vent hole (6.0 mm × 3.0 mm) in the upper part of the PCP film (Fig. 1a, b4, b5) similar to Test 3 (Tab. 1).

For Tests 1-5, photocurrent generation (photocurrent tests) was performed by connecting the external parallel circuit for one route via a voltmeter and another route via an ammeter with a resistance of 0.5 Ω (Fig. 1a). Both catalysts were exposed for 30 min to UV–visible light and then kept in the dark for 30 min, and this process was repeated five-seven times. The current and voltage between the two electrodes were monitored.

1.3 Current–Voltage Dependence Tests

The current (i)–voltage (V) characteristics were also measured for several types of PFC using a similar parallel circuit (Fig. 1a). After the photocurrent test for Figure 2 (2.5, 3.5 and 3.5 h of irradiation of UV–visible light for Tests 1, 2 and 4, respectively), the resistance was gradually decreased from 500 kΩ to 0.3 Ω over 20 min to plot the i–V dependence. For comparison, a PFC in the conditions of Test 1

(Sect. 1.2.1) was also prepared and *i–V* characteristics were also measured similar to the experiments described above, but in darkness.

1.4 Product Analyses during Photocurrent Tests at the Anode

1.4.1 Analysis of O_2 Formed at the Anode

56.2 mg of iron(III) chloride hexahydrate (*Kanto Chemical*, purity $\geq 99\%$) was dissolved in 90 mL of HCl aqueous solution in a quartz flask. The pH of this solution was adjusted to 2.0 by the addition of HCl solution. 151 mg of TiO_2 powder (P25, *Degussa*) was suspended in this solution for 10 min. Then, dissolved O_2 in the solution was removed by N_2 bubbling for 30 min. Next, N_2 bubbling was stopped and the flask was isolated. The solution suspended with TiO_2 was irradiated with UV–visible light from model SX-UID502-XAM. The distance between the light exit (ϕ = 60 mm) and the center of the flask was 120 mm. The sample was exposed to the light for 15 min and kept in the dark for 15 min. This step was repeated five times. The solution was magnetically stirred at a rate of 1 000 rpm and the temperature was 298-301 K. The formed O_2 as dissolved was monitored *in situ* by a Dissolved Oxygen (DO) electrode (Model 9520-10D, *Horiba*).

1.4.2 Analysis of Hypochlorous Acid Formed at the Anode by the Photooxidation of Cl^-

The possibility of the photooxidation of Cl^- ions at the anode of the PFC instead of the photooxidation of water was checked. If Cl_2 is formed by the photoxidation at the anode (Eq. 2), it should immediately decompose into hypochlorous acid and hydrochloric acid in the HCl solution of the anode (Eq. 3):

$$2Cl^- + 2h^+ \rightarrow Cl_2 \quad (2)$$

$$Cl_2 + H_2O \rightarrow HCl + HClO \quad (3)$$

The HClO formed was evaluated by the *N,N*-Diethyl-*p*-PhenyleneDiamine (DPD) method [22].

20 mg of DPD reagent (0.8 mg of DPD sulfate salt (H_2SO_4) and 19.2 mg of Na_2SO_4) were dissolved in 1.0 mL phosphate buffer solution (pH 6.5). This solution was mixed with 5.0 mL of 0-39.0 μmol L^{-1} of NaOCl aqueous solutions and used as a standard solution. The mixed solutions (6.0 mL) were used to obtain a linear regression relationship between the absorbance at 515 nm and the concentration of ClO^- ions.

Next, 20 mg of DPD reagent were dissolved in 4.0 mL phosphate buffer solution (pH 6.5). This solution was mixed with 2.0 mL of sample solution (HCl aqueous solution) after a photocurrent Test 1 using a PFC (Fig. 1) for 3 h at pH 2.0. The UV–visible spectrum of the mixed solution (6.0 mL) was measured by UV–visible spectrometer (Model V650, *JASCO*). The concentration of HClO in the sample was determined based on the absorbance at 515 nm and the relationship obtained using NaOCl standard solutions. The reason for the volume difference for the phosphate buffer and sample solutions was to adjust the pH within 6.4 and 6.6 for mixed solutions (6.0 mL).

2 RESULTS AND DISCUSSION

2.1 Photocurrent Tests

The time courses of the photocurrent tests are summarized in Figure 2. For Test 1, in response to the UV–visible irradiation, photocurrents gradually increased from 43.7 to 60.5 μA in five cycles at pH 2.0. In contrast, the photocurrents increased and converged to a constant value (26.7 μA) in the test at pH 3.0 (not shown) [11]. Thus, the active metallic Ag^0 species for photoreduction of O_2 at the cathode should gradually increase, transformed from Ag_2O by the reaction with protons, in the five cycles at pH 2.0. The transformation of the active Ag species (Ag^0, Ag_2O and AgCl) was reported based on an extended X-ray absorption fine-structure study [11].

At pH 4.0, a PFC current of 3 μA was obtained (not shown) for 2.5 h irradiated by UV–visible light (0.28 μmol-e^- generated in total) [11]. The pH values starting from 4.0 for the HCl solution (40 mL, 4.0 μmol-H^+) remained constant after the PFC test at both the anode and cathode. The pH change due to each half reaction was calculated as ±0.03. As this value is nearly the accuracy of pH monitoring under these conditions, we cannot judge that the photooxidation of water (at the anode) and photoreduction of O_2 (at the cathode) exclusively took place. The possibility of a by-reaction, *e.g.* the consumption of acid, remained. For the test at pH 2.0, the initial proton concentration was 400 μmol in 40 mL of electrolyte solution at both the anode and cathode, while the PFC current for 2.5 h was 4.7 μmol-e^- in total. The pH change due to each half reaction is calculated as ±0.005. It was even more difficult to judge the possibility of photooxidation of water and/or anions of acid under the conditions.

In Test 2, the photocurrent in the first cycle (34.7 μA) was similar to that in Test 1 (43.7 μA). In response to the UV–visible irradiation, photocurrents gradually increased to 35.8 μA in three cycles. In contrast to the gradual increase in photocurrents throughout 5 h of the study in Test 1, the photocurrents peaked at 35.8 μA in the third cycle and gradually decreased to 32.2 μA in the seventh cycle in Test 2 (Fig. 2). This decrease by 3.6 μA suggested partial deactivation of the TiO_2 photocatalyst at the anode. One of the major reasons is the reverse reaction of produced O_2 which remained dissolved in the HCl solution to water (Eq. 1). In contrast to the gas-flowing Test 1, the anode compartment of TiO_2/ITO/Pyrex was purged with N_2 gas and produced O_2 remained in the compartment for Test 2 (Tab. 1).

The photocurrent in the first cycle (54.8-40.3 μA) was clearly higher in Test 3 than that in Tests 1 and 2. The photocurrent decreased to 36.5 μA at the end of the second cycle (Fig. 2) and continued to decrease gradually to 31.0 μA at the end of the seventh cycle. The photocurrents in the first cycle were probably higher because N_2 was purged in the anode compartment more completely than in Tests 1 and 2 by the effect of hexane to preferably dissolve a contamination level of O_2 (~0.04 ppb). This evaluation was based on the impurity of O_2 (< 1 ppm) in the N_2 gas used and the equilibrium of dissolution. Conversely, the decrease in the photocurrent in seven cycles suggested an even smaller amount of formed O_2 remained in the anode and inhibited the PFC performance considerably, and also unexpected reverse diffusion of O_2 from the cathode into the anode was not excluded through the vent hole.

In Test 4, the photocurrent in the first cycle (69.7 μA) was clearly higher than that in Tests 1-3. However, in response to the UV–visible irradiation, photocurrents gradually decreased to 51.9 μA in seven cycles (Fig. 2). The photocurrent in Test 4 was always higher than that in Tests 2 and 3 throughout the photocurrent test. Thus, the positive effects of hexane on PFC were confirmed (Tab. 1), but unexpectedly, the vent hole in Test 3 contributed negatively to the photocurrent generation. Furthermore, the linear decrease of the photocurrent as a function of time that elapsed during Test 4 was more gradual than the exponential-like decrease as a function of time in Test 3.

In Test 5, the photocurrent in the first cycle (41.0-33.0 μA) was similar to that in Test 2 (Fig. 2). Then, it decreased to 29.2 μA in the second cycle. Later than the third cycle, the photocurrents became stable at 28.4 μA. This stabilized current was lowest among the Tests 1–5, as the result of produced O_2 which remained dissolved in the HCl solution to water, and the unexpected reverse diffusion of O_2 in the cathode into the anode through the vent hole, as in Test 3 (Tab. 1).

In Test 2, 1.11 μmol of O_2 were generated during the 7-h photocurrent test based on the total photocurrent for 3.5 h and assuming the next reaction (Eq. 1 is the reverse reaction of Eq. 4):

$$2H_2O + 4h^+ \rightarrow O_2 + 4H^+ \tag{4}$$

All of them were dissolved in the anode electrolyte based on the O_2 solubility. Thus, the O_2 concentration in the anode electrolyte was 27.8 μmol L^{-1}. On the other hand, in Test 4, 2.01 μmol of O_2 were generated after the 7-h photocurrent test based on the total photocurrent for 3.5 h and assuming reaction 4. In this test, the O_2 should be distributed between the HCl solution and hexane. Based on the O_2 solubility values in the literature to hydrochloric acid solution (1.27 mmol L^{-1}) and to hexane (15.0 mmol L^{-1}) [23], the distribution of O_2 to hexane was calculated as:

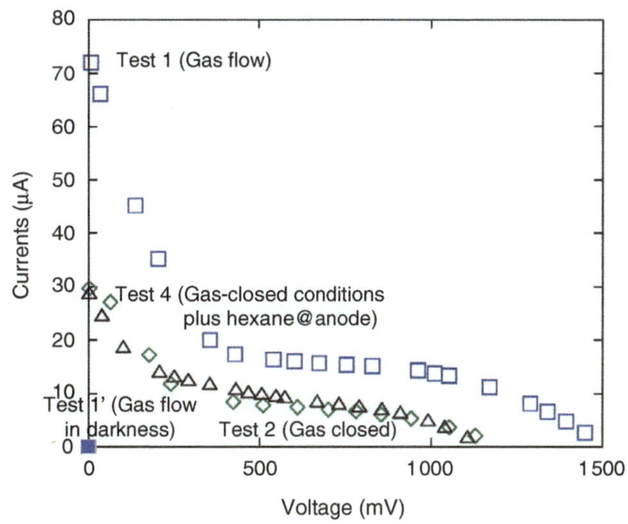

Figure 3

i–V characteristics of the PFC comprising TiO_2 and Ag-TiO_2 photocatalysts in Test 1 (gas flow), Test 2 (gas closed), Test 4 (gas-closed conditions plus hexane@anode), and control Test 1' in darkness (gas flow).

$$\frac{0.055\,\mathrm{L} \times 15\,\mathrm{mmol\,L^{-1}}}{0.055\,\mathrm{L} \times 1.27\,\mathrm{mmol\,L^{-1}} + 0.055\,\mathrm{L} \times 15\,\mathrm{mmol\,L^{-1}}} = 0.922 \tag{5}$$

Thus, 1.85 and 0.157 μmol of O_2 should be distributed to the hexane and HCl solutions, respectively, in equilibrium. Thus, the O_2 concentration in the anode electrolyte (HCl solution) was calculated as 2.85 μmol L^{-1}. Hence, the O_2 concentration at the end of Test 4 was only 10% of that at the end of Test 2. This difference accounted for the higher current density in Test 4 by a factor of two as compared with Test 2 (Fig. 2).

2.2 *i–V* Characteristic Tests

The obtained *i–V* characteristics are shown in Figure 3. Furthermore, the properties of PFC calculated based on *i–V* data in Tests 1, 2 and 4 are summarized in Table 2. In Test 1, the *i* value gradually increased as the cell voltage decreased, starting from open circuit voltage (V_{OC}) at 1.59 V, which is a similar trend to the *i–V* dependence of the SC [24] or the Ohmic

TABLE 2
PFC characteristics defined by i–V characteristics in Tests 1 (gas flow), 2 (gas closed), and 3 (gas circulating inside the PFC)

Test	I_{SC} (μA)	V_{OC} (V)	P_{Max} (μW)
1 (Gas flow)	73.7	1.59	14.0
2 (Gas closed)	29.8	1.23	5.5
4 (Gas-closed conditions plus hexane@anode)	29.2	1.18	6.1
1' (Gas flow; control in darkness)	< 0.1	< 0.002	< 0.0002

loss region of a Polymer Electrolyte Fuel Cell (PEFC). When the voltage became less than 0.45 V, the current increased linearly from 20 to 74 μA (short-circuit current, I_{SC}). This increasing trend was different from that in a SC or the transportation limit region of a PEFC [7, 8].

In Test 2, the shape of the i–V characteristic curve was similar to that for Test 1. However, the increase in currents in the lower voltage region (≥ 0.45 V) was less steep than the slope in Test 1. The values of I_{SC} (29.8 μA), V_{OC} (1.23 V) and P_{Max} (5.51 μW) were 40, 77 and 39% of the corresponding values obtained in Test 1. The lower photocurrents in Test 2 suggested that the reverse reaction 1 that occurred in the anode seriously affected the performance of the PFC.

In Test 4, the curve of i–V characteristics was similar to that for Test 2; the increase in photocurrents in the low-voltage region (≥ 0.45 V) was less steep than the slope in Test 1. The values of I_{SC} (29.2 μA) and V_{OC} (1.18 V) were slightly lower than the corresponding values in Test 2, whilst the value of P_{Max} (6.10 μW) was clearly higher than that in Test 2 (5.51 μW). This low performance (Fig. 3) was inconsistent with the better performance in photocurrent Test 4 than that in Tests 1 and 2 (Fig. 2). This trend cannot be rationalized based on the remaining amount of produced O_2 dissolved in the anode. One of the possibilities is the polarization of photoelectrodes during the i–V measurements at high voltage to modify the band bending (charge balance at the surface) at the surface of photocatalysts [25-29].

In a control Test 1' in darkness, the value of the current was always less than the detection limit by changing the resistance from 500 kΩ to 0.3 Ω. The corresponding voltage was less than 0.002 V. Therefore, the values of I_{SC}, V_{OC} and P_{Max} were less than 0.1 μA, 0.002 V and 0.0002 μW, respectively. This indicated electricity of the PFC was generated exclusively by irradiation with UV–visible light.

2.3 Product Analyses during Photocurrent Tests at the Anode

2.3.1 Analysis of O_2 Formed at the Anode

The time course of dissolved O_2 concentrations is shown in Figure 4. When TiO_2 was irradiated with UV–visible light,

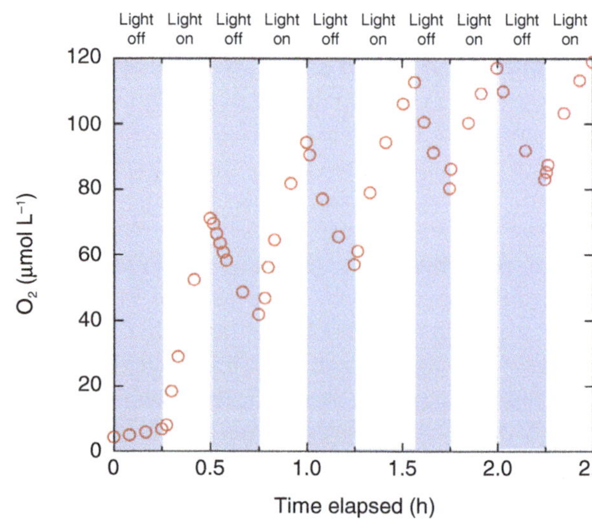

Figure 4

The time course of dissolved O_2 concentrations during the photoirradiation test on TiO_2. UV–visible light and $FeCl_3{\cdot}6H_2O$ (2.31 mmol L^{-1}) were used as excitation light and a sacrificial reagent, respectively. The initial pH value was 2.0.

the dissolved O_2 concentration linearly increased immediately in all five cycles. For 1.25 h irradiation, dissolved O_2 was increased from 7.8 μmol L^{-1} to 119 μmol L^{-1}. If the irradiation was stopped, dissolved O_2 decreased gradually in all cycles. This indicated that the formed O_2 dissolved in acidic water and was transported to the gas phase in the quartz flask. In this photoreaction test, 151 mg of TiO_2 was used, significantly greater than the 5 mg for the photocurrent test (Sect. 2.1, 2.2). Accordingly, a greater amount of O_2 was generated from the water (63.1 μmol L^{-1} per 0.25 h) in this photoreaction test than in photocurrent Test 2 (27.8 μmol L^{-1} per 3.5 h). Thus, excessive O_2 greater than the dissolution amount would be transported to the gas phase.

When the photocatalyst was irradiated, the rate would be the balance between O_2 formation and the transportation to the gas phase. The increase in dissolved O_2 in the first cycle was the quickest among all the cycles. The increase in dissolved O_2 was 63.1 μmol L^{-1} per 0.25 h. The slope of the dissolved O_2 gradually became less steep as the cycles were repeated. As a result, in the fifth cycle, it was 36.2 μmol L^{-1} per 0.25 h. This trend is contradicted by the gradual increase in the photocurrent in Figure 2, Test 1 as the cycles repeated. We suspect that the major reason was that the transportation of dissolved O_2 to the gas phase became faster as the concentration of dissolved O_2 increased, as the cycles were repeated (Fig. 4).

The increase rate of dissolved O_2 in the first cycle was:

$$63.1\ \mu\text{mol L}^{-1} \times 0.090\ \ \text{L}/0.25\ \ \text{h}/0.151\ \ g_{cat} = 151\ \mu\text{mol-O}_2\ \text{h}^{-1}\ g_{cat}^{-1} \quad (6)$$

In comparison, the total photocurrent in Test 2 was average current:

$$34.0\ \mu\text{A} \times 3\,600\ \text{s}\,\text{h}^{-1}/9.65 \times 10^4\ \text{A}\,\text{s}\,\text{mol-}e^{-1}/0.0050\ g_{cat} = 253\ \mu\text{mol-}e^{-}\text{h}^{-1}\ g_{cat}^{-1} \quad (7)$$

The electron flow rate corresponded to the O_2 generation rate of 63 μmol-O_2 h^{-1} $g_{cat}{}^{-1}$. The net O_2 generation rate in Figure 4 should be even higher because the O_2 transported to the gas phase was not included for the evaluation in Figure 4. This difference in the amount based on the PFC current was not inconsistent. In the presence of a sacrificial oxidation agent, Fe^{3+}, photoexcited electrons to the conduction band of TiO_2 easily diffuse with sufficient Fe^{3+} and react in contrast so that the electrons need to travel to the cathode to react with holes at the valence band of [Ag-]TiO_2 in the PFC. The concentration of holes was determined by the balance of charge separation by light and the cathode reaction rate of O_2 reduction [11, 21].

2.3.2 Analysis of Hypochlorous Acid Formed at the Anode from Cl^- Ions

The absorbance at 515 nm for standard mixed solutions (6.0 mL) was plotted as a function of NaOCl concentrations in the solution (6.0 mL). Then, a standard curve was calculated as follows (Fig. 5):

$$\text{Absorbance@515 nm} = 0.0166 \times [\text{ClO}^-]\,(\mu\text{mol L}^{-1}) + 0.007 \quad (8)$$

20 mg of DPD reagent in 4.0 mL phosphate buffer solution (pH 6.5) were mixed with 2.0 mL of sampled solution from the anode solution (40 mL) of the PFC after Test 1 for 3 h. The absorbance of this mixed solution was 0.018 ± 0.008. Therefore, the amount of HClO in the mixed solution (6.0 mL) was 3.98 nmol based in Equation (8).

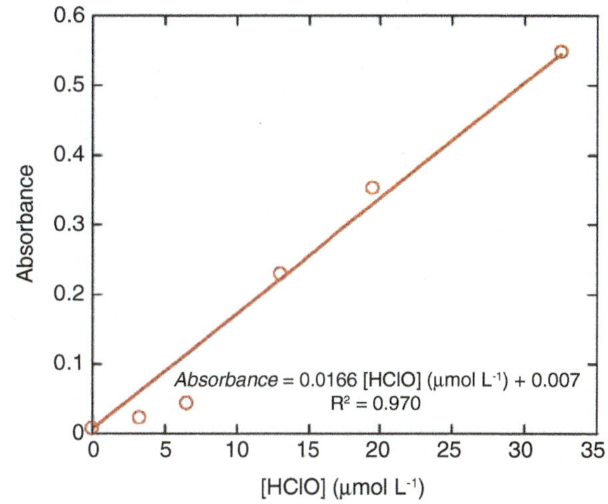

Figure 5

Absorbance at 515 nm for standard mixed solutions as a function of the concentration of HClO standard solutions.

Thus, the amount of HClO in the anode solution was calculated as:

$$3.98\,\text{nmol} \times 40\,\text{mL}/2.0\,\text{mL} = 0.080 \pm 0.056\,\mu\text{mol} \quad (9)$$

For 3 h of Test 1 (pH 2.0), total electron flow was 3.92 μmol based on the integration of photocurrents. Two holes reacted with Cl^- ions (Eq. 2) correspond to one HClO formed from Cl_2 (Eq. 3). As the concentration of HClO was the impurity level for the anode solution in Test 1, the amount of HClO was nearly the detection limit. Therefore, the upper limit of selectivity to proceed with Equation (2) for photogenerated holes was:

$$2 \times 0.080\,\mu\text{mol-HClO}/3.92\,\mu\text{mol-}h^{+} = 0.041 \pm 0.031 \quad (10)$$

The participation of chlorine ions in the anode reaction cannot be denied; however, the ratio was as low as 4.1% at most.

CONCLUSIONS

First, the anode reaction in the PFC was demonstrated exclusively to be the photooxidation of water over TiO_2 based on the O_2 evolution test, and negligible (or no) photooxidation of Cl^- occurred based on the detected amount of HClO. The selectivity of photooxidation of Cl^- was 4.1% at most as compared with the photooxidation of water. Next, a product gas-trapping PFC was demonstrated to achieve a photocurrent value of 69.7 μA per 1.3 cm^2 of photocatalysts (53.6 μA cm^{-2}). Gas trapping was enabled by accommodating an organic solvent layer over the aqueous electrolyte solution phase in the anode and utilizing the solubility difference of produced O_2. However, gas circulation by a vent hole in the upper part of the PCP film due to pressure differences was not well achieved. To transfer generated O_2 to an anode, a greater pressure difference was needed. The performance of the gas-closed PFC equipped with a hexane layer on the anode (Test 4) was the best at the initial stage of the photocurrent test due to the O_2 solubility effect of hexane rather than the other conditions. However, the photocurrents gradually decreased during the measurements for 7 h. The small amount of remaining O_2 in the anode electrolyte (2.85 μmol L^{-1}) reduced the PFC performance considerably in Test 4. The I_{SC}, V_{OC} and P_{Max} values of the gas-circulating PFC were 29.2 μA (22.5 μA cm^{-2}), 1.18 V and 6.10 μW (4.69 μW cm^{-2}), respectively. These values were similar to corresponding values for the gas-closed PFC.

The feasibility of a gas-trapping PFC was successfully demonstrated in this study. Further, the improvement of power efficiency is required by the optimization of photocatalysts and the thickness and density of photocatalyst layers in order to be accommodated in portable electronic devices, remote-control light gliders, and environmental cameras/sensors.

ACKNOWLEDGMENTS

The authors are grateful for financial support from the Feasibility Study Stage of A-STEP (AS262Z00159L, AS251Z00906L, AS231Z01459C) from the Japan Science and Technology Agency, the Iwatani Naoji Foundation (2011-2012), and a Grant-in-Aid for Scientific Research C (26410204, 22550117) from the Japan Society for the Promotion of Science. X-ray absorption experiments were conducted under the approval of the Photon Factory Proposal Review Committee (2013G159).

REFERENCES

1 Lewis N.S., Nocera D.G. (2006) Powering the Planet: Chemical Challenges in Solar Energy Utilization, *Proc. Natl. Acad. Sci. USA* **103**, 15729.

2 Izumi Y. (2013) Recent Advances in the Photocatalytic Conversion of Carbon Dioxide to Fuels with Water and/or Hydrogen Using Solar Energy and Beyond, *Coord. Chem. Rev.* **257**, 171.

3 Bolton J.R. (1996) Solar Photoproduction of Hydrogen: A review, *Sol. Energy* **57**, 37.

4 Grätzel M. (2001) Photoelectrochemical Cells, *Nature* **414**, 338.

5 Gasteiger H.A., Marković N.M. (2009) Just a Dream or Future Reality? *Science* **324**, 48.

6 Schlapbach L. (2009) Hydrogen-Fuelled Vehicles, *Nature* **460**, 809.

7 Oka K., Ogura Y., Izumi Y. (2014) X-Ray Evaluation of the Boundary between Polymer Electrolyte and Platinum and Carbon Functionalization to Conduct Protons in Polymer Electrolyte Fuel Cell, *J. Power Sources* **258**, 83.

8 Oka K., Shibata Y., Itoi T., Izumi Y. (2010) Synthesis and Site Structure of a Replica Platinum–Carbon Composite Formed Utilizing Ordered Mesopores of Aluminum-MCM-41 for Catalysis in Fuel Cells, *J. Phys. Chem. C* **114**, 1260.

9 Joya K.S., Joya Y.F., Ocakoglu K., van de Krol R. (2013) Water-Splitting Catalysis and Solar Fuel Devices: Artificial Leaves on the Move, *Angew. Chem. Int. Ed.* **52**, 10426.

10 Seery M.K., George R., Floris P., Pillai S.C. (2007) Silver Doped Titanium Dioxide Nanomaterials for Enhanced Visible Light Photocatalysis, *J. Photochem. Photobio. A* **189**, 258.

11 Ogura Y., Okamoto S., Itoi T., Fujishima Y., Yoshida Y., Izumi Y. (2014) A photofuel Cell Comprising Titanium Oxide and Silver(I/0) Photocatalysts for Use of Acidic Water as a Fuel, *Chem. Comm.* **50**, 3067-3070.

12 Huang S.Y., Ganesan P., Park S., Popov B.N. (2009) Development of a Titanium Dioxide-Supported Platinum Catalyst with Ultrahigh Stability for Polymer Electrolyte Membrane Fuel Cell Applications, *J. Am. Chem. Soc.* **131**, 13898.

13 Orilall M.C., Matsumoto F., Zhou Q., Sai H., Abrũna H.D., DiSalvo F.J., Wiesner U. (2009) One-Pot Synthesis of Platinum-Based Nanoparticles Incorporate into Mesoporous Niobium Oxide–Carbon Composites for Fuel Cell Electrodes, *J. Am. Chem. Soc.* **131**, 9389.

14 Sun Y., Zhuang L., Lu J., Hong X., Liu P. (2007) Collapse in Crystalline Structure and Decline in Catalytic Activity of Pt Nanoparticles on Reducing Particle Size to 1 nm, *J. Am. Chem. Soc.* **129**, 15465.

15 Wang L.L., Johnson D.D. (2007) Shear Instabilities in Metallic Nanoparticles: Hydrogen-Stabilized Structure on Pt_{37} on Carbon, *J. Am. Chem. Soc.* **129**, 3658.

16 Izumi Y., Nagamori H., Kiyotaki F., Masih D., Minato T., Roisin E., Candy J.P., Tanida H., Uruga T. (2005) X-ray Absorption Fine Structure Combined with X-ray Fluorescence Spectrometry. Improvement of Spectral Resolution at the Absorption Edges of 9–29 keV, *Anal. Chem.* **77**, 6969.

17 Hanna M.C., Nozik A.J. (2006) Solar Conversion Efficiency of Photovoltaic and Photoelectrolysis Cells with Carrier Multiplication Absorbers, *J. Appl. Phys.* **100**, 074510.

18 Morikawa M., Ogura Y., Ahmed N., Kawamura S., Mikami G., Okamoto S., Izumi Y. (2014) Photocatalytic Conversion of Carbon Dioxide into Methanol in Reverse Fuel Cells with Tungsten Oxide and Layered Double Hydroxide Photocatalysts for Solar Fuel Generation, *Catal. Sci. Technol.* **4**, 1644.

19 Heller A. (1981) Conversion of Sunlight into Electrical Power and Photoassisted Electrolysis of Water in Photoelectrochemical Cells, *Acc. Chem. Resear.* **14**, 154.

20 Chen F., Liu H., Bagwasi S., Shen X., Zhang J. (2010) Visible-Light Responsive Photocatalytic Fuel Cell Based on WO_3/W Photoanode and Cu_2O/Cu Photocathode for Simultaneous Wastewater Treatment and Electricity Generation, *Environ. Sci. Technol.* **46**, 11451.

21 Fujishima Y., Okamoto S., Yoshiba M., Itoi T., Kawamura S., Yoshida Y., Ogura Y., Izumi Y. (2015) Photofuel Cell Comprising Titanium Oxide and Bismuth Oxychloride ($BiO_{1-x}Cl_{1-y}$) Photocatalysts for Use of Acidic Water as a Fuel, *J. Mater. Chem. A.* **3**, 8389.

22 Moberg L., Karlberg B. (2000) An Improved *N*, *N*'-Diethyl-*p*-Phenylenediamine (DPD) Method for the Determination of Free Chlorine Based on Multiple Wavelength Detection, *Anal. Chim. Acta* **407**, 127.

23 Battino R., Rettich T.R., Tominaga T. (1983) The Solubility of Oxygen and Ozone in Liquids, *J. Phys. Chem. Ref. Data* **12**, 163.

24 Anderson A.Y., Barnes P.R.F., Durrant J.R., O'Regan B.C. (2011) Quantifying Regeneration in Dye-Sensitized Solar Cells, *J. Phys. Chem. C* **115**, 2439.

25 Hu Z.T., Liu J., Yan X., Oh W.D., Lim T.T. (2015) Low-Temperature Synthesis of Grapheme/$Bi_2Fe_4O_9$ Composite for Synergistic Adsorption-Photocatalytic Degradation of Hydrophobic Pollutant under Solar Irradiation, *Chem. Eng. J.* **262**, 1022.

26 Wu W., Huang Z.H., Lim T.T. (2014) Recent Development of Mixed Metal Oxide Anodes for Electrochemical Oxidation of Organic Pollutants in Water, *Appl. Catal. A.* **480**, 58.

27 Nath R.K., Zain M.F.M., Kadhum A.A.H. (2014) Artificial Photosynthesis using $LiNbO_3$ as Photocatalyst for Sustainable and Environmental Friendly Construction and Reduction of Global Warming: A Review, *Catal. Rev. Sci. Eng.* **56**, 175.

28 Swierk J.R., Mallouk T.E. (2013) Design and Development of Photoanodes for Water-Splitting Dye-Sensitized Photoelectrochemical Cells, *Chem. Soc. Rev.* **42**, 2357.

29 Wenyi T., Qin Z., Han Y., Xiufang Z., Hongyi L. (2012) Deactivation Anode Catalyst $La_{0.75}Sr_{0.25}Cr_{0.5}Mn_{0.5}O_{3+\delta}$ in SOFC with Fuel Containing Hydrogen Sulfur: The Role of Lattice Oxygen, *Int. J. Hydrogen Energy* **37**, 7398.

4

A Novel Method for Performance Analysis of Compartmentalized Reservoirs

Mohammad Sadeq Shahamat[1]*, Hamidreza Hamdi[2], Louis Mattar[1] and Roberto Aguilera[2]

[1] *IHS Canada, Sun Life Plaza, East Tower Suite 800, 112 –4th Avenue S.W., Calgary, AB, T2P 0H3, Calgary - Canada*
[2] *University of Calgary, 2500 University Dr. NW, Calgary, AB, T2N 1N4, Calgary - Canada*
e-mail: sadeq.shahamat@ihs.com

* Corresponding author

Abstract — *This paper presents a simple analytical model for performance analysis of compartmentalized reservoirs producing under Constant Terminal Rate (CTR) and Constant Terminal Pressure (CTP). The model is based on the well-known material balance and boundary dominated flow equations and is written in terms of capacitance and resistance of a production and a support compartment. These capacitance and resistance terms account for a combination of reservoir parameters which enable the developed model to be used for characterizing such systems. In addition to considering the properties contrast between the two reservoir compartments, the model takes into account existence of transmissibility barriers with the use of resistance terms.*

The model is used to analyze production performance of unconventional reservoirs, where the multistage fracturing of horizontal wells effectively creates a Stimulated Reservoir Volume (SRV) with an enhanced permeability surrounded by a non-stimulated region. It can also be used for analysis of compartmentalized conventional reservoirs. The analytical solutions provide type curves through which the controlling reservoirs parameters of a compartmentalized system can be estimated. The contribution of the supporting compartment is modeled based on a boundary dominated flow assumption. The transient behaviour of the support compartment is captured by application of "distance of investigation" concept. The model shows that depletion of the production and support compartments exhibit two unit slopes on a log-log plot of pressure versus *time for CTR. For CTP, however, the depletions display two exponential declines. The depletion signatures are separated by transition periods, which depend on the contribution of the support compartment (* i.e. *transient or boundary dominated flow).*

The developed equations can be implemented easily in a spreadsheet application, and are corroborated with the use of a numerical simulation. The study provides a new and non-conventional insight into the use of production data as an aid in the understanding of hydraulically fractured tight formations and compartmentalized conventional reservoirs.

Résumé — Nouvelle méthode d'analyse de la performance des réservoirs compartimentés — Cet article présente un modèle analytique simple pour analyser la performance des réservoirs compartimentés produisant un Débit Final Constant (DFC) et une Pression Finale Constante (PFC). Ce modèle est basé sur un équilibre bien connu des matières et sur des équations d'écoulement soumis aux limites et est symbolisé par les critères de capacité et de résistance d'un compartiment de production et de réserve. Ces critères de capacité et de résistance constituent un ensemble de paramètres du réservoir qui permettent d'utiliser le modèle développé pour caractériser de tels systèmes. En plus de considérer les différentes propriétés des deux compartiments du réservoir, ce modèle prend en compte l'existence de barrières de transmissibilité en se servant des critères de résistance.

Ce modèle peut être utilisé pour analyser la performance de production de réservoirs non conventionnels où la fracturation multi-étape de puits horizontaux crée de manière efficace un Volume de Réservoir Stimulé (VRS) avec une perméabilité accrue, et entouré d'une zone non stimulée. Il peut également être utilisé pour l'analyse de réservoirs compartimentés conventionnels. Les solutions analytiques fournissent des courbes types, grâce auxquelles il est possible d'évaluer les paramètres de contrôle des réservoirs d'un système compartimenté. La contribution du compartiment de réserve est modélisée sur la base d'une hypothèse d'écoulement soumis aux limites. Le comportement transitoire du compartiment de réserve est pris en compte par l'application du concept de « distance d'étude ». Le modèle montre que l'épuisement des compartiments de production et de réserve est représenté sur un graphique logarithmique par deux courbes distinctes de pression et de temps pour le DFC. Pour la PFC cependant, les signes d'épuisement sont représentés par deux courbes de déclin exponentiel. Les signes d'épuisement sont séparés par des périodes de transition qui dépendent de la contribution du compartiment de réserve (c'est-à-dire écoulement transitoire ou écoulement soumis aux limites).

Les équations développées peuvent être facilement mises en œuvre dans un tableur, et sont corroborées grâce à l'utilisation d'une simulation numérique. Cette étude fournit une perspective nouvelle et non conventionnelle de l'utilisation des données de production en vue de mieux comprendre les réservoirs compacts traités par fracturation hydraulique et les réservoirs conventionnels compartimentés.

NOMENCLATURE

BDF	Boundary Dominated Flow
c	Compressibility, 1/psi
C	Fluid flow capacitance, Stb/psia
C^*	Fluid flow capacity, Stb/(ft.psia)
CTP	Constant Terminal Pressure
CTR	Constant Terminal Rate
Der	Bourdet derivative
F_c	Ratio of capacitance of the production to support compartment
F_R	Ratio of resistance of the production to support compartment
$F_c{}^*$	Ratio of capacity of the production to support compartment
$F_R{}^*$	Ratio of resistivity of the production to support compartment
h	Net pay, ft
k	Permeability, md
p_{avg}	Average reservoir pressure within the compartment, psi
p_D	Dimensionless pressure used in classical well testing literature
p_{DM}	Dimensionless pressure defined in this work
$p_{DM,BDF}$	Dimensionless pressure at BDF
$p_{DM}{}^c$	Dimensionless pressure based on Δp^c
$p_{DM}{}^{cs}$	Dimensionless pressure based on Δp^{cs}
p_i	Initial reservoir pressure, psi
p_{wf}	Wellbore flowing pressure, psia
q	*In situ* well-production rate, bbl/D
q_D	Dimensionless rate used in classical well testing literature
q_{Dd}	Dimensionless decline rate used in production data analysis literature
q_{DM}	Dimensionless rate defined in this work
ODE	Ordinary Differential Equation
R	Flow resistance, psi/bbl
s	(Subscription) support
SRV	Stimulated Reservoir Volume
t	Time, days
t_D	Dimensionless time used in classical well testing literature
t_{DM}	Dimensionless time
$t_{DM,BDF}$	Dimensionless time at BDF
V	Volume, bbl
WFT	Wireline Formation Tester
x_f	Fracture half length, ft
y_e	Reservoir size in y-direction, ft
y_{eD}	Dimensionless distance
y_{inv}	Distance of investigation, ft
Δp^c	Pressure depletion of a dummy production compartment neglecting the support, psia
Δp^{cs}	Pressure depletion of a dummy support compartment, psia

α Constant used for defining the distance of investigation

β_1 Constant used for determining q_D and p_D (equal to $2\pi \times 141.2$ in oil field units)

β_2 Constant used for determining t_D (equal to 0.00633 in oil field units)

λ_1 First eigenvalue of the matrix of coefficients, A

λ_2 First eigenvalue of the matrix of coefficients, A

μ Viscosity, cp

ϕ Porosity, fraction

SI CONVERSION FACTORS

bbl $\times$ 1.589873E-1 = m^3
cp $\times$ 1E-3 = Pa.s
ft $\times$ 3.048E-1 = m
mD $\times$ 9.869223E16 = m^2
psi $\times$ 6.894757E3 = Pa
* Conversion factors are exact

INTRODUCTION

Compartmentalized reservoirs are the result of segregation of a petroleum accumulation into a number of individual fluid/pressure compartments. They occur when flow is restricted across some form of boundaries in the reservoir. These boundaries are caused by a variety of geological features and processes and can be either completely sealed, across which no flow can occur, or partially sealed in which case the boundaries exhibit low to very low permeability (permeability barriers). The latter allows fluids and pressures of the compartments to equilibrate over geologic time-scales, and retards cross-flow between compartments during production. As a result, there will be pressure discontinuities across the reservoir, which according to Stewart and Whaballa (1989) can occur over near horizontal events such as shales or micaceous streaks and/or near vertical events such as partially communicating faults or turbidite lobe interfaces.

Evidence of compartmentalization has been observed in both oil and gas reservoirs (Junkin *et al.*, 1992). A number of compartmentalized reservoirs have been discovered in different parts of the world including those in the North Sea (Fox *et al.*, 1988), Texas Gulf Coast (Junkin *et al.*, 1992) and Australia (Malavazos and McDonough, 1991). Stewart (2011) presents a well-test case study in a Middle-East carbonate reservoir, where the derivative unit slope buildup indicates a charging-recharge mechanism. The compartmentalized behaviour can also occur in particular meandering fluvial reservoirs. In these environments, the depletion and the lateral cross-flow between the laterally avulsing point bar deposits separated by the semi-permeable channel fills can effectively create a compartmentalized system (Hamdi, 2014; Hamdi *et al.*, 2014). Corbett *et al.* (2012) present an interesting well-test response in a meandering channelized system, where numerical well-test simulations and a multi-point statistics approach support the analytical well-test interpretation and the existence of effective sand bar compartments. Malavazos and McDonough (1991) demonstrate another channelized reservoir example where there is communication between the stacked channel compartments.

An important aspect of reservoir compartmentalization is a stacked channel realization where two sand bodies communicate hydraulically with each other through an overlapping area (interface). Chilingarian *et al.* (1992) give an account of geologic and engineering aspects of compartmentalization in carbonate reservoirs and point out that reservoir compartmentalization may result not only from vertical and lateral lithofacies changes, but also from spatial variations stemming from processes of post-depositional diagenesis that create secondary porosity in the rocks.

Reservoir compartmentalization can also be considered in unconventional reservoirs, where multi-frac completion of horizontal wells create an SRV region of enhanced permeability around the horizontal well that is surrounded by an unstimulated region of much lower permeability (Stalgorova and Mattar, 2012). It is therefore reasonable to model the production behaviour of such a reservoir using a model where the production and support tanks are attributed to the SRV and unstimulated reservoir sections, respectively.

1 COMPARTMENT DETECTION AND ANALYSIS

Reservoir compartmentalization is detected primarily from observations of pressure discontinuities both areally and vertically (Stewart and Whaballa, 1989). Such reservoir pressure variations over a producing field can be monitored using Wireline Formation Testers (WFT) and are due to differential depletion of the system caused by permeability barriers or reservoir heterogeneities of one kind or another (Stewart, 2011). Bradley and Powley (1994) indicate other ways for detection of reservoir compartmentalization, including the changes in mineralogy and drilling rates and also using electrical resistivity, sonic velocity and density of shales. Reservoir simulation is another method which can be used to detect compartmentalization. Because of its complexity and associated degrees of freedom, though, this method is not justified in most cases before identifying and quantifying the major flow units and barrier resistances using other methods (Rahman and Ambastha, 2000; Stewart and Whaballa, 1989; Aguilera, 2013).

There are also well-testing and production data analysis methods which can contribute to an improvement of the understanding of geology. Analytical models and related solutions can lead to a simplified description of the geological heterogeneities around the well and can provide good insights into the significance of various parameters affecting the transport phenomena (Massonnat *et al.*, 1993). Stewart and Whaballa (1989) argue that an extended drawdown test should be performed to detect the limits of the producing compartment. They develop type curves to identify the extent of faulted boundaries between compartments. By matching the late time data, they estimate the resistance to flow of fluid at the boundary (which they call transmissibility). Their approach is based on the use of single-phase material balance equations, the concepts of boundary pressure time delay and desuperposition to predict pressures over different flow regimes. Along the same lines, Rahman and Ambastha (2000) develop analytical 1D, 2D and 3D solutions for constant-rate transient flow in compartmentalized systems.

Stewart (2011) presents a mathematical method for well-test analysis of a two-cell compartmentalized system. He presents the material balance equations in dimensionless terms and discusses the existence of two derivative unit slopes in logarithmic scales indicating depletion of the well and support compartments. He considers the effect of compartment volumes, neglecting the variations in other reservoir parameters, *e.g.* permeability, between the compartments. This work does not involve the mathematical complexities of Rahman and Ambastha (2000) and is appealing from a practical point of view.

The primary objective of this paper is to obtain information on the contrast of rock and fluid properties from analysis of compartmentalized or composite systems using the capacitance and resistance terms introduced by Shahamat *et al.* (2014). Two tanks (and therefore two capacitance and resistance terms) are considered in the model – production and support. The support tank feeds the production tank; therefore, its contribution starts after investigating the production tank. The length component in the capacitance and resistance of the support tank is first assumed to be constant, equal to the size of the support tank in the y-direction. This assumption implies instantaneous depletion (*i.e.* BDF) of the support tank. Based on this assumption and by using the fluid-flow and material balance equations for both tanks, simple analytical solutions are derived for constant-rate and constant-pressure production of a compartmentalized system. Using the distance of investigation equation in the capacitance and resistance terms, it is shown that the solutions can be easily modified to account for transient flow within the support tank. The derived mathematical solutions enable the generation of production type curves which can be used to characterize a compartmentalized system and obtain meaningful information related to the contrast in rock and fluid properties between the compartments. To the authors' best knowledge, no analytical model is available in the literature that discusses the constant-pressure production of compartmentalized systems, and that can be used for long-term production forecasting. The mathematical simplification described in this paper has the same spirit as the work documented by many authors attempting to replace a complex reservoir model by a simpler equivalent "upscaled" system (Valko *et al.*, 2000; Kaviani *et al.*, 2008; Sayarpour *et al.*, 2008; Yousef *et al.*, 2009; Noetinger, 2015).

2 MATHEMATICAL DEVELOPMENT

The mathematical framework of this study is based on the capacitance-resistance approach for a single-phase fluid. Adapted from Shahamat and Aguilera (2010), the configuration of the different sections of a compartmentalized reservoir is depicted using a tank model (Fig. 1a). The production tank is directly connected to the wellbore through

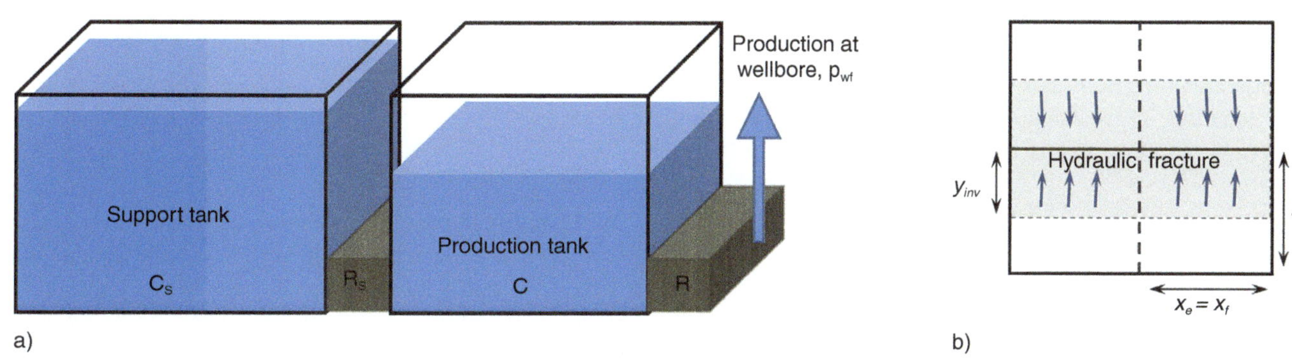

Figure 1

a) Tank representation of a compartmentalized reservoir; b) a hydraulically fractured well in the center of a rectangular reservoir.

the sand pack which implies a flow resistance (R) to the wellbore. In addition, there is a support tank which is connected to the production tank through another barrier with a flow resistance, R_s. Depending on the values of flow capacitance (C and C_s) and resistance (R and R_s) there are various degrees of pressure communication between these compartments. Figure 1b shows the base reservoir geometry of each resistance. This geometry implies pure transient linear flow and is chosen because linear flow has widespread application in unconventional reservoirs analysis.

It is assumed that the contribution of the support tank starts after investigating the production tank. This means that by the time the support tank starts feeding the production tank, the characteristics of the production tank (*i.e.* its C, R and t_{BDF}) are known (Shahamat, 2014). These properties are used for modeling the compartmentalized reservoir behaviour.

Using the fluid-flow equation for relating the production rate to the average pressure of the production tank, one gets:

$$q = \frac{1}{R}\left(p_{avg} - p_{wf}\right) \tag{1}$$

Resistance, R, in the above equation is equivalent to the inverse of the productivity index. Knowing that the pressure at the adjoining sides of the production tank is average pressure (p_{avg}), one can use the Darcy's equation for linear flow to analytically derive an equation for resistance according to $R = \left(\frac{\beta_1 B\mu}{kh}\right)\left(\frac{y_e}{x_f}\right)$. In this equation, β_1 is a constant which is equal to $2\pi \times 141.2$ in oilfield units and B is the dimensionless formation volume factor.

For relating the average pressure of the support tank with the average pressure of the production tank we use Darcy's equation through the following formula:

$$q_s = \frac{1}{R_s}\left(p_{avg,s} - p_{avg}\right) \tag{2}$$

Linear flow regime within the support resistance leads to the following definition, $R_s = \left(\frac{\beta_1 B\mu}{kh}\right)_s \times \left(\frac{y_e}{x_f}\right)_s$. In the above two equations, q and q_s are *in situ* rates.

In addition to Equations (1) and (2) for describing the behaviour of a compartmentalized system, the material balance equations for both tanks are employed. Material balance equation (*i.e.* compressibility equation) for the production tank is written in the following form:

$$C\frac{dp_{avg}}{dt} = q_s - q \tag{3}$$

Similar to Equation (3), the compressibility equation for the support tank takes the following form:

$$C_s\frac{dp_{avg,s}}{dt} = -q_s \tag{4}$$

In the above two equations, capacitances (C and C_s) are expressed in Stb/psia. It is noted that capacitance is the multiplication of pore volume and total compressibility.

Equations (1) through (4) are the building blocks of the proposed methodology. Solutions to different production constraints (CTR or CTP) are obtained by defining different dimensionless parameters, as shown in the following sections.

2.1 Constant Terminal Rate Production (Reservoir Limit Testing)

For the CTR production, the dimensionless pressure and time, p_{DM} and t_{DM}, are defined as follows (see Appendix C):

$$p_{DM} = \frac{p_i - p}{q \times R} \tag{5}$$

and,

$$t_{DM} = \frac{t}{C \times R} \tag{6}$$

It is noted that the dimensionless pressure defined in Equation (5) is different from the definition used in the classical well-testing literature. In well-testing literature, the dimensionless pressure is defined based on wellbore pressure, whereas Equation (5) is written in terms of average reservoir pressure (to be used in Eq. 1 through Eq. 4).

Moreover, two additional dimensionless parameters are introduced to characterize reservoirs with two communicating compartments. The first parameter, F_R, is the ratio of resistances, and reflects the combined effect of contrast in permeability, viscosity, formation thickness and skin factor, between the production and support tanks:

$$F_R = \frac{R}{R_s} \tag{7}$$

The second parameter, F_c, is the ratio of flow capacitances, and reflects the combined effect of changes in reservoir pore volume and compressibility of the production tank with respect to the support tank:

$$F_c = \frac{C}{C_s} \tag{8}$$

While normally one would expect to define the ratios of resistance and capacitance with respect to the production tank (*i.e.* $\frac{R_s}{R}$ and $\frac{C_s}{C}$), defining F_R and F_c according to Equations (7) and (8) simplify the formulations and obtained solutions in a concise form. As such, smaller values of F_R (or F_c) indicate larger values of resistance (or capacitance) of the support tank compared with the production tank.

Using the dimensionless variables defined in Equation (5) through (8), and upon some mathematical manipulations, Equations (3) and (4) take the following dimensionless forms, respectively:

$$\frac{dp_{DM}}{dt_{DM}} = F_R(p_{DMs} - p_{DM}) - 1 \tag{9}$$

$$\frac{dp_{DMs}}{dt_{DM}} = -F_R F_c(p_{DMs} - p_{DM}) \tag{10}$$

In these equations, p_{DM} and p_{DMs} are the dimensionless pressures of the production and support compartments. Equations (9) and (10) construct a system of linear differential equations, which can be solved to obtain the pressure of the support and production tanks. This system can be solved using matrix algebra routines. The detailed procedure for the solution is given in Appendix A. The resulting dimensionless pressure of the production tank (p_{DM}) is given by the following closed form:

$$\begin{aligned} p_{DM} = {} & p_{DM,BDF} + \left(1 + \frac{F_R}{\lambda_2}\right)(t_{DM} - t_{DM,BDF}) \\ & + \frac{F_R}{\lambda_2}\left(\frac{1}{\lambda_2} + p_{DM,BDF}\right)\left(1 - e^{\lambda_2\left(t_{DM} - t_{DM,BDF}\right)}\right) \end{aligned} \tag{11}$$

and the solution for the support tank (p_{DMs}) is:

$$\begin{aligned} p_{DMs} = {} & \left(1 + \frac{F_R}{\lambda_2}\right)\left\{(t_{DM} - t_{DM,BDF}) + \left(\frac{1}{\lambda_2} + p_{DM,BDF}\right)\right. \\ & \left. \times \left(1 - e^{\lambda_2\left(t_{DM} - t_{DM,BDF}\right)}\right)\right\} \end{aligned} \tag{12}$$

where, $\lambda_2 = -F_R(1 + F_c)$ is the smallest eigenvalue of the matrix of coefficients for the system of Equations (9) and (10). In these equations, $t_{DM,BDF}$ and $p_{DM,BDF}$ are the dimensionless time and pressure at the beginning of the BDF for the production tank.

2.2 Constant Terminal Pressure Production (Rate Transient Analysis)

For the CTP production, the dimensionless rate variable, q_{DM}, is defined as follows (see Appendix C):

$$q_{DM} = \frac{q \times R}{p_i - p_{wf}} \tag{13}$$

The dimensionless rate defined in this equation is based on the average reservoir pressure and therefore is the equivalent linear flow form of the Fetkovich rate decline parameter (Fetkovich, 1980). Employing the dimensionless rate (Eq. 13) and time (Eq. 6) in Equations (1) through (4), the constant-pressure behaviour of the compartmentalized system is obtained based on the solution of a system of ordinary differential equations:

$$\frac{dq_{DM}}{dt_{DM}} = -q_{DM} + q_{DMs} \tag{14}$$

$$\frac{dq_{DMs}}{dt_{DM}} = F_R q_{DM} - (F_R + F_c F_R) q_{DMs} \tag{15}$$

In these equations, q_{DM} and q_{DMs} are dimensionless rates for the production and support compartments. Similar to the constant terminal rate solution, these equations can be solved using matrix algebra routines. The detailed procedure for the solution is outlined in Appendix B. The resulting dimensionless rate of the production tank (q_{DM}) is given as following closed form:

$$\begin{aligned} q_{DM} = q_{DM,BDF} & \left\{\left(\frac{\lambda_2 + 1}{\lambda_2 - \lambda_1}\right) \times e^{\lambda_1\left(t_{DM} - t_{DM,BDF}\right)}\right. \\ & \left. - \left(\frac{\lambda_1 + 1}{\lambda_2 - \lambda_1}\right) \times e^{\lambda_2\left(t_{DM} - t_{DM,BDF}\right)}\right\} \end{aligned} \tag{16}$$

Similarly, the solution for the support tank (q_{DMs}) is obtained according to Equation (17):

$$\begin{aligned} q_{DMs} = {} & q_{DM,BDF}\left(\frac{(\lambda_1 + 1)(\lambda_2 + 1)}{(\lambda_2 - \lambda_1)}\right) \\ & \times \left\{e^{\lambda_1\left(t_{DM} - t_{DM,BDF}\right)} - e^{\lambda_2\left(t_{DM} - t_{DM,BDF}\right)}\right\} \end{aligned} \tag{17}$$

We note that in these two equations, $\lambda = \frac{-(1+F_R(1+F_c))\pm\sqrt{(1-F_R(1+F_c))^2+4F_R}}{2}$, adopting the convention that λ_1 and λ_2 correspond to positive (+) and negative (−) signs, respectively. In these equations, $t_{DM,BDF}$ and $q_{DM,BDF}$ are the dimensionless time and rate at the beginning of the BDF for the production tank.

3 RESULTS AND DISCUSSION

In order to discuss the behaviour of the obtained equations in a systematic manner, first the equations obtained for the CTR production are elaborated and after that the CTP production behaviour is discussed.

3.1 CTR Production

The CTR behaviour of a compartmentalized reservoir can be better understood using the principle of superposition. The pressure change in the production tank is expressed

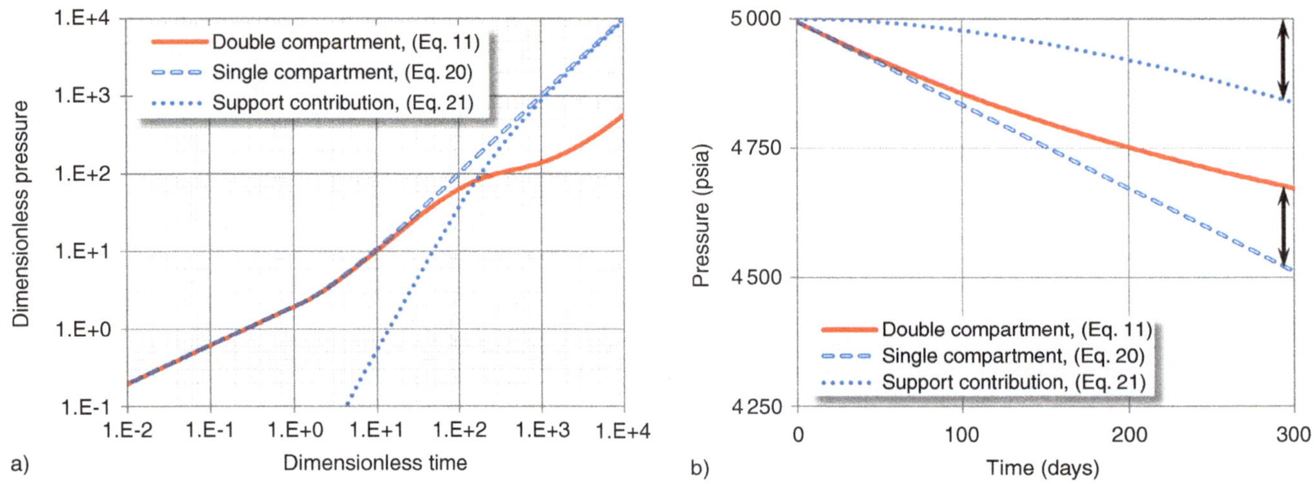

Figure 2

a) Log-log and b) Cartesian plots showing comparison of pressure behaviour of a single tank model *versus* a 2-compartment reservoir, $F_R = 0.01$ and $F_c = 0.1$. Also shown is the contribution of the support tank.

as the depletion of a single tank without any support (Δp^c) subtracted by the pressure maintenance caused by the support tank (Δp^{cs}). This means that the pressure change of a single tank can be expressed as below:

$$\Delta p = \Delta p^c - \Delta p^{cs} \tag{18}$$

Writing the above equation in dimensionless form, we obtain:

$$p_{DM}{}^{cs} = p_{DM}{}^{c} - p_{DM} \tag{19}$$

As evident from the definition of $p_{DM}{}^{c}$, an equation for its behaviour can be obtained by using Equation (11), when R_s approaches infinity and therefore $F_R = 0$. This leads to the following relation:

$$p_{DM}{}^{c} = p_{DM,BDF} + \left(t_{DM} - t_{DM,BDF}\right) \tag{20}$$

Using Equations (20) and (11) inside Equation (19), we get:

$$p_{DM}{}^{cs} = -\frac{F_R}{\lambda_2}\left(t_{DM} - t_{DM,BDF}\right) - \frac{F_R}{\lambda_2}\left(\frac{1}{\lambda_2} + p_{DM,BDF}\right) \times \left(1 - e^{\lambda_2\left(t_{DM} - t_{DM,BDF}\right)}\right) \tag{21}$$

Note that this equation is different from Equation (12) and is useful for illustrating the contribution of the support tank. Figure 2 demonstrates the behaviour of a compartmentalized reservoir and the support contribution compared with a single tank reservoir model. The plot on the left hand side of this figure shows dimensionless pressure *versus* dimensionless time in logarithmic scales while the right hand side of the figure gives the Cartesian plot of pressure *versus* time. It is evident from these figures that during transient flow, the behaviour of both single and double tank models are identical. However, the effects of the support tank are felt after reaching the boundaries of the production tank. During this period, the log-log plot gives a straight line of unit slope for the single tank model which corresponds to a straight line on the Cartesian plot shown on the right hand side of Figure 2. Deviation from this behaviour (unit slope on the log-log plot or late time straight line on the Cartesian plot) is due to the pressure contribution of the support tank. The double-sided black arrows on the Cartesian plot clearly demonstrate that the amount of deviation is equal to the support contribution. It is noted that at late times, *i.e.* total system pseudo-steady state, the two compartments are depleted at the same rate with a constant pressure difference between them. Subtracting Equation (12) from Equation (11) and evaluating the result at late time gives the pressure difference according to the following relation:

$$p_{DM} - p_{DMs} = -\frac{1}{\lambda_2} \tag{22}$$

One can also determine the points where there are major changes in the plot of pressure *versus* time. These points include the start of BDF for the production tank, the time where deviation from the first unit slope on the log-log plot of p_{DM} *versus* t_{DM} takes place, and the time attributed to the start of the total system behaviour. The equation for the distance of investigation is used to identify the time at which these changes occur in the behaviour of a compartmentalized reservoir:

$$y_{inv} = \alpha_1 \sqrt{\frac{\beta_2 kt}{(\phi \mu c)_t}} \tag{23}$$

In this equation, k is the permeability in md, μ is the viscosity in cp, ϕ is the porosity, c is the compressibility in 1/psia and time is in days. $\beta_2 = 0.00633$ and α_1 is a constant which depends on the criterion used for defining the distance of investigation. α_1 depends on the type of production at the wellbore. Wattenbarger *et al.* (1998) proposed $\alpha_1 = 1.42$ and 2 for constant-rate and constant-pressure production, respectively. Later, Nobakht and Clarkson (2012) suggested a modification of the distance of investigation equation over that of Wattenbarger *et al.* (1998) for constant-pressure production, giving $\alpha_1 = 2.55$. In this study and based on simulation studies, we use $\alpha_1 = 1.76$ and 2.23 for constant-rate and constant-pressure production, respectively. It should be mentioned that the definition of y_{inv} in the above equation is applicable only for $t < t_{BDF}$, where t_{BDF} is the time to reach BDF. This upper limit of time for y_{inv} ensures that the calculated reservoir distance in the y-direction is limited to the size of the reservoir, y_e.

Using Equation (23), one can identify the time at which these changes occur in the behaviour of a compartmentalized reservoir. Writing this equation in terms of dimensionless variables defined in this work, one can obtain the time at which BDF of the production tank starts:

$$t_{DM,BDF} = \frac{\pi}{2} \frac{1}{\alpha_1{}^2 \alpha_2} \tag{24}$$

The time where deviation from the single tank behaviour happens is determined from the following equation:

$$t_{DM} = t_{DM,BDF} \left(\frac{F_R + 1}{F_R} \right) \tag{25}$$

and the time at which the total system BDF occurs can be determined using the distance of investigation equation for the total system:

$$t_{DM} = t_{DM,BDF} \left(\frac{F_R F_c + 1}{F_R F_c} \right) \tag{26}$$

In addition to the above characteristic points, we can calculate the well-testing or semilog derivative (*Der*) of Equation (11) according to the following equation:

$$Der = \left(1 + \frac{F_R}{\lambda_2}\right) t_{DM} - F_R \left(\frac{1}{\lambda_2} + p_{DM,BDF} \right) \times t_{DM} e^{\lambda_2 (t_{DM} - t_{DM,BDF})} \tag{27}$$

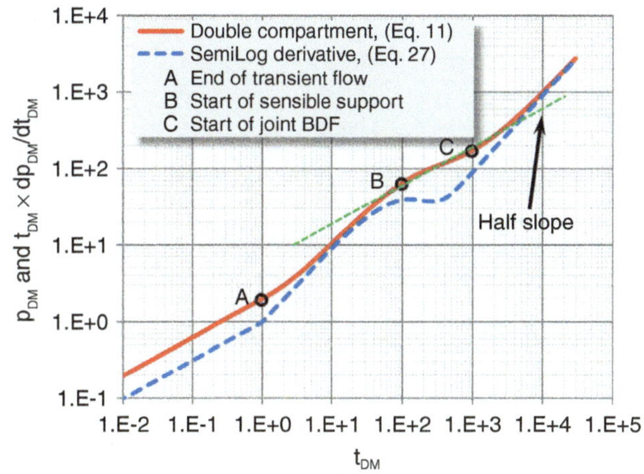

Figure 3

Log-log plot of p_{DM} and well-testing derivative (*Der*) *versus* t_{DM}, along with the position of the characteristic points for the behaviour of the compartmented reservoir.

Figure 3 shows a plot of dimensionless pressure and the well-testing derivative *versus* dimensionless time with inclusion of the points at which the transient flow in the production tank ends (point A), the support tanks becomes active (B) and the BDF for the composite system begins (C). This figure demonstrates the powerful ability of the well-testing derivative for detecting different flow regimes. Compared with the dimensionless pressure plot, the well-testing derivative shows the start of joint BDF at a much earlier time. In addition, it is noted that while the dimensionless pressure curve manifests a half slope during the transition period between points B and C, the welltesting derivative does not exhibit a half slope and therefore indicates that the flow regime during this period is not linear flow. This powerful diagnostic feature of derivative plot is used to advantage in well-test analysis for the detection of different flow regimes. In the context of production data analysis, however, the sensitive nature of the derivative amplifies the inherent noise in the production data so much that it makes the derivative plot essentially useless. Accordingly, it is reasonable to use the developed pressure solutions (rather than their derivatives) to make acceptable production analysis and therefore use the points with significant changes in the production behaviour to simply characterize a compartmentalized system (*via* Eqs. 24-26).

Evident from Equation (11), the behaviour of the double compartment model is a function of the value of F_c and F_R. These parameters determine the time at which the straight line attributed to the behaviour of the total system takes place. Figure 4 shows the effect of F_c and F_R on the behaviour of the double tank model. As mentioned previously,

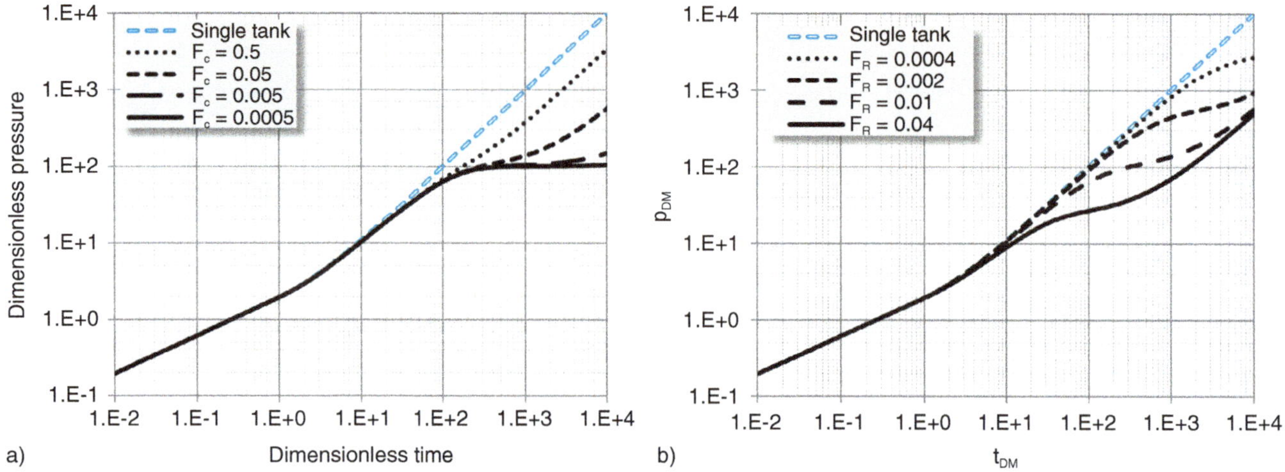

Figure 4

a) Effect of F_c on the CTR response of a compartmentalized system with $F_R = 0.01$; b) effect of F_R on the CTR response of a compartmentalized system with $F_c = 0.05$.

smaller values of F_R (or F_c) indicate larger resistance (or capacitance) of the support tank compared with the production tank and therefore later appearance of the total system unit slope. The plot on the left-hand side of this figure shows the effect of F_c for a constant value of F_R and *vice versa* for the plot on the right hand side of this figure.

Using Equation (11), type curves for different values of F_c and/or F_R can be easily generated which, through a simple type curve matching approach, can be used for estimating the contrast of capacitance (F_c) and resistance (F_R) between production and support compartments. These parameters can be used to compare the hydraulic diffusivity of the two reservoir sections.

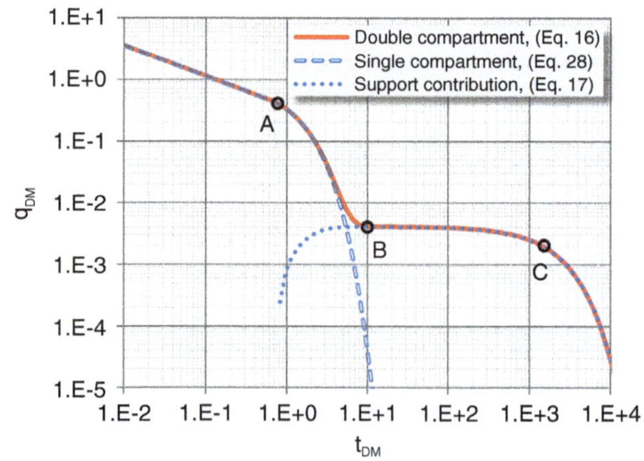

Figure 5

Log-log plot of q_{DM} *versus* t_{DM} showing the behaviour of a double tank model, $F_R = 0.01$ and $F_c = 0.05$, the position of its characteristic points and its comparison with the single tank model.

3.2 CTP Production

CTP production from a two-compartment reservoir yields a double exponential decline in rate; each is attributed to depletion of one compartment. A log-log plot of dimensionless rate *versus* dimensionless time (Eq. 16) for typical values of $F_R = 0.01$ and $F_c = 0.05$ is given in Figure 5. This figure demonstrates the difference between the behaviour of a single tank and that of a two-tank reservoir. Depletion of a single tank results in the well-known exponential decline (Fetkovich, 1980) which can be obtained by substitution of $\lambda_1 = 0$ and $\lambda_2 = -1$ in Equation (16):

$$q_{DM} = q_{DM,BDF} \times e^{-\left(t_{DM} - t_{DM,BDF}\right)} \tag{28}$$

Based on these parameters and using the equation for the distance of investigation, the points where major changes take place in the plot of rate *versus* time can be determined. The first point is the start of the BDF for the production tank, which as in the case of CTR can be determined with the use of Equation (24). The second point is the time where deviation from the single tank behaviour happens. This can be determined according to the equation (Shahamat, 2014):

$$t_{DM} = t_{DM,BDF} + \frac{1}{\lambda_1 + 1} \times ln\left(\frac{\lambda_2 - \lambda_1}{\lambda_2 + 1}\right) \tag{29}$$

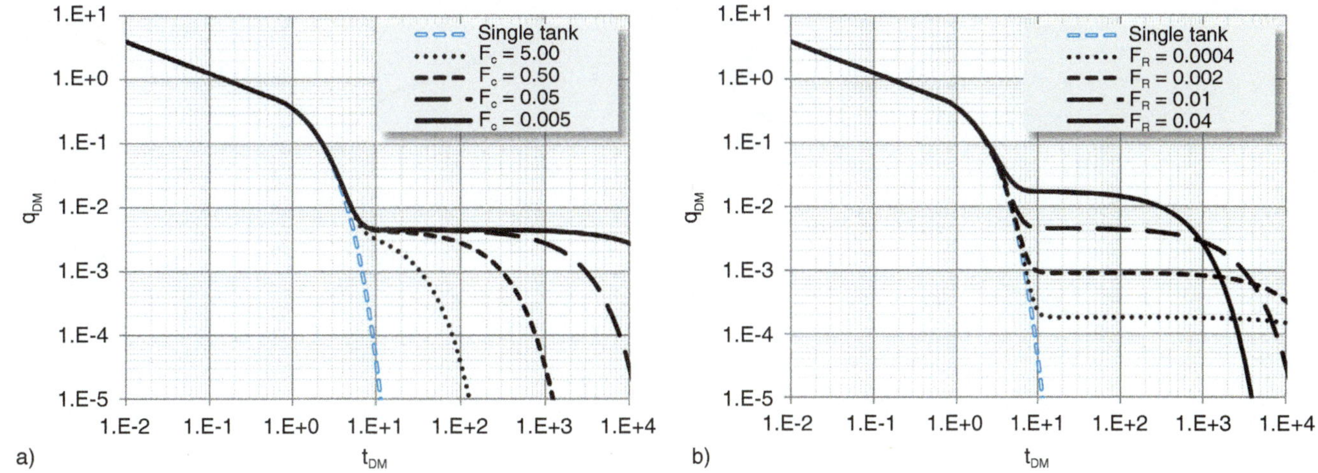

Figure 6

a) Effect of F_c on the CTP response of a compartmentalized system with $F_R = 0.01$; b) effect of F_R on the CTP response of a compartmentalized system with $F_c = 0.05$.

The time for the total system BDF can be determined using the distance of investigation equation for the total system, similar to the case of CTR production, Equation (26).

Figure 5 shows the location of the characteristic points (end of transient flow in the production tank (A), start of sensible support tank contribution (B) and start of BDF for the composite system (C)) on the log-log plot of q_{DM} *versus* t_{DM} for $F_R = 0.01$ and $F_c = 0.1$. It depicts that the rate decline associated with such a compartmentalized system exhibits a period following depletion of the production tank where the flow rate is constant. q_{DM} for this period can be obtained from the second exponential term in Equation (16).

Similar to the CTR, deviation from the single tank behaviour is because of the pressure contribution of the support tank and is a function of the values of F_c and F_R. These parameters determine the behaviour of the system. Figure 6 shows the effect of F_c and F_R on the behaviour of the double tank model. The plot on the left-hand side of this figure shows the effect of F_c for a constant value of F_R and *vice versa* for the plot on the right of this figure.

Using Equation (16), type curves for different values of F_c and/or F_R can be easily generated which can be used through a simple type curve matching approach to estimating the contrast of capacitance (F_c) and resistance (F_R) between production and support compartments. These parameters reflect the effective variation of reservoir and fluid properties across the compartments.

4 TRANSIENT FLOW WITHIN THE SUPPORT TANK

The equations developed so far are based on the premise of instantaneous (or relatively instantaneous) depletion of the support tank. This means that in the above equations the transient flow in the support tank is ignored. As a result, immediately after the start of support tank contribution, its flow regime is assumed to be BDF. This is a reasonable assumption for cases where the size of the support and production tanks and also their permeabilities are very close to each other. However in cases where the size of the support tank is bigger than that of the production compartment, the support permeability should be higher in order to satisfy its relative instant depletion. In practice, this may not be the case and the production tank may be surrounded by a number of lower (instead of higher) permeability sections. Moreover, having a lower-permeability support tank would require extended transient flow to be considered.

In order to extend the application of the proposed model for capturing transient flow within the support tank, the distance of investigation concept is employed. The length components in the capacitance and resistance parameters of the support tank are allowed to change with time according to the distance of investigation equation, Equation (23), until reaching its BDF. This implies that parameters F_c and F_R decrease with time and become constant only at late times, therefore they cannot be used for type curve generation. For the purpose of producing the appropriate type curves, we define $F_c{}^*$ and $F_R{}^*$ where $F_c{}^* = F_c \times \frac{y_{inv,s}}{y_e}$ and $F_R{}^* = F_R \times \frac{y_{inv,s}}{y_e}$. These parameters indicate the ratios of

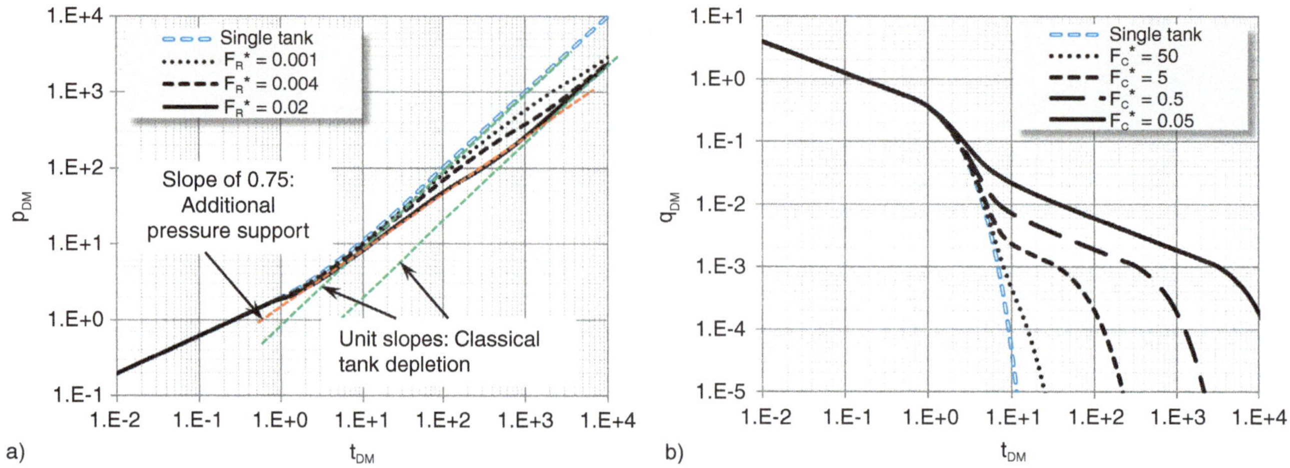

Figure 7

a) Effect of $F_R{}^*$ on the CTR response of a compartmentalized system with $F_c{}^* = 0.1$; b) effect of $F_c{}^*$ on the CTP response of a compartmentalized system with $F_R{}^* = 0.005$.

the capacity and resistivity of the production to support tank, rather than their capacitance and resistance, and exclude the effect of variable distance introduced by y_{inv} in the support compartment (Shahamat *et al.*, 2014). Using these parameters and just for demonstration purposes, the obtained CTR and CTP type curves for typical values of $F_R{}^*$ and $F_c{}^*$ are shown in Figure 7.

Figure 7a is the CTR response of a compartmentalized system considering the effect of transient flow within the support tank. It shows that contribution of the support tank can considerably alter the production signature in such a way that the classic half or unit slopes may not exist for long times. In this figure, there is one case with a slope of 0.75 during the transition period for almost 3 log cycles. Figure 7b depicts the effect of $F_c{}^*$ on the CTP response of a reservoir. It illustrates two straight line of slope −0.5 on the log-log plot of q_{DM} *versus* t_{DM} indicating linear flow regime. Each of these linear periods is followed by significant drops in dimensionless rate that correspond to exponential declines signifying depletion in the associated compartment.

The proposed tank model and developed type curves presented in this work can be used to characterizing a compartmentalized or composite reservoir. In the following section, two different numerical simulation models with known compartment parameters are presented. Using the developed analytical models in this paper, we then estimate the compartments properties and compare them with the numerical models input data. Consistent results between the numerical simulation and the proposed analytical

TABLE 1

Reservoir and fluid properties used for constructing Case 1

Parameter	Region 1	Region 2
k (md)	0.5	0.01
Φ (fraction)	0.04	0.06
c_f (psi^{-1})	3×10^{-6}	9×10^{-6}

approach shows the validity and application of our developed solutions.

5 SIMULATION EXAMPLE, CASE 1

Using a commercial numerical simulator, Case 1 is a single-phase liquid composite reservoir with contrasting reservoir properties shown in Table 1 and reservoir schematic depicted in Figure 8. The production well is in the center of the higher permeability reservoir section, is hydraulically fractured and is put on production for about 12 years.

Constant-rate and constant-pressure production of the well results in pressure and rate data whose logarithmic plots are shown in Figure 9. Using the developed type curves in Figure 7 and through a matching process, the parameters are determined to be $F_c{}^* = 0.28$ and $F_R{}^* = 0.02$. Calculation of these parameters based on the simulation input data give identical values. This consistency validates the approach and developed equations for production data analysis of a composite reservoir.

6 SIMULATION EXAMPLE, CASE 2

In this case, a compartmentalized reservoir with similar properties and reservoir configuration as the previous example is used. Compartmentalization is imitated using a thin lower permeability region between the two reservoir sections. Such a lower permeability region (which in this case is 5% of the lower permeability region) represents a fault or a low-permeability shale barrier and is used to simulate a compartmentalized reservoir. The production well in the center of the higher permeability region is hydraulically fractured and is put on production for 12 years under both constant-rate and constant-pressure production.

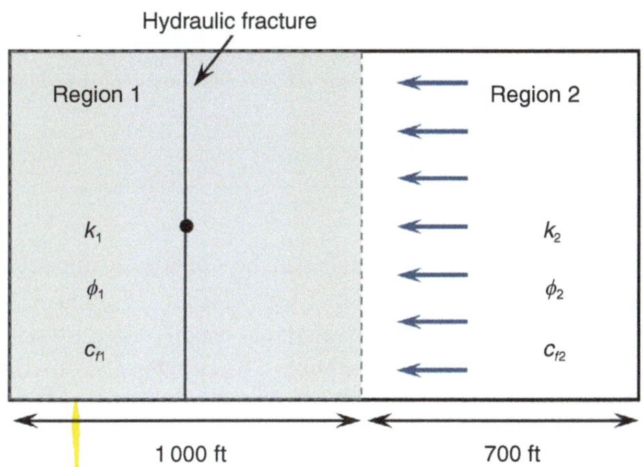

Figure 8

Reservoir geometry used for simulation of a composite reservoir.

Existence of the low-permeability barrier delays the production from the lower permeability region in such a way that its transient flow (half slope) is absent from the well's production signature. As a result, one can assume instantaneous support and simply use Equations (11) and (16) for its production analysis. For the CTR production data and using the type curves in Figure 4, we obtain a perfect match as illustrated in Figure 10 showing $F_R = 1.2 \times 10^{-3}$ and $F_c = 0.2$. It is noted here that F_R represents the ratio of the flow resistance between the production and support compartments. For the simulation case defined in this example, it fuses the effect of a flow barrier (low-permeability region) in addition to the contrast of permeability between the two reservoir sections. This is an advantage of this method as it combines the effect of different factors affecting production at the wellbore in a single term to be used in the type curve analysis. On the other hand, F_c represents the ratio of capacitance of production and support compartments. This term represents the combined effects of the drainage volume and compressibility associated with each compartment.

Analyzing the CTP production data and using Figure 6, the type curve matching (Fig. 10b) yields $F_R = 2.3 \times 10^{-3}$ and $F_c = 0.2$. Comparison of the matching parameters for CTP and CTR indicates different values of F_R for the same compartmentalized system. The difference is attributed to the effect of transient flow within the support compartment for each of these production scenarios. In other words, as indicated in Equation (23), the propagation of pressure disturbance in the reservoir as a result of

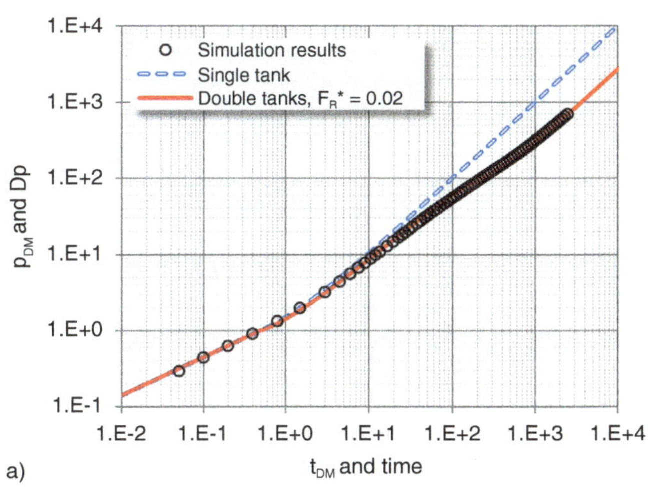

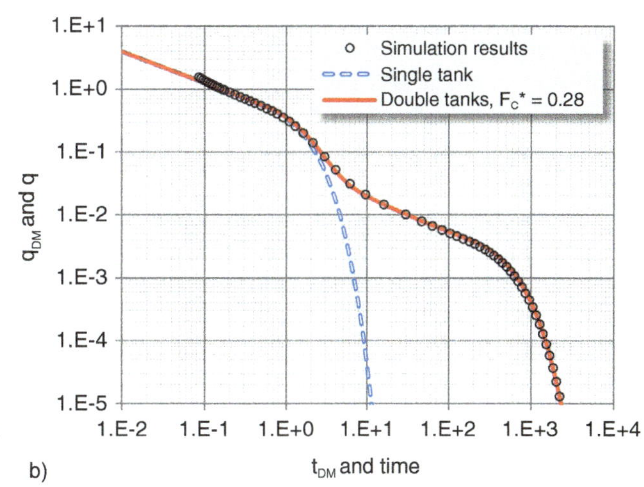

Figure 9

Type curve match results for Case 1; a) constant rate, b) constant pressure.

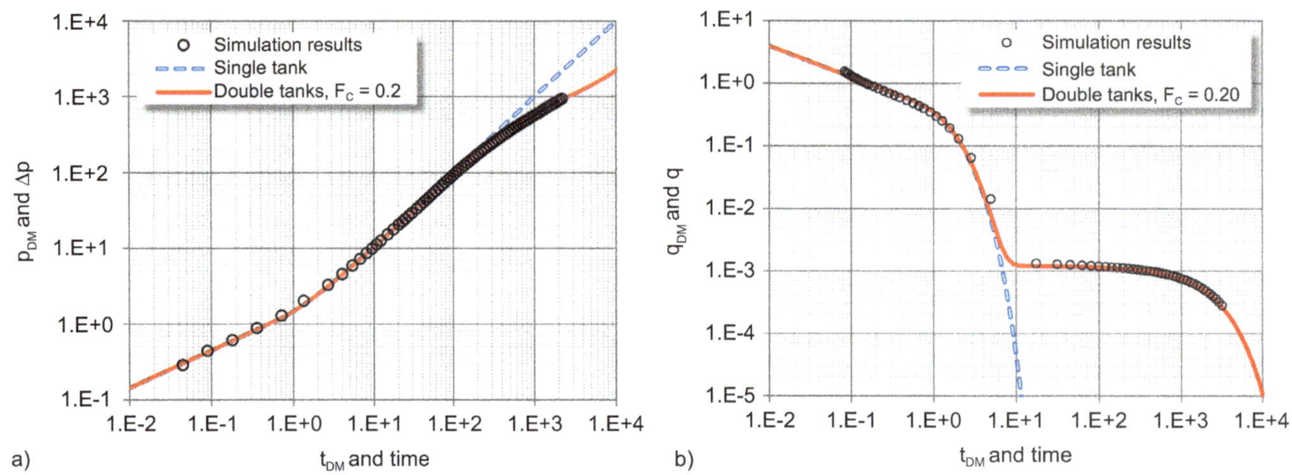

Figure 10

Type curve match results for Case 2; a) constant rate, b) constant pressure.

constant terminal rate and constant terminal pressure is different for systems that exhibit linear flow. This leads to different values of F_R obtained from analysis of constant-rate and constant-pressure solutions.

7 DISCUSSION

Comparison of the results for the composite and compartmentalized reservoir models in the above examples shows that CTR and CTP of a composite model yield similar values for $F_c{}^*$ and $F_R{}^*$. However, for the compartmentalized model with BDF within the support tank, type curve matching of the CTR and CTP gives different values of F_R for the same reservoir. This reveals that F_R is not only dependent on the reservoir and fluid characteristics of the two compartments, but also on the production mechanisms. Intuitively, it has resemblance to the skin factor that is used extensively in well-testing analysis. The skin factor explains the additional pressure drop at the wellbore and has different effects for CTR compared with CTP. While the skin factor is defined as the ratio of the permeabilities and the sizes of damaged and undamaged zones, F_R is the ratio of the permeability, viscosity and size of the production and support compartments.

Our extensive numerical experiments with different contrast in properties and with low-permeability barrier showed that not only the permeability and size of the support tank, but also its compressibility and porosity greatly influence the performance behaviour of a compartmentalized reservoir. For smaller values of hydraulic diffusivity of the support compartment, lower permeability barrier results in a production signature which can be adequately modeled using the simplified solution (with constant F_c and F_R). In general, for a system with relatively equally-sized production and support regions and with a barrier that has smaller than 1% permeability of the lower permeability region, the BDF assumption of the support compartment is valid and its performance can be reasonably modeled using the developed solution with constant F_c and F_R.

This study illustrates that the parameters F_c and F_R (or $F_c{}^*$ and $F_R{}^*$) can be used as parameters to characterize a compartmentalized/composite reservoir. Type curves can be generated for different values of F_c and F_R (or $F_c{}^*$ and $F_R{}^*$). Here it should be stressed that when analyzing production data, these parameters are not known *a priori*. Therefore type curve matching process may not be very practical. Although a type curve matching process was used for obtaining these parameters, a history matching process can be employed instead to determine F_c and F_R (or $F_c{}^*$ and $F_R{}^*$).

It is worth mentioning that even though F_c and F_R are based on physical properties (*e.g.* permeability, porosity, compressibility etc.) of the system, the history matched F_c and F_R may not reflect these exact properties but are more likely a combination of the various complexities in the system (transient flow, existence of barrier, constant-rate *versus* constant-pressure production, etc.). Nevertheless, because they are based on system properties, they are meaningful, and their value obtained from the history matching should be much more useful than a purely empirically derived match (as in the Arps curves). In essence, it can be stated that these parameters are matching parameters that qualitatively have physical meaning as they are related to physical properties of the system, but quantitatively can have different values under different production scenarios (as in the case of F_R).

The focus of this paper was on modeling of capacitance and resistance for the single-phase fluid systems. However, the approach can be extended for multiphase flow situations by employing proper pseudo variable (Behmanesh *et al.*, 2015) definitions to linearize the equations. Therefore the single-phase liquid analogy can still be applied. This is the subject of another paper and is not discussed in this work.

CONCLUSIONS

In this work, the long-term performance of compartmentalized reservoirs under constant terminal rate and pressure production has been studied. Using the capacitance and resistance concepts, a mathematical model has been developed for compartmentalized reservoirs and detailed analytical solutions have been presented. Different scenarios for BDF and transient flow within the support cell are considered. The model allows for existence of leaky boundaries as well as different reservoir properties across the compartments. Key conclusions are as follows:

- the analytical solutions are obtained in real time domain and therefore can be easily implemented in a spreadsheet application through automatic fitting procedures and inversion loops for parameter estimation;
- depletion of each compartment is revealed by two unit slopes on a log-log plot of pressure *versus* time for CTR. For CTP, however, the depletions display two exponential declines. These depletion signatures are separated by transition periods which depend on the support cell contribution (*i.e.* BDF or transient);
- expressed in dimensionless forms, the solutions provide type curves which can be used through a matching process for estimating the contrast of capacitance and resistance (*i.e.* F_c and F_R or F_c^* and F_R^*). These parameters reflect the combined effect of variation of reservoir and fluid properties across the compartments and control the flow performance of compartmentalized systems;
- the basic equations are developed based on the BDF as the prevailing conditions. However, the distance of investigation concept can be used to model the transient effect of the supporting compartment. This is considered as a generalization of two-region linear composite model with inclusion of a leaky interface (permeability barrier);
- the difference in distance of investigation for the CTR and CTP leads to different values of F_R (for instantaneous depletion of the support) which are obtained from the type curve matching process of pressure and rates for the same system.

REFERENCES

Aguilera R. (2013) Flow Units: From Conventional to Tight Gas to Shale Gas to Tight Oil to Shale Oil Reservoirs, *SPE Paper* 165360 presented at the *SPE Western Regional & AAPG Pacific Section Meeting, Joint Technical Conference*, Monterey, CA, USA, 19-25 April.

Behmanesh H., Hamdi H., Clarkson C.R. (2015) Production data analysis of tight gas condensate reservoirs, *Journal of Natural Gas Science and Engineering* **22**, 0, 22-34.

Bradley J.S., Powley D.E. (1994) Pressure Compartments in Sedimentary Basins: A Review, in *Basin Compartments and Seals*, Ortoleva P.J. (ed.), The American Association of Petroleum Geologists.

Chilingarian G.V., Mazzullo S.J., Rieke H.H. (1992) Introduction, in *Carbonate Reservoir Characterization: A Geologic-Engineering Analysis, Part I*, Elsevier Science Publishers B.V.

Corbett P.W.M., Hamdi H., Gurav H. (2012) Layered fluvial reservoirs with internal fluid cross flow: a well-connected family of well test pressure transient responses, *Petroleum Geoscience* **18**, 219-229.

Fetkovich M.J. (1980) Decline Curve Analysis Using Type Curves, *Journal of Petroleum Technology*, **32**, 6, 1065-1077, *SPE*-4629-PA.

Fox M.J., Chedburn A.C.S., Stewart G. (1988) Simple Characterization of Communication Between Reservoir Regions, *SPE Paper* 18360 presented at the *European Petroleum Conference*, London, United Kingdom, 16-19 Oct.

Hamdi H. (2014) Well-test response in stochastic permeable media, *Journal of Petroleum Science and Engineering* **119**, 169-184.

Hamdi H., Ruelland P., Bergey P., Corbett P.W.M. (2014) Using geological well testing for improving the selection of appropriate reservoir models, *Petroleum Geoscience* **20**, 4, 353-368.

Junkin J.E., Sippel M.A., Collins R.E., Lord M.E. (1992) Well Performance Evidence for Compartmented Geometry of Oil and Gas Reservoirs, *SPE Paper* 24356 presented at the *SPE Rocky Mountain Regional Meeting*, Casper, Wyoming, USA, 18-21 May.

Kaviani D., Jensen J.L., Lake L.W., Fahes M. (2008) Estimation of Interwell Connectivity in the Case of Fluctuating Bottomhole Pressures, *SPE Paper* 117856 presented at the *International Petroleum Exhibition and Conference,* Abu Dhabi, 3-6 Nov, DOI: 10.2118/117856-MS.

Malavazos M., McDonough R.C.M. (1991) Pressure-Transient Response in Compartmentalised Gas Reservoirs: A South Australian Field Example, *SPE Paper* 23009 presented at the *SPE Asia-Pacific Conference,* Perth, Australia, 4-7 Nov.

Massonnat G.J., Norris R.J., Chalmette J.-C. (1993) Well Test Interpretation in Geologically Complex Channelized Reservoirs, *SPE Paper* 00026464 presented at the *SPE Annual Technical Conference and Exhibition,* Houston, Texas, 3-6 Oct.

Nobakht M., Clarkson C.R. (2012) A New Analytical Method for Analyzing Linear Flow in Tight/Shale Gas Reservoirs: Constant-Flowing-Pressure Boundary Condition, *SPE Reservoir Evaluation & Engineering* **15**, 3, 370-384, *SPE*-143989-PA.

Noetinger B. (2015) About the Determination of Quasi Steady State Storativity and Connectivity Matrix of Wells in 3D Heterogeneous Formations, *Mathematical Geosciences* (submitted).

Rahman N.M.A., Ambastha A.K. (2000) Generalized 3D Analytical Model for Transient Flow in Compartmentalized Reservoirs, *SPE Journal* **5**, 3, 276-286, *SPE*-65106-PA.

Sayarpour M., Kabir C.S., Lake L.W. (2008) Field Applications of Capacitance-Resistive Models in Waterfloods, *SPE Paper* 114983 presented at the *SPE Annual Technical Conference and Exhibition*, Denver, 21-24 Nov, DOI: 10.2118/114983-MS.

Shahamat M.S. (2014) Production Data Analysis of Tight and Shale Reservoirs, *PhD Thesis*, University of Calgary, Calgary.

Shahamat M.S., Aguilera R. (2010) A New Method for Production Decline Analysis of Tight Gas Formations, *SPE Paper* 138149 presented at the *Canadian Unconventional Resources and International Petroleum Conference,* Calgary, Alberta, Canada, 19-21 Oct.

Shahamat M.S., Mattar L., Aguilera R. (2014) A Physics-Based Method for Production Data Analysis of Tight and Shale Petroleum Reservoirs Using Succession of Pseudo-Steady States, *SPE Paper* 167686 presented at the *SPE/EAGE European Unconventional Resources Conference and Exhibition*, Vienna, Austria, 25-27 Feb.

Stalgorova E, Mattar L. (2012) Practical Analytical Model To Simulate Production of Horizontal Wells With Branch Fractures, *SPE Paper* 162515 presented at the *SPE Canadian Unconventional Resources Conference,* Calgary, Alberta, Canada, 30 Oct.-1 Nov.

Stewart G. (2011) *Well Test Design & Analysis*, PennWell.

Stewart G., Whaballa A.E. (1989) Pressure Behavior of Compartmentalized Reservoirs, *SPE Paper* 19779 presented at the *SPE Annual Technical Conference and Exhibition*, San Antonio, Texas, USA, 8-11 Oct.

Valko P., Doublet L.E., Blasingame T.A. (2000) Development and Application of the Multiwell Productivity Index (MPI), *SPE Journal* **5**, 1, 21-31, DOI: 10.2118/51793-PA.

Wattenbarger R.A., El-Banbi A.H., Villegas M.E., Maggard J.B. (1998) Production Analysis of Linear Flow Into Fractured Tight Gas Wells, *SPE Paper* 39931 presented at the *SPE Rocky Mountain Regional/Low-Permeability Reservoirs Symposium*, Denver, Colorado, USA, 5-8 April.

Yousef A.A., Jensen J.L., Lake L.W. (2009) Integrated Interpretation of Interwell Connectivity Using Injection and Production Fluctuations, *Mathematical Geosciences* **41**, 1, 81-102, DOI: 10.1007/s11004-008-9189-x.

Appendix A

Equations (9) and (10), which are the result of simple combination of material balance and deliverability equations construct a coupled system of two ordinary differential equations for two unknown (p_{DM} and p_{DMs}) since knowledge of p_{DM} is required in order to find p_{DMs} and *vice versa*. Using the notation that $X_1 = p_{DM}$ and $X_2 = p_{DMs}$, we can solve this system using matrix algebra, where the system is written in the following manner:

$$X' = AX + g \tag{A.1}$$

In this equation, $X = \begin{bmatrix} X_1 \\ X_2 \end{bmatrix}$, $X' = \frac{dX}{dt_{DM}}$, A is the matrix of coefficients $A = \begin{bmatrix} -F_R & F_R \\ F_c F_R & -F_c F_R \end{bmatrix}$ and $g = \begin{bmatrix} 1 \\ 0 \end{bmatrix}$.

As evident, this is a linear nonhomogeneous system of differential equations. In order to solve the above system and find the general solutions, we need to obtain the homogeneous solution (*i.e.* solution to the homogeneous system of ODE). This can be achieved by first evaluating the eigenvalues of Equation (A.1). The eigenvalues are obtained through finding the roots of the characteristic polynomial of the matrix of coefficients, *i.e.* $det(A - \lambda I) = 0$. Since A in this case is a 2×2 matrix, then its characteristic polynomial is a second degree polynomial. Upon finding the roots, the following eigenvalues are obtained:

$$\begin{cases} \lambda_1 = 0 \\ \lambda_2 = -F_R(1 + F_c) \end{cases} \tag{A.2}$$

Then we need to find the eigenvectors $\vec{\alpha} = \begin{bmatrix} \alpha_1 \\ \alpha_2 \end{bmatrix}$ and $\vec{\beta} = \begin{bmatrix} \beta_1 \\ \beta_2 \end{bmatrix}$ corresponding to the first and second eigenvalues, respectively. These eigenvectors can be obtained according to the following equations:

$$\vec{\alpha} = \begin{bmatrix} 1 \\ \frac{\lambda_1}{F_R} + 1 \end{bmatrix} \tag{A.3}$$

$$\vec{\beta} = \begin{bmatrix} 1 \\ \frac{\lambda_2}{F_R} + 1 \end{bmatrix} \tag{A.4}$$

After finding the eigenvectors, the homogeneous solution is obtained as below:

$$X_h(t_{DM}) = C_1 \, \vec{\alpha} \, e^{\lambda_1 t_{DM}} + C_2 \, \vec{\beta} \, e^{\lambda_2 t_{DM}} \tag{A.5}$$

Upon substitution and simplification, we obtain:

$$X_h(t_{DM}) = C_1 \begin{bmatrix} 1 \\ \frac{\lambda_1}{F_R} + 1 \end{bmatrix} e^{\lambda_1 t_{DM}} + C_2 \begin{bmatrix} 1 \\ \frac{\lambda_2}{F_R} + 1 \end{bmatrix} e^{\lambda_2 t_{DM}} \tag{A.6}$$

After finding the homogeneous solution, the particular solution can be obtained using the method of variation of parameters. For this purpose, we first construct the matrix M using coefficients of C_1 and C_2 of the homogeneous solution (A.6):

$$M = \begin{bmatrix} 1 & e^{\lambda_2 t_{DM}} \\ 1 & \left(\frac{\lambda_2}{F_R} + 1\right) e^{\lambda_2 t_{DM}} \end{bmatrix} \tag{A.7}$$

Here we substituted the first eigenvalue ($\lambda_1 = 0$) to simplify the resulting matrix. Then we find the inverse of M which upon simplification yields:

$$M^{-1} = \frac{F_R}{\lambda_2} \begin{bmatrix} \left(\frac{\lambda_2}{F_R} + 1\right) & -1 \\ -e^{-\lambda_2 t_{DM}} & e^{-\lambda_2 t_{DM}} \end{bmatrix} \tag{A.8}$$

The particular solution is then obtained by finding the below integration:

$$X_p(t_{DM}) = M\int M^{-1}g\,dt_{DM} \tag{A.9}$$

The resulting particular solution is then determined to be according to below equation:

$$X_p(t_{DM}) = \begin{bmatrix} \left(\frac{\lambda_2+F_R}{\lambda_2}\right)t_{DM} + \frac{F_R}{\lambda_2^2} \\ \left(\frac{\lambda_2+F_R}{\lambda_2}\right)t_{DM} + \frac{F_R}{\lambda_2^2}\left(\frac{\lambda_2}{F_R}+1\right) \end{bmatrix} \tag{A.10}$$

After finding the particular solution, the general solution is determined by adding the homogeneous and particular solutions:

$$X_g(t_{DM}) = C_1\begin{bmatrix}1\\1\end{bmatrix} + C_2\begin{bmatrix}1\\ \frac{\lambda_2}{F_R}+1\end{bmatrix}e^{\lambda_2 t_{DM}} + \begin{bmatrix} \left(\frac{\lambda_2+F_R}{\lambda_2}\right)t_{DM} + \frac{F_R}{\lambda_2^2} \\ \left(\frac{\lambda_2+F_R}{\lambda_2}\right)t_{DM} + \frac{F_R}{\lambda_2^2}\left(\frac{\lambda_2}{F_R}+1\right) \end{bmatrix} \tag{A.11}$$

We can find the constants C_1 and C_2 using the initial conditions. We note here that the obtained results are for the average pressure. We also note that for the case of production at a constant rate and after reaching the boundaries of the first tank, the wellbore pressure tracks the average pressure. As a result we can set the initial condition as the starting wellbore pressure during BDF ($p_{DM,BDF}$, as obtained by the conventional methods of pressure transient analysis). Therefore the initial conditions are:

$$X_g(t_{DM,BDF}) = \begin{bmatrix} p_{DM,BDF} \\ 0 \end{bmatrix} \tag{A.12}$$

Substitution of these initial conditions and simplification of the results give the following equation for C_1 and C_2:

$$C_1 = \left(\frac{\lambda_2+F_R}{\lambda_2}\right)(p_{DM,BDF} - t_{DM,BDF}) \tag{A.13}$$

$$C_2 = \left(-\frac{F_R}{\lambda_2^{\,2}} - \frac{F_R}{\lambda_2}p_{DM,BDF}\right)e^{-\lambda_2 t_{DM,BDF}} \tag{A.14}$$

Using these constants, the general solution $X_g(t_{DM})$ is therefore obtained. The first element of the matrix $X_g(t_{DM})$ is p_{DM} which upon simplification yields:

$$p_{DM} = p_{DM,BDF} + \left(1+\frac{F_R}{\lambda_2}\right)(t_{DM} - t_{DM,BDF}) + \frac{F_R}{\lambda_2}\left(\frac{1}{\lambda_2} + p_{DM,BDF}\right)\left(1 - e^{\lambda_2(t_{DM}-t_{DM,BDF})}\right) \tag{A.15}$$

And the first element of the matrix $X_g(t_{DM})$ is p_{DMs} which is according to Equation (A.16):

$$p_{DMs} = \left(1+\frac{F_R}{\lambda_2}\right)\left\{(t_{DM} - t_{DM,BDF}) + \left(\frac{1}{\lambda_2} + p_{DM,BDF}\right)\left\{1 - e^{\lambda_2(t_{DM}-t_{DM,BDF})}\right\}\right\} \tag{A.16}$$

Appendix B

In this case, Equations (14) and (15), construct a coupled system of two ordinary differential equations for two unknown (q_{DM} and q_{DMs}), since knowledge of q_{DM} is required in order to find q_{DMs} and *vice versa*. Using the notation that $X_1 = q_{DM}$ and $X_2 = q_{DMs}$, we can solve this system using matrix algebra, where the system is written in the following manner:

$$X' = AX \tag{B.1}$$

In this equation, $X = \begin{bmatrix} X_1 \\ X_2 \end{bmatrix}$, $X' = \frac{dX}{dt_{DM}}$ and A is the matrix of coefficients given by $A = \begin{bmatrix} -1 & 1 \\ F_R & -(F_R + F_c F_R) \end{bmatrix}$.

This is a linear homogeneous system of differential equations. In order to solve the above system and find the general solutions, we need to obtain the homogeneous solution (*i.e.* solution to the homogeneous system of ODE). This can be achieved by first evaluating the eigenvalues of Equation (B.1). The eigenvalues are obtained through finding the roots of the characteristic polynomial of the matrix of coefficients, *i.e.* $det(A - \lambda I) = 0$. Since A in this case is a 2×2 matrix, its characteristic polynomial is of second degree, as below equation shows:

$$\lambda^2 + (1 + F_R(1 + F_c))\lambda - F_R F_c = 0 \tag{B.2}$$

Depending on the discriminant for this equation, we can have real or imaginary roots. The discriminant for Equation (B.2) is as below:

$$\Delta = (1 + F_R(1 + F_c))^2 - 4F_R F_c \tag{B.3}$$

This equation can be manipulated to yield the following relation, which shows that we have a positive discriminant and therefore (B.3) has two real roots:

$$\Delta = (1 - F_R(1 + F_c))^2 + 4F_R \tag{B.4}$$

The roots are then obtained as below:

$$\lambda = \frac{-(1 + F_R(1 + F_c)) \pm \sqrt{(1 - F_R(1 + F_c))^2 + 4F_R}}{2} \tag{B.5}$$

We adopt the convention that λ_1 and λ_2 correspond to the positive (+) and negative (−) signs in Equation (B.5), respectively.

Then we need to find the eigenvectors $\vec{\alpha} = \begin{bmatrix} \alpha_1 \\ \alpha_2 \end{bmatrix}$ and $\vec{\beta} = \begin{bmatrix} \beta_1 \\ \beta_2 \end{bmatrix}$ corresponding to the first and second eigenvalues, respectively. These eigenvectors can be obtained according to the following equations:

$$\vec{\alpha} = \begin{bmatrix} 1 \\ 1 + \lambda_1 \end{bmatrix} \tag{B.6}$$

$$\vec{\beta} = \begin{bmatrix} 1 \\ \lambda_2 + 1 \end{bmatrix} \tag{B.7}$$

Similar the previous case and after finding the eigenvectors, the homogeneous solution is obtained using Equation (A.5). Upon substitution and simplification, the homogeneous solution, which in this case is equal to the general solution gives:

$$X_g(t_{DM}) = X_h(t_{DM}) = C_1 \begin{bmatrix} 1 \\ \lambda_1 + 1 \end{bmatrix} e^{\lambda_1 t_{DM}} + C_2 \begin{bmatrix} 1 \\ \lambda_2 + 1 \end{bmatrix} e^{\lambda_2 t_{DM}} \tag{B.8}$$

We can find the constants C_1 and C_2 using the following initial conditions:

$$X_g(t_{DM,BDF}) = \begin{bmatrix} q_{DM,BDF} \\ 0 \end{bmatrix} \tag{B.9}$$

Substitution of these initial conditions and simplification of the results give the following equation for C_1 and C_2:

$$C_1 = \left(\frac{\lambda_2 + 1}{\lambda_2 - \lambda_1}\right) q_{DM,BDF} \times e^{-\lambda_1 t_{DM,BDF}} \tag{B.10}$$

$$C_2 = -\left(\frac{\lambda_1 + 1}{\lambda_2 - \lambda_1}\right) q_{DM,BDF} \times e^{-\lambda_2 t_{DM,BDF}} \tag{B.11}$$

Therefore from the obtained general solution we get the following equation for q_{DM}:

$$q_{DM} = q_{DM,BDF}\left\{\left(\frac{\lambda_2 + 1}{\lambda_2 - \lambda_1}\right) \times e^{\lambda_1 \left(t_{DM} - t_{DM,BDF}\right)} - \left(\frac{\lambda_1 + 1}{\lambda_2 - \lambda_1}\right) \times e^{\lambda_2 \left(t_{DM} - t_{DM,BDF}\right)}\right\} \tag{B.12}$$

And for q_{DMs} we obtain:

$$q_{DMs} = q_{DM,BDF}\left(\frac{(\lambda_1 + 1)(\lambda_2 + 1)}{(\lambda_2 - \lambda_1)}\right)\left\{e^{\lambda_1 \left(t_{DM} - t_{DM,BDF}\right)} - e^{\lambda_2 \left(t_{DM} - t_{DM,BDF}\right)}\right\} \tag{B.13}$$

Appendix C

Dimensionless parameters used in well testing and production data analysis literature are as follows:

$$t_D = \frac{\beta_2 k t}{\phi \mu c_t} \tag{C.1}$$

$$p_D = \frac{2\pi k h}{\beta_1 q \mu}\left(p_i - p_{wf}\right) \tag{C.2}$$

$$q_D = \frac{\beta_1 q \mu}{2\pi k h\left(p_i - p_{wf}\right)} \tag{C.3}$$

$$y_{eD} = \left(\frac{y_e}{x_f}\right) \tag{C.4}$$

Using the definitions of capacitance and resistance for linear flow, we have:

$$cV = \left(\frac{4 x_f c_t h \phi}{5.615}\right) y_e \tag{C.5}$$

$$R = \left(\frac{\beta_1 \mu}{kh}\right)\left(\frac{y_e}{x_f}\right) \tag{C.6}$$

Using the above equations for capacitance and resistance in definition of t_{DM}, p_{DM} and q_{DM} (Eq. 5, 6 and 13) and upon simplification we obtain the following relation:

$$t_{DM} = \frac{t_D}{{y_{eD}}^2} \tag{C.7}$$

$$q_{DM} = q_D \times y_{eD} \tag{C.8}$$

$$p_{DM} = \frac{p_D}{y_{eD}} \tag{C.9}$$

It has to be mentioned that in addition to the difference between p_{DM} and p_D shown in Equation (C.9), p_{DM} is defined based on average pressure while the p_D is in terms of the wellbore pressure. In addition, q_{DM} is the linear equivalent of the dimensionless decline rate (q_{Dd}) defined by Fetkovich (1980).

5

Use of Computational Fluid Dynamics for Absorption Packed Column Design

Yacine Haroun* and Ludovic Raynal

IFP Energies nouvelles, Rond-point de l'échangeur de Solaize, BP 3, 69360 Solaize - France
e-mail: yacine.haroun@ifpen.fr - ludovic.raynal@ifpen.fr

* Corresponding author

Abstract — *Computational Fluid Dynamics (CFD) is today commonly used in a wide variety of process industries and disciplines for the development of innovative technologies. The present article aims to show how CFD can be used as an effective analysis and design tool for the development and design of packed gas/liquid absorption columns. It is first shown how CFD can be used for the characterisation of packings. The different hydrodynamic and mass transfer design parameters are investigated and adapted CFD methods are suggested. Secondly, column distribution internal development is discussed to show how CFD simulations should be performed to improve the design of gas and liquid distributors. An example of the development of new distribution technologies for floating installation of a reactive absorption column is also presented.*

Résumé — Utilisation de la dynamique des fluides numérique pour le design des colonnes d'absorption à garnissages — La dynamique des fluides numérique (*Computational Fluid Dynamics*, CFD) est aujourd'hui couramment utilisée pour le développement de technologies innovantes dans de nombreux domaines et procédés industriels. Cet article a pour objectif de montrer comment la CFD peut être utilisée comme outil d'analyse et de dimensionnement pour le développement des colonnes d'absorption gaz-liquide à garnissages. En premier, il est présenté comment la CFD peut être utilisée pour caractériser les différents paramètres de dimensionnement d'un garnissage (hydrodynamique et transfert). Les méthodes CFD appropriées sont suggérées et discutées. En second, l'utilisation de la CFD pour le développement et l'optimisation des internes de distribution gaz-liquide est exposée. Un exemple de développement par CFD d'une nouvelle technologie de distribution pour une colonne d'absorption réactive sur une installation flottante offshore est présenté.

INTRODUCTION

Gas/liquid absorption operation constitutes an important aspect in several industrial applications using packed columns such as distillation, reactive distillation, acid gas removal and solvent-based post-combustion CO_2 capture.

Reactive absorption packed columns consist of packing elements (structured or random packing) contacted by cocurrent or counter-current gas-liquid flow. Liquid mostly trickles over the packing's wall as film flow, with the gas being a continuous phase flowing within the core of the packing elements. The absorption of gases in the liquid solution is often accompanied by chemical reactions. The distribution of the liquid and gas phases in packing is ensured by the use of gas-liquid tray distributors situated at the top of each packed bed and the bottom gas distributor at the bottom of the column (Billet, 1995; Kister, 1990; Olujic *et al.*, 2004; Stemich and Spiegel, 2011).

To meet the requirements of size optimisation of the column (in terms of height and diameter) and pressure drop limitation, efficient and high-capacity packings and appropriate design of the gas and liquid distributor are needed (Billet, 1995). The selection of the column internals requires a deep understanding of the purpose of the reactive absorption application and precise knowledge of the characteristics of the gas and liquid solvent flows throughout the packed bed in terms of hydrodynamics (pressure drop, flooding), mass transfer (gas-side and liquid-side mass transfer coefficients) and indirectly, kinetics and thermodynamics (reaction regime, reaction acceleration factor). These parameters are used in process simulations to achieve optimum design. For example, in the particular case of post-combustion CO_2 capture, Raynal *et al.* (2013) showed that the interfacial area is the key parameter for the absorber design. This leads to promoting structured packings for such a reactive absorption application that offer higher geometric area per unit volume, from at least 250 m^2/m^3 and preferentially more than 350 m^2/m^3, and a large void fraction (porosity), around 90%, which induce high mass transfer efficiency and low pressure drop, respectively (Lassauce *et al.*, 2014). For high-pressure acid gas treatment, H_2S removal efficiency will be more sensitive to gas-side mass transfer, since H_2S reacts instantaneously with amine-based solvent, and, due to high-pressure operation, the column design will consider diameter optimisation as more important than height and resulting pressure drop optimisation; the corresponding choice of packing would thus differ from ambient pressure CO_2 post-combustion absorption.

Much work and many experimental tests have been conducted in order to measure the packing characteristics and to develop adapted models (liquid holdup, effective area, gas-side and liquid-side mass transfer, pressure drop) for commercial packings (Billet, 1995; Bravo *et al.*, 1985; Fair and Bravo, 1990) and design rules for distribution technologies (Kister, 1990). However, developing understanding is still needed, especially for the development of new reactive absorption packings and internals, which involves various physical and geometrical parameters that cannot be easily investigated experimentally and are still not well understood. Besides, the corresponding experimental work is quite time-consuming and parametric optimisation is almost impossible.

In this context, Computational Fluid Dynamics (CFD) is seen as a powerful tool in complement to experimental work to investigate performance characteristics and the development of original reactive absorption columns (Charpentier, 2009). In the last decade, CFD has been used more and more to calculate flow characteristics in packed beds (Petre *et al.*, 2003; Raynal and Royon-Lebeaud, 2007; Mohamed Ali *et al.*, 2003). Indeed, the use of numerical simulation can provide a significant gain in time and could limit the number of experiments. Another reason to use CFD is the possibility of accessing information on a local scale which is not measurable with experimental methods.

The present article aims to show how CFD is used for the development of reactive absorption packed columns. Different CFD approaches and simulations are presented, addressing hydrodynamic and mass transfer characterisation of packings as well as distributing tray evaluation and development.

In the following first section, the use of CFD for the characterisation of packing is presented. The second section focuses on the development of gas and liquid distribution internals by using CFD.

1 STRUCTURED PACKING CHARACTERISATION BY USING CFD

As discussed in Raynal *et al.* (2013), the choice of the most adequate packing is linked to its performances in terms of pressure drop and mass transfer efficiency. There is no best packing which would offer high capacity, homogeneous distribution and good mass transfer efficiency, since from one case to another the expectations and sensitivity of the process differ. The capacity of the packing, giving the maximum gas and liquid mass flow rates, is used to determine the diameter of the column. This characteristic is often given by the pressure drop, and more exactly it corresponds to the flooding limit over which operation is no longer possible. The mass transfer efficiency is used to determine the height of the column. This latter characteristic is much more difficult to determine since, for reactive absorption application studied in the framework of the two-film theory, the mass transfer flux is linked to five parameters (Danckwerts, 1970). Three of these parameters are directly linked to the packing contactor geometry and the gas/liquid operating conditions. These parameters are the following: the liquid-side mass transfer coefficient, k_L (m/s), the gas-side mass transfer coefficient, k_G (m/s), and the interfacial effective area, a_e (m^{-1}). The two remaining parameters correspond to the thermodynamic and kinetic performances of the solvent given, respectively, by the Henry coefficient, H_e ($Pa.m^3.mol^{-1}$), and the acceleration factor coefficient, E (-), both varying with pressure and temperature conditions and with solvent loading. All these parameters are linked in the following relationship giving the flux for the chemical component j from the gas to the liquid:

$$\Phi_j(\text{mol/m}^3/\text{s}) = \left(\frac{1}{1/k_G a_e + H_e/E k_L a_e}\right) \times (P_j - H_e C_{j,bulk})$$

where: P_j (Pa): the partial pressure of the component j in the gas phase $C_{j,bulk}$ (mol/m^3): the concentration of the component j in the liquid bulk.

The characterisation of packing contactors consists of determining and modeling all parameters dealing with hydrodynamics and mass transfer. In recent years, several works can be found in the literature which deal with the CFD investigation of hydrodynamic (pressure drop, liquid hold-up, wetting, liquid distribution) and mass transfer design parameters (liquid- and gas-side mass transfer, acceleration factor, effective area mass transfer). Petre *et al.* (2003) were among the first authors to use CFD in order to calculate dry pressure drop in structured packings. They proposed an original approach by considering the structured packing by a combination of four Representative Elementary three-dimensional Units (REU). The total pressure drop in the packed bed was determined by combining CFD simulation results obtained in the different REU domains. The choice of the REU was further discussed by Raynal *et al.* (2004), who proposed considering the flow domain between two consecutive plates for gas flow calculations, or preferentially the smallest periodic element. More recently, Said *et al.* (2011) show that the pressure drop per unit length for a small number of REU is the same as for the entire column, which simplifies its computational modeling. Nikou and Ehsani (2008) tested different turbulence models to study wet and dry pressure drop and investigated heat and mass transfer in structured packings. Fernandes *et al.* (2009) presented a pseudo two-phase CFD model for determination of the wet pressure drops in a *Sulzer* EX packing.

Szulczewska *et al.* (2000) used the Volume Of Fluid (VOF) method to perform 2D simulations of counter-current gas-liquid flow on the structured packing Mellapak 250Y, first, to determine the effect of flow rates on the interfacial effective area, and second, to specify the lowest phase flow rate at which the packing becomes fully wetted. The interfacial area was determined by assuming that the 2D picture of flow in any cross-section of the packing is identical. Raynal *et al.* (2004) performed 2D CFD simulations with the VOF method to determine the liquid film thickness on a wavy plate similar to the packing surface, and the results were used to derive the liquid hold-up. Raynal and Royon-Lebeaud (2007) used the same VOF method for gas-liquid flow simulations on a small scale and the results are further used on a large scale in three-dimensional calculations with a geometry corresponding to a complete packed column. Ataki and Bart (2006) investigated the wetting in the structured packing elements of Rombopack using the same VOF method and used simulation results to develop correlations for the effective area and liquid hold-up. They also described the effect of liquid properties on wetting. Shojaee *et al.* (2011) investigated wet pressure drop and the effective area in Gempak 2A structured packing also by using the VOF method. The authors compare the evaluated effective area with existing correlations. They found good agreement with Brunazzi *et al.* (1995) model. More recently, Haroun *et al.* (2014) applied a VOF numerical model to investigate wetting phenomena, the effective interfacial area and liquid hold-up within a set of actual Mellapak 250X REU. The values of liquid-solid contact angles were varied to mimic the geometrical imperfections of the plate surface. The results showed significant variations in the wetted area depending on the liquid-solid contact characteristics and liquid flow rate. The authors pointed out the fact that experimental studies on wetting in structured packing are still needed to verify the numerical results.

Few works can be found in the literature which deal with CFD modeling of mass transfer in structured packings. Haroun *et al.* (2010a, b, 2012) extended hydrodynamic studies by further implementing liquid-side mass transfer and reactive mass transfer calculations by considering a 2D approach assuming full wetting of the packing using the VOF approach. They show how the liquid flow rate and the complex geometry of packing affect the liquid film flow topology and interfacial mass transfer. The authors reported good agreement between CFD results and the Higbie penetration theory when the adequate exposure time is used. For reactive mass transfer, their results show a good match between the numerical predictions and the theoretical relationship between the Hatta number and the enhancement factor for both first- and second-order reactions. More recently, Sebastia-Saez *et al.* (2013, 2014) developed a three-dimensional VOF model on a small scale for hydrodynamics and implemented physical mass transfer as a source term using the Higbie approach. The authors performed CFD simulation to study transient behaviour and the impact of the liquid load on the absorption rate in structured packing.

Regarding gas-side mass transfer investigation using CFD, only a few studies can be found in the literature;

indeed, most of the studies focused only on hydrodynamics or liquid-side mass transfer. Erasmus (2004) was one of the first to use CFD to simulate gas-side mass transfer in structured packing. The author performed CFD simulations of naphthalene sublimation in simplified sections of structured packing. He found that using a BSL turbulence model gives satisfactory agreement between CFD calculations and experimental measurements. He points out, however, the sensitivity of the results to the placement of inlet and outlet boundaries. He found that a distance equal to the corrugation base was found to be adequate. Recently, Lautenschleger *et al.* (2015) used CFD to analyse the impact of flow behaviour on gas-side mass transfer in new and conventional structured packings. The authors reported good agreement between numerical simulation and literature modeling for *Sulzer* BX and *Montz* B1-500 commercial structured packings.

It is known that liquid dispersion in packing is strongly linked with the packing geometry (Fourati *et al.*, 2012). The quality of distribution inside the packing is crucial for the mass transfer and thus the efficiency of the reactive absorption column. The gas-liquid distribution behaviour inside packed beds was mostly given by industrial experience and little comes from more scientific explanations. Recently, the large-scale two-fluid CFD simulation approach has been developed and used more and more to study such aspects. Mahr and Mewes (2007) proposed a CFD model for simulation of the macroscopic gas-liquid two-phase flow field in a column with structured packing. The model is based on the elementary cell model extended to be used on anisotropic porous structures such as corrugated structured packings. The numerical models are tested against X-ray radiographic measurements on a quasi-two-dimensional segment of structured packing and good agreement is reported. More recently, Fourati *et al.* (2013) developed a porous media CFD model in the framework of Eulerian two-fluid flows taking into account liquid dispersion in the packed zone. The authors compared CFD prediction of liquid spreading in the Mellapak 250X commercial packing previously studied by means of gamma-ray tomography, good agreement being observed. Soulaine *et al.* (2014) developed CFD modeling for simulation of gas-liquid flow on bistructured porous media representing structured packing. The set of macroscale equations that makes up the model combines three continuity equations and three multiphase Darcy's laws and is directly based on the Darcy generalised model. The authors compared the CFD results with the tomography imaging of Fourati *et al.* (2012) and show qualitative good agreement of the liquid distribution behaviour.

From the literature review, it clearly appears that most of the CFD work focused on structured packings. Indeed, achieving a representative loading domain of a random packing bed remains a complex challenge. However, with the development of the Discrete Element Simulation (DEM) loading simulation for complex geometry (Bai *et al.*, 2009), the perspective of CFD simulation with random packing will soon become accessible. In this article, we will focus only on the structured packing contactors.

In the following, typical CFD approaches and simulations performed to determine physical design parameters in terms of hydrodynamics and mass transfer for structured packing geometry are presented. The different CFD approaches used for each physical parameter are introduced and discussed.

1.1 Pressure Drop

In this section, the ability of CFD to predict dry pressure (gas flow only) drop in different types of commercial structured packings (*Sulzer* Mellapak type or equivalent, such as Flexipac from *Koch-Glitsch* or B1-series from *Montz*) are presented and the results compared with experimental data. The purpose of this section is to present an appropriate approach to simulate dry pressure drop and thus to validate the straightforward CFD method to simulate gas flow in structured packing. As discussed in Petre *et al.* (2003) and Said *et al.* (2011), the dry pressure drop is an important design parameter, since it gives insight into the trend of the wet pressure drop performance. It thus leads to a first evaluation of the capacity of the packing.

Structured packings are made of corrugated metal sheets arranged side by side with opposing channel orientation. Channels are oriented with an angle from horizontal, θ, which varies from 45° for the *Y* type of Mellapak packing to 60° for the *X* type. Figure 1a shows the triangular-based channel geometry of packing.

Structured packings are inherently periodic. Therefore, Representative Periodic Elements (RPE) were mostly used for the simulations, to reduce the CPU computational time. Different types of representative elements for corrugated sheet structured packings can be found in the literature. Petre *et al.* (2003) suggested a criss-crossing channel element. Raynal and Royon-Lebeaud (2007) and Said *et al.* (2011) proposed a geometry with periodic conditions at all cut boundaries. In this work, the second type of REU was considered. Figures 1b, c show the REU of Mellapak 250Y (M250Y) and M250X packing, respectively. In Figure 1a, it can be seen that the metal sheet of structured packing is not smooth, but is perforated and composed of small-scale texture (embossed metal sheet). For computing resource reasons, the wall texture of real packing and the perforation are not represented in simulated geometry. Indeed, in this work, the packing wall considered is smooth. A similar approach was considered by Raynal *et al.* (2004), and Nikou and Ehsani (2008) for the simulation of pressure drop in

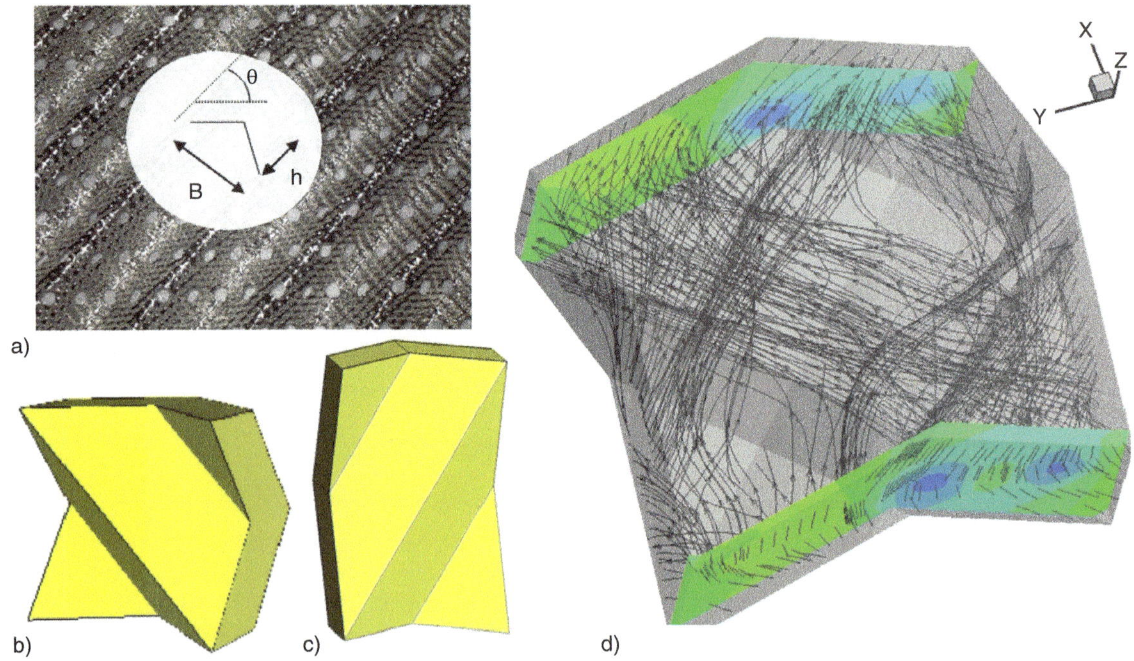

Figure 1

a) Corrugation and surface geometries; b, c) illustration of the geometries considered for simulation, b) the periodic REU of M250Y, c) the REU of M250X, d) illustration of gas flow streamline in the REU of M250Y.

structured packings. As boundary conditions, opposite faces are considered periodic and a pressure gradient is imposed in the flow direction. The CFD simulations presented in this section were carried out with the commercial CFD software ANSYS Fluent 14.

One critical aspect of CFD simulations of gas flow for pressure drop in structured packings is choosing the appropriate turbulent model, especially when the range of gas Reynolds numbers is within laminar and fully turbulent flow, as discussed in Raynal *et al.* (2004). In order to determine the best CFD approach to simulate aerodynamics in structured packings, the results of two CFD turbulent models are presented in this work: Large-Eddy Simulation (LES) modeling and Reynolds Average Navier-Stokes models k-ω SST.

LES explicitly solves the large turbulent scales and modeling assumptions are restricted to small scales only. The subgrid model used is based on the Wall-Adapting Local Eddy-viscosity (WALE) model formulation of Nicoud and Ducros (1999). The assumptions of this model remain the same as in Smagorinsky's model but involve a local rotation rate in the expression of the turbulent viscosity; this is justified since dissipative scales are also characterised by a high rotation rate. The WALE model thus allows the prediction of the correct wall behaviour with an implicit damping effect. The Shear Stress Transport (SST) k-ω low Reynolds is based on model transport equations for turbulence kinetic energy (k) and the specific dissipation rate (ω). This model blends the robust and accurate formulation of the k-ω model in the near-wall region with the free-stream independence of the k-ε model in the far field. For more details about this turbulent modeling, the reader is referred to the ANSYS user's guide.

In the simulations presented here, we paid attention to the validation of the mesh resolution. For near-wall treatment, the adapted mesh grid resolution near the wall region was adopted for each turbulent model; for example, for the k-ω models, to take full advantage of this formulation, the viscous sublayer is meshed with the $y+ < 1$. The mesh size used is such that the REU domain contains 600 000 cells.

Figure 1d shows an example of simulation results with Mellapak 250Y structured packing (M250Y in the following). It shows gas flow streamlines in the REU considered. The picture shows the flow pattern in the packing channels. The streamline shapes confirm the preferential flow direction of the gas in the packed bed; indeed, the gas flow is oriented through the channel inclination angle. However, at the interaction face with the opposite channel orientation, part of the flow deviates the direction because of flow collisions.

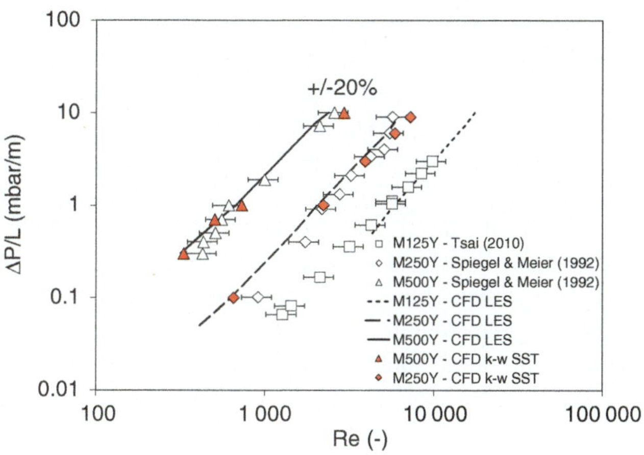

Figure 2

Experimental and simulated dry pressure drops *versus Re* number for the Mellapak 250Y REU.

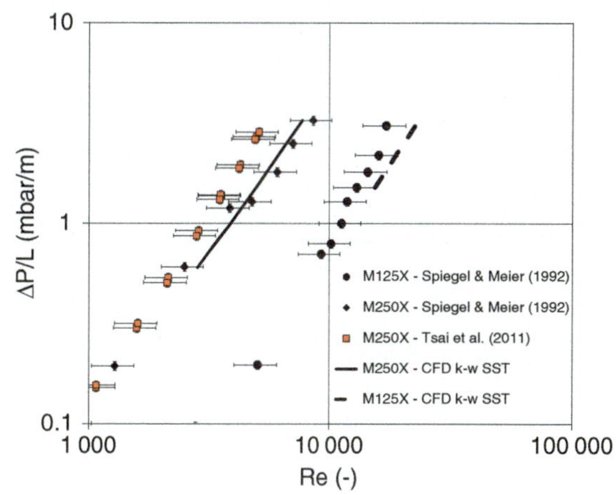

Figure 3

Experimental and simulated dry pressure drops *versus Re* number for the Mellapak 250X REU.

Figure 2 shows the predicted dry pressure drop *versus* Reynolds number for Mellapak 125Y, 250Y and 500Y, simulated using k-ω SST and LES turbulent models compared with the experimental results obtained by Spiegel and Meier (1992), the Reynolds number being calculated as follows:

$$Re = \frac{\rho v_{SG} d_{eq}}{\mu} \text{ with } d_{eq} = 4/a_G \qquad (1)$$

where ρ (kg/m^3) is the gas density, μ (Pa.s) the gas dynamic viscosity, a_G (m^2/m^3) the geometric area of the packing and v_{SG} (m/s) the superficial velocity.

The results show that the LES and k-ω SST approaches are well adapted to determine dry pressure drop in structured packings. Indeed, the relative deviation between these two approaches and experimental data is less than 20%. These results highlight that both LES and k-ω SST modeling is satisfactory, even if k-ω SST model results seem less precise at high Reynolds numbers than LES. However, taking into account the CPU resources necessary to run simulations using the LES model, using k-ω SST appears to be a very good compromise between simulation time and result accuracy. In Figure 3, only the k-ω SST low Reynolds model is used to determine pressure drop in M125X and M250X. The results are compared with the experimental data of Spiegel and Meier (1992) and Tsai *et al.* (2011). The simulated pressure drop in the REU is found to be in good agreement with the experimental results of Spiegel and Meier (1992). However, less good agreement is observed with the results of Tsai *et al.* (2011). This highlights the difficulty of the comparison exercise related to the fact that there are discrepancies in the literature data. However, good agreement is obtained with the experimental data of Spiegel and Meier (1992) for the Mellapak *Y* packing type as well as for the *X* packing type.

To sum up, these studies showed that the basic gas flow behaviour and the dry pressure drop of the packing can be predicted fairly well by only one REU with the k-ω SST low Reynolds model. Furthermore, this work shows that CFD based on the REU element can be used as a fast and straightforward means for the investigation and optimisation of new packing geometries or benchmark pressure drop performance of different existing packing technologies.

1.2 Physical and Reactive Liquid-Side Mass Transfer

In recent years, with the development of numerical methods for simulating a deformable interface, the direct numerical investigation of interfacial reactive mass transfer in complex geometry has become possible. Haroun *et al.* (2010a, b) developed VOF modeling for physical and reactive mass transfer study in two-dimensional geometries corresponding to a slice of a structured packing. The computational method is based on the VOF method (Hirt and Nichols, 1981; Scardovelli and Zaleski, 1999), where the chemical species concentration equation is solved coupled with the Navier–Stokes equations by using the JADIM code developed at IMFT (Haroun *et al.*, 2010a, b). The resolved equations are:

$$\nabla \cdot U = 0 \qquad (2)$$

$$\frac{\partial U}{\partial t} + U \cdot \nabla U = -\frac{1}{\rho}\nabla P + g + \frac{1}{\rho}\nabla \cdot T - F \qquad (3)$$

$$\frac{\partial \alpha}{\partial t} + (V \cdot \nabla)\alpha = 0 \tag{4}$$

$$\frac{\partial C_j}{\partial t} + \nabla \cdot (UC_j) = \nabla \cdot \left(D_j \nabla C_j - \left(D_j \frac{C_j(1 - He_j)}{\alpha + He_j(1-\alpha)} \nabla \alpha \right) \right) + W_j \tag{5}$$

where U, P, ρ, g, T, F and α are the local velocity, pressure, density, dynamic viscosity, gravity, viscous deformation tensor, capillary force and volume fraction, respectively. The location of each phase is given by a scalar α (corresponding in most cases to the volume fraction). Cj is the chemical species concentration. D_j is the molecular diffusivity, and W_j is the production term related to the chemical reaction. For the calculation of capillary force, a continuum method (CSF) is employed, as proposed by Brackbill *et al.* (1992). An option to specify a wall adhesion angle in conjunction with the surface tension model is available in the VOF method. A contact angle, γ, can be imposed at the wall. This latter is used to adjust the surface normal in cells near the wall (Haroun *et al.*, 2014).

This approach takes into account the solubility of chemical species at the interface (Eq. 5) described by Henry's law (Haroun *et al.*, 2010a, b). This formulation allows interpreting the jump of concentration at the interface as a continuous phenomenon, the Henry solubility law being converted into a solubility flux. This original modeling allows, from the gradient of concentration and flux of solubility, the determination of the local liquid mass transfer coefficient simultaneously with the evolution of the gas-liquid interface. This approach differs from recent studies of Sebastia-Saez *et al.* (2013, 2014) where the physical mass transfer is implemented in the VOF approach *via* a source term using Higbie modeling.

In Figure 4, the CFD results of Haroun *et al.* (2010a, b) on physical and reactive mass transfer obtained with the VOF approach are presented. For physical mass transfer, Haroun *et al.* (2010a, b, 2012) show that despite the complex geometry of structured packing, the CFD liquid-side mass

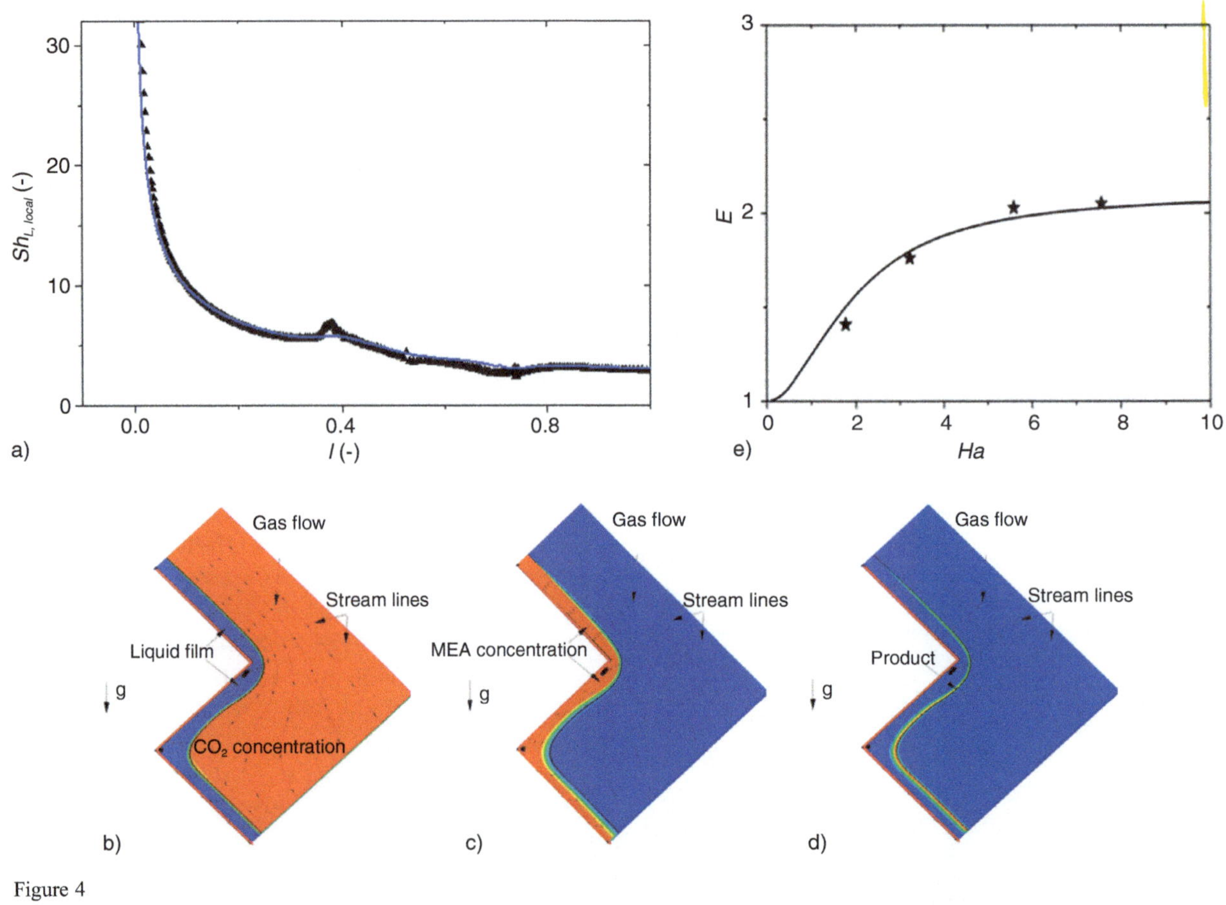

Figure 4

a) Local evolution of the liquid-side Sherwood number with the dimensionless distance from the inlet. Comparison between numerical results (triangle) and the Higbie solution (line). b) CO_2, c) MEA and d) product reaction concentration contours in gas/liquid flow on structured packing. e) Enhancement factor E evolution with Hatta number (Ha) for different values of second-order rate constants. Physical properties and operating conditions: $\rho_L/\rho_G = 780$, $\mu_L/\mu_G = 50$, $Fr = 5$, $We = 8$, $Pe_L = 104$, $Q = 45$ m^3/m^2/h.

transfer coefficient is well reproduced by the Higbie model when the adequate velocity and length scales are considered for exposure time determination. Figure 4a shows a comparison between the non-dimensional local liquid-side mass transfer coefficient, that is a Sherwood number, with the Higbie (1935) theory (line). It is found that the exposure time of the fluid element at the interface corresponds to the ratio between the curvilinear distance between two periodic corrugation contact points and the interface velocity, as further discussed in Haroun *et al.* (2012). For reactive mass transfer, the example of the transfer of CO_2 in MEA (MonoEthanol-Amine, a reference solvent for post-combustion carbon capture) solvent accompanied by second-order reaction in the liquid phase is presented in Figures 4b-e. The 2D plots (Fig. 4b-d) show concentration contours of CO_2, MEA and the reaction product in the gas-liquid flow down a two-dimensional slice of the structured packing at steady state. The results show that at steady state, the depth of penetration of CO_2 in the liquid phase is very thin, indicating that the reaction occurs at the gas-liquid interface. The simulation results also show that the MEA concentration is depleted in the neighbourhood of the interface and two regions appear, the region with no reactant CO_2 in the bottom of the liquid film and the region with a low concentration of MEA close to the interface. The concentration of the reaction product is mainly restricted at the gas-liquid interface. Finally, the CFD acceleration factor (Haroun *et al.*, 2012) is compared with the approximate solution presented by Brian *et al.* (1961). Figure 4e shows the evolution of the enhancement factors E between the gas and liquid phases with the Hatta number. Very good agreement with the solution of Brian *et al.* (1961) is obtained, indicating that the VOF approach is well adapted to deal with problems of reactive absorption in complex geometries.

As discussed in Raynal *et al.* (2013), such CFD modeling can be very useful to understand the impact of the packing geometry, physical parameters of fluids and the shape of flow on physical and reactive mass transfer and the development of physical models as well.

1.3 Gas-Side Mass Transfer

In Section 1.1, it is shown that, by using the REU of structured packing and the k-ω SST low Reynolds model with periodic boundary conditions, the gas flow behaviour and the dry pressure drop can be predicted with good precision. Similarly, it is shown in the present section that the REU element can also be used to estimate the impact of the flow behaviour on the gas-side mass transfer. This parameter is of prime importance in some reactive absorption applications such as the reactive absorption of H_2S from the acid natural gas in amine solvent; the main resistance of mass transfer is usually concentrated in the gas phase and, thus, the gas-phase flow analysis provides important information.

In this approach, the liquid film flow is not considered in the calculation; the walls of structured packings are considered as the gas-liquid interface. It is assumed that the thin mass boundary layer of gas next to the wall is equivalent to the mass boundary layer next to the gas-liquid interface. This simplification is valid as long as there is little gas/liquid interaction. Indeed, it is well known that waves can form on the surface of a free-falling liquid film (Tailby and Portalski, 1962; Pierson and Whitaker, 1977; Ishimatsu *et al.*, 1990) that can change the transfer mechanism compared with a smooth interface. However, it is also well known that the effective area for high liquid loads is found to be very close to the geometric area, while waves would lead to an efficient area significantly higher than the geometric area. Moreover, the experimental work of Wang (2012) showed that gas-side mass transfer is independent of the liquid flow rate. The assumption of a flat wall to represent the gas layer next to the liquid interface is therefore thought to be reasonable. A similar approach was used by Erasmus (2004) for the calculation of gas-side mass transfer in a simplified geometry of structured packings. More recently, with the same idea by simulating gas flow only, Lautenschleger *et al.* (2015) used CFD to analyse the impact of flow behaviour on gas-side mass transfer in novel and conventional structured packing. For the determination of gas-side mass transfer, the authors consider a fast reaction of H_2S from N_2 by NaOH. The phase interface was represented by a packing wall. The authors reported very good agreement between the numerical simulation and literature modeling for *Sulzer* BX and *Montz* B1-500 commercial structured packings.

In this work, the analogy of Chilton-Colburn (Chilton, 1934) is used to calculate the gas-side mass transfer coefficient from heat transfer simulations from the packing wall. The energy equation is solved with the mass and momentum equation. The dimensionless numerical gas-side mass transfer coefficient, the Sherwood number, is then calculated from the CFD Nusselt number computed from the thermal gradient at the wall and Chilton-Colburn analogy as follows:

$$Nu = \frac{1}{S}\iint_S \frac{k\nabla T}{k\,\frac{T_w - T_b}{d_{eq}}}\,dS \tag{6}$$

$$\frac{Sh}{Sc^{1/3}} = \frac{Nu}{Pr^{1/3}} \tag{7}$$

where k, T_w, T_b, Sh, Sc, Nu and Pr are the effective conductivity, wall and bulk temperatures, Sherwood numbers, Schmidt numbers, Nusselt numbers, and Prandtl numbers, respectively. The CFD simulations presented in this section

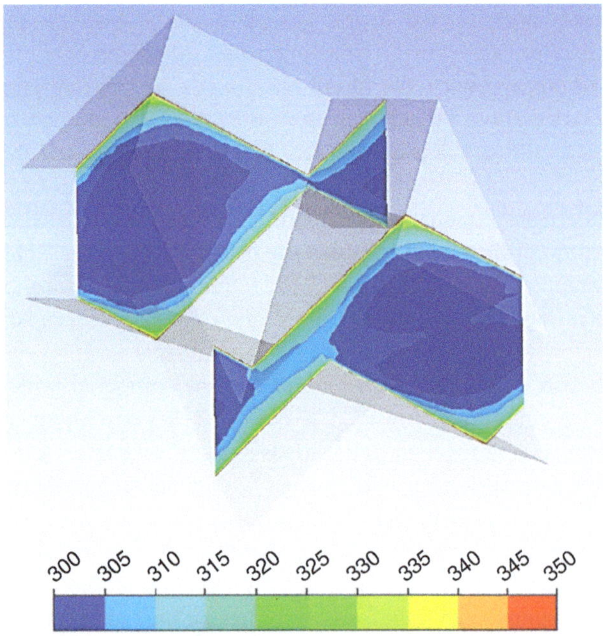

Figure 5

Temperature contour in the REU element of Mellapak 250X structured packing at Reynolds equal to 15 000.

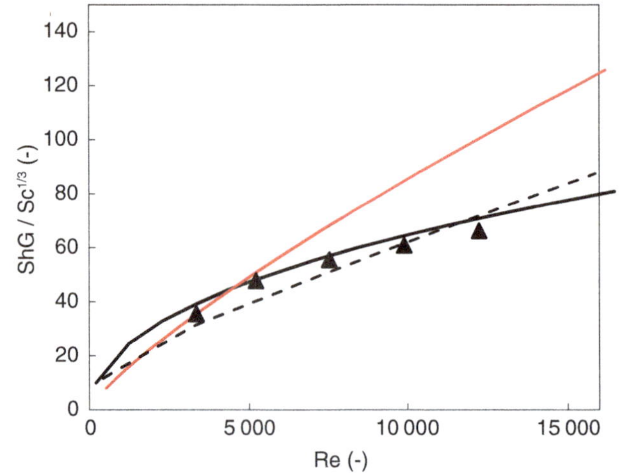

Figure 6

Comparison between gas-side mass transfer in Mellapak 250X packing resulting from REU CFD simulations (dotted line) and the experimental measurements of Wang 2012 (symbol), the model of Wang 2012 (black line) and the model of Brunazzi 1997 (red line) for a liquid load of approx. 40 $m^3/m^2/h$.

were carried out with the commercial CFD software ANSYS Fluent 14.5.

Figure 5 shows an illustration of the temperature contour plot obtained in the REU corresponding to M250X structured packing. As expected, there exists a temperature drop from the wall to the bulk, due to the lower temperature value near the bulk. The CFD results are compared with the model and experimental data of Wang (2012) and the correlation of Brunazzi (1997) for different gas Reynolds numbers and shown in Figure 6. Very good agreement between REU CFD simulations and the experimental data and physical model of Wang (2012) is observed. The relative deviation between CFD and experimental data is about 10%. However, a higher deviation is obtained with the correlation proposed by Brunazzi (1997) at large Reynolds numbers with relative deviation of about 25%.

Nevertheless, the presented gas-side mass transfer calculation approach appears to be applicable for the investigation of the gas-side mass transfer in structured packing and the comparison of different packing geometries.

1.4 Effective Area and Liquid Hold-Up

The effective area, a_e, in structured packing is defined as the effective interfacial area developed by the liquid phase normalised by the packing volume. This parameter is of determinant importance for reactive absorption applications. Indeed, in post-combustion CO_2 capture processes, for example, Raynal *et al.* (2013) show that the most important mass transfer parameter is the interfacial area, the gas and liquid mass transfer coefficients having almost no influence.

In recent years, the VOF CFD approach has been used more and more to investigate wetting, the effective area and liquid hold-up in structured packing (Ataki and Bart, 2006; Hoffmann *et al.*, 2006; Shojaee *et al.*, 2011; Haroun *et al.*, 2014; Sebastia-Saez *et al.*, 2013; Iso *et al.*, 2013; Iso and Chen, 2011). Indeed, this method is adapted to simulate wetting phenomena since it is possible to specify a wall adhesion angle in conjunction with the surface tension model (Raynal *et al.*, 2009). A given contact angle γ can be imposed at the wall to specify the adherence characteristic of the packing surface. This latter is then used to adjust the surface normal in cells near the wall. The combination of this contact angle with the calculated surface normal one cell away from the wall determines the local curvature of the surface, and this curvature is used to adjust the body force term in the surface tension calculation.

Recently, Haroun *et al.* (2014) applied a VOF method to investigate the effective area and liquid hold-up within the REU of Mellapak 250X structured packing. The values of liquid-solid contact angles were varied to mimic the geometrical imperfections of the plate surface. The results showed significant variations of the wetted area depending on the liquid-solid contact characteristics and liquid flow rate.

Figure 7

Typical liquid film flow on the surface of the REU of structured packing for $Re_L = 100$, $Re_G = 3\,300$ and $\gamma = 10°$ with different channel orientations from horizontal θ, a) $\theta = 60°$; b) $\theta = 70°$; c) $\theta = 30°$. The coloured surface corresponds to the iso-contour surface $\alpha = 0.5$.

The comparison between the CFD effective area and the experimental data of Tsai *et al.* (2011) shows that the best agreement with experimental data is obtained with a static angle of $\gamma = 10°$, the deviation being about 20%.

For the liquid hold-up prediction, the authors obtain good agreement between CFD and literature modeling.

Despite the interesting results with a static angle $\gamma = 10°$, the authors pointed out the lack of experimental studies on the wetting in structured packing to check the reliability of the numerical results. Nevertheless, the method appears to be appropriate for a reasonable estimation of interfacial area and liquid hold-up and for the comparison between different packing geometries. In this work, an illustration of such CFD calculations carried out with the commercial CFD software ANSYS Fluent 14 is presented.

The CFD VOF approach is applied to calculate the effective interfacial area in REU structured packings with different channel angle orientation from horizontal θ for a fixed packing geometric area $a_G = 250$ m^2/m^3. Computational options as well as boundary conditions and fluid system property details can be found in Haroun *et al.* (2014).

Characteristic shapes of the liquid film flow on the surface of corrugated sheet are shown in Figure 7 for different channel angle orientation. The simulation results show that an increase in channel angle orientation from the horizontal leads to a decrease in the packing wetted area and thus in the interfacial effective area. Indeed, for $\theta = 70°$ the liquid flows more in the hollow of the packing corrugation, further resulting in a decrease in the wetted area. On the contrary, for low channel angle orientation from the horizontal, the liquid flows over all the packing surface, thus increasing the wetted area. This behaviour is quantified in Figure 8, where the CFD effective interfacial area evolution with the liquid flow rate is presented for the different channel orientation angles considered. The results show clearly that a decrease in channel angle orientation from the horizontal leads to an increase in effective area values for different Reynolds numbers.

Such a parametric CFD effective interfacial area calculation combined with pressure drop and mass transfer analysis appears to be a powerful tool for first structured packing characterisation, as well as for the optimisation of new packing geometry. These parameters could be further used in process simulation on a larger scale for the development and design of efficient reactive absorption columns.

2 ABSORBER INTERNAL DEVELOPMENT USING CFD

All previous results implicitly assume that local velocities correspond to the process operating conditions, which implies perfectly homogeneous flow distribution. To make sure that perfect distribution is reached in packed towers, one has first to develop efficient internals, that are gas and liquid distributing devices, and second to make sure that this good distribution is maintained along the bed, which is known to be impossible and requires splitting a packed column into multiple packed beds with collecting and distributing devices in between.

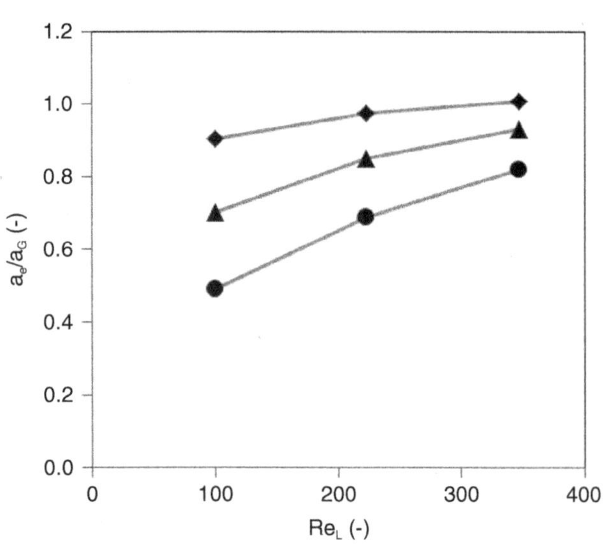

Figure 8

Dimensionless interfacial effective area *versus* liquid Reynolds number for different channel orientations: θ. (●) $\theta = 70°$; (▲) $\theta = 60°$; (♦) $\theta = 30°$.

Concerning phase distribution evolution in packed columns, one is interested in determining the maximum bed length that can be designed before considering redistributing phases. This knowledge is mostly given by experience from the field since it is quite difficult to perform experimental tests on very large-diameter columns. A maximum bed height is thus given either in terms of maximum absolute height or in terms of maximum number of theoretical stages per bed (respectively, from 8 m to 12 m or from 10 to 20 stages depending on the packing considered and the vendor); it is sometimes given in terms of the height to diameter ratio. This upscaling issue has recently been addressed with original CFD developments *via* the integration of liquid phase dispersion modeling (Fourati *et al.*, 2013); however, the simulation tool has not yet been used either for identifying the relationship between operating conditions, bed height and maldistribution phenomena or for design purposes; this would require further work and is thus not further discussed in the present paper. On the contrary, CFD can be used as a design tool for distributing devices, as is discussed in the next paragraph.

The uniform distribution of gas and liquid above packings is indeed very important to achieve optimum performance of a reactive absorption column. Poor distribution reduces the effective wetted packing and promotes liquid channelling and poor gas/liquid contact efficiency, resulting in a higher HETP (Height Equivalent to a Theoretical Plate). Consequently, the gas-liquid distributors play a major role to ensure utilising the full potential of the packed bed.

Most of the work done (Moore and Rukovena, 1987; Olujic and de Graauw, 1989, 1990; Olujic *et al.*, 2004) shows that the sources and the nature of large-scale liquid and gas maldistribution are often due to insufficient quality of the initial distribution. In this context, CFD has become a very frequently used tool to improve the design of gas-liquid distributors and to ensure the scale-up to industrial size. With this purpose, Stemich and Spiegel (2011) and Raynal *et al.* (2013) showed how CFD can be used to evaluate the vapour velocity distribution at the entrance to the packed bed above the vapour feed, and define design rules. Mohamed Ali *et al.* (2003) used CFD as an effective tool for analysing and comparing the distribution performance of internals encountered in packed columns, such as initial gas distributors, liquid distributors and liquid collectors. The comparison of measured and predicted profiles for single-phase gas flow conditions indicates good agreement. By using the CFD VOF approach, Heggemann *et al.* (2007) investigated liquid outflow through the distributor

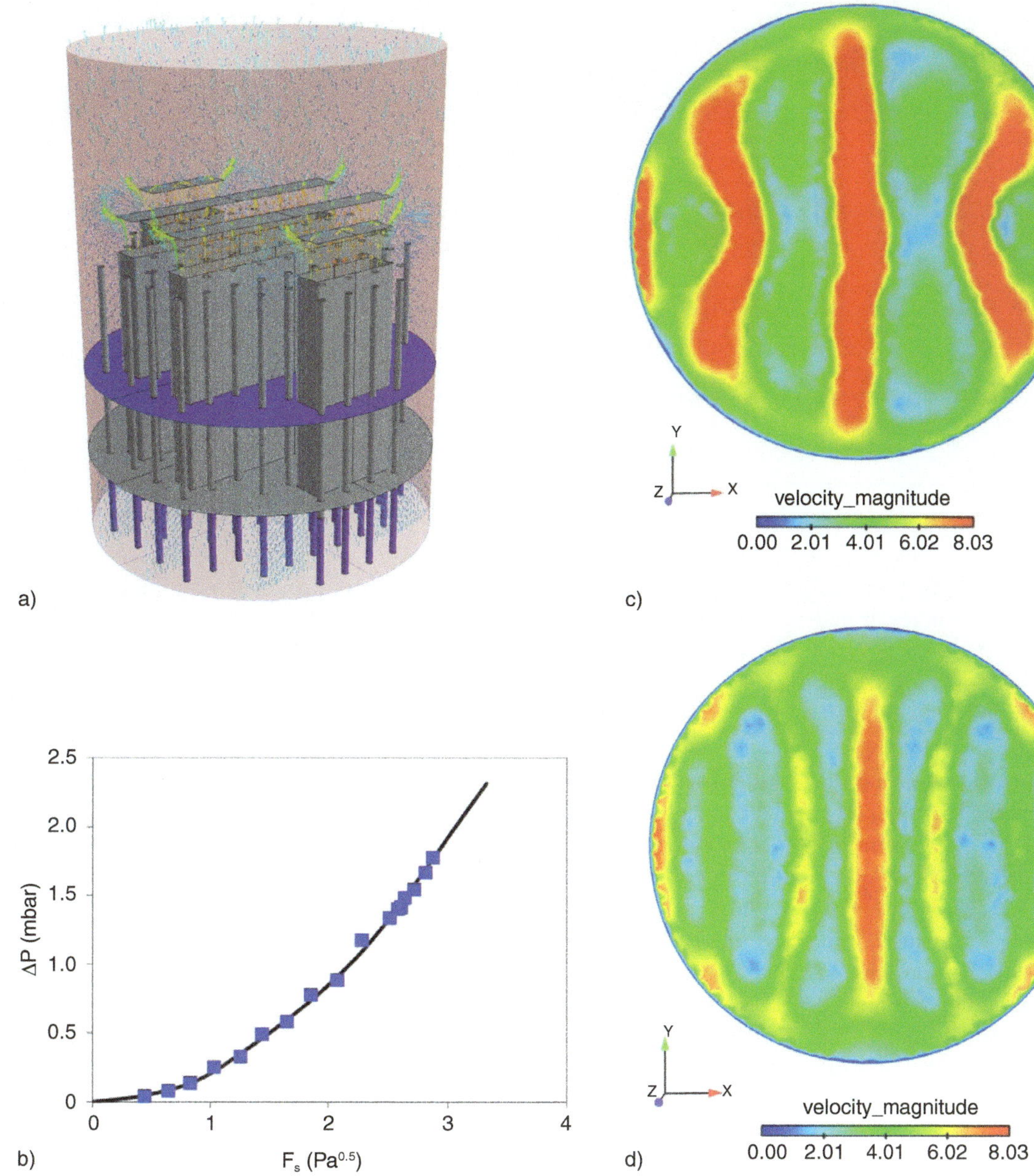

Figure 9

Picture of a 1-m-diameter tray with gas velocity vectors obtained from a gas-only simulation; b) pressure drop curve across the distributor, comparison between CFD calculations (line) and experiments (squares); c) velocity contours in a plane above the tray with 4 chimneys for gas flow; d) velocity contours in a plane above the tray with 6 chimneys for gas flow.

orifices and the influence of the lateral fluid velocity on the orifice coefficient in the liquid channel distributor tray.

In the following, typical CFD calculations performed to improve liquid and gas tray distributors are presented. Moreover, the use of CFD to build or improve the design rules is discussed. CFD simulations presented in this section were carried out with the commercial CFD software Fluent 5.4 and ANSYS Fluent 14.

2.1 Liquid Distributors

As discussed above, the liquid distribution tray plays an important part in the efficient operation of a reactive absorption column. Different types of commercial distributor trays can be found: perforated pipe, orifice distributors, and tunnel or channel distributors (Kister, 1990). Figure 9a shows an illustration of a common deck-type liquid distributor.

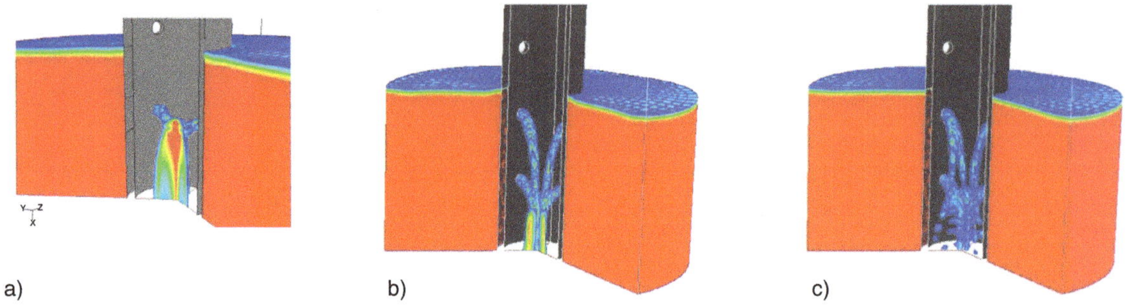

Figure 10

VOF simulations of the liquid flow through a drip tube with the liquid flow rate and surface tension's respective influence, only 2/3 of the geometry being shown; a) air/water system for $Q_L = 6\ m^3/m^2/h$, b) $Q_L = 14\ m^3/m^2/h$, c) similar to b) but with $\sigma = 20$ mN/m.

The latter consists of a deck equipped with covered rectangular chimneys for gas flow and drip tubes for liquid flow. This figure (Fig. 9a), where gas velocity vectors are shown, corresponds to a gas-only *k-ω* simulation, the aim of which is first to determine the pressure drop across the distributor, and second to optimise the number and shape of gas chimneys. It is seen from Figure 9b that the comparison between CFD calculations and experimental measurements is excellent, the measurements being performed on a 1-m-diameter column with air at atmospheric pressure. This good agreement implies a good velocity distribution description and further allows design optimisation. Figures 9c, d show velocity contours in a plane 0.1 m above the gas chimney covers for two different designs, one with 4 gas chimneys, the other one with 6. High velocity zones can clearly be identified with velocities as high as 8 m/s and even higher, the superficial gas velocity being 3 m/s only. It is clearly seen that, when increasing the number of gas chimneys from 4 to 6, the zones of high velocity strongly decrease, which may avoid any liquid re-entrainment to the upper packed bed, leading to efficiency loss.

Depending on plugging risks, liquid operating ranges and operating conditions, one would consider drip tubes with side circular or slotted orifices, or with overflow weirs; perforations in the tray can also be used instead (Kister, 1990). In some high-performance designs corresponding to operation with a high turndown ratio, the distributors can be enhanced by using perforated drip tubes with orifices at different levels on the tubes. Such a solution must be carefully engineered to ensure a good distribution quality (Bazer-Bachi *et al.*, 2013). For this case, the design rules have to take into account several parameters, such as the liquid height at nominal and turndown flow conditions, the distance between the gas-liquid interface and the drip tube orifice(s), the number and the area of orifices, the friction coefficient at the orifice, the resistance to the tray unlevelness, etc. Once again, CFD can be of great help to design the shape and number of orifices. Figures 10a, b show liquid contours, obtained with the VOF approach described in Raynal and Harter (2001), for two liquid flow conditions corresponding to air/water physical properties and liquid flow rates of 6 and 14 $m^3/m^2/h$, respectively. The main difference between the two cases is of course the liquid height on the tray and the fact that in the higher flow rate case two levels of orifices dispatch liquid, while there is only one level for the lower flow rate case. One also observes that, in both cases, the liquid jets from the orifices meet along the axis into one liquid jet further going down to the packed bed downstream, which is the flow configuration which would be expected for proper operation in counter-current mode for avoiding too much liquid entrainment. Figure 10c shows a case similar to case b, that is, with the same liquid flow rate but for different physical properties, the surface tension being changed from 73 mN/m, corresponding to water, to 20 mN/m, corresponding to a light naphtha, which covers the case of amine-based solutions characterised by surface tension in the range of 40-50 mN/m (Asprion, 2005). One observes in the latter case that the liquid jets break up into droplets, which means, in the case of a counter-current flow, a risk of liquid entrainment if the gas velocity inside the tube is too high, which calls for good design of gas chimneys, as previously discussed.

For offshore applications, where the reactive absorption unit is on a floating vessel such as a Floating Liquefied Natural Gas vessel (FLNG) or a Floating Production Storage and Offloading vessel (FPSO), the column undergoes wave motions and is subject to significant accelerations which might affect transfer efficiency and fluid distribution. From a design point of view, this could lead to an increase in absorption unit size (higher packed bed heights, cumbersome distribution internal) to avoid any risk of off-spec gas being sent to the liquefied gas production unit. Therefore,

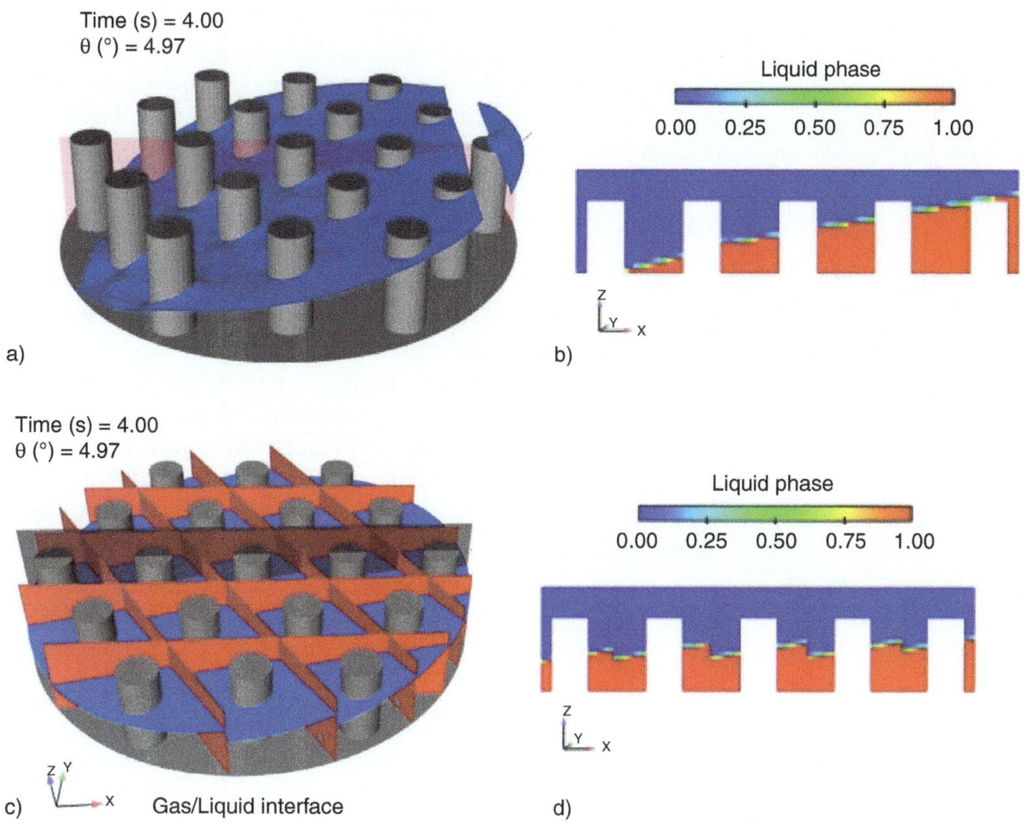

Figure 11

Illustration of the liquid guard level at the maximum oscillating angle with the standard orifice pan distributor (a, b) and partitioned distributor tray (c, d).

development of efficient and compact distribution technologies that are less sensitive to motion is crucial to reduce floating absorption unit investment cost.

For offshore applications, the use of the CFD VOF approach is particularly interesting for the evaluation and design of high-performance distributors, since experiments in dynamic floating conditions are less accessible and are much more expensive than a classical static experimental setup. Moreover, this method has been validated upon experimental data on sloshing liquid in moving tanks. (Wacławczyk and Koronowicz, 2008; Li *et al.*, 2014) Recently, in the patent US 20130277868 (Haroun *et al.*, 2013), the inventors used CFD to develop a new distributor tray for a packed column on a floating installation (FPSO, FLNG) submitted to wave oscillating motion.

The inventors used the VOF method in floating conditions to evaluate the performance of a standard orifice pan distributor tray and new partitioned distributor tray. The motion effect on reactive absorption columns has been taken into account in the calculation through body force terms (Celebi and Akyildiz, 2002) derived from a moving coordinate system and introduced, in the framework of the moving distributor, as source terms *via* an in- house user-defined function using Fluent CFD software. These forces introduce into the calculation the effect of rolling motion and acceleration of fluids. The rolling conditions considered correspond to distributors positioned at the top of the column, about 50 m above the gyration centre of the boat. The roll oscillation amplitude and period used are ±5° and 15 s, respectively.

Figure 11 shows at the maximum oscillating angle the liquid guard shape obtained by CFD VOF simulations for the conventional orifice pan distributor (Fig. 11a, b) and partitioned distributor tray (Fig. 11c, d). Figure 11a, c shows the three-dimensional shape of the gas-liquid interface and Figure 11b, d gives the contour of the liquid phase on a two-dimensional slice of the distributors. The results show that with the conventional onshore distributors (Fig. 11a, b) the liquid interface is greatly disturbed in floating motion conditions, resulting in a poor liquid distribution. In contrast,

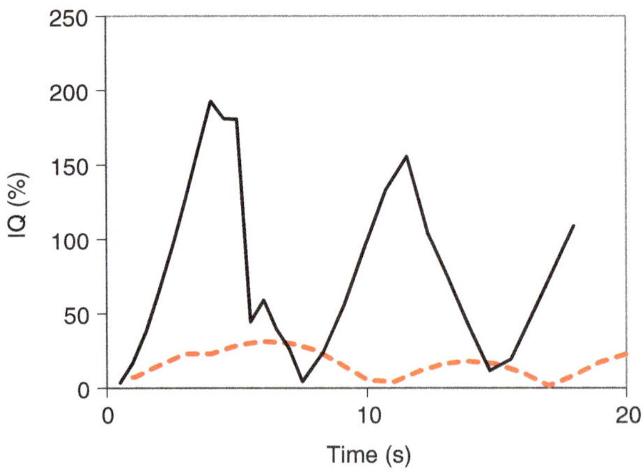

Figure 12

Evolution of the distribution Quality Index (IQ) (see Eq. 8) with floating motion time; (line) onventional orifice pan distributor; (dotted line) partitioned distributor tray.

in the partitioned distributor tray (Fig. 11c, d), the baffles allow the reduction of liquid inertia induced by wave motion and thus ensure a homogeneous liquid level on the distributor. This results in appropriate liquid distribution on the packed bed. This is well illustrated in Figure 12, which shows the evolution of the distribution Quality Index, IQ, with floating motion time for conventional and novel distributor trays. A low value of the IQ indicates a good distribution quality and low sensitivity of the distributor to the oscillating motion. By contrast, a high value of the IQ points to a strong imbalance of distribution.

From the plot, a high imbalance of distribution is obtained with the conventional distributor with a maximum IQ as high as 150% and even higher. It is clearly seen that with the partitioned distributor tray, the distribution IQ strongly decreases, which indicates a good distribution performance.

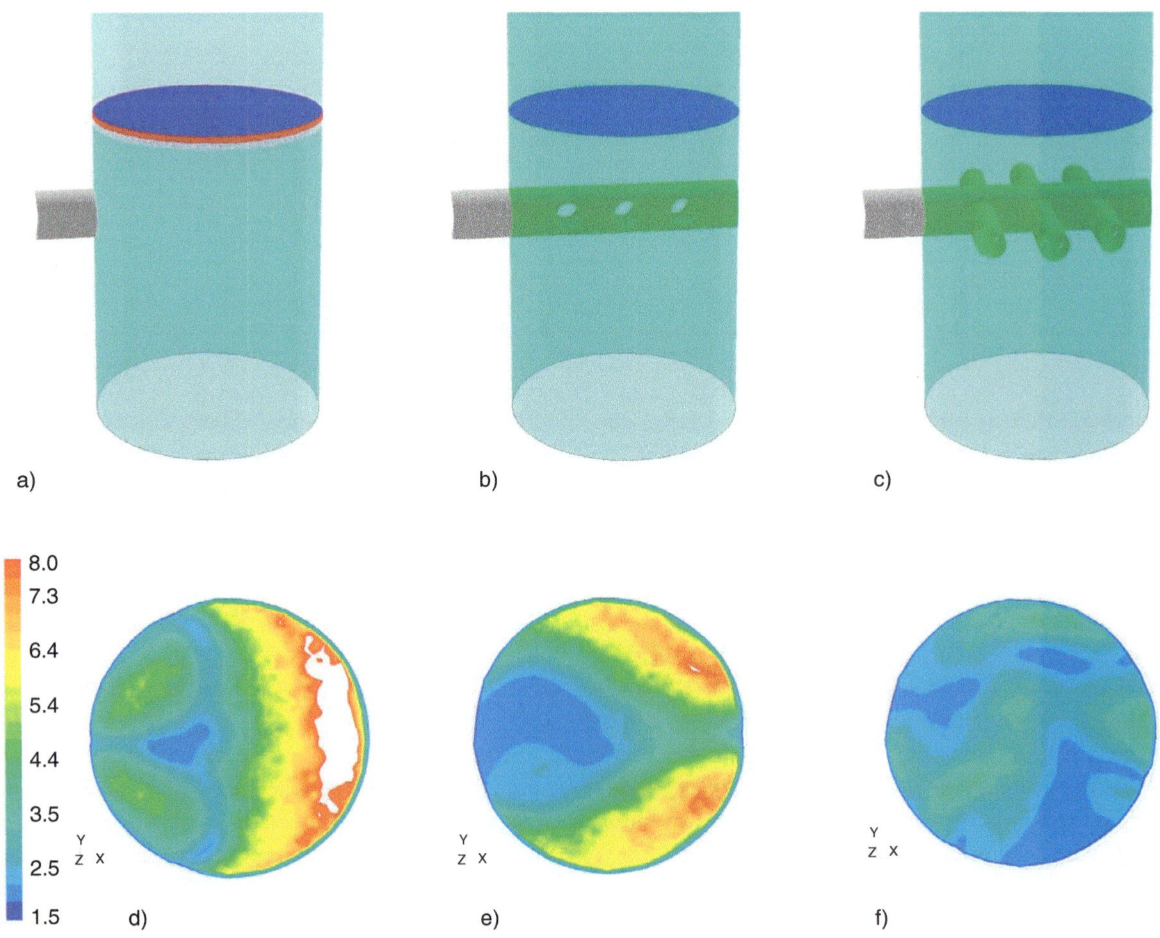

Figure 13

Gas distributor configurations and corresponding velocity contours at the packed bed inlet; a, d) no gas distributor; b, e) simple perforated tube; c, f) perforated pipe distributor.

Further developments such as parametric CFD simulations will optimise this design and improve the distribution quality index:

$$IQ = \frac{F_{max} - F_{min}}{F_{average}} \quad (8)$$

where F_{max} and F_{min} are, respectively, the maximum and minimum liquid flows through the pan orifice distributors, and $F_{average}$ represents the average liquid flow.

2.2 Gas Distributors

Gas or vapour distributors are used to achieve a uniform gas flow at the packed bed inlet. Depending on the service, pressure drop can be a specification and must be minimised; in natural gas treatment, this would not be an issue, while for CO_2 post-combustion capture this would be critical. Gas flow simulations are quite easy to perform and can be used for gas distributing device testing and design (Mohamed Ali *et al.*, 2003; Stemich and Spiegel, 2011; Raynal *et al.*, 2013). Indeed, as can be seen from Figure 13, it is possible to evaluate the quality distribution at the packed bed inlet, downstream of the distributor, for different distributors. We considered cases without any distributor (Fig. 13a), with a simple large perforated tube (Fig. 13b) or with a multiple perforated pipe distributor (Fig. 13c). Calculations were performed in the framework of CO_2 capture in an 8-m-diameter column with a standard *k*-ε turbulent model. The use of a simple tube, which was believed to provide some improvement, does not make any change with the case without any internal, the maximum velocity still being above 8 m/s for a superficial gas velocity of 2 m/s. The ladder-type distributor provides a significant improvement for moderate complexity; the quality index, as defined in the previous section, was improved by a factor of 10, decreasing from 269% to 20%. However, this improvement has a cost since the associated pressure drop increases from 104 mbar up to 200 mbar.

CONCLUSION

The aim of this paper was to illustrate on a large spectrum how CFD is used for packed absorption tower optimum design. It is shown how the simulation tool can complete experiments either by providing supplementary information that experiments can provide too, at least partially, such as mass transfer coefficient determination, or by providing supplementary information that experiments cannot address, or with limiting inputs, in particular linked to scale-up issues, such as internal design.

It is first important to underline that both tools really complete each other: on the one hand, CFD calculations can provide sensitivity analysis or extrapolation to operating conditions and corresponding physical properties outside the accessible range of experiments; on the other hand, CFD calculations will still require validation since the development of new closure models, for example concerning wetting of phase dispersion phenomena, are still not fully understood and integrated into physical models one can easily use. Second, CFD can address gas only or gas/liquid flows over a very large range of characteristic lengths *via* the use of different approaches in quite a time-efficient way so that it can be used for the development or optimisation of new design rules, from asperities on the packing wall to a full industrial-size 3D column solving the flow within the packed bed and through column internals.

For future developments, further CFD investigations and validation should first focus on the wetting phenomena and on improving the modeling of the contact angle in the roughness surface. This could lead to improving the prediction of the effective area in structured packing and also to extending the prediction of pressure drop to gas-liquid flow conditions.

Second, regarding the upscaling issue, the future development of the two-fluid models with porous medium description (Fourati *et al.*, 2013) will allow identifying the relationship between operating conditions, bed height and maldistribution phenomena on an industrial column scale. Moreover, hydrodynamic simulations using a two-fluid model could be coupled with chemical species transport and reactive absorption modeling. Such an approach would allow quantifying the impact of hydrodynamic distribution on the packing transfer performance and thus better definition of design criteria for distribution technologies.

It is thus believed that CFD should be a more and more useful tool for R&D chemical engineers who develop new solutions to optimise absorption column design.

ACKNOWLEDGMENTS

The authors would like to acknowledge Dr. M. Fourati and M. Morin for their collaboration in CFD modeling of gas-side mass transfer and P. Ancelet for her collaboration in CFD modeling of effective area in structured packing.

REFERENCES

Ataki A., Bart H.J. (2006) Experimental and CFD simulation study for the weeting of s structured packing element with liquids, *Chem. Eng. Technol.* **29**, 3, 336-347.

Asprion N. (2005) Surface tension models for aqueous amine blends, *Ind. Eng. Chem. Res.* **44**, 7270-7278.

Bai H., Theuerkauf J., Gillis P.A. (2009) A Coupled DEM and CFD Simulation of Flow Field and Pressure Drop in Fixed Bed Reactor with Randomly Packed Catalyst Particles, *Ind. Eng. Chem. Res.* **48**, 4060-4074.

Bazer-Bachi F., Haroun Y., Augier F., Boyer C. (2013) Experimental evaluation of distributor technologies for trickle-bed reactors, *Ind. Eng. Chem. Res.* **52**, 11189-11197.

Billet R. (1995) *Packed Towers*, VCH Eds, Weinheim.

Bravo J.L., Rocha J.A., Fair J.R. (1985) Mass transfer in gauze packings, *Hydrocarbon Processing* **64**, 1, 91-95.

Brackbill J.U., Khote D.B., Zemach C. (1992) A continuum method for modelling surface tension, *J. Comput. Phys.* **100**, 335-354.

Brian P.L.T., Hurey J.F., Hassettine E.H. (1961) *AIChE J.* **7**, 226.

Brunazzi E., Nardini G., Paglianti A., Petraca L. (1995) Interfacial area of Mellapak packing Absorption of 1,1,1-trichloroethane by Genosorb 300, *Chem. Eng. Technol.* **18**, 248.

Brunazzi E.P. (1997) Liquid-Film Mass-Transfer Coefficient in a Column Equipped with Structured Packings, *Ind. Eng. Chem. Res.* **36**, 3792-3799.

Celebi M.S., Akyildiz H. (2002) Nonlinear modeling of liquid sloshing in a moving rectangular tank, *Ocean Eng.* **29**, 1527-1553.

Charpentier J.C. (2009) Perspective on multiscale methodology for product design and engineering, *Comp. Chem. Eng.* **33**, 936-946.

Chilton T.C. (1934) Mass Transfer Coefficient, Prediction from Data on Heat Transfer and Fluid Friction, *Ind. Eng. Chem.* 1183-1187.

Danckwerts P.V. (1970) *Gaz-Liquid Reaction*, McGraw-Hill, New York.

Erasmus A.B. (2004) Mass Transfer in Structured Packing, *Ph.D. Dissertation Chem. Eng.*, University of Stellenbosch, South Africa.

Fair J.R., Bravo J.L. (1990) Distillation columns containing structured packings, *Chem. Eng. Prog.* **86**, 19-29.

Fernandes J., Lisboa P.F., Simoes P.C., Mota J.P.B., Saatdjian E. (2009) Application of CFD in the study of supercritical fluid extraction with structured packing: wet pressure drop calculations, *J. Supercrit. Fluids* **50**, 61.

Fourati M., Roig V., Raynal L. (2012) Experimental study of liquid spreading in structured packings, *Chem. Eng. Sci.* **80**, 1-15.

Fourati M., Roig V., Raynal L. (2013) Liquid dispersion in packed columns: experiments and numerical modelling, *Chem. Eng. Sci.* **100**, 266-278.

Haroun Y., Legendre D., Raynal L. (2010a) Direct numerical simulation of reactive absorption in gas–liquid flow on structured packing using interface capturing method, *Chem. Eng. Sci.* **65**, 351-356.

Haroun Y., Legendre D., Raynal L. (2010b) Volume of fluid method for interfacial reactive mass transfer: application to stable liquid film, *Chem. Eng. Sci.* **65**, 2896-2909.

Haroun Y., Raynal L., Legendre D. (2012) Mass transfer and liquid hold-up determination in structured packing by CFD, *Chem. Eng. Sci.* **75**, 342-348.

Haroun Y., Raynal L., Alix P. (2013) Partitioned distributor tray for offshore gas/liquid contact column, Patent US 20130277868.

Haroun Y., Raynal L., Alix P. (2014) Prediction of effective area and liquid hold-up in structured packings by CFD, *Chem. Eng. Res. Des.* **92**, 2247-2254.

Heggemann M., Hirschberg S., Spiegel L., Bachmann C. (2007) CFD Simulation and Experimental Validation of Fluid Flow in Liquid Distributors, *Chem. Eng. Res. Des.* **85**, 59-64.

Higbie R. (1935) The rate of absorption of a pure gas into a still liquid during short periods of exposure, *Trans. AIChE* **35**, 365.

Hirt C.W., Nichols B.D. (1981) Volume of Fluid method for the dynamics of free boundaries, *J. Comput. Phys.* **39**, 201-225.

Hoffmann A., Ausner I., Repke J.-U., Wozny G. (2006) Detailed investigation of multiphase (gas–liquid and gas–liquid–liquid) flow behaviour on inclined plates, *Chem. Eng. Res. Des.* **84**, 147-154.

Ishimatsu M., Nosoko T., Nagata T. (1990) Flow patterns and wave characteristics of falling liquid films, *Heat Trans. Japanese Res.* **19**, 602-615.

Iso Y., Chen X. (2011) Flow transition behavior of the wetting flow between the film flow and rivulet flow on an inclined wall, *J. Fluids Eng.* **133**, 091101-91111.

Iso Y., Huang J., Kato M., Matsuno S., Takano K. (2013) Numerical and experimental study on liquid film flows on packing elements in absorbers for post-combustion CO_2 capture, *Energy Procedia* **37**, 860-868.

Kister H.Z. (1990) *Distillation Operation*, McGraw-Hill, New York, USA.

Lautenschleger A., Olenberg A., Kenig E.Y. (2015) Systematic CFD-based method to investigate and optimise novel structured packings, *Chem. Eng. Sci.* **122**, 452-464.

Lassauce A., Alix P., Raynal L., Royon-Lebeaud A., Haroun Y. (2014) Pressure Drop, Capacity and Mass Transfer Area Requirements for Post-Combustion Carbon Capture by Solvents, *Oil Gas Sci. Technol.* **69**, 6, 1021-1034.

Li H.-L., Li J., Zong Z., Chen Z. (2014) Numerical studies on sloshing in rectangular tanks using a tree-based adaptive solver and experimental validation, *Ocean Eng.* **82**, 20-31.

Mahr B., Mewes D. (2007) CFD Modelling and calculation of dynamic two phase flow in columns equipped with structured packing, *Chem. Eng. Res. Des.* **85**, 1112-1122.

Moore F., Rukovena F. (1987) Liquid and gas distribution in commercial packed towers, *Chem. Plants and Processing* No. 8, 11.

Mohamed Ali A., Jansens P.J., Olujic Z. (2003) Experimental Characterization and Computational Fluid Dynamics Simulation of Gas Distribution Performance of Liquid (Re)Distributors and Collectors in Packed Columns, *Chem. Eng. Res. Des.* **81**, 1, 108-115.

Nikou K., Ehsani M.R. (2008) Turbulence models application on CFD simulation of hydrodynamics, heat and mass transfer in a structured packing, *Int. Commun. Heat Mass Transfer* **35**, 1211.

Nicoud F., Ducros F. (1999) Subgrid-Scale Stress Modelling Based on the Square of the Velocity Gradient Tensor, *Flow Turb. Comb.* **62**, 3, 183-200.

Olujic Z., de Graauw J. (1989) Appearence of maldistribution in distillation columns equiped with hight performance packings, *Chem. Biochem. Eng. Q.* **3**, 181-196.

Olujic Z., de Graauw J. (1990) Experimental studies on interaction between the initial liquid distribution and the performance of structured packing, *Separation Sci. Tech.* **25**, 1723-1735.

Olujic Z., Mohamed Ali A., Jansens P.J. (2004) Effect of the initial gas maldistribution on the pressure drop of structured packings, *Chem. Eng. Process.* **43**, 465-476.

Petre C.F., Larachi F., Illiuta I., Grandjean B.P.A. (2003) Pressure drop through structured packings: breakdown into the contributing mechanisms by CFD modelling, *Chem. Eng. Sci.* **58**, 163-177.

Pierson F.W., Whitaker S. (1977) Some theoretical and experimental observations of the wave structure of falling liquid film, *Ind. Eng. Chem. Fundam.* **16**, 401-408.

Raynal L., Harter I. (2001) Studies of Gas-Liquid flow through distributing devices using VOF-CFD simulations, *Chem. Eng. Sci.* **56**, 6385-6391.

Raynal L., Boyer C., Ballaguet J.P. (2004) Liquid holdup and pressure drop determination in structured packing with CFD simulation, *Can. J. Chem. Eng.* **82**, 871-879.

Raynal L., Royon-Lebeaud A. (2007) A multi-scale approach for CFD calculations of gas–liquid flow within large size column equipped with structured packing, *Chem. Eng. Sci.* **62**, 7196-7204.

Raynal L., Ben Rayana F., Royon-Lebeaud A. (2009) Use of CFD for CO_2 absorbers optimum design: from local scale to large industrial scale, *Energy Procedia* **1**, 917-924.

Raynal L., Gomez A., Caillat B., Haroun Y. (2013) CO_2 capture cost reduction: Use of a multiscale simulations strategy for a multiscale issue, *Oil Gas Sci. Technol.* **68**, 1093-1108.

Said W., Nemer M., Clodic D. (2011) Modeling of dry pressure drop for fully developed gas flow in structured packing using CFD simulations, *Chem. Eng. Sci.* **66**, 2107-2117.

Scardovelli R., Zaleski S. (1999) Direct numerical simulation of free-surface and interfacial flow, *Annu. Rev. Fluid Mech.* **31**, 567-603.

Sebastia-Saez D., Gu S., Ranganathan P., Papadikis K. (2013) 3D modeling of hydrodynamics and physical mass transfer characteristics of liquid film flows in structured packing elements, *Int. J. Greenhouse Gas Control* **19**, 492-502.

Sebastia-Saez D., Gu S., Ranganathan P., Papadikis K. (2014) Micro-scale CFD study about the influence of operative parameters onphysical mass transfer within structured packing elements, *Int. J. Greenhouse Gas Control* **28**, 180-188.

Shojaee S., Hosseini S.H., Rafati A., Ahmadi G. (2011) Prediction of the effective area in structured packings by computational fluid dynamics, *Ind. Eng. Chem. Res.* **50**, 10833-10842.

Soulaine C., Horgue P., Franc J., Quintard M. (2014) Gas–liquid flow modeling in columns equipped with structured packing, *AIChE Journal* **60**, 10, 3665-3674.

Stemich C., Spiegel L. (2011) Characterization and quantification of the quality of gas flow distributions, *Chem. Eng. Res. Design* **89**, 1392-1396.

Spiegel L., Meier W. (1992) A generalized pressure drop model for structured packings, Distillation and Absorption, *IChemE Symp.* **128**, B85-B94.

Szulczewska B., Sek J., Gorak A., Zbicinski I. (2000) CFD calculation of two-phase flow on elements of structured packing, *Chisa Int. Conf.*, Praha, Czech Republic, Aug. 27-37, Paper #565, 1-10.

Tailby S.R., Portalski S. (1962) Wave inception on a liquid film flowing down a hydrodynamically smooth plate, *Chem. Eng. Sci.* **19**, 283-290.

Tsai R.E., Seibert A.F., Eldridge R.B., Rochelle G.T. (2011) A dimensionless model for predicting the mass-transfer area of structured packing, *AIChE J.* **57**, 1173-1184.

Wacławczyk T., Koronowicz T. (2008) Comparison of CICSAM and HRIC high-Resolution schemes for interface capturing, *J. Theo. App. Mech.* **46**, 2, 325-345.

Wang C. (2012) Measurement of Packing Effective Area and Mass Transfer Coefficients, Luminant Carbon Management Program and Process Science and Technology Center, Austin, Texas, pp. 1-9.

6

Permeability Variation Models for Unsaturated Coalbed Methane Reservoirs

Yumin Lv[1,2*], Zhiping Li[1], Dazhen Tang[1], Hao Xu[1] and Xiaozhi Chen[2]

[1] School of Energy Resources, China University of Geoscience, Beijing 100083 - China
[2] New Energy Research Center, CNOOC Research Institute, Beijing 100027 - China
e-mail: yale1210@163.com

* Corresponding author

Abstract — *A large number of models have been established to describe permeability variation with the depletion of reservoir pressure to date. However, no attempt has been made to draw enough attention to the difference in the effect of various factors on permeability variation in different production stages of unsaturated CoalBed Methane (CBM) reservoirs. This paper summarizes the existing and common permeability models, determines the relationship between various effects (effective stress effect, matrix shrinkage effect and Klinkenberg effect) and desorption characteristics of the recovery of unsaturated CBM reservoirs, then establishes two improved models to quantificationally describe permeability variation, and finally discusses the effects of various factors (gas saturation, cleat porosity, Poisson's ratio and shrinkage coefficient) on permeability variation. The results show that permeability variation during the recovery of unsaturated CBM reservoirs can be divided into two stages: the first one is that permeability variation is only affected by the effective stress effect, and the second is that permeability variation is affected by the combination of the effective stress effect, matrix shrinkage effect and Klinkenberg effect. In the second stage, matrix shrinkage effect and Klinkenberg effect play much more significant role than the effective stress effect, which leads to an increase in permeability with depletion of reservoir pressure. Sensitivity analysis of parameters in the improved models reveals that those parameters associated with gas saturation, such as gas content, reservoir pressure, Langmuir volume and Langmuir pressure, have a significant impact on permeability variation in the first stage, and the important parameters in the second stage are the gas content, reservoir pressure, Langmuir volume, Langmuir pressure, Poisson's ratio, Young's modulus and shrinkage coefficient during the depletion of reservoir pressure. A comparative study of the improved models indicates that the improved SD model has a greater sensitivity to various parameters than the improved PM model and the improved models describe permeability dynamic variation more exactly than the original ones.*

Résumé — **Modèles de variation de perméabilité pour des réservoirs de gaz de houille insaturé** — Un grand nombre de modèles ont été établis à ce jour pour décrire la variation de perméabilité en fonction de l'épuisement de la pression du réservoir. Toutefois, aucune tentative n'a été faite pour attirer suffisamment l'attention sur la différence d'effet des différents facteurs sur la variation de perméabilité dans différentes phases de production des réservoirs de gaz de houille insaturés (*CoalBed Methane*, CBM). Le présent article résume les modèles de perméabilité existants et usuels, détermine la relation entre les différents effets (effet de contrainte effective, effet du rétrécissement de la matrice et effet Klinkenberg) et les caractéristiques de désorption pour la récupération des réservoirs de CBM insaturé, puis établit deux modèles améliorés pour décrire de manière quantitative la variation de perméabilité, et enfin discute des effets de différents facteurs (saturation en gaz, porosité traversante, coefficient de Poisson et coefficient de retrait) sur la variation de la perméabilité. Les résultats montrent que la variation de perméabilité pendant la récupération des réservoirs de CBM insaturé peut être divisée en deux phases : la première pour laquelle la variation de perméabilité n'est affectée que par l'effet de contrainte effectif et la seconde pour laquelle la variation de perméabilité est affectée par la combinaison de l'effet de contrainte effectif, de l'effet de retrait de matrice et de l'effet Klinkenberg. Dans la seconde phase, l'effet de retrait de matrice et l'effet Klinkenberg jouent un rôle plus significatif que l'effet de contrainte effectif, qui mène à une augmentation de la perméabilité lors de l'épuisement de la pression du réservoir. L'analyse de sensibilité des paramètres dans ces modèles améliorés révèle que les paramètres associés à la saturation en gaz, tels que la teneur en gaz, la pression du réservoir, le volume de Langmuir et la pression de Langmuir, ont un impact significatif sur la variation de perméabilité dans la première phase. Les paramètres importants dans la seconde phase pendant l'épuisement de la pression du réservoir sont la teneur en gaz, la pression du réservoir, le volume Langmuir, la pression Langmuir, le coefficient de Poisson, le module de Young et le coefficient de retrait. Une étude comparative des modèles améliorés indique que le modèle SD amélioré a une sensibilité aux différents paramètres supérieure au modèle PM amélioré et les modèles améliorés décrivent la variation dynamique de perméabilité de manière plus exacte que les modèles d'origine.

NOMENCLATURE

V_L	Langmuir storage capacity (m^3/t)
V	Gas storage capacity in the condition of pressure p (m^3/t)
P_L	Langmuir pressure (MPa)
P	Pressure (MPa)
p_i	Initial reservoir pressure (MPa)
p_c	Critical desorption pressure (MPa)
k	Reservoir permeability at the pressure of p (mD)
k_i	Initial reservoir permeability (mD)
k_g	Gas-phase permeability (mD)
k_∞	Absolute permeability (mD)
φ_i	Initial porosity (-)
φ_c	Porosity at the critical desorption pressure (-)
c_m	Coal compressibility (1/MPa)
c_f	Cleat volume compressibility (1/MPa)
ε_l	Langmuir-type matrix shrinkage constant (-)
σ_{h1}	Stress in the $h1$ direction (MPa)
σ_{h2}	Stress in the $h2$ direction (MPa)
b	Klinkenberg coefficient (MPa)
b_c	Klinkenberg coefficient at the critical desorption pressure (MPa)
K	Bulk modulus (MPa)
M	Constrained axial modulus (MPa)
ϑ	Poisson's ratio (-)
C	Proportionality factor (-)
λ	Mean free path of gas molecules (m)
r	Pore size radius (m)

INTRODUCTION

Coal reservoir physical properties (particularly permeability) are affected by several factors and show dynamic variation through the recovery of CoalBed Methane (CBM). These factors can be summarized as follows: the effective stress effect [1-7], matrix shrinkage effect [8-19] and Klinkenberg effect [20]. To date, the impacts of the first two effects on permeability variation in CBM reservoirs have been intensively studied [10, 13, 14, 20-44]. However, these two effects on permeability dynamic variation in different development stages of CBM reservoirs, especially in unsaturated CBM reservoirs, have not been investigated in depth. Additionally, the third Klinkenberg effect has seldom been considered in permeability variation of the recovery of CBM reservoirs.

It is widely thought that coal can be substantially unsaturated with thermogenic gas if not augmented by hydrocarbon migration or late-stage bacterial methanogenesis [45-58]. In consideration of this fact, we improved the existing permeability models in this paper to make them more suitable for describing permeability dynamic variation in unsaturated CBM reservoirs based on the production performance of those reservoirs, and discuss the sensitivity of various parameters, including the gas content, initial reservoir pressure, Langmuir volume and pressure, Poisson's ratio, Young's modulus, Langmuir strain, and Klinkenberg coefficient, to permeability dynamic variation in the recovery of unsaturated CBM reservoirs.

1 RECOVERY MECHANISM OF UNSATURATED CBM RESERVOIRS

Gas saturation in coal reservoirs is defined as the percentage of adsorbed gas content relative to adsorption capacity in the *in situ* reservoir conditions. Thus, unsaturated CBM reservoirs refer to those reservoirs in which gas content is less than their maximum gas adsorption in the reservoir conditions. Most coal reservoirs can be substantially unsaturated with thermogenic gas [45-58]. Gas saturation of coal reservoirs in the Weibei CBM field ranges from 39.2%~87.2% [55, 56]. The micro-pilot test results showed that the gas saturation is 40.8% in the anthracitic coals of the South Qinshui basin, China [58]. Seam 4 in the Lower Cretaceous Gates Formation, Inner Foothills of Alberta showed gas saturation levels of 90% in the *in situ* reservoir conditions [53]. Unsaturated CBM reservoirs initially produce CBM only when the reservoir pressure is reduced below the critical desorption pressure [59, 60]. A small degree of unsaturation can necessitate prolonging dewatering period before a large reservoir volume can reach the critical desorption pressure [57]. Therefore, most discovered CBM reservoirs are unsaturated, which often leads to significant differences in production performance from high-saturated CBM reservoirs.

A variety of models have been established to describe the sorption process [61-67]. However, the most commonly used model for the gas adsorption capacity in coal is the Langmuir isotherm model, which can be used for both single-component and multicomponent gases, given by the equation below:

$$V = V_L \frac{p}{p + P_L} \tag{1}$$

where: V is the gas storage capacity in the condition of pressure p, m^3/t; V_L is the Langmuir storage capacity, m^3/t; p is the pressure, MPa; P_L is the Langmuir pressure, MPa.

The area below the Langmuir isotherm curve refers to unsaturated CBM reservoirs, while the other area indicates oversaturated ones (Fig. 1). An important term established from the Langmuir isotherm curve is the critical desorption pressure, which a well situated in unsaturated reservoirs must be reduced to in order for CBM desorption to occur. Once the well pressure is higher than the critical desorption pressure, there will not be any appreciable desorption.

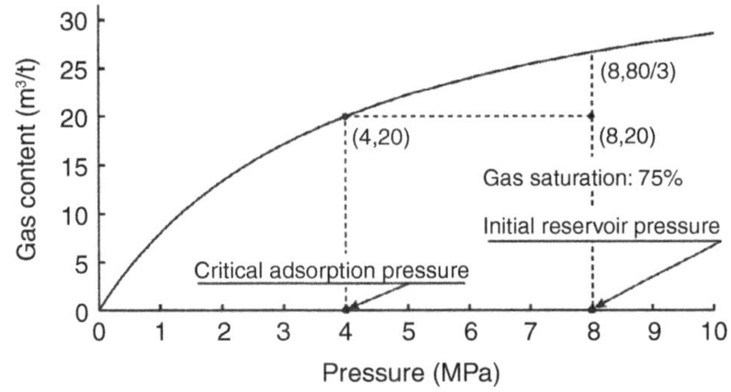

Figure 1

The Langmuir isotherm curve and the recovery of unsaturated CBM reservoirs.

Therefore, unsaturated CBM reservoirs do not produce CBM at the dewatering stage and CBM initially desorbs from the internal surface of the coal matrix only after the reservoir pressure is reduced to the critical desorption pressure and goes on [59, 60]. Thus, a small degree of unsaturation can necessitate widening the pressure drop to produce gas for CBM reservoirs and prolong dewatering before a large reservoir volume can reach the critical desorption pressure [57].

2 DYNAMIC MODELS OF PERMEABILITY VARIATION FOR UNSATURATED CBM RESERVOIRS

Permeability is an important reservoir parameter determining whether the CBM reservoir can be commercially developed. Thus, it is necessary to research reservoir permeability, especially its variations with pressure. To date, a substantial number of models [4-6, 10, 13, 14, 21-28, 30, 32, 33, 36-38, 40, 44] for permeability variations in coal have been established to attempt to account for the effect of effective stress as well as matrix shrinkage on permeability variation during the recovery of CBM reservoirs. The most widely used permeability models are the Seidle and Huitt model (SH), Palmer and Mansoori model (PM) and Shi and Durucan model (SD), as shown in Table 1.

However, all the models do not consider the effect of adsorbed gas saturation on permeability variations. In fact, adsorbed gas saturation controls which effect will affect the permeability variation in the different production stages according to the development of unsaturated CBM reservoirs (Fig. 1). Additionally, almost no models take the Klinkenberg effect into consideration for describing permeability variation in CBM reservoirs, which are typical low-permeability reservoirs.

TABLE 1
List of the three most widely used coalbed permeability models

Authors	Permeability equation	Notes
Seidle *et al.* (1992) [4]	$k = k_i \exp\left(-3c_f(\sigma - \sigma_i)\right)$	Effective stress effect only
Seidle and Huitt (1995) [23]	$k = k_i\left(1 + \left(1 + \frac{2}{\varphi_i}\right)\varepsilon\left[\frac{p_i}{p_i + P_L} - \frac{p}{p + P_L}\right]\right)^3$	Matrix shrinkage effect only
Palmer and Mansoori (1996, 1998) [24, 25]	$k = k_i\left(1 + \frac{c_m}{\varphi_i}(p - p_i) + \left(\frac{K}{M} - 1\right)\frac{\varepsilon_L}{\varphi_i}\left[\frac{p_i}{p_i + P_L} - \frac{p}{p + P_L}\right]\right)^3$	Derived from a rock mechanic approach
Shi and Durucan (2004, 2005) [26, 28]	$k = k_i \exp\left(3c_f\left(\frac{v}{1 - v}(p - p_i) - \frac{E}{3(1 - v)}\varepsilon_l\left[\frac{p_i}{p_i + P_L} - \frac{p}{p + P_L}\right]\right)\right)$	Uniaxial strain and constant vertical stress conditions

2.1 Improved Permeability Models above the Critical Desorption Pressure

During the recovery of unsaturated CBM reservoirs, it is indispensable to dewater the coalbed to release the reservoir pressure. CBM never desorbs from the matrix before the reservoir pressure is reduced to the critical desorption pressure. Permeability variation is controlled by the effective stress effect, rather than the matrix shrinkage or the Klinkenberg effect. Thus, the models of permeability variation during this process must exclude the impact of matrix shrinkage and the Klinkenberg effect.

As a result, the mathematical form of the PM model at this stage can be converted to the following form:

$$\frac{k}{k_i} = \left[1 + c_m\left(\frac{p - p_i}{\varphi_i}\right)\right]^3 \tag{2}$$

where k is the reservoir permeability at the pressure of p, mD; k_i is the initial reservoir permeability, mD; c_m is the coal compressibility, 1/MPa; φ_i is the initial porosity, dimensionless; p_i is the initial reservoir pressure, MPa.

Likewise, the mathematical form of the SD model is also converted to the following form:

$$\frac{k}{k_i} = \exp\left[3c_f\frac{\vartheta}{1 - \vartheta}(p - p_i)\right] \tag{3}$$

where ϑ is the Poisson's ratio, dimensionless; c_f is the cleat volume compressibility, 1/MPa.

Seidle *et al.*'s model (1992) [4], which only considered the effect of stress on permeability, is adapted for this single-phase dewatering stage:

$$\frac{k}{k_i} = \exp\left[-3c_f(\sigma_{h2} - \sigma_{h1})\right] \tag{4}$$

$$\sigma_{h2} - \sigma_{h1} = \frac{1}{3}\frac{1 + \vartheta}{1 - \vartheta}(p - p_i) \tag{5}$$

where σ_{h1} or σ_{h2} is the stress in the $h1$ or $h2$ direction, MPa.

2.2 Improved Permeability Models below the Critical Desorption Pressure

Once the reservoir pressure has decreased to the critical desorption pressure, the coal matrix starts to shrink due to the desorption of adsorbed CBM from the internal surface of the coal matrix, and additionally the Klinkenberg effect begins to take effect during the seepage flow of free gas in the ultra-low permeable coalbeds. These two effects may eventually counteract the effective stress effect and lead to an increase in permeability. Thus, the models of permeability variation during this process include not only the effective stress effect, but also the matrix shrinkage effect and the Klinkenberg effect.

As a result, the mathematical form of the PM model at this stage should be converted to the following form:

$$\frac{k}{k_i} = \frac{\left(1 + \frac{b}{p}\right)}{\left(1 + \frac{b_c}{p_c}\right)}\left[1 + c_m\left(\frac{p - p_i}{\varphi_i}\right) + \frac{\varepsilon_l}{\varphi_c}\left(\frac{K}{M} - 1\right)\left(\frac{p_c}{P_L + p_c} - \frac{p}{P_L + p}\right)\right]^3 \tag{6}$$

where b is the Klinkenberg coefficient, MPa; b_c is the Klinkenberg coefficient at the critical desorption pressure, MPa; p_c is the critical desorption pressure, MPa; ε_l is the Langmuir-type matrix shrinkage constant, dimensionless; φ_c is the porosity at the critical desorption pressure, dimensionless; K is the bulk modulus, MPa; M is the constrained axial modulus, MPa.

Similarly, the mathematical form of the SD model for the gas-producing stage is converted to the following form:

$$\frac{k}{k_i}=\frac{\left(1+\frac{b}{P}\right)}{\left(1+\frac{b_c}{P_c}\right)}\exp\left[-3c_f\right.$$

$$\left.\times\left[-\frac{\vartheta}{1-\vartheta}(p-p_i)+\frac{E}{3(1-\vartheta)}\varepsilon_l\left(\frac{p_c}{P_L+p_c}-\frac{p}{P_L+p}\right)\right]\right] \quad (7)$$

Additionally, the Seidle *et al.*'s model, which only considers the matrix shrinkage, can be used to compare the above two models:

$$\frac{k}{k_i}=\frac{\left(1+\frac{b}{P}\right)}{\left(1+\frac{b_c}{P_c}\right)}\left[1+\left(1+\frac{2}{\varphi_c}\right)\varepsilon_l\left(\frac{p_c}{P_L+p_c}-\frac{p}{P_L+p}\right)\right]^3 \quad (8)$$

3 RESULTS AND DISCUSSION

3.1 Comparisons of the Improved Models with the Original Ones

The essence of the improved models is piecewise functions with the boundary of the critical desorption pressure to describe permeability variations of unsaturated CBM reservoirs, which is different from the original models. Permeability variations are controlled only by the effective stress effect when the reservoir pressure is above the critical desorption pressure, while its variations are controlled not only by the effective stress effect but also by the matrix shrinkage effect and the Klinkenberg effect when the reservoir pressure is below the critical desorption pressure. In order to distinguish the differences among the new models and the original ones, some work was carried out based on the same input parameters. The major parameters of the coal reservoirs used in the previous studies and chosen for our study are summarized in Table 2.

A plot of the differences in the original and improved PM and SD models for a coalbed reservoir with the same parameters is shown in Figure 2. On the onset of production, the permeability calculated from the improved PM and SD models decreases with the depletion of reservoir pressure. Once reservoir pressure decreases below the critical desorption pressure, the permeability simulated by the improved PM and SD models does not decrease again and maintains a low value. However, since the reservoir pressure further decreases, permeability starts to rebound. Obviously, the rebound rate of permeability in the improved PM model is smaller than that in the SH model, which the rate in the improved SD model approaches. However, in the rebound process, the rate under high pressure in the improved SD model is a bit lower than that in the SH model and the rate under low pressure is a bit higher than that in the SH model. The reason for the difference in those two models is that the SH model only considers the matrix shrinkage effect and the improved SD model considers not only the matrix shrinkage effect but also the effective stress effect and the Klinkenberg effect. Permeability simulated in the original PM and SD models decreases only slightly with depletion of pressure, which cannot reflect the only impact of the effective stress effect in the initial recovery stage of unsaturated CBM reservoirs, because those original models also consider the impact of the matrix shrinkage effect on permeability, which conflicts with the initial production performance of unsaturated CBM reservoirs.

TABLE 2

Summary of parameters and their values used in the models

Parameters	Previous values	Values for this study
Langmuir volume (V_L), (m^3/t)	-	30
Langmuir pressure (P_L), (MPa)	-	2
Gas content (v), (m^3/t)	-	20
Initial reservoir pressure (p_i), (MPa)	-	8
Cleat volume compressibility (c_f), (MPa^{-1})	0.14 [27]	0.140
Initial cleat porosity (φ_i), (dimensionless)	0.001~0.06 [68] 0.01~0.05 [69]	0.008
Poisson (υ), (dimensionless)	0.35 [27] 0.39 [70]	0.35
Young's modulus (E), (MPa)	2 900 [27] 2 070~4 140 [13]	3 000
Langmuir curve-match for shrinkage (ε_l), (dimensionless)	0.00507~0.04389 [71] 0.01264 [72] 0.00754 [13] 0.01075 [39]	0.008
Klinkenberg coefficient (b), (MPa) $b_c = 2b$	~ 0.0528 for 1 mD [73] ~ 0.1269 for 0.1 mD [73] ~ 0.3050 for 0.01 mD [73] ~ 0.7334 for 0.001 mD [73]	0.15

3.2 Sensitivity Analysis of the Improved Permeability Models

3.2.1 Gas Content

Gas content determines gas saturation, affecting the critical desorption pressure in coal reservoirs with similar reservoir pressure and isothermal adsorption properties. A slight unsaturation can necessitate prolonging the dewatering time [57].

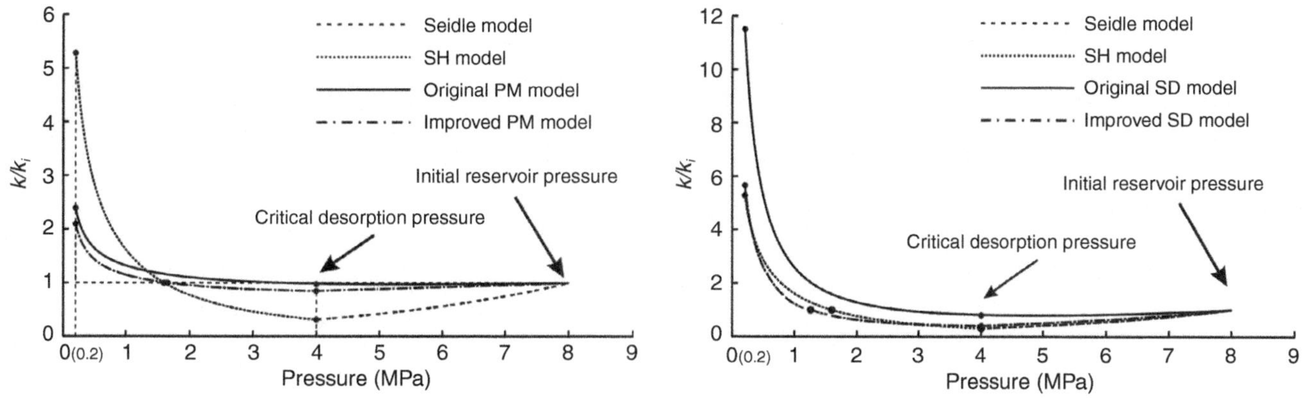

Figure 2

Comparisons of the original and improved models.

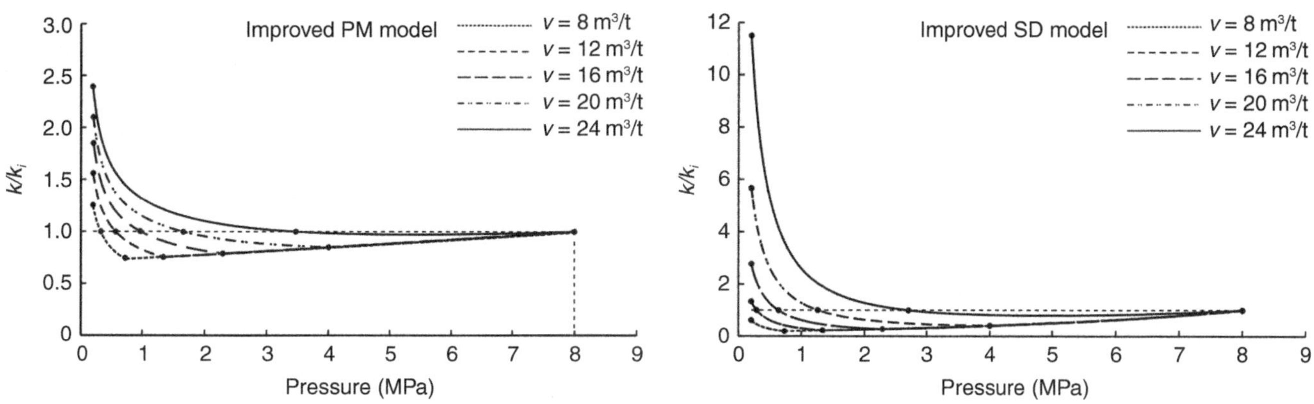

Figure 3

Sensitivity analysis of the improved permeability variation models with gas content.

The dynamic variation of coal seam permeability and its rebound effect are obviously affected by gas content/gas saturation in Figure 3. In the case of the similar other parameters, both the improved PM and SD models with gas content of 8, 12, 16 and 20 m^3/t show that coal seam permeability is damaged only by effective stress in the dewatering stage. The lower the gas content, the more serious the damage of permeability caused by effective stress and the poorer the rebound effect in the late gas-producing stage. If gas content in the coal seam is 24 m^3/t, in other words, gas saturation reaches 100%, permeability does not cause damage or improve significantly in a long time of the initial recovery, because it is affected comprehensively both by the effective stress and coal matrix shrinkage effect. Until the reservoir pressure is further reduced, permeability starts to improve significantly. Overall, high gas content highlights permeability rebound by matrix shrinkage and suppresses permeability damage by effective stress. The sensitivity of various gas contents to permeability in the improved SD model is stronger than that in the improved PM model.

3.2.2 Initial Reservoir Pressure

The characteristic of permeability dynamic variation caused by the initial reservoir pressure in the recovery of unsaturated CBM reservoirs is shown in Figure 4. In the conditions of similar other parameters, both the improved PM model and SD models with initial reservoir pressure of 6, 8, 10 and 12 MPa show that coal reservoir permeability is damaged only by effective stress. The larger the initial reservoir pressure, the more serious the permeability damage and the poorer rebound effect in the late production stage.

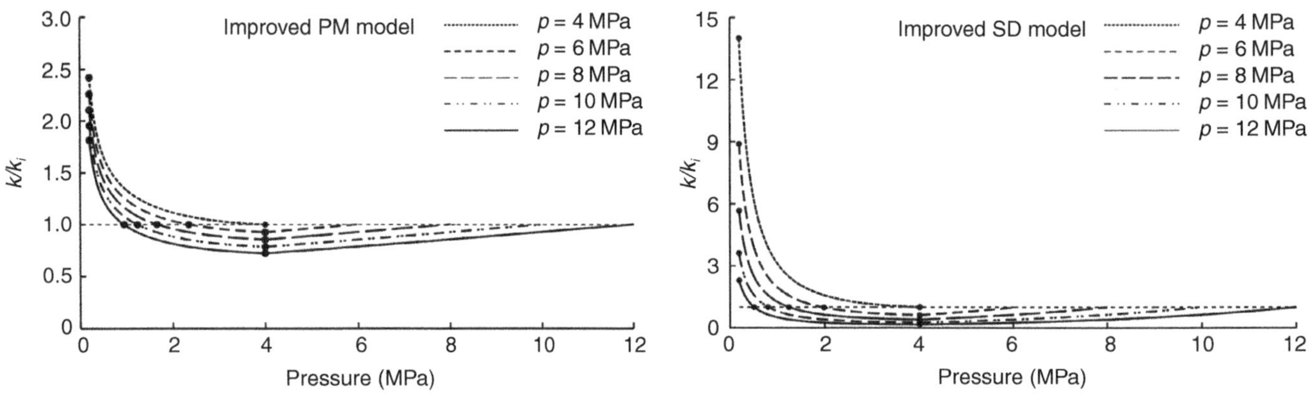

Figure 4

Sensitivity analysis of the improved permeability variation models with reservoir pressure.

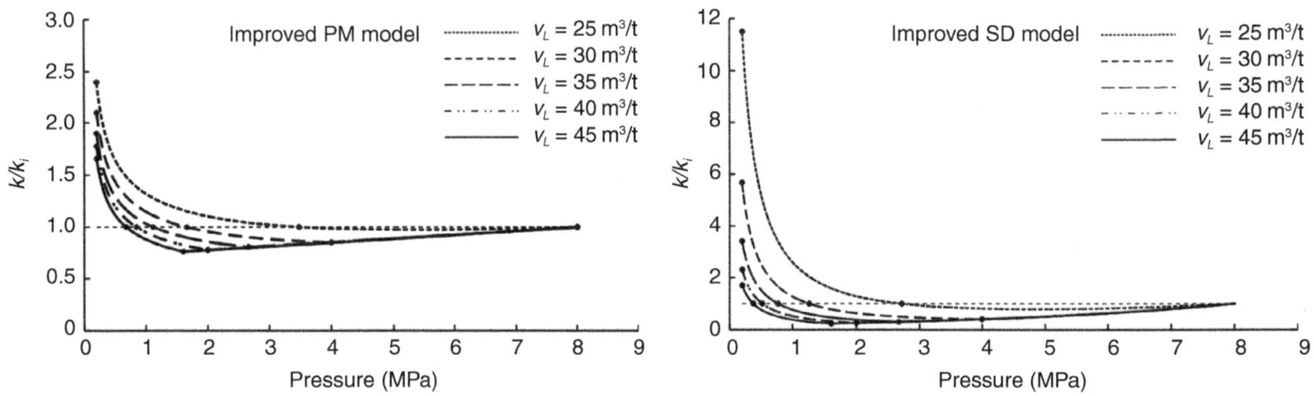

Figure 5

Sensitivity analysis of the improved permeability variation models with Langmuir volume.

The main reason of this phenomenon is that coal seam adsorption ability was controlled by initial coal reservoir pressure under the same other parameters. The higher the initial coal pressure, the greater the adsorption capacity. Therefore, gas saturation in a coal seam with high coal reservoir pressure is obviously low even with the same gas content. Permeability with low gas saturation is often seriously damaged by effective stress in the dewatering stage, which leads to an insignificant rebound of permeability in the late production stage. However, permeability with an initial coal reservoir pressure of 4 MPa (namely, gas saturation of 100%) is almost unaffected by effective stress, and shows that the impact of the coal matrix shrinkage effect and Klinkenberg effect on permeability exceeds that of the effective stress effect. A comparative study proves the sensitivity of the initial coal reservoir pressure to permeability in the improved SD model is greater than that in the improved PM model.

3.2.3 Langmuir Parameters of Adsorption

The coal seam has a strong physical adsorption capacity, and it is usually characterized by the isothermal adsorption curve. During the recovery of CBM reservoirs, CBM is desorbed from the inner face of pores in the coal seam by reducing pressure, which causes coal matrix shrinkage. This shrinkage is the main effect causing the increase in coal permeability with the depletion of pressure. The influences of the Langmuir volume and Langmuir pressure on unsaturated CBM reservoirs' permeability variation are shown in Figures 5 and 6.

The difference in coal reservoir permeability dynamic variation is obvious among Langmuir volumes from 25 to 45 m^3/t, especially in the improved SD model. In general, in the case of the same other parameters, the larger the Langmuir volume, the smaller the gas saturation, and the longer the dewatering stage at the same depressurization rate. Therefore, under the same conditions, a large Langmuir

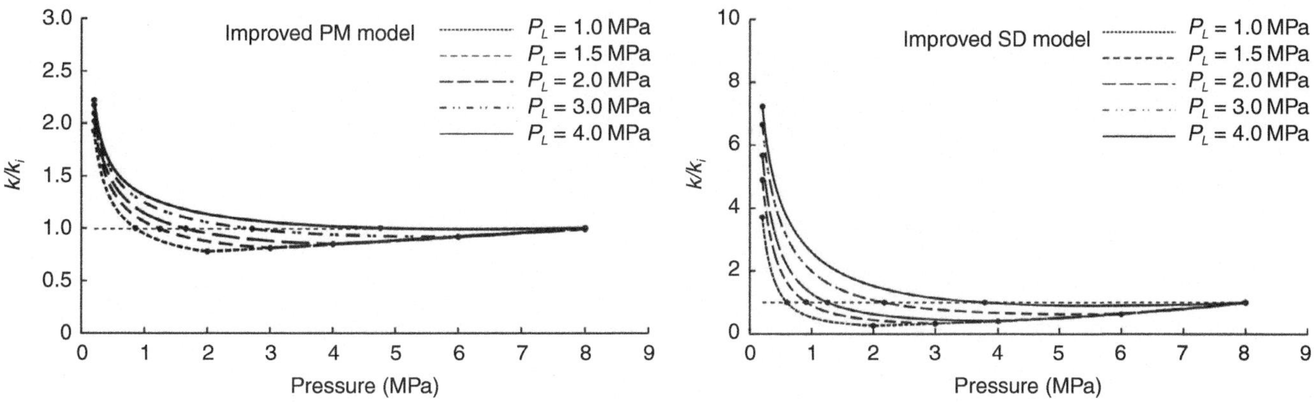

Figure 6

Sensitivity analysis of the improved permeability variation models with Langmuir pressure.

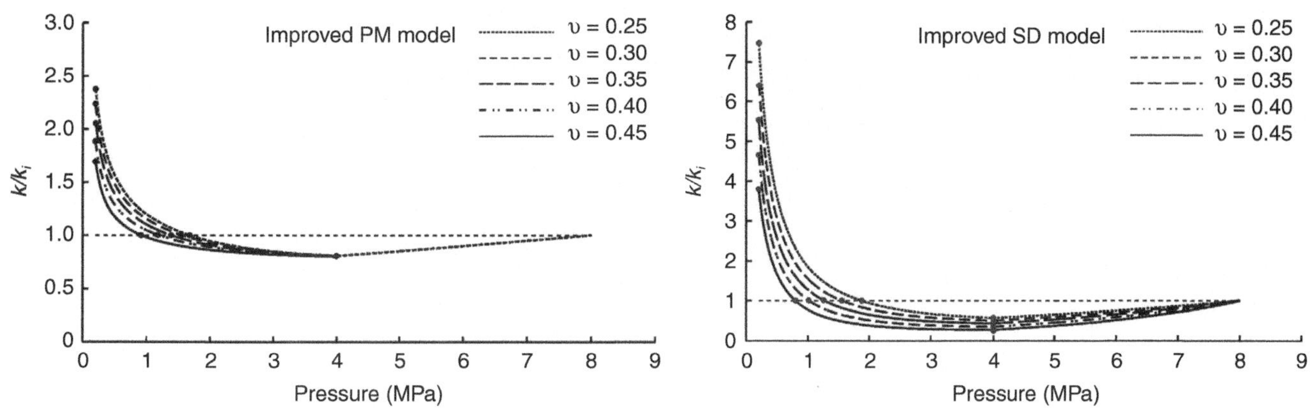

Figure 7

Sensitivity analysis of the improved permeability variation models with Poisson's ratio.

volume is apt to damage permeability by effective stress in the dewatering stage, which is unfavorable for permeability rebound in the late production stage. Coal permeability dynamic variation is also influenced by the Langmuir pressure. In the improved PM model, permeability dynamic variation caused by various Langmuir pressures is obvious in the middle production stage, but permeability remains almost the same in the end with reservoir pressure of 0.2 MPa. In the improved SD model, permeability with a Langmuir pressure of 4 MPa is approximately twice that with a Langmuir pressure of 1 MPa when the reservoir pressure drops to 0.2 MPa.

3.2.4 Poisson's Ratio

Poisson's ratio refers to the ratio of a material's transverse strain and longitudinal strain. As the cleats growing perpendicularly to the horizontal bedding are main flow channels, the increase in effective stress results in the lateral strain increasing and the cleat fissure gradually narrowing during the depletion of pressure, and finally causes permeability reduction. A plot of the impact of various Poisson's ratios on dynamic variation of permeability is shown in Figure 7. In the improved PM model, Poisson's ratio does not take effect in the stage of permeability damage because Equation (2) does not contain Poisson's ratio. Once the reservoir pressure is below the critical desorption pressure, the increase in permeability at high Poisson's ratio is always less than that at low Poisson's ratio. In the improved SD model, the higher the Poisson's ratio, the greater the damage of effective stress, and finally, the worse the ultimate rebound of permeability. At the same time, the ultimate permeability at 0.25 Poisson's ratio is about twice as much as the permeability at 0.45.

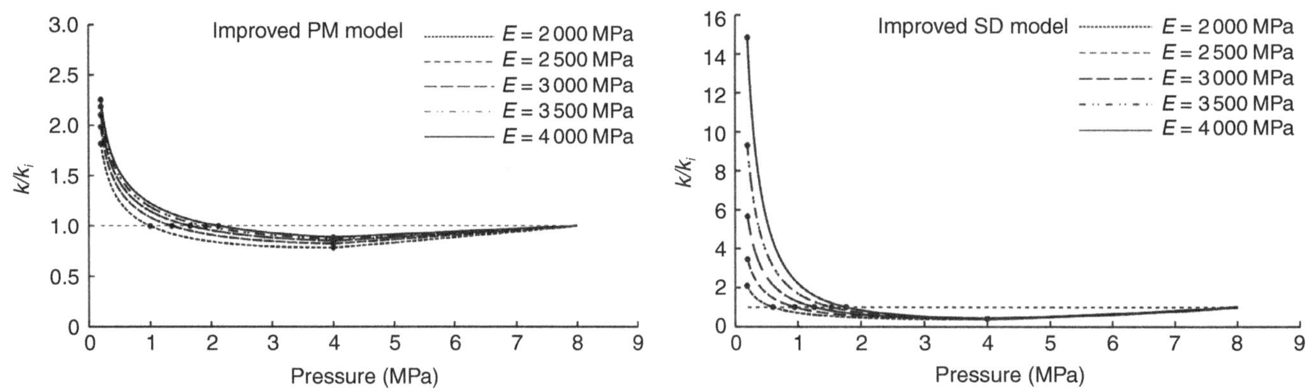

Figure 8
Sensitivity analysis of the improved permeability variation models with Young's modulus.

3.2.5 Young's Modulus

Young's modulus is one of the most important mechanics parameters in coal permeability. In order to compare the effect of Young's modulus on the dynamic variation of permeability in the improved PM and SD models, permeability is calculated under the range of modulus from 2 000 MPa to 4 000 MPa (Fig. 8). Because of the differences between introduction of Young's modulus into the improved PM and SD models, dynamic variation of permeability is obviously different in those two improved models. In the improved PM model, permeability in the whole CBM production process is affected by Young's modulus, but the impact of Young's modulus from 2 000 MPa to 4 000 MPa on the dynamic variation of permeability is limited, especially at the initial stage of recovery. In the improved SD model, the dynamic variation of permeability is slightly affected by Young's modulus when the reservoir pressure is greater than the critical desorption pressure, while it is obviously affected when the reservoir pressure is smaller than the critical desorption pressure. The ultimate improved permeability at 4 000 MPa is more than 7 times that at 2 000 MPa in the improved SD model, indicating that the impact of Young's modulus on permeability increases as the reservoir pressure decreases gradually.

3.2.6 Langmuir Curve-Match for Shrinkage Coefficient

The swelling/shrinkage behavior in coal due to gas adsorption/desorption follows the same form as the Langmuir isotherm, thus explained by a Langmuir-like equation [12, 13, 30]. Since Levine (1996) [13] used a Langmuir form of equation to describe the swelling and achieved good agreement with the experimental measurements, many authors have applied the Langmuir curve-match for shrinkage coefficient to describe the effect of matrix shrinkage on permeability variation [25-27]. In this study, a series of Langmuir curve-match for shrinkage coefficients ranging from 0.004 to 0.020 were applied to calculate the dynamic variation of coalbed reservoirs in the improved PM and SD models (Fig. 9). Shown in Figure 9, these parameters do not lead to dynamic variation of permeability before reservoir pressure decreases to the critical desorption pressure. These parameters start to take effect after the reservoir pressure decreases and reaches the critical desorption pressure and then plays a more and more significant role in permeability variation when the pressure continues to decrease. The ultimate permeability with a shrinkage coefficient of 0.020 is 2.5 times that of permeability with a shrinkage coefficient of 0.004 in the improved PM model. The ultimate permeability with a shrinkage coefficient of 0.012 is 2.5 times that of permeability with shrinkage coefficient of 0.004 in the improved SD model. It indicates that the shrinkage coefficient is an important factor affecting permeability variation as the reservoir pressure decreases.

3.2.7 Klinkenberg Coefficient

The Klinkenberg effect is very obvious in reservoirs with low porosity and low permeability, which is rarely considered in permeability models, even though a coal reservoir is a typical low poro-perm reservoir. In this paper, in order to analyze the characteristics of permeability variation in low-porosity and low-permeability coal reservoirs, the Klinkenberg effect is cited in the improved PM and SD models. According to the study of Tanikawa and Shimamoto (2006) [73], Klinkenberg coefficient b is closely related to the magnitude of permeability and it doubles as the magnitude of permeability decreases each time, which is listed in Table 2.

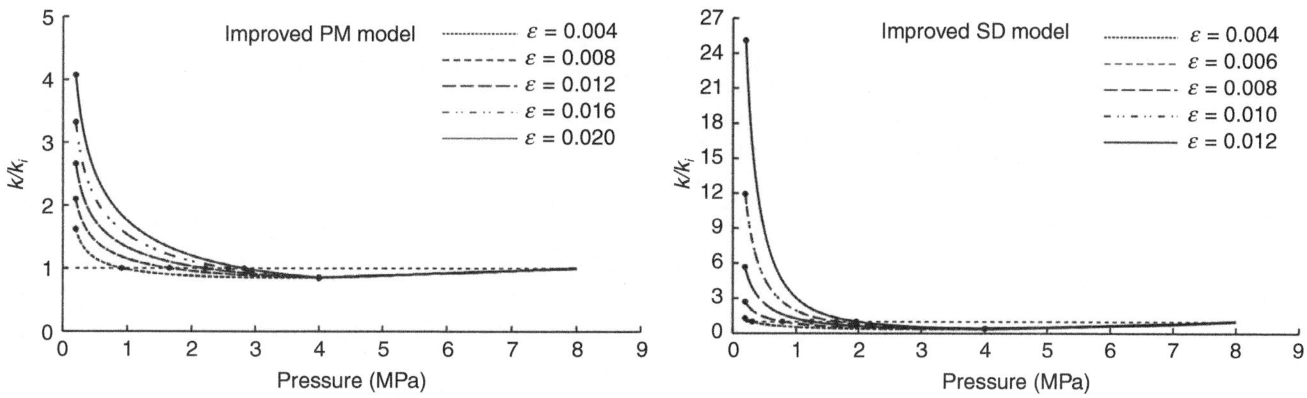

Figure 9

Sensitivity analysis of the improved permeability variation models with the shrinkage coefficient.

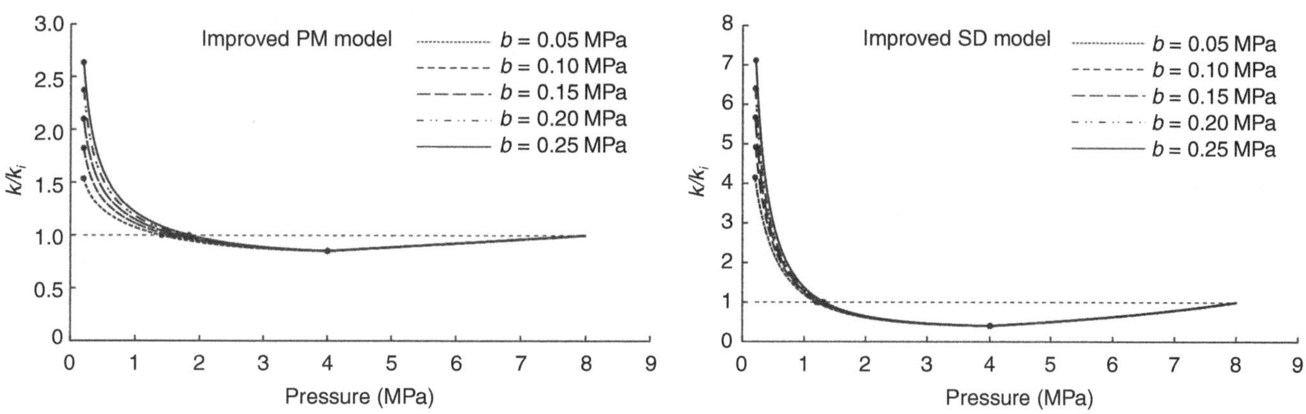

Figure 10

Sensitivity analysis of the improved permeability variation models with the Klinkenberg coefficient.

Thus, the Klinkenberg coefficient at the critical desorption pressure (b_c) was twice the Klinkenberg coefficient b in our paper. Before a CBM reservoir's pressure drops to its critical desorption pressure, especially when the CBM has not been desorbed, the Klinkenberg effect does not exist because there is no flow of gas. The permeability is affected slightly by the Klinkenberg effect most of the time after dropping below the critical desorption pressure. Only when the reservoir pressure drops to 1.5 MPa does the Klinkenberg effect begin to take effect and is highlighted with the decreasing reservoir pressure. From Figure 10, the impact of the Klinkenberg effect on permeability shows no difference between the improved PM model and the improved SD model, which can also be reflected in the formulae of these two models. Reservoir permeability with a slippage coefficient of 0.25 MPa is approximately 1.8 times that with a slippage coefficient of 0.05 MPa.

CONCLUSIONS

Associating the existing permeability variation models with the production performance of unsaturated CBM reservoirs, the improved PM and SD models are established by considering the Klinkenberg effect. The improved models are much more suitable than the original models for describing permeability dynamic variation of unsaturated CBM reservoirs due to gas unsaturation. Sensitivity analysis of various parameters on permeability variation show that these parameters, including the gas content, reservoir pressure, Langmuir volume, Langmuir pressure, Poisson's ratio, Young's modulus, shrinkage coefficient and Klinkenberg coefficient have a significant impact on permeability variation of unsaturated CBM reservoirs. These parameters show obvious different impacts on permeability variation at different production stages. The gas content, reservoir pressure,

Langmuir volume and Langmuir pressure have an impressive impact on permeability variation in the first stage in which the reservoir pressure decreases but does not reach critical adsorption pressure. Except for the above-mentioned parameters, Poisson's ratio, Young's modulus and the shrinkage coefficient have a significant impact in the second stage in which the reservoir pressure is depleted below the critical adsorption pressure. A comparative study of the improved PM and SD models indicates the improved SD model has a greater sensitivity to various parameters than the improved PM model.

ACKNOWLEDGMENTS

This work was financially supported by the Key Project of the National Science & Technology Pillar Program (Grant No. 2011ZX05038-001), the National Natural Science Foundation Project (41272175) and the China Postdoctoral Science Foundation-funded project (2014M561020).

REFERENCES

1 Somerton W.H., Söylemezoglu I.M., Dudley R.C. (1975) Effect of stress on permeability of coal, *International Journal of Rock Mechanics and Mining Sciences & Geomechanics Abstracts* **12**, 5-6, 129-145.

2 Durucan S., Edwards J.S. (1986) The Effect of stress and fracturing on permeability of coal, *Mining Science and Technology* **3**, 205-216.

3 McKee C.R., Bumb A.C., Koenig R.A. (1987) Stress-dependent permeability and porosity of coal, *Proceedings of the 1987 Coalbed Methane Symposium*, Tuscaloosa, Alabama, pp. 183-190.

4 Seidle J.P., Jeansonne M.W., Erickson D.J. (1992) Application of matchstick geometry to stress dependent permeability in coals, *Rocky Mountain Regional Meeting of the Society of Petroleum Engineers*, Casper, Wyoming, 18-21 May, *SPE* 24361.

5 Connell L.D., Lu M., Pan Z. (2010) An analytical coal permeability model for tri-axial strain and stress conditions, *International Journal of Coal Geology* **84**, 2, 103-114.

6 Connell L.D., Pan Z., Lu M., Heryanto D., Camilleri M. (2010) Coal permeability and its behavior with gas desorption, pressure and stress, *SPE Asia Pacific Oil & Gas Conference and Exhibition*, Brisbane, Australia, 18-20 Oct., *SPE Paper* 133915.

7 Jasinge D., Ranjith P.G., Choi S.K. (2010) Effects of effective stress changes on permeability of latrobe valley brown coal, *Fuel* **90**, 1292-1300.

8 Harpalani S., Zhao X. (1989) An investigation of the effect of gas desorption on coal permeability formation, *Proceedings of the 1989 Coalbed Methane Symposium*, Tuscaloosa, Alabama, pp. 57-64.

9 Harpalani S., Schraufnagel R.A. (1990) Shrinkage of coal matrix with release of gas and its impact on permeability of coal, *Fuel* **69**, 551-556.

10 Harpalani S., Chen G.L. (1997) Influence of gas production induced volumetric strain on permeability of coal, *Geotechnical and Geological Engineering* **15**, 4, 303-325.

11 Bodden W.P., Ehrlich R. (1998) Permeability of coals and characteristics of desorption tests: Implications for coalbed methane production, *International Journal of Coal Geology* **35**, 333-347.

12 Chikatamarla L., Cui X., Bustin R.M. (2004) Implications of volumetric swelling/shrinkage of coal in sequestration of acid gases, *Proceedings of the 2004 Coalbed Methane Symposium*, Tuscaloosa, Alabama.

13 Levine J.R. (1996) Model study of the influence of matrix shrinkage on absolute permeability of coal bed reservoirs, in Coalbed Methane and Coal Geology, Gayer R., Harris I. (eds), *Geological Society Special Publication*, **109**, London, pp. 197-212.

14 Robertson E.P., Christiansen R.L. (2006) A permeability model for coal and other fractured, sorptive-elastic media, *SPE Eastern Regional Meeting, Society of Petroleum Engineers*, Society of Petroleum Engineers, Canton, Ohio, USA.

15 Pan Z., Connell L.D. (2007) A theoretical model for gas adsorption-induced coal swelling, *International Journal of Coal Geology* **69**, 243-252.

16 Viete D.R., Ranjith P.G. (2007) The mechanical behavior of coal with respect to CO_2 sequestration in deep coal seams, *Fuel* **86**, 2667-2671.

17 Lin W., Tang G.-Q., Kovscek A.R. (2008) Sorption-induced permeability change of coal during gas-injection processes, *SPE Reservoir Evaluation & Engineering* **11**.

18 Balan H.O., Gumrah F. (2009) Assessment of shrinkage–swelling influences in coal seams using rank-dependent physical coal properties, *International Journal of Coal Geology* **77**, 203-213.

19 Clarkson C.R., Pan Z., Palmer I.D., Harpalani S. (2010) Predicting sorption-induced strain and permeability increase with depletion for coalbed-methane reservoirs, *SPE Journal* **15**, 1, 152-159.

20 Zhu W.C., Liu J., Sheng J.C., Elsworth D. (2007) Analysis of coupled gas flow and deformation process with desorption and Klinkenberg effects in coal seams, *Journal of Rock Mechanics and Mining Sciences* **44**, 971-980.

21 Gray I. (1987) Reservoir engineering in coal seams, part 1—the physical process of gas storage and movement in coal seams, *SPE Reservoir Engineering* **2**, 1, 28-34.

22 Sawyer W.K., Zuber M.D., Kuuskraa V.A. (1987) Using reservoir simulation and field data to define mechanisms controlling coalbed methane production, *Proceedings of the 1987 Coalbed Methane Symposium*, Alabama, 295-307.

23 Seidle J.R., Huitt L.G. (1995) Experimental measurement of coal matrix shrinkage due to gas desorption and implications for cleat permeability increases, *International Meeting on Petroleum Engineering*, Beijing, China.

24 Palmer I., Mansoori J. (1996) How permeability depends on stress and pore pressure in coalbeds, a new model, *SPE Annual Technical Conference and Exhibition*, Denver, Colorado.

25 Palmer I., Mansoori J. (1998) Permeability depends on stress and pore pressure in coalbeds, a new model, *SPE Reservoir Evaluation and Engineering* **1**, 6, 539-544.

26 Shi J.Q., Durucan S. (2004) Drawdown induced changes in permeability of coalbeds: a new interpretation of the reservoir response to primary recovery, *Transport in Porous Media* **56**, 1-16.

27 Shi J.Q., Durucan S. (2004) A numerical simulation study of the Allison Unit CO_2 – ECBM pilot: the effect of matrix shrinkage and swelling on ECBM production and CO_2 injectivity, *Proceedings of the 7th International Conference on Greenhouse Gas Control Technologies*, Sept. 5-9, Vancouver, Canada **1**, 431-442.

28 Shi J.Q., Durucan S. (2005) A model for changes in coalbed permeability during primary and enhanced methane recovery, *SPE Reservoir Evaluation and Engineering* **8**, 4, 291-299.

29 Zhao Y., Hu Y., Wei J., Yang D. (2003) The experimental approach to effective stress law of coal mass by effect of methane, *Transport in Porous Media* **53**, 3, 235-244.

30 Cui X., Bustin R.M. (2005) Volumetric strain associated with methane desorption and its impact on coalbed gas production from deep coal seams, *AAPG Bulletin* **89**, 9, 1181-1202.

31 Cui X., Bustin R.M., Chikatamarla L. (2007) Adsorption-induced coal swelling and stress, implications for methane production and acid gas sequestration into coal seams, *Journal of Geophysical Research-Solid Earth* **112**, B10202.

32 Gu F., Chalaturnyk R.J. (2006) Numerical simulation of stress and strain due to gas sorption/desorption and their effects on *in situ* permeability of coalbeds, *Journal of Petroleum Science and Engineering* **45**, 10, 52-62.

33 Gu F., Chalaturnyk R.J. (2010) Permeability and porosity models considering anisotropy and discontinuity of coalbeds and application in coupled simulation, *Journal of Petroleum Science and Engineering* **74**, 3-4, 113-131.

34 Connell L.D. (2009) Coupled flow and geomechanical processes during gas production from coal seams, *International Journal of Coal Geology* **79**, 1-2, 18-28.

35 Connell L.D., Detournay C. (2009) Coupled flow and geomechanical processes during enhanced coal seam methane recovery through CO_2 sequestration, *International Journal of Coal Geology* **77**, 1-2, 222-233.

36 Wang G.X., Massarotto P., Rudolph V. (2009) An improved permeability model of coal for coalbed methane recovery and CO_2 geosequestration, *International Journal of Coal Geology* **77**, 1-2, 127-136.

37 Liu H.H., Rutqvist J. (2010) A new coal-permeability model, internal swelling stress and fracture-matrix interaction, *Transport in Porous Media* **82**, 1, 157-171.

38 Liu J., Chen Z., Elsworth D., Miao X.X., Mao X.B. (2010) Linking gas-sorption induced changes in coal permeability to directional strains through a modulus reduction ratio, *International Journal of Coal Geology* **83**, 1, 21-30.

39 Liu S., Harpalani S., Mallikarjun P. (2012) Laboratory measurement and modeling of coal permeability with continued methane production: Part 2 – Modeling results, *Fuel* **94**, 117-124.

40 Wu Y., Liu J., Elsworth D., Miao X.X., Mao X.B. (2010) Development of anisotropic permeability during coalbed methane production, *Journal of Natural Gas Science and Engineering* **2**, 4, 197-210.

41 Wei Z., Zhang D. (2010) Coupled fluid-flow and geomechanics for triple-porosity/dual- permeability modeling of coalbed methane recovery, *International Journal of Rock Mechanics and Mining Sciences* **47**, 8, 1242-1253.

42 Izadi G., Wang S., Elsworth D., Liu J., Wu Y., Pone D. (2011) Permeability evolution of fluid-infiltrated coal containing discrete fractures, *International Journal of Coal Geology* **85**, 202-211.

43 Ma Q., Harpalani S., Liu S. (2011) A simplified permeability model for coalbed methane reservoirs based on matchstick strain and constant volume theory, *International Journal of Coal Geology* **85**, 1, 43-48.

44 Pan Z., Connell L.D. (2011) Modelling of anisotropic coal swelling and its impact on permeability behaviour for primary and enhanced coalbed methane recovery, *International Journal of Coal Geology* **85**, 257-267.

45 Rice D.D. (1993) Composition and origins of coalbed gas, *AAPG* **38**, 159-184.

46 Scott A.R. (1993) Composition and origin of coalbed gases from selected basins in the United States, *Proceedings of the 1993 Coalbed Methane Symposium*, Tuscaloosa, Alabama, pp. 207-222.

47 Scott A.R., Kaiser W.R., Ayers W.B. Jr (1994) Thermogenic and secondary biogenic gases, San Juan Basin, Colorado and New Mexico—implications for coalbed gas producibility, *AAPG Bulletin* **78**, 1186-1209.

48 Scott A.R. (2002) Hydrogeologic factors affecting gas content distribution in coal beds, *International Journal of Coal Geology* **50**, 363-387.

49 Diamond W.P., Schatzel S.J. (1998) Measuring the gas content of coal: a review, *International Journal of Coal Geology* **35**, 311-331.

50 Stricker G.D., Flores R.M. (2002) Coalbed methane content in the Powder River Basin, Wyoming: saturation by coal rank and depth, *2002 International Pittsburgh Coal Conference*, Sept. 23-27.

51 Lamarre R. (2006) Under-saturation in coals: How does it happen and why is it important: Search and Discovery Article 40195, Available at: http://www.searchanddiscovery.net/documents/2006/06034/amarre/index.htm.

52 Bustin A.M.M., Bustin R.M. (2008) Coal reservoir saturation: impact of temperature and pressure, *American Association of Petroleum Geologists Bulletin* **92**, 77-86.

53 Gentzis T., Goodarzi F., Cheung F.K., Laggoun-Défargec F. (2008) Coalbed methane producibility from the Mannville coals in Alberta, Canada: A comparison of two areas, *International Journal of Coal Geology* **74**, 237-249.

54 Mares T.E., Moore T.A., Moore C.R. (2009) Uncertainty of gas saturation estimates in a subbituminous coal seam, *International Journal of Coal Geology* **77**, 320-327.

55 Yao Y.B., Liu D.M., Tang D.Z., Huang W.H. (2009) Preliminary evaluation of the coalbed methane production potential and its geological controls in the Weibei Coalfield, Southeastern Ordos Basin, China, *International Journal of Coal Geology* **78**, 1, 1-15.

56 Yao Y.B., Liu D.M., Qiu Y.K. (2013) Variable gas content, saturation, and accumulation characteristics of Weibei coalbed methane pilot-production field in the southeastern Ordos Basin, China, *AAPG Bulletin* **97**, 8, 1371-1393.

57 Pashin J.C. (2010) Variable gas saturation in coalbed methane reservoirs of the Black Warrior Basin: Implications for exploration and production, *International Journal of Coal Geology* **82**, 135-146.

58 Wong S., Macdonald D., Andrei S., Guntera W.D., Denga X., Lawc D., Yed J., Fengd S., Fand Z., Hoe P. (2010) Conceptual economics of full scale enhanced coalbed methane production and CO_2 storage in anthracitic coals at South Qinshui basin, Shanxi, China, *International Journal of Coal Geology* **82**, 280-286.

59 Ayers W.B. (2002) Coalbed gas systems, resources, and production and a review of contrasting cases from the San Juan and Powder River Basins, *AAPG Bulletin* **86**, 1853-1890.

60 Hamelinck C., Faaij A., Turkenburg W., van Bergen F., Pagnier H.J.M., Barzandji O.H.M., Wolf K.-H.A.A., Ruijg G.J. (2002) CO_2 enhanced coalbed methane production in the Netherlands, *Energy* **27**, 7, 647-674.

61 Langmuir I. (1918) The adsorption of gases on plane surfaces of glass, mica and platinum, *Journal of the American Chemical Society* **40**, 1361.

62 Dubinin M.M., Astakhov V.A. (1971) Description of adsorption equilibria of vapors on zeolites over wide ranges of temperature and pressure, *Advances in Chemistry* **102**, 69-85, American Chemical Society Publications, Washington, DC.

63 DeGance A.E. (1992) Multicomponent high-pressure adsorption equilibria on carbon substrates: theory and data, *Fluid Phase Equilibria* **78**, 99-137.

64 Zhou C., Hall F., Gasem K.A.M., Robinson R.L. Jr. (1994) Predicting gas adsorption using two-dimensional equations of state, *I&EC Research* **33**, 1280-1289.

65 Clarkson C.R., Bustin R.M. (2000) Binary gas adsorption/desorption isotherms: Effect of moisture and coal composition upon carbon dioxide selectivity over methane, *International Journal of Coal Geology* **42**, 241-271.

66 Fitzgerald J.E., Pan Z., Sudibandriyo M., Robinson R.L. Jr, Gasem K.A.M. (2005) Adsorption of methane, nitrogen, carbon dioxide and their mixtures on wet Tiffany coal, *Fuel* **84**, 2351-2363.

67 Pan Z.J., Connell L.D. (2009) Comparison of adsorption models in reservoir simulation of enhanced coalbed methane recovery and CO_2 sequestration in coal, *International Journal of Greenhouse Gas Control* **3**, 77-89.

68 Gas Research Institute (GRI) (2002) A Guide to Coalbed Methane Reservoir Engineering, Report GRI-94/ 0397.

69 Mavor M.J., Gunter W.D. (2004) Secondary porosity and permeability of coal: gas composition and pressure, *SPE Annual Technical Conference and Exhibition*, Houston, Texas, USA, 26-29 Sept.

70 Jones S.C. (1972) A rapid accurate unsteady-state klinkenberg permeameter, *SPE Formation Evaluation* **12**, 5, 383-397.

71 Zahner B. (1997) Application of material balance to determine ultimate recovery of a San Juan Fruitland coal wel, *SPE Annual Technical Conference and Exhibition*, 5-8 Oct., San Antonio, Texas, *SPE* 38858.

72 Mavor M.J., Vaughn J.E. (1998) Increasing coal absolute permeability in the San Juan basin Fruitland formation, *SPE Reservoir Evaluation and Engineering* **1**, 3, 201-206.

73 Tanikawa W., Shimamoto T. (2006) Klinkenberg effect for gas permeability and its comparison to water permeability for porous sedimentary rocks, *Hydrology & Earth System Sciences Discussions* **3**, 4, 1315-1338.

74 Klinkenberg L.J. (1941) The permeability of porous media to liquids and gases, *Drilling and production Practice*, American Petroleum Inst., pp. 200-213.

Appendix: Derivation of the Permeability Variation due to the Klinkenberg Effect

According to Klinkenberg (1941) [74], effective gas permeability at a finite pressure is given by:

$$k_g = k_\infty \left(1 + \frac{b}{p}\right) \tag{A-1}$$

where k_g is the gas phase permeability, mD; k_∞ is the absolute permeability under very large gas-phase pressure in which conditions the Klinkenberg effects are negligible, mD; and b is the Klinkenberg factor, dependent on the pore structure of the medium and temperature for a given gas, MPa:

$$b = \frac{4C\lambda p}{r} \tag{A-2}$$

where C is the proportionality factor, dimensionless; λ is the mean free path of gas molecules, m; r is the radius of a capillary or a pore, m.

Permeability variations are due to the Klinkenberg effect when the CBM reservoir pressure varies from p_i to p with the depressurization of CBM reservoirs. Thus:

$$k_i = k_\infty \left(1 + \frac{b_i}{p_i}\right) \tag{A-3}$$

$$k = k_\infty \left(1 + \frac{b}{p}\right) \tag{A-4}$$

and then,

$$\frac{k}{k_i} = \frac{k_\infty \left(1 + \frac{b}{p}\right)}{k_\infty \left(1 + \frac{b_i}{p_i}\right)} = \frac{1 + \frac{b}{p}}{1 + \frac{b_i}{p_i}} \tag{A-5}$$

Considering that the unsaturated CBM reservoirs start to desorb coalbed methane only when the pressure declines to or below the critical desorption pressure, the mathematical form of the permeability variation model has been converted to:

$$\frac{k}{k_i} = \frac{1 + \frac{b}{p}}{1 + \frac{b_c}{p_c}} \tag{A-6}$$

7

CFD Applied to Process Development in the Oil and Gas Industry

L. Raynal*, F. Augier, F. Bazer-Bachi, Y. Haroun and C. Pereira da Fonte

IFP Energies nouvelles, Rond-point de l'échangeur de Solaize, BP 3, 69360 Solaize - France
e-mail: ludovic.raynal@ifpen.fr - frederic.augier@ifpen.fr - frederic.bazer-bachi@ifpen.fr
yacine.haroun@ifpen.fr - claudio.fonte@ifpen.fr

* Corresponding author

Abstract — *Computational Fluid Dynamics (CFD) is increasingly used in the oil and gas industry. The present article aims to show how CFD can be used at all steps of the development of a new process, with a focus on refining technologies. Those different steps consist first of setting up tools that will be used during the development phase, second of obtaining data in complement with experiments required for the process development, and finally, of troubleshooting actions or technology developments that will make the process even more efficient. A large number of applications corresponding to various flow configurations, single-phase or gas-liquid, gas-solid or even gas-liquid-solid, characterised by significantly different scales and requiring adapted simulation approaches, are discussed based on original results and a review of the literature. Perspectives are given in particular on the multi-scale approach and physical phenomena coupling.*

Résumé — La simulation numérique des écoulements appliquée au développement de procédés dans le monde de l'industrie du pétrole et du gaz – une revue — La simulation numérique des écoulements, ou *Computational Fluid Dynamics* (CFD), est un outil de plus en plus employé dans les secteurs pétrolier et gazier. Cet article permet d'illustrer comment l'outil numérique est utilisé lors des différents stades de développement d'un nouveau procédé, en particulier dans le monde du raffinage. Ces différentes phases correspondent tout d'abord à la mise au point des équipements qui seront utilisés pour le développement, puis au complément des essais lors de la phase de développement proprement dite et enfin à des actions support pour la résolution de problèmes opératoires ou l'amélioration de technologies qui permettront d'optimiser le procédé. De nombreuses applications correspondant à différents types d'écoulements, monophasique ou gaz-liquide, gaz-solide et gaz-liquide-solide, caractérisées par des échelles de résolution différentes et requérant des approches de simulations adaptées sont discutées sur la base de résultats originaux et d'une revue de travaux déjà publiés. Des perspectives sont aussi données tant sur les aspects approche multi-échelles que sur le couplage de phénomènes.

ACRONYMS

CAD	Computer-Aided Design
CCR	Continuous Catalytic Reforming
CDU	Crude Distillation Unit
CFD	Computational Fluid Dynamics
CPU	Central Processing Unit
CR	Catalytic Reforming
DNS	Direct Numerical Simulation
FCC	Fluid Catalytic Cracking
FT	Fischer-Tropsch
GPU	Graphics Processing Unit
HCK	HydroCracKing
HDS	HydroDeSulphurisation unit
HDT	HyDroTreating unit
LES	Large-Eddy Simulations
MEA	MonoEthanolAmine
MDEA	Methyl-DiEthanolAmine
MRF	Multiple Reference Frame
PIV	Particle Image Velocimetry
RANS	Reynolds Averaged Navier-Stokes
SGS	SubGrid Scale
TBR	Trickle-Bed Reactor
VDU	Vacuum Distillation Unit
VOF	Volume Of Fluid

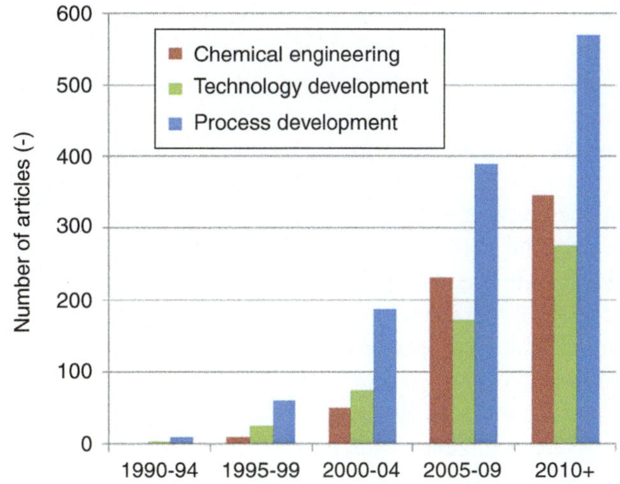

Figure 1

Time evolution of the number of articles including CFD and various keywords in the topic.

INTRODUCTION

Computational Fluid Dynamics (CFD) has been used more and more in the oil and gas industry since the 90s for many purposes. Indeed, since the original expectations of Trambouze (1993) and one of the first reviews on the subject (Kuipers and van Swaaij, 1997), one can see from Figure 1 that the number of scientific papers concerning CFD and chemical reaction engineering is booming. This figure is based on research performed on the World of Science core collection by Thomson Reuters on Nov. 28th 2014. It shows the number of papers containing the word CFD and another keyword (chemical engineering, technology development or process development) in the topic over periods of 5 years. From this, and whatever the keyword used in addition to CFD, it can be considered that the early 90s correspond to the dawn of CFD in chemical reaction engineering, and the late 90s show a real start, the boom being obvious in the 2000s.

CFD is first used to complement experimental data for obtaining results either experimentally difficult to obtain or deduced from sensitivity studies that would be time-consuming experimentally. The objective here is to develop a better comprehension of physical phenomena and to provide further understanding of what happens in chemical reactors. It is also used for designing equipment which has to be linked with Computer-Aided Design (CAD) use in the mechanical industry. More recently, it has been considered as an efficient tool to be used before performing any experiments and of great help for designing the corresponding experimental set-ups. CFD is thus applied at all steps of process development and the present paper aims to illustrate it with an original approach proposing an analogy between process development steps and CFD calculation steps in the particular framework of the oil and gas sector.

The first step in CFD modeling is called pre-processing. It consists of setting up the case, defining the geometry and most importantly, since it is less visible, choosing the adequate boundary conditions and the appropriate mesh. With the increasing Central Processing Unit (CPU) performances and some novel meshing approaches, young research engineers tend to pay less attention to the mesh quality. However, it is well known that a carefully prepared mesh with low skewness factor cells and oriented in the main flow direction will contribute to a more efficient convergence, and that minimum numbers of cells are required in zones with high shear. Boundary condition choice is of even greater importance since it gives the physical meanings to the calculations. The choice is often not obvious since the flow characteristics at inlets (phase distributions, particle sizes, pressure, etc.) are not always well known. They can thus be the subject of sensitivity analysis during the calculation step. Concerning the process development, this step consists of the definition of the tools that will be used for laboratory elementary studies or during the pilot test phase. For example, if one develops a new catalyst material or a new catalytic process, the development and catalyst screening phase would probably be performed on tools with hydrodynamics that significantly differ from those encountered in industrial catalytic reactors. One has consequently to make sure that what will be obtained on the screening tool is meaningful and that there is no artifact, such as a gas-liquid mass transfer limitation, for example, that will affect the catalytic performances. As another example, if one wants to develop mass-transfer relationships in a trickle-bed reactor, one must be sure that the column used for the test is large enough so that wall effects are negligible. Thus, this pre-processing step is the period of time in which one has to invest to make the real development phase more efficient, and it is sometimes neglected.

Step 2 consists of the calculations themselves, which corresponds, in the present analogy, to the process development step. This is the hard part, concentrating most of the work. In CFD simulations, appropriate models (in terms of physical modeling and mathematical solvers) will have to be chosen. Calculations will be performed directly or *via* a multi-scale approach so as to develop better understanding and physical correlations that could be used in process flow sheet simulations or in process design rules. CFD is used as a complementary tool to experiments which require appropriate sensors and can be performed on a laboratory scale or on a large pilot scale. The purpose of step 2 is to develop the appropriate knowledge to design the chemical reactors, and especially the reactive section, and further bring corresponding innovations to the market.

The third step in CFD modeling is called post-processing or post-treatment. It consists of making the most of the calculations; it can either be used to show colourful illustrations or provide quantitative information that would increase the value of simulations. Concerning the process development, the analogy would correspond to diagnostics, trouble-shooting or equipment design. The main objective is a better design and optimised technologies, making the flow as ideal as possible for achieving the desired yields or performances.

This article presents original work and already published results which illustrate these different steps with examples almost all taken from the oil refining industry. Simulations of refining chemical reactors are indeed quite challenging. The range of scales from the reactor main dimensions to the thickness of the liquid diffusion layer on the catalyst is about 10^6 orders of magnitude. In most cases, one has to deal with very complex and unstable multi-phase flows with extreme operating conditions, not always resulting in well-known physical properties and, last but not least, many physical phenomena are involved and coupled with each other: fluid mechanics with strong gas/liquid/solid multi-phase interactions, mass and heat transfer and chemical reactions. As a result, a full CFD simulation of a process is impossible, and may still remain impossible for some years to come, whatever the progress of CPU or GPU computing may be. Simplification is thus required; either by decoupling phenomena or by considering a multi-scale strategy combining different simulations on different scales, and often by doing both.

The following section describes the main refining processes and associated simulation approaches that are considered further in the paper. Section 2 provides examples of CFD use for pre-processing/pre-development studies. Development and post-development case studies are respectively discussed in Sections 3 and 4. The paper ends with a conclusion and tries to show some perspectives in terms of both CFD development and CFD use.

1 REFINING PROCESSES AND CFD APPROACHES

1.1 Refining Processes and Associated Flow Conditions

Crude oil is a very complex mixture of thousands of compounds which need to be separated, upgraded and eventually rearranged before being transformed into valuable products, mostly as intermediate products for the petrochemical industry or final products for the automotive and transportation industry. Petroleum refining is thus based on a smart combination of separation and reactive processes to yield various valuable products which have to comply with more and more severe specifications.

Figure 2 shows a simplified block flow diagram of a typical refinery. Crude oil enters the Crude Distillation Unit (CDU), which consists of an atmospheric distillation column in which vapour and liquid flow countercurrently through internals. Usually, five products are generated from the CDU; namely gas, naphtha, which is usually further split into light and heavy naphtha, kerosene or jet fuel, light straight run gas oil, and the atmospheric residue being further split in the Vacuum Distillation Unit (VDU) into a vacuum gas oil and a vacuum residue. Distillation is sometimes considered of less importance than reactive processes since it does not directly contribute to the heart of the process, that is the expected chemical transformation. However, it is a key elementary separation operation without which the standard distillation scheme would not be operational. Besides, it is interesting to note that, in refineries, distillation columns are far more numerous than reactor vessels. Optimisation of distillation columns can thus significantly contribute to refinery optimisation. Distillation columns are either equipped with trays, or random or structured packings. Packings have become more and more popular, in particular since they offer significant pressure loss reduction. However, for most column diameters, trays are considered of lower

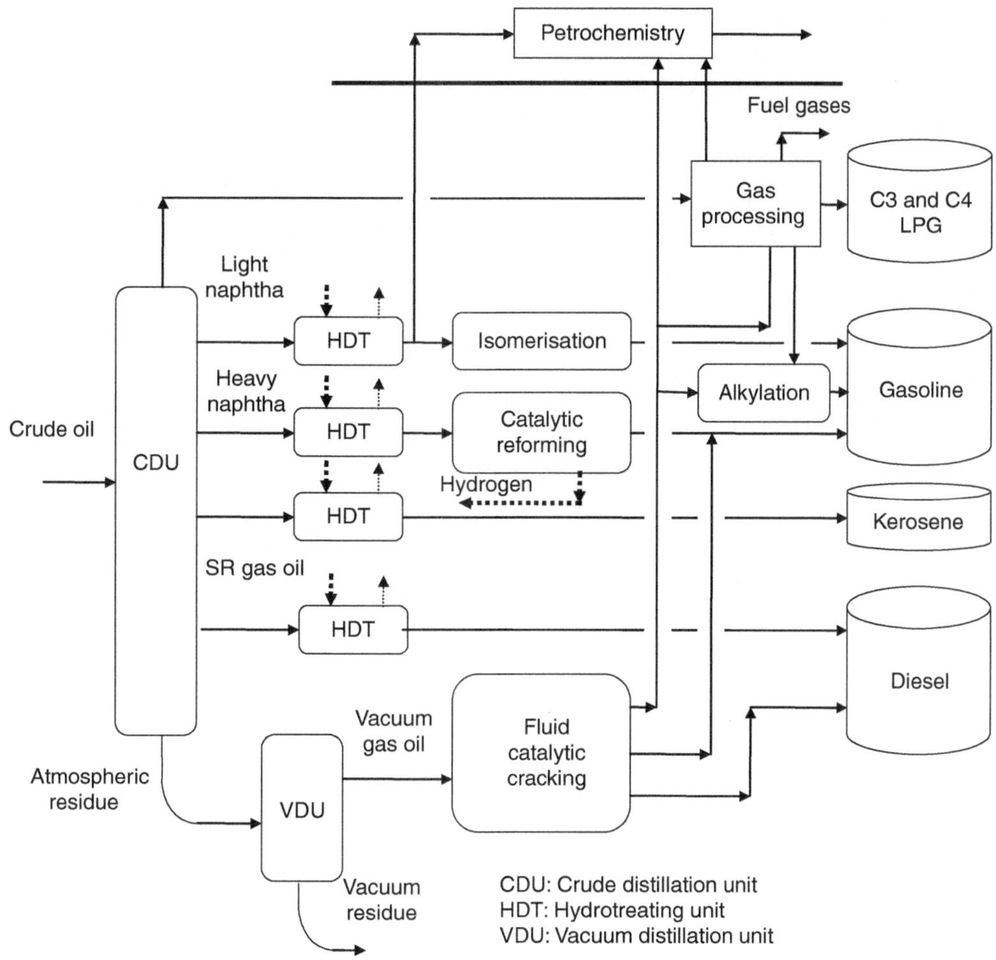

Figure 2

Simplified refining block flow diagram.

cost than packings and, in the case of high-pressure operation (above 6-10 bar), they still offer good separation efficiency while back-mixing may be encountered with packings, further resulting in column efficiency drop and distillation failure. Trays are still of high importance; CFD application to this type of internal is discussed in Section 3.1. Gas-liquid flow in packings has also been investigated, on different scales either for closure law developments, in particular for liquid-side mass transfer parameters, or for pressure drop determination (Haroun *et al.*, 2010; Petre *et al.*, 2003), for studies on a large scale, in particular for distribution device development and liquid dispersion description (Raynal *et al.*, 2013; Fourati *et al.*, 2013), and even for packing geometry optimisation (Luo S.J. *et al.*, 2008). No further details on this type of flow are given in the present paper since it is discussed by Haroun and Raynal (2015).

Most products coming out from the CDU contain impurities, such as sulphur or nitrogen, and need to be hydrotreated. This process operation, represented here as a HyDroTreating unit (HDT), involves fixed-bed reactors with catalysts with the shape of spherical particles and most often of extrudates, a few millimetres in diameter (typically trilobes of 1.2-2.5 mm in diameter). Depending on the operating conditions and the range of boiling temperatures of the oil cuts considered, single or two-phase flows are present in the reactors. In the last case, various hydrodynamic behaviours are possible, namely trickling, bubbles, and pulsed or atomising regimes (Ng and Chu, 1987). However, in practice, two types of fixed-bed reactors are frequently encountered in a refinery: single-phase gas flows and mainly liquid-gas trickling flows. Bubbly flow regimes in fixed-bed reactors are specifically involved in the case of selective hydrogenation units of steam-cracking products, not discussed in the present paper.

Among the most important processes shown in Figure 2, it is interesting to focus on Catalytic Reforming (CR) and Fluid Catalytic Cracking (FCC), which operate with continuous catalyst regeneration and with two very different gas/solid flows. CR is a key refining process whose principal aims are first to transform the naphtha feed into a high-octane reformate for gasoline blending, second to produce high-value aromatics for the petrochemical industry, and finally, to produce hydrogen, a high-value by-product which is used in other processes, in particular in the HDT described above (Antos and Aitani, 2004). The latest technology, called Continuous Catalytic Reforming (CCR), consists of several radial moving bed reactors mounted in series, preceded by furnaces to compensate for the reactions' endothermy, the whole being in connection with a continuous *in situ* regenerator. In each capacity, the millimetric catalyst flows downwards by gravity (velocity ~ 10^{-4} $m.s^{-1}$), while gas naphtha feed flows radially inward (velocity ~ $m.s^{-1}$). FCC is a very complex process made of two reactor zones. In the riser, the heavy oil feedstock is introduced as a liquid, and is atomised and vaporised due to immediate contact with the high-temperature catalyst, in which it further cracks into lighter valuable products, in particular gasoline, but also propylene and diesel. The catalyst is made of solid particles 70 μm in diameter which belong to group A in the Geldart classification; they thus behave as a fluid with strong solid/solid interactions. The catalyst particles are continuously transported from the riser zone to the regenerator, a turbulent fluidised bed reactor where the coke formed during cracking is burnt with air. Once regenerated, it goes back to the riser. The solid entrainment is possible due to the high gas velocities in the riser, which can be up to 20 $m.s^{-1}$.

In parallel to "standard refining", it is worth mentioning the development of the Fischer-Tropsch (FT) process, which, from synthesis gas, produces high-quality diesel and for which a slurry bubble column type of reactor is now considered as the preferred reactor technology (Dry, 2002). This catalytic process, initially developed for producing fuels from coal, has been extensively studied and further developed in the last 20 years. This can be explained in particular by the large reserves of coal and natural gas and the recent use of biomass, which are all growing carbon resources in competition with crude oil and which can be used in the gasification plant upstream of the FT unit. The advantage of the FT process is also the fact that it can provide high-quality diesel fuel with no sulphur, no aromatics and a high cetane number. Since the FT reactions are highly exothermic, it is important to efficiently remove the heat of reaction from the catalyst particles, which is in favour of a slurry reactor technology in comparison with multi-tubular fixed-bed technology. This has motivated the study of slurry bubble columns with the particular objective of extrapolation to very large-diameter columns in order to increase the capacity and reduce costs. The slurry bubble column is the technology used in the recent and huge FT plant in Qatar (Oryx GTL, 2015). The simulation of bubble columns for FT development purposes is also discussed in Section 3.

1.2 CFD Approaches

The purpose of the present paper is not to describe CFD approaches that are used in the examples discussed hereafter. In all cases the required references where simulation details can be found are given. The idea is just to give a quick overview of the simulation approaches underlining the possibility they offer.

With a classification from the less to the more complex flow, one would first describe the case of the reforming process. As previously described, a significant difference in velocity exists in the radial gas and the axial solid flow. The consequence for CFD modeling is that the catalyst may be considered as an immobile material compared with the gas flow, at least if the CFD simulations do not focus on the interactions between the gas flow and catalyst that can provoke catalyst flow disturbances (Ginestra and Jackson, 1985). Depending on the objective, that is either being able to determine local gas flow and heat transfer properties, as performed *e.g.* by Nijemeisland and Dixon (2004), or determining flow behaviour at column scale, as discussed in Section 4.1, one would either consider a local approach with no assumption which can be considered as a Direct Numerical Simulation (DNS) limited to a few number of particles, or a porous-media approach, which makes simulation at column scale possible. In both cases, CFD simulations are thought to be an adequate tool from which quantitative results are expected.

In the case of distillation, once again, two approaches can be considered. The first one would focus on a local scale with a Volume Of Fluid (VOF) or similar approach, that is, an approach with no interpenetrating phases and calculation of the interface. This makes possible the study of the gas-liquid flow behaviour at an orifice. The second approach would consist of Euler/Euler simulations, that is, a macroscopic averaged approach considering phases as interpenetrating media with momentum equations containing interaction terms, the latter requiring adapted closure terms. As discussed in this article, these approaches can provide very interesting results with close agreement with experiments and are thus of great interest for extrapolation purposes. However, they require closure terms, which often result from complementary work not always performed in fully representative conditions.

For trickle-bed reactors, the gas/liquid/flow is even more complex. The gas/liquid flow is superimposed on a fixed bed and simulations must take into account the interactions with the solid in terms of momentum exchanges, but should also integrate reactant and product diffusion in the catalyst and consideration of reaction effects *via* heat sinks or sources.

Due to this complexity, multi-scale approaches and decoupling phenomena must be considered. Local or global approaches are discussed in Section 3.2.

FCC is probably one of the most challenging processes to simulate. Feed atomization and vaporization, reaction and coke deposition, liquid/solid and strongly coupled group A solid particles/gas flow are all complex phenomena, each of them still being not well understood. On top of this, one has to consider a reaction scheme implying a huge number of chemical compounds, which has to be coupled with heat transfers and phase change between solid, gas and liquid. Once again, decoupling and simplification must be considered, which also means that results must be cautiously considered. One would thus conclude that qualitative trends only can be deduced from CFD simulations; however, this is still important, as discussed in Section 3.1.

In most cases, laminar and mostly Reynolds Averaged Navier-Stokes, (RANS) k-ε, turbulent models are respectively used for low or high Reynolds number flows. Since the flows considered are quite far from the well-known reference cases (homogeneous jet, mixing layer, wall boundary layer or wake flow), some model constants, and most frequently, the interaction terms are adapted or tuned *via* a validation step against experimental results. Note that the latter experiments are not always that representative, but at least they are close to what people want to simulate. The *a priori* more robust, for relying on less assumptions, Large-Eddy Simulations (LES) are scarcely used in chemical engineering. This can be explained by two main reasons. First, such an approach requires adapted SubGrid Scale (SGS) models, and most of them have been developed for fully turbulent homogeneous flow only and still require further development for two-phase flows; this is work in progress (Boivin *et al.*, 2000; Deen *et al.*, 2007). Second, this approach requires intensive CPU resources which are not always accessible in the industrial sector, and limit its use to the applied chemical engineering world.

In the following, examples, implying different types of modeling approaches, will be given for all steps encountered in process development.

2 CFD FOR PRE-DEVELOPMENT STUDIES

2.1 Impeller Choice for Proper Stirring

Before being adopted in an industrial process, the catalysts being developed for the refining and petrochemical industries must undergo a laboratory screening step in order to evaluate their intrinsic kinetics. It can either be done with powder or crushed catalyst or with the real shape catalyst.

In the case of powder, which is considered for preliminary catalyst screening tests, it is important to make sure that all the reactor volume is homogeneously filled with the catalyst. To that purpose, Spogis and Nunhez (2009) studied the effect of geometry on solid suspension in stirred tank reactors and were able to make recommendations on operating conditions for proper stirring. It is also possible to consider more complex phenomena such as phase change and solid precipitation (Gavi *et al.*, 2010). CFD developments are quite important in this area since this type of reactor is commonly used in the specialty chemicals sector and has recently gained in importance with the development of biochemistry with issues concerning reactant concentration and shear-stress distributions (Mehmood *et al.*, 2010; Hong *et al.*, 2014).

Once preliminary screening is performed, it is important to check that similar performances are reached with "real" industrial size and shape catalysts. This is discussed in the following section.

2.2 Catalyst Screening Tool

Catalysts in the refining industry are usually manufactured in the shape of beads or extrudates with diameters ranging from 1 to 3 mm for fixed-bed technologies. During the screening step, it is essential to assess the catalyst's intrinsic kinetics in its commercial form, maintaining the spatial distribution of the active sites in the particle, and external and internal mass transfer rates. Only in this way can laboratory measurements be representative of how the catalyst will perform during industrial operation.

Laboratory-scale stirred tank reactors containing a stationary basket have been proposed since the late 1970s for the kinetic testing of commercially shaped catalysts (Mahoney *et al.*, 1978). In these reactors, also known as Robinson-Mahoney reactors, the catalyst is supported and immobilised inside an annular basket manufactured from a metallic net or a perforated metal sheet, and placed inside the reactor (Fig. 3). The gap between the side walls of the annular basket is generally narrow, where the catalyst particles form a shallow packed bed with a length equivalent to ~3-10 particles, from wall to wall. An axial flow impeller is placed in the centre of the annular basket, simultaneously promoting the flow through the basket walls and catalyst bed, and quite strong turbulent mixing of the liquid or gas-liquid mixture bulk. These catalyst testing devices need to fulfil some essential requirements for allowing a true assessment of the reactions' kinetics (Perego and Peratello, 1999): isothermal operation; appropriate flow distribution and blending times; and the absence of interphase mass transfer limitations.

Robinson-Mahoney reactors are known for having a higher performance compared with laboratory fixed-bed reactors, also frequently used for the testing of shaped catalysts. Due to the strong turbulent mixing promoted by the

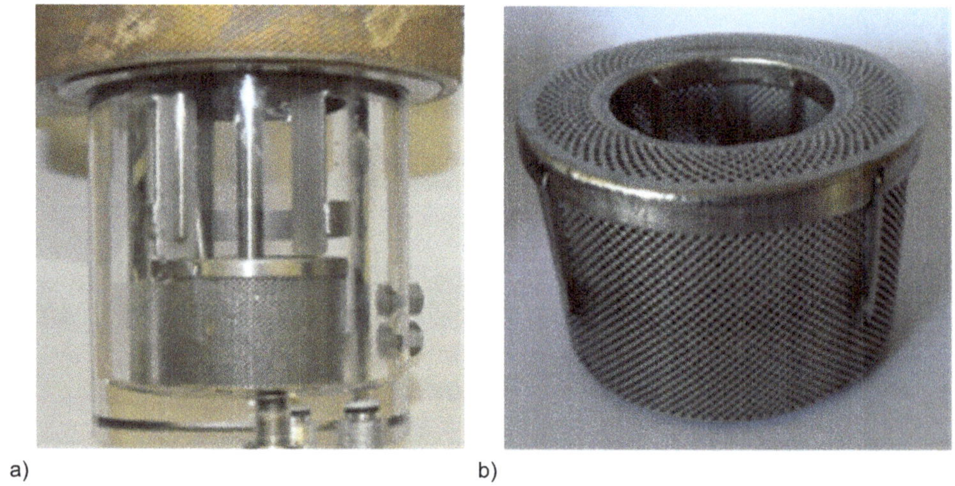

Figure 3

Photographs of a Robinson-Mahoney reactor; a) reactor with a transparent vessel (Braga, 2013), b) stationary catalytic basket.

use of an impeller, mass transfer coefficients between the phases (gas-liquid and solid-liquid) are usually higher (Pitault *et al.*, 2004). Also, due to the strong agitation, these reactors present lower temperature and concentration gradients (Forni, 1997). However, the reactor performance can be affected by operational parameters and the physical properties of the reactants. Geometrical changes to the basket or impeller designs, or the selection of a different catalyst filling, of different shape or size, can also lead to different reactor performances, even when similar operating conditions are chosen. With the continuous development of newer and more efficient catalysts and the growing interest in the use of complex fluids such as heavy crude feedstock or biomass-derived fluids, the catalyst testing and selection task in lab-scale reactors may become hindered by interphase mass transport limitations, external to the particles, or by mixing limitations. In these cases, measuring the mass transfer coefficients prior to the catalyst kinetic assessment is recommended (Pitault *et al.*, 2004).

In spite of the concerns regarding the reliability of the kinetic measurements in more extreme and non-tested conditions, only a reduced number of studies for the characterisation of Robinson-Mahoney-type reactors' hydrodynamics and mass transfer rates has been reported in the scientific literature. The few examples are the works of Pitault *et al.* (2005), Mitrovic *et al.* (2005) and Braga (2013), whose reactor set-ups have similar dimensions but different impellers and basket geometries, and catalyst shapes. This small number of reported results occurs mostly because an extensive experimental study of Robinson-Mahoney reactors is neither practical nor economical when a large number of geometrical or operational parameters need to be studied. As a consequence, the range of applicability of these devices on the kinetic testing of shaped catalysts and the reactor geometry adapted for each application is still not completely known.

Recently, since the pioneering work of Magnico and Fongarland (2006), and later of Santos-Moreau *et al.* (2012), CFD has been providing an important contribution to the study of Robinson-Mahoney reactors. A CFD model of the reactor can indeed provide a detailed tri-dimensional description of the flow field, with special interest in the flow inside the catalytic basket, which is very difficult to visualise experimentally. The numerical results can then be used to estimate mass transfer coefficients and search for operational regimes of kinetic control, where reaction rates can be truly measured. The complexity of the flow in Robinson-Mahoney reactors poses a difficult challenge when developing a computational model. This is mainly due to the existence of flow zones with quite different turbulent intensities – high turbulence in the region close to the impeller, and low turbulence intensity in the outer region of the basket – and to a strong hydrodynamic interaction between the impeller blades and the basket structure. However, the computational strategy being adopted is the development of low computationally-demanding models, both in computation time and machine requirements, but able to describe the most essential aspects of turbulent mixing and the flow in the catalyst porous medium inside the basket. The Multiple Reference Frame (MRF) approach has been chosen to deal with impeller motion, and RANS modeling to describe the turbulence generated in the reactor, allowing running steady-state simulations. The packed bed of catalyst particles is modeled as a homogeneous porous medium, which allows reducing greatly the number of mesh

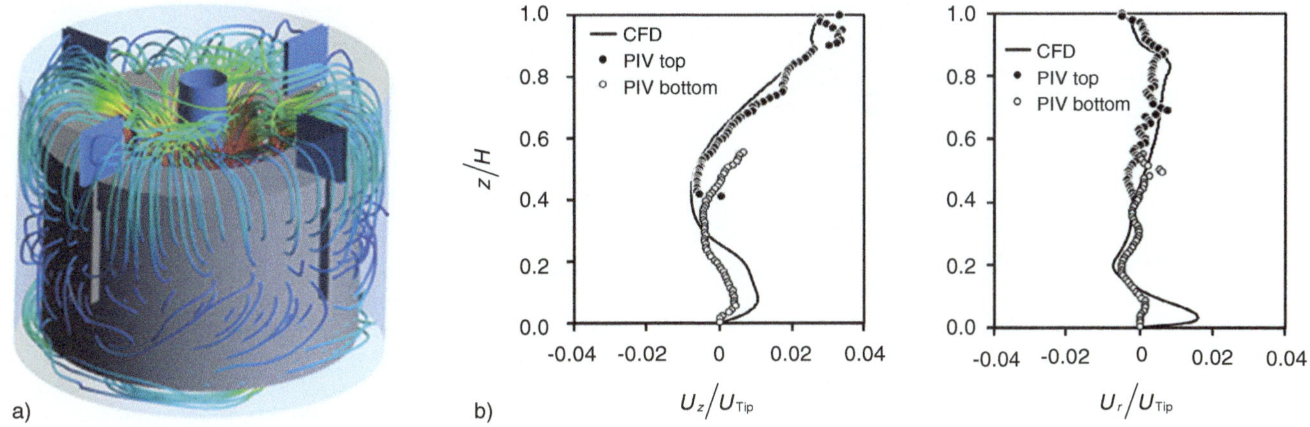

Figure 4

Computational results and model validation; a) time-averaged fluid path lines in the Robinson-Mahoney reactor obtained from CFD, b) comparison between time-averaged velocity profiles obtained from CFD and the PIV results, Braga (2013).

elements necessary to discretise the basket domain. This less computationally-demanding modeling strategy allows obtaining flow results (Fig. 4a) in only a few hours.

In spite of the adopted simplifications, the developed CFD models are able to produce quite satisfactory results. Figure 4b shows, for illustration, the good agreement that can be obtained between time-averaged radial and axial velocity components between the outer basket walls of the catalytic basket and the tank walls obtained from CFD simulations and the PIV experiments of Braga (2013). The axial and radial velocity components, U_z and U_r respectively, are normalised by the speed at the tip of the impeller blades, U_{Tip}. The axial coordinate, z, is normalised by the height of the liquid in the reactor, H.

After experimental validation, these CFD models can then be used as a numerical tool for the design of optimised devices. CFD can be used, as an example, to study the effect of the impeller clearance from the bottom of the stirred tank on the flow field and mass transfer rates (Fig. 5). The results from this parametric study allowed finding the optimal positioning of the impeller in the reactor, avoiding the actual need to build testing prototypes and further experimentation.

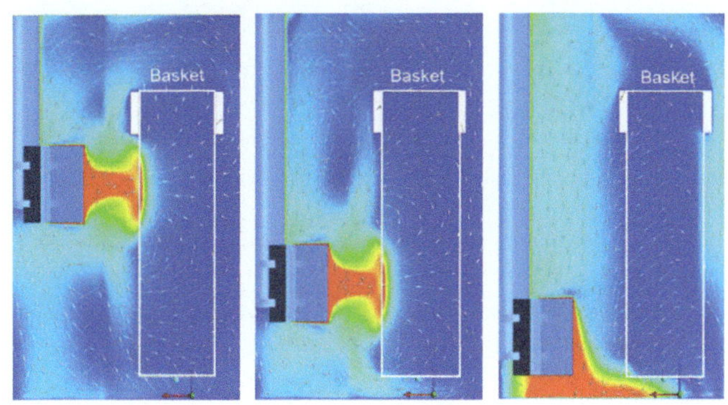

Figure 5

Velocity contour plots and vector maps in the Robinson-Mahoney reactor at different impeller clearances from the bottom of the tank.

2.3 Solvent Screening Tool

Similarly to catalyst screening, it is of interest to discuss solvent screening, which also requires dedicated tools. In spite of not being shown in Figure 1, amine-based gas treatment processes, often combined with other processes, are required downstream of the HydroDeSulphurisation units (HDS – a specialised HDT) for treating the sour gas containing high hydrogen sulphide (H_2S) content, thus limiting sulphur emissions into the atmosphere. Usually, robust standard amine-based processes are used with MEA or MDEA (MonoEthanolAmine and Methyl-DiEthanolAmine) as solvents. With energy efficiency constraints, refiners may benefit from developments performed in the energy sector on post-combustion capture processes where CO_2 contained in the flue gas is absorbed in low-energy-requirement solvents. Such developments need dedicated and precise laboratory tools, in particular thermodynamic cells to obtain Vapour-Liquid equilibrium data and kinetic cells to determine the solvent reactivity. In the case of fast chemical reaction, a wetted-wall column or equivalent string of discs are often considered for kinetic measurements (Chen *et al.*, 2011; Luo X. *et al.*, 2012). In these experimental set-ups, a very thin well-controlled liquid film flows down over a wall of well-known geometry and is surrounded by the acid

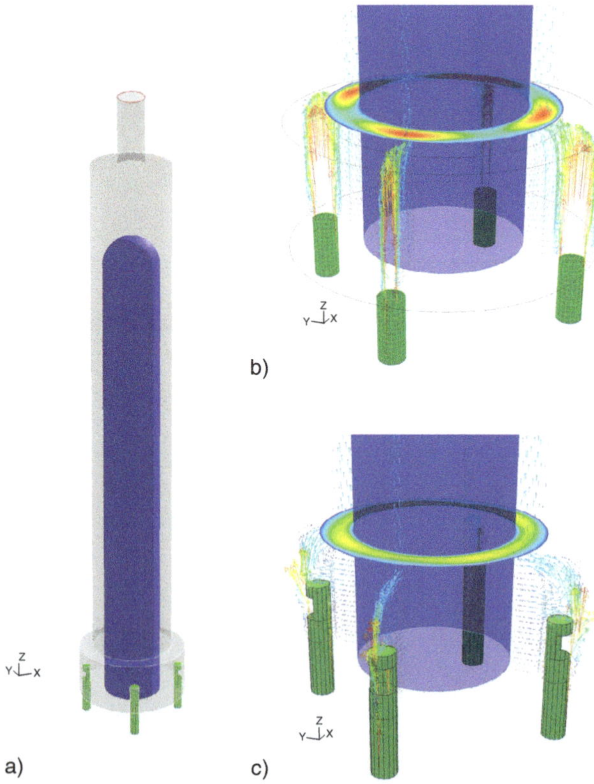

Figure 6

Wetted-wall column flow design; a) general view, b) close view to gas injection – base case, c) optimised injection.

gas in such a way that flow distribution must be perfect with a smooth gas/liquid interface. A few years ago, *IFP Energies nouvelles* (*IFPEN*) designed its own wetted-wall column with the help of CFD concerning special items. As illustrated in Figure 6, the aim of CFD simulations was to check the quality of the gas phase distribution around the central cylinder over which the liquid flows. From simulations, considering various gas injections, it was possible to set a minimum number and type of injection points. For the same number of injection points, two types of injectors were tested. In Figure 6b, one can observe significant non-homogeneities in the velocity contours; in Figure 6c, one can observe a much better flow distribution, although not yet perfect, indicating that four injection points were not enough. Finally, a special multiple injection system, as described by Servia *et al.* (2014), was retained.

Finally, once the catalyst, the separation agent or solvent, has been selected, the most important phase (that is the development step both on the catalytic side and on the reactor design side) has to be addressed. CFD is rarely used today for the catalyst development and for its different production steps (mixing, spraying, impregnation, drying, etc.); the following examples thus focus on the chemical reactor design, showing different types of inputs that can be obtained on different scales.

3 DEVELOPMENT

3.1 Trickle-Bed Reactors

3.1.1 Trickle-Bed Reactors on a Local Scale

As mentioned in Section 1, trickle-bed reactors are frequently used for hydrodesulphurisation, hydrocracking (HCK) and in general for HDT processes of various oil cuts. For decades, catalysts involved in HDT or HCK reactors have been continuously enhanced both in terms of intrinsic reactivity and physical access to active sites throughout the particle volume. At the same time, new catalyst shapes are proposed by licensors, such as trilobes, quadrilobes, asymmetric shapes and so on. Although arguments relative to the catalyst effectiveness are proposed, it is not trivial for the final user to understand the exact effect of the catalyst shape and size on its own process.

Actually, the purpose is complex. It involves several physical phenomena on various scales. On a macroscopic (reactor) scale, catalyst geometry impacts active solid loading, pressure drop and global holdup of different fluid phases. On a meso-scale from tens to thousands of particles, the degree of homogeneity of the product concentrations and temperature may be affected, leading to undesirable hot spots and heterogeneity of particle reactivity. On the particle scale, the shape can modify external interphase transfers and wetting of particles (in the case of trickling flows). Inside the catalyst particles, the diffusion lengths to be travelled by molecules also depend on the particle geometry. Finally, mechanical strength has to be considered, as catalyst particles are generally gravity-loaded several times during their lifetime. Considering the problem from a one-sided point of view may lead to erroneous conclusions (Cooper *et al.*, 1986).

In this context, CFD can be a powerful tool to deal with (multi-)fluid-solid interactions and their consequences on local and global transport properties. Realistic particle loadings can be simulated by mechanical models (Zou, 1995; Augier *et al.*, 2010a) or characterised by non-intrusive analysis such as MRI or X-Tomography, leading to the extraction of usable meshing domains for simulation (Sederman and Gladden, 2001; Larachi *et al.*, 2014). Otherwise, in the case of multi-fluid flows, interface reconstructive models are of special interest. In particular, VOF models are popular as they are easy to use even in complex geometries. They can give information on the wetting efficiency of the catalyst and its dynamics (Heidari and Hashemabadi, 2013).

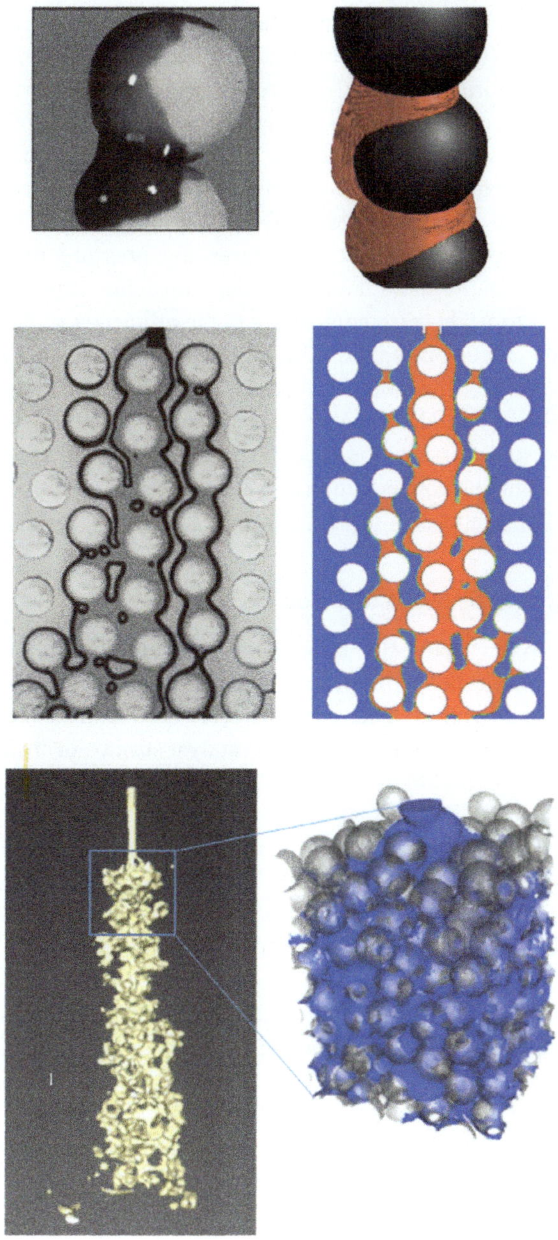

a)

b)

c)

Figure 7

Comparison between experiments (left) and CFD-VOF simulations (right) on different scales: a) on the particle scale (Augier *et al.*, 2010b); b) on the meso-scale for 2D flow (Horgue *et al.*, 2013); c) on the scale of hundreds of particles.

This kind of calculation was initially limited to small calculation domains, of several particles, because of their high CPU usage and being time-consuming. Figure 7 reports different CFD calculations of trickling flows between spherical particles, operated on different scales. A qualitative comparison with experiments is proposed:

– 3-particle-scale 3D simulation (Augier *et al.*, 2010a);
– 36-particle-scale pseudo-2D simulation (Horgue *et al.*, 2013);
– over 300-particle-scale simulation performed in 2014 on a 64-processor cluster.

More quantitative comparisons have also been reported in the cited papers. Similar calculations are possible with more complex shapes of particles. Given the possibility to perform direct simulation on hundreds of particles, statistics and averaging procedures (Whitaker, 1999) can be applied to representative elementary volumes in order to develop rigorous closure laws for macroscopic porous models. Friction forces for pressure drop analysis and apparent viscosity are, for instance, terms that need to be specified in macroscopic models.

Another use of local CFD simulations consists of additionally solving the transport of passive scalars (such as chemical species or temperature) in established flows. Different scalar boundary conditions can be applied at the domain inlet or at particle surfaces in order to calculate axial/transversal dispersion properties or liquid/solid transfer coefficients, respectively (Augier *et al.*, 2010a). As an example, the *L*/*S* mass transfer coefficient ($k_{ls} \times S$) around single particles was calculated in two dimensions for different Reynolds numbers and shapes of particles. Figure 8a shows the results corresponding to 3 different shapes of extruded catalyst particles: cylinder, perfect trilobe and smoothed trilobe. In the same figure, the associated concentration profiles of liquid reactants are also reported to illustrate the differences between particle shapes. One can see the significant differences in the overall transfer coefficient ($k_{ls} \times S$) between shapes because of dead volumes between lobes and the heterogeneity of diffusive film thickness at the particle surface. On a lower scale, CFD solvers can also be used to calculate transport of chemical species by diffusion inside catalyst particles, and their consumption or production by reaction. Associated partial differential equations are solved in complex geometry including particle shape, heterogeneous wetting of catalysts, if any, and phase change inside particles in the case of vaporisation (Kirillov *et al.*, 2002; Bazer-Bachi *et al.*, 2011). In the case of a single reactant, the problem can generally be tackled by one-dimensional approximations, and classical approaches based on the Thiele modulus, Biot or Prater numbers are usable through some adaptation of characteristic lengths. However, in more complex cases, involving several reactants in gas and liquid phases, for example, CFD-based multidimensional approaches have to be used to calculate the effect of transport phenomena on the catalyst effectiveness factor (van Houwelingen *et al.*, 2010; Schwidder and Schnitzlein, 2012).

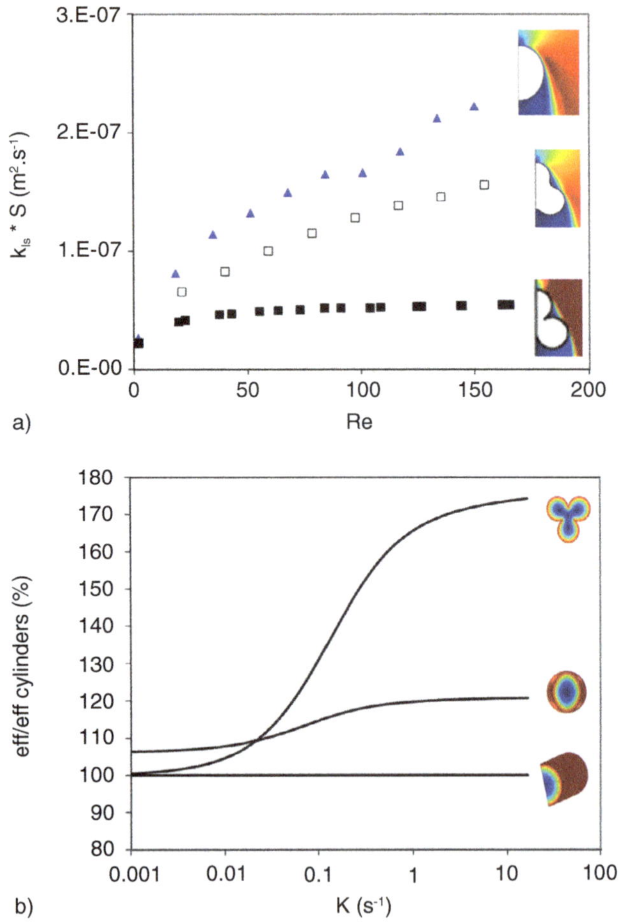

Figure 8

Effect of catalyst shape on: a) liquid-solid mass transfer ($k_{ls} \times S$), and b) global reactor efficiency depending on the kinetic constant K. Results are normalised by the efficiency corresponding to the cylindrical shape.

From the final user's viewpoint, a good catalyst is a more efficient one without increasing pressure drop inside existing reactors. A comparison is proposed between 3 simple catalyst shapes: sphere, cylinder and trilobe, to illustrate the topic. External transfer limitations are neglected and in this simple example, pressure drops are calculated using the model of Tosun, including particle size and shape effects (Tosun, 1984). The 3 considered particle shapes lead to different loading density, pressure drops and diffusion limitations. In Figure 8b, a comparison between overall catalyst efficiencies is presented, calculated by cubic metres of the catalyst bed, at iso-pressure drop, and as a function of the considered 1st-order reaction kinetics K. The higher the reaction rate K, the stronger the diffusion limitation. Calculations are normalised by the efficiency obtained for cylinders. The results show that the best shape depends on the diffusion regime. At low values of K, the Thiele modulus is small (chemical regime); the best shape is the one leading to the more dense catalyst loading, which is the sphere (the solid fraction is up to 66%). Inversely in the diffusion regime, the best shape is the most irregular as it provides more surface at the iso-catalyst volume and thus decreases diffusion lengths inside solids. The lower loading density associated with trilobes decreases the pressure drop and allows one to use smaller particles to reach a targeted pressure drop. This also contributes to the efficiency of catalysts in the diffusion regime. This simple illustration shows how multi-scaled phenomena interact. With the development of massively parallel computing, more realistic fluid dynamics representation may be used in the future, including wetting phenomena, concentration and flow heterogeneities between particles, and thus for different reaction kinetics.

3.1.2 Trickle-Bed Reactors on the Bed Scale

Catalytic Trickle-Bed Reactors (TBR) consist of a fixed bed of solid catalyst particles contacted by a co-current gas-liquid downflow. The reaction occurs between the dissolved gas and the liquid phase at the catalyst surface or inside the catalyst pellet. In the last decade, the increasing environmental standards of fuels have led to the development of high-performance units with improved catalyst activities. However, an efficient catalyst relies on a perfect liquid distribution over the catalyst bed's cross-section. Thus, the hydrodynamic distribution is of prime importance in HDT/HDS operations, especially due to the increased demand on catalyst performance. Indeed, the uniform distribution of gas and liquid above the catalyst is very important to achieve the optimum performance of the catalyst.

Hydrodynamic distribution study inside the catalytic bed is often carried out in a laboratory column of moderate size (diameter lower than 1 m, since it is quite difficult to perform experimental tests on very large-diameter columns) with ideal horizontal trays and model system fluids (water/air). However, in the industrial reactors of several meters in diameter, trays are often not perfectly levelled (Maiti and Nigam, 2007; Alvarez *et al.*, 2007; Bazer-Bachi *et al.*, 2013) because of their weight or because of column deviation from verticality. In this case, the horizontality default of the distributor tray becomes an important factor that must be carefully considered. Indeed, when the distribution of liquid reactants is not perfect on catalyst beds, preferential liquid flows appear, causing radial temperature gradients and loss of conversion. This phenomenon leads to increasing the temperature at the inlet of reactors, with associated negative consequences concerning cocking and deactivation of catalysts.

In this context, CFD simulation has become a very interesting tool to study the effect of maldistribution on reactor efficiency and to ensure the scale-up to industrial reactor size.

As discussed above, in recent years, many interesting CFD approaches have been considered to solve the two-phase flow hydrodynamics in trickle-bed reactors. For the bed-scale simulation, the Eulerian two-fluid approach appears to be very interesting since the real physical geometry of the bed is represented by an effective porous medium, which leads to significantly reduced calculation time when compared with interface capturing methods (such as VOF or Level Set methods). The Eulerian two-fluid approach is based on an averaging method of local gas and liquid mass and momentum conservation equations, the packed bed being represented as an "effective porous medium" (Wang *et al.*, 2013; Solomenko *et al.* 2015). Since these models do not directly simulate the flow over the real physical geometry, one should deal with a closure problem where interactions with the solid surface of particles, as well as fluid-fluid interactions, should be accounted for through specific closure laws. Recently, Solomenko *et al.* (2015) proposed adapted closure laws for fluid-solid interactions, gas-liquid interaction, and dispersion mechanisms that contribute to the liquid dispersion inside the catalyst bed, based on capillary and mechanical forces as respectively proposed by Attou and Ferschneider (2000) and Lappalainen *et al.* (2009). The authors show that these dispersion mechanisms should be taken into account in the simulations to reproduce gas-liquid hydrodynamic flow in TBR well. They validated the CFD model upon various experimental data of liquid distribution obtained by Marcandelli *et al.* (2000) and Boyer *et al.* (2005). Figures 9 and 10 show simulation results obtained by Solomenko *et al.* (2015) in order to evaluate CFD modeling. The simulation results were compared with the experimental data of Marcandelli *et al.* (2000). The experimental work consists of liquid distribution in a 0.3 m-diameter and 1.3 m-high column packed with 2-mm glass beads using air and water. The gas phase was fed through four 25-mm-internal-diameter chimneys, while liquid was injected through two 2.5-mm-internal-diameter orifices in the co-current mode. The liquid flow rate distribution in the bed was measured using 9 collectors of equal surface at the bottom of the column. Simulations were carried out considering a 3D grid that represents half of the column. The simulation domain was chosen so that the symmetry of the flow about plane *xy* is fulfilled. Gas and liquid, flowing in the co-current mode, were introduced into the domain through velocity inlet conditions and the gauge pressure was specified at the trickle-bed outlet. The superficial velocities of gas and liquid used for simulation are, respectively, 3.10^{-3} m.s^{-1} and 5.10^{-3} m.s^{-1} (Marcandelli *et al.*, 2000). As discussed in Solomenko *et al.* (2015), relatively small time steps, varying between 1.10^{-5} s and 1.10^{-2} s, were required for this simulation.

Figure 9 shows liquid saturation contours in the simulated trickle-bed for the configuration of two liquid injection points considered in Marcandelli *et al.* (2000) experiments. The plot shows the liquid spreading along the column. The dissymmetry of liquid patterns in the top of the column is gradually attenuated at lower positions inside the trickle-bed.

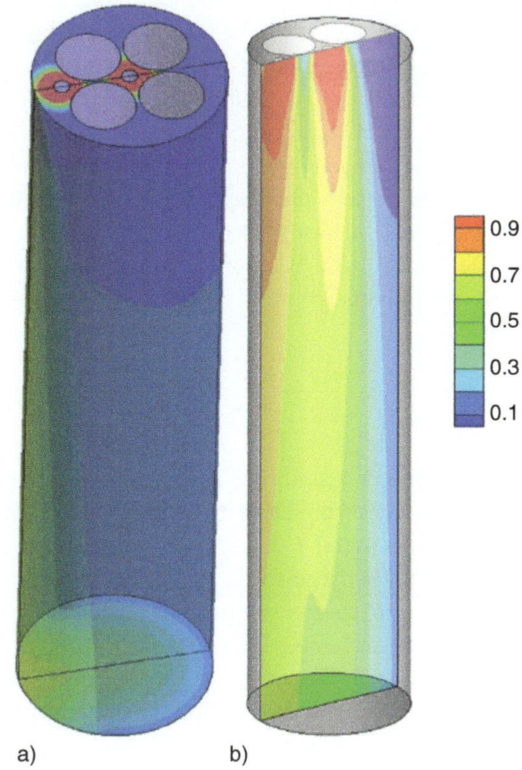

Figure 9

Contours of liquid saturation obtained by simulation in the case of Marcandelli *et al.* (2000) experiment. a) On the entire domain, b) on the symmetry plane, Solomenko *et al.* (2015).

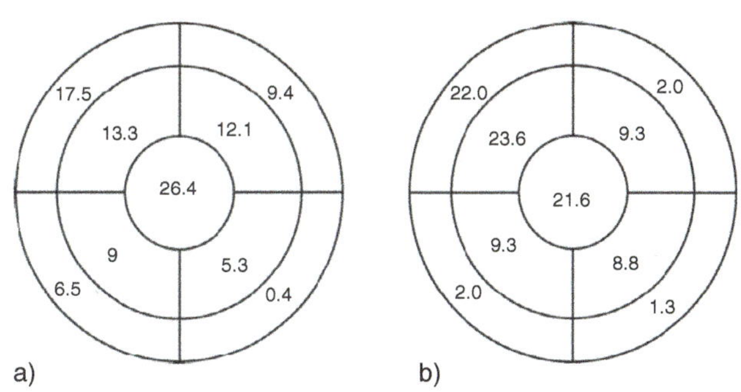

Figure 10

Liquid distribution at the bottom of the trickle-bed in terms of % of total liquid flow rate. a) Experimental results (Marcandelli *et al.*, 2000), b) simulation results, Solomenko *et al.* (2015).

Figure 10 shows, respectively, the experimental and the CFD liquid distribution at the bed outlet; the ratio of the liquid flow rate in each collector to the total liquid flow rate is given. One can observe very good agreement between the experimental and numerical data, bearing in mind that relative errors on measured liquid flow rates are evaluated between 8% and 10%. Moreover, the simulation predicts a global maldistribution factor (as defined in Eq. 1) of 27%, which is quite close to that reported by Marcandelli *et al.* (2000), whose value is 23%:

$$M_f = \sqrt{\frac{1}{N(N-1)}\sum_i \left(\frac{Q_i - Q_{mean}}{Q_{mean}}\right)^2} \qquad (1)$$

These studies show that the bed-scale liquid distribution in TBR can indeed be well predicted by CFD. Such CFD modeling can thus be very useful to understand the impact of maldistribution on reactor efficiency well, and to ensure the scale-up of TBRs to industrial size. For further studies, such as hydrodynamic, simulations can be coupled with chemical species transport and chemical reaction to relate the quality of distribution to the catalytic performance and thus define the design criteria for distribution technologies better.

3.2 Bubbly Flows

As discussed in Section 1, bubbly flow is mostly encountered in distillation columns equipped with distillation trays. Sieve trays or valve trays are preferred to old bubble-cap trays, but in all cases the flow configuration, at least just above the tray, is mainly of churn turbulent bubbly flow type. This type of flow is also encountered in many other chemical processes and deserves significant attention within the framework of the FT process. In both cases illustrated hereafter, the Euler/Euler approach is considered for modeling, the bubbles being the dispersed phase in the liquid phase.

3.2.1 Case of Distillation Column Trays

CFD can help to understand the flow configuration and characteristics better, especially in terms of liquid holdup and flow dispersion. Among the numerous articles dealing with that subject, the paper by Krishna *et al.* (1999a) can be considered as a pioneering work. The authors showed that, provided an appropriate drag law is selected for the interphase momentum exchange term, the Euler/Euler model is appropriate to determine the clear liquid height above the tray. The calculations were performed on a simplified rectangular tray and the results were compared with experimental data obtained on a set-up with similar geometry. The clear liquid height determined by CFD was found to be overestimated in comparison with experimental data. However, its trends as a function of the liquid flow rate, the gas flow rate and the weir height are in good agreement with what was experimentally observed. It was thus concluded that the use of the CFD tool should be amplified. Among several other papers from the same group, from the University of Amsterdam, the article by Krishna and van Baten (2003) is of particular interest since first, the geometry can be considered as more realistic, the column shape being circular, and more importantly, because a comparison is made between configurations with and without catalyst containers placed on the tray with application to reactive distillation. Figure 11 shows two snapshots of the flow with a view from above, the liquid flowing from left to right; the colour map corresponds to liquid holdup values and the arrows to liquid velocity vectors. Such a simulation shows the quite chaotic character of the flow, with local flow recirculation patterns and local flow directions somewhat different from the mean flow. This behaviour shows that a simple 1D analysis of the flow would not be sufficient to describe the liquid phase flow. The study illustrates the important differences observed in terms of flow structure and liquid holdup between the cases with and without gas injection. Moreover, it shows that liquid residence time calculation is possible, which is of high importance when a catalytic reaction takes place. More recently, it has been shown that CFD could also be used for the design and optimisation of tray orifices. Indeed, Li *et al.* (2014) performed calculations to study the effects of an original fixed-valve tray geometry by combining the use of Euler/Euler and VOF approaches. Comparison with experimental measurements for clear liquid height determination is made, and a detailed description of the flow at the valve is presented, which would be very difficult to perform experimentally.

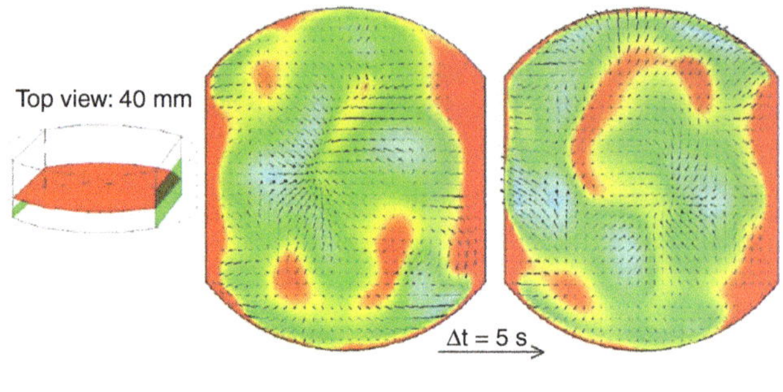

Figure 11

Snapshots of the liquid volume fraction contour, Krishna and van Baten (2003).

3.2.2 Case of Bubble Columns for Fischer-Tropsch Process Development

More than 20 years have been required to be able to match CFD simulations with the reference measurement made by

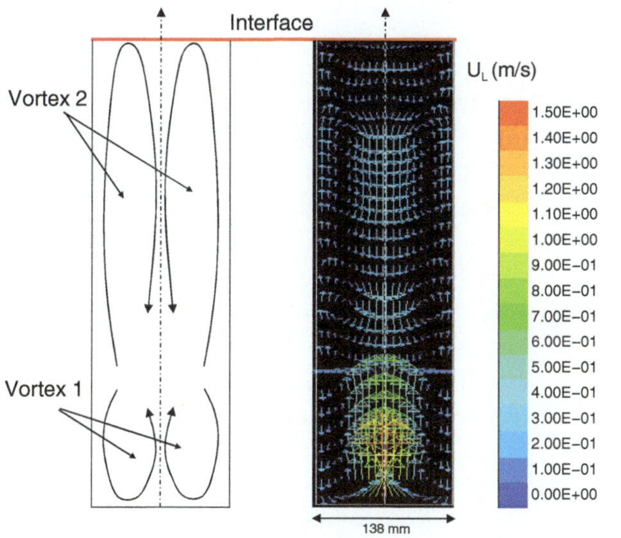

Figure 12

Sketch of the two-phase flow inside a bubble column before steady state is reached with corresponding liquid velocity vectors.

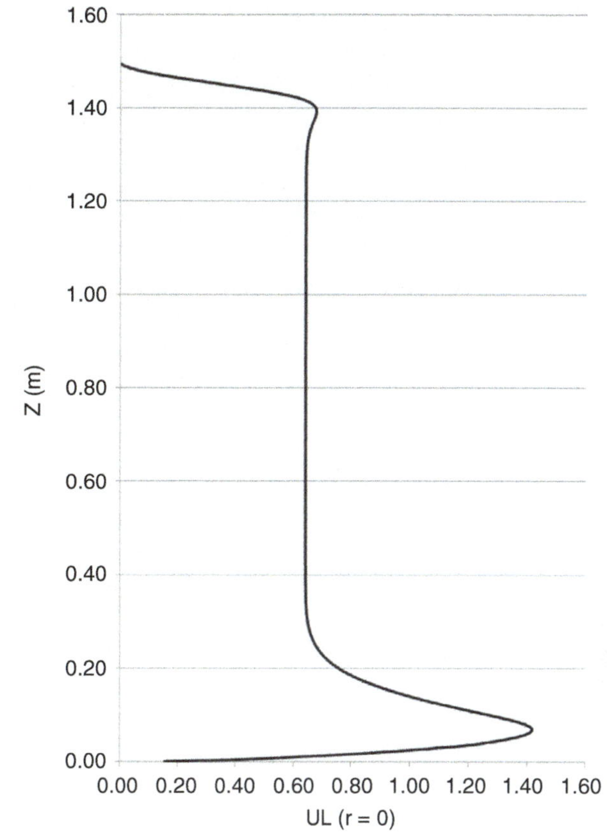

Figure 13

Liquid velocity profile along the axis – D = 138 mm, V_{SG} = 16.9 cm/s.

Hills in 1974 (Hills, 1974). Among the first studies, Lappin and Lübbert (1994) and Boisson and Malin (1996) showed that it was possible to be able to obtain the main flow behaviour, that is a large liquid recirculation pattern in the shape of an elongated torus with a maximum liquid velocity along the central axis. This liquid recirculation is induced by ascending bubbles, since bubble drag corresponds to a transfer of momentum from the gas to the liquid. Satisfactory agreement was obtained between simulations and experiments for the radial void fraction and radial liquid velocity profiles. However, some difficulties have to be faced. First of all, simulations are quite unstable and to obtain the expected recirculation pattern, one has to proceed to long unsteady calculations before a steady state is reached. During the non-stationary process, several recirculation cells, or large-scale vortices, can form, making the calculation procedure long and complex if no simplifications are assumed; such as, for example, gas injection on a central portion only, as discussed by Raynal *et al.* (2001). As an illustration, Figure 12 shows both a simplified flow pattern with two vortices and the corresponding liquid velocity vectors. Second, one needs adapted closure laws, both on gas/liquid interaction terms and on the bubble diameter. Boisson and Malin (1996) showed that, in this special gas-induced flow, the drag law is prominent when compared with other interactions (lift, added mass or Basset effects), and Raynal (2005) discussed the sensitivity toward the bubble diameter for a given set of drag laws and showed that the latter could not be neglected but was not too important. Indeed, an increase of 25% in the bubble size results in a decrease of the centreline liquid velocity along the axis of only 8%. Once steady state is reached, the axial velocity profile, as can be seen in Figure 13, shows a region of fully established flow with a plateau value for the liquid velocity. This is this latter centreline velocity, V_{CL}, which is compared with experiments. Despite the good agreement between simulations and experiments observed on small-diameter columns, here of 138 mm as used by Hills (1974), data for large diameters was lacking; the process development indeed aims at designing columns of diameters closer to 10 m. Several studies have been dedicated to this extrapolation issue, with an important input from the group of Professor Krishna. Krishna *et al.* (1999b, 2000) discussed the large number of existing correlations for the centreline liquid velocity with a huge discrepancy observed when applied to large diameters above 5 m, ranging from 1 to almost 5 m.s^{-1}. The potential of CFD was to discriminate between the existing correlations or possibly propose a new one. Simulations were performed for diameters up to 0.63 m, which made possible a first choice among all correlations identified in the literature. Two correlations were considered since they were found to be in better agreement with CFD and with some experimental measurements. However, since some discrepancies were still observed, it was mentioned that further experimental work was needed on larger-diameter columns to confirm the strong scale dependence observed with the Euler/Euler simulations. This work was completed by van Baten and Krishna (2004) with simulations up to 10 m; 3D simulations were shown to be more representative of the flow, in particular in terms of void fraction radial distribution, little discrepancy being observed regarding liquid velocity. This is in good agreement with what was later obtained by Chen *et al.* (2011). From experimental measurements performed for diameters up to 1 m and CFD simulations performed for diameters up to 10 m, the correlation proposed by Zehner (1986) was chosen to be the most adapted. Similar simulations were performed at *IFPEN* with a 2D approach for diameters ranging from 0.15 up to 5 m and for high gas velocities ranging from 2 to 20 $cm.s^{-1}$ (Raynal, 2001). The objective was quite similar to the one described above; that is, performing CFD simulations to identify the most adequate correlation that would be used for design purposes (heat exchange transfer coefficient determination, catalyst attrition conditions, vibration calculations, etc.). The results were compared with correlations suggested by Krishna *et al.* (2000) and also with some other results obtained in the group of Prof. Miyauchi, who had the opportunity to perform measurements of gas holdup, liquid velocity and mass transfer in an industrial column of 5.5 m in diameter and 14 m in height (Koide *et al.*, 1979; Kojima *et al.*, 1980; Kataoka *et al.*, 1979). Among those correlations, one was preferred to give the reference value for the liquid centreline velocity, V_{CL_ref}. Despite these numerical results, it was considered that the risk of extrapolating to diameters up to 10 m was too high for making safe designs and, as a consequence, original measurements were performed at *IFPEN* on large-diameter columns. The experimental data obtained by Forret *et al.* (2003) on a 1-m-diameter column and by Schweitzer (2003) on a 3-m-diameter column are shown in Figure 14 with a comparison with the *a priori* CFD-chosen reference correlation. Normalised liquid centreline velocity values are plotted *versus* the gas superficial velocity. One can observe that the agreement is not fully satisfactory; ideally all points should be on the $y = 1$ line, and more reasonably equally distributed between the dashed lines, corresponding to a ±20% discrepancy. It is, however, observed that, in most cases, the reference correlation underpredicts the measured values. Nonetheless, most experimental points deviate from the reference correlation by less than +30%, the average deviation being +19% for the $D = 3$ m case, the "newly obtained" experimental values being especially well represented by the CFD-based reference at high gas superficial velocities, which confirms the potential of such an approach for extrapolation purposes.

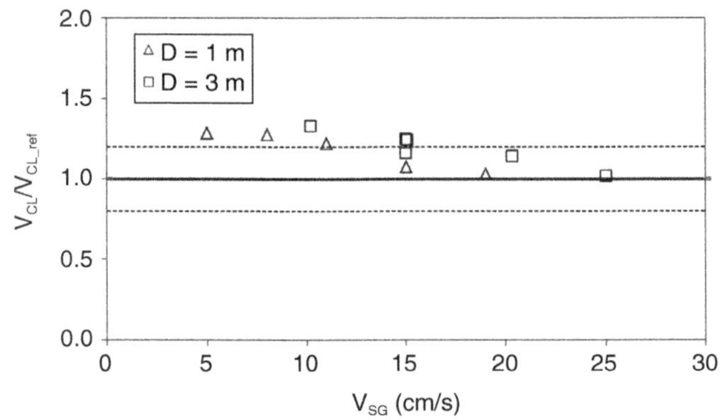

Figure 14

Comparison between predicted and experimental values for the liquid velocity profile along the axis for large-diameter columns, normalised liquid centreline velocity *versus* superficial gas velocity. The dashed lines correspond to 0.8 and 1.2.

The scale-up in diameter is a first important step to go beyond; however, other parameters are also of importance, namely physical properties of fluids, such as gas density, liquid viscosity and surface tension, which have been partially addressed by Krishna and van Baten (2001, 2002). In fact, those influences are mainly imposed *via* bubble size, which can be considered as mono-dispersed (work performed at *IFPEN*) bi-dispersed (work performed at the University of Amsterdam) or poly-dispersed (Chen *et al.*, 2005). However, to develop and implement coalescence and breakup complex models, such as those described by Olmos *et al.* (2001) and Sanyal *et al.* (2005), further information is required. Indeed, such models use closure laws, or kernels, based on theoretical approaches, as given by Luo H. and Svendsen (1996), for example, or on experiments, which are rather scarce for high void fraction bubbly turbulent flows.

Another difficulty to surmount is the influence of the heat exchanger that will have to be inside the vessel to control the reactor temperature. Larachi *et al.* (2006) performed a first study on that difficult aspect. They showed that the presence of bundles of vertical heat-exchanger tubes could significantly affect the well-known liquid circulation pattern. Different internal geometries were tested and, in some cases, the ascending central structure was found not to exist any longer; on the contrary, a central downward flow was observed. Since quite different and not fully explained results were obtained depending on the choice of heat-exchanger tube configurations, one can consider that this subject is still open for further studies.

From this very complex example, it can be seen that work is still needed, first on a local scale to develop a better comprehension and modeling of bubble interaction and properties. Second, Euler/Euler approaches also need to be further developed to be able to consider internals within gas-liquid bubbly flow. It is believed that there will be many other applications which will drive such developments, in particular due to the recent development of biomass use for large-scale industrial processes for which gas-liquid flow modeling will face similar challenges to those that FT development is facing (Jiang *et al.*, 2014).

3.3 FCC Gas-Solid Flow

As described in Section 1, FCC is a very important process for the refinery with a high complexity, and some similarities with gas bubbles in liquid discussed in the preceding section. In both cases, a dispersed phase strongly interacts with a continuous phase in such a way that the Euler/Euler approach with a two-way coupling is well adapted. In the FCC riser a huge number of solid particles are interacting with a hydrocarbon mixture which, due to low or moderate pressure and high temperature conditions, vaporises and cracks to form lighter compounds and behaves almost like a gaseous phase. Professor Gidaspow proposed an original approach for modeling the solid as a fluid *via* a kinetic approach and a granular temperature concept (Gidaspow, 1994), which has been seen to be successful for global flow patterns in risers (Neri and Gidaspow, 2000; Jiradilok *et al.*, 2006). This approach is different from what can be obtained with a multi-scale approach as used by van der Hoef *et al.* (2004), but it can be seen as an efficient approach to perform more applied process design calculations. Indeed, it makes it possible to consider not only mass transfer (Chalermsinsuwan *et al.*, 2009a, b) but also cracking chemical reactions (Lopes *et al.*, 2011). Lopes *et al.* (2011) used a relatively simple lumped kinetic scheme considering only four main compounds (gas oil, gasoline, coke and light gases) and were able to obtain close agreement with data obtained by Derouin *et al.* (1997), as can be seen in Figure 15a. The reaction kinetics modeling can of course not be as precise as that which can be done in more standard 1D chemical engineering reactor modeling (Gupta *et al.*, 2007), which considers a larger number of pseudo-components with a more detailed kinetic mechanism scheme. However, and as can be seen in Figure 15b, CFD offers the opportunity to study some 3D effects which can be important, for example, for predicting local hot spots whose occurrences depend on some injection configurations that can be optimised.

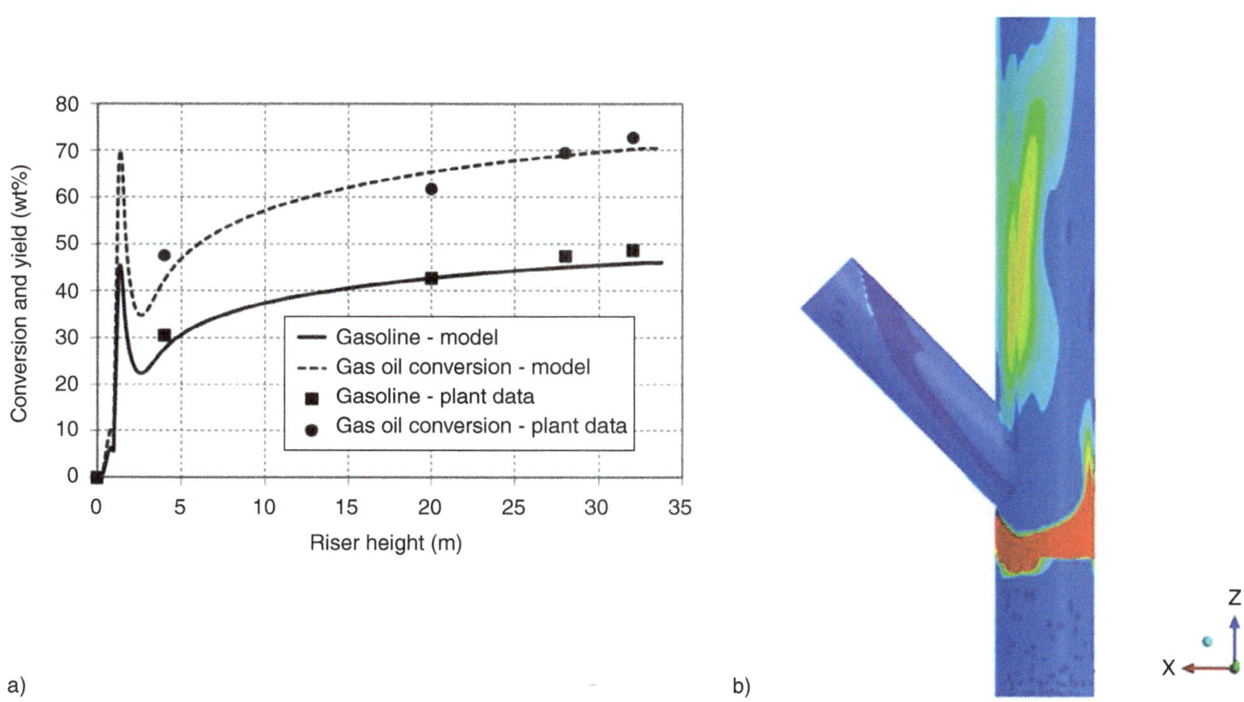

Figure 15

FCC riser simulations by Lopes *et al.* (2011). a) Axial conversion profile, comparison between CFD and experimental results, and b) temperature contours at the bottom of the riser.

Note that, even if CFD simulations are mostly focused on risers, some deal with regenerators with a focus on coke combustion modeling (Moula *et al.*, 2013). Such developments are important since the regenerator is of high capital cost and its optimisation could be addressed by CFD simulations in terms of catalyst and air injection systems and associated mixing.

4 DIAGNOSTICS, TROUBLE-SHOOTING AND EQUIPMENT DESIGN

In many cases, process development assumes perfect flow distribution and reaction section designs are made accordingly. However, it is well known from feedback from the field that industrial plant operation needs technical support to understand phenomena which were not expected; hot spots observed with thermocouples in fixed bed reactors, local high pressure drop, bad separation performances, etc. Another important role of CFD is to solve these operational issues by providing new technologies that make possible an operation as close as possible to that expected. This role is increasingly considered during the development step. However, in some cases, the demand arises only once the problem is observed in the industrial plant and must be solved as quickly as possible. Hopefully, the required time for CFD studies' start-up is very short. In this section, three examples are given; they illustrate how CFD can be used for reactor internal development for optimal flow and temperature distribution.

4.1 Gas Solid Flow in a Catalytic Reformer

Although CCR could appear to be a mature refining process, constant improvements are still required in order to optimise catalysts and reactor/regenerator internals. The role of CFD is illustrated by an example consisting of the optimisation of the oxychlorination internals in CCR regenerators.

The Continuous Catalytic naphtha Reforming process implements a reaction zone with typically four reactors and a catalyst regeneration zone with several stages: a combustion stage, an oxychlorination stage followed by a calcination stage, and finally, a reduction stage (Antos and Aitani, 2004).

In the regenerator, the catalyst circulates in a downward direction. In the bottom of this regenerator, a calcination gas is injected under the calcination annular zone, generally at the periphery using external baffles. Then, it goes through the calcination moving bed in countercurrent compared with the catalyst. This gas is then mixed with an oxychlorination gas, directly in the moving bed at the bottom of the oxychlorination section: the mixture crosses the oxychlorination zone, again in countercurrent. The role of the oxychlorination gas is to redisperse the metal crystallite particles, by the addition of chlorine.

For old regenerators, both calcination and oxychlorination gases were injected at the periphery of the annular zones, which generates a non-homogeneous velocity profile at the bottom of both the calcination and oxychlorination zones. Moreover, this particular mode of injection can cause a bad mixing of the calcination and oxychlorination gases, because of the poor radial mixing properties of catalytic moving beds. Some solutions were proposed to mix both gases outside the reactor by removing the calcination gas before the oxychlorination zone (Greenwood, 1987). Although some other technologies have recently been developed to have a better mixing between the two gases, such as different internal injection systems (Capelle *et al.*, 2000), there is still a challenge to:

- correctly mix calcination and oxychlorination gases in order to reach the desired radial repartition of chlorine in the oxychlorination zone;
- avoid any disturbance in the downward catalyst flow;
- do all these steps inside the regenerator, for a minimum size of the whole device.

Both hydrodynamic and transport models are standardised ones, and can be developed easily. The model used for characterising the hydrodynamics, called a Brinkman-Forchheimer model, is a modification of the laminar Navier-Stokes equations in order to take into account the friction between the liquid and the catalyst particles (Augier *et al.*, 2008). These hydrodynamic equations are solved coupled with the transport equations of chemical species, which consider the radial dispersion of the components in the porous media (Peclet numbers are evaluated using the correlation of Guedes de Carvalho and Delgado, 2001).

Using this CFD model, a new internal was recently proposed to enhance the chlorine distribution in the oxychlorination zone (Fig. 16). A compact chamber was added between the oxychlorination and calcination zones, with an internal space located between two plates. A plurality of legs passes through the chamber to allow the downward movement of the catalyst. The calcination gas crosses the chamber using another series of tubes. The oxychlorination gas is introduced using an injection line in the internal space of the chamber, and orifices around the tubes enable the mixing of this oxychlorination gas with the calcination gas in the tubes, before their injection into the oxychlorination zone. A significant number of options is available to optimise this chamber: density of tubes and legs, means of injection of the mixing gas into the oxychlorination zone, size of the elements, position and size of the orifices, etc. The developed CFD model allows one to study:

- the behavior of the gases in the chamber (Fig. 17 of a hydrodynamic model);

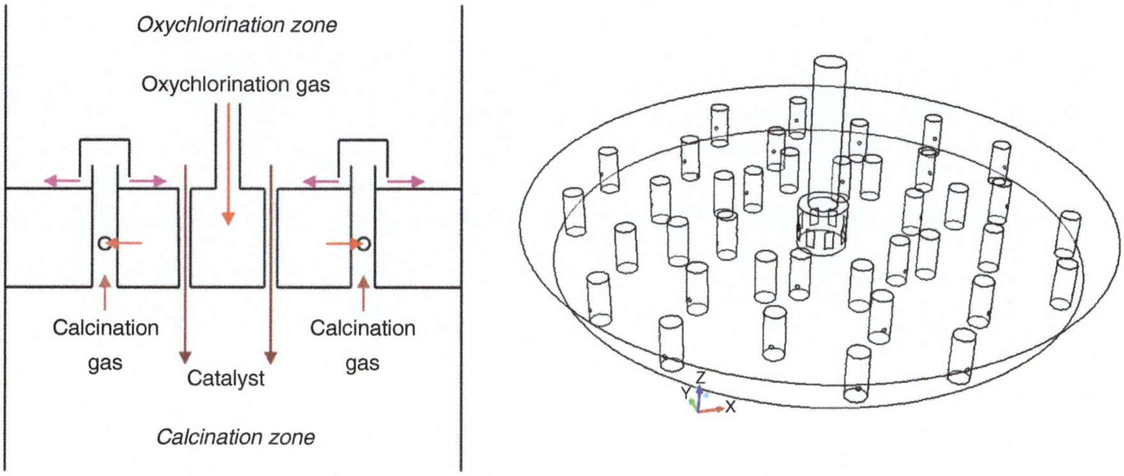

Figure 16

2D sketch and corresponding 3D view of the distributing device in the regenerating section of a catalytic reformer, Decoodt *et al.* (2014).

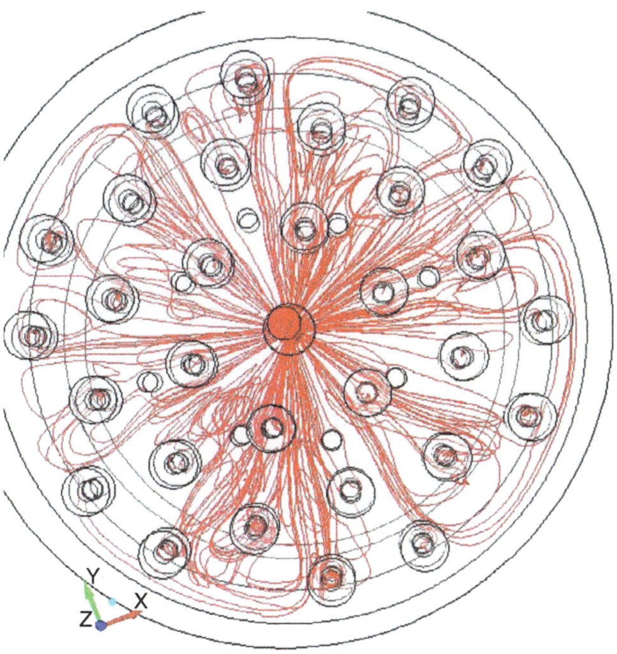

Figure 17

Top view of the streamlines of the oxychlorination gas entering the mixing chamber.

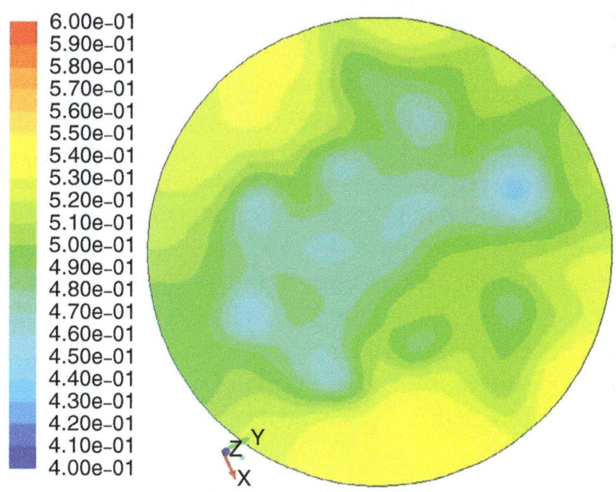

Figure 18

Chlorine concentration contours 10 cm above the mixing chamber, in the oxychlorination zone.

- the injection of the mixing gas into the oxychlorination bed and its radial dispersion above the internal (hydrodynamic and transport model using a chlorine tracer, Fig. 18).

As an example, the area of catalyst legs must be sufficiently large to prevent any fluidization of the catalyst in the legs, *i.e.* by reducing the local gas velocities in these legs, also considering that the bypass of calcination gas in the legs must be avoided by creating an adequate pressure drop. This modeling approach produced results for the optimization of this chamber in a few weeks. The choice of geometric parameters enables a perfect distribution of the oxychlorination gas streamlines in the chamber (Fig. 17), leading to an improved repartition of this gas in the mixing tubes (±1% difference in the flow rate). As a consequence, chlorine was quasi-equally distributed in the above oxychlorination bed, the small differences being due to the bypass of the calcination gas in the catalyst legs (Fig. 18). Other simulations have shown that a perfect distribution of chlorine

(±1% difference of concentration) was reached only 30 cm above the chamber.

This new internal was recently implemented in several industrial units, greatly enhancing the radial temperature profiles in the oxychlorination bed, and extending the catalyst lifetime.

This clearly demonstrates the potential of CFD for the development of internals: these simple models generated remarkable results with an immediate impact in the industrial units, at a reasonable development cost.

4.2 Internals for Hydrotreating Units

As described in Section 1, hydrotreatment is required, first to end with products in compliance with quality specifications, and second because sulphur, nitrogen and metals act as poisons for most catalysts that will be used in processes downstream of the VDU. This operation takes place in high-pressure high-temperature reactors where there are multiple injections of hydrogen (H_2). The purpose of the latter injection is first to maintain the H_2 partial pressure, and second to control the temperature profile and avoid excessively high temperature values inside the reactor that would damage the catalyst, hydrogenations being exothermic. The reactor section is thus divided into multiple beds with quench and distributing devices between them. These internals must be as efficient as possible in terms of flow and temperature homogeneity entering the following bed and be as compact as possible so that the catalyst volume inside the vessel is maximised.

CFD is used not only to perform technology screening but also to check and eventually adapt the preferred design to real industrial conditions which are impossible to reproduce by experiments.

4.2.1 Quench Apparatus

The quench box is one of the key internal elements of a hydroprocessing reactor. It is fixed between two beds and has a specific role for mixing the quench fluid (commonly cold hydrogen) and the hot reaction mixture. During the mixing process, the quench hydrogen should dissolve rapidly in the liquid phase through gas-liquid mixing in providing fresh hydrogen and lowering the temperature of the reaction mixture (Zhenyuan *et al.*, 2013). CFD is today used more and more for the development of innovative quench box technologies. Recently, in the US patent 2014/0224707 by Haroun and Augier (2014), the inventors used a CFD Euler-Euler approach to develop new quench box technology. Figure 19 shows an illustration of CFD calculation in the quench box and the quench fluid injection system, respectively. The plots show the liquid guard shape on the quench pane and the gas-liquid stream lines in the quench injection system obtained by CFD Euler-Euler simulations. The inventors show that the quench injection system plays an important role in the mixing efficiency. This study shows that CFD is a powerful tool that can be used to optimise or develop new quench box technologies.

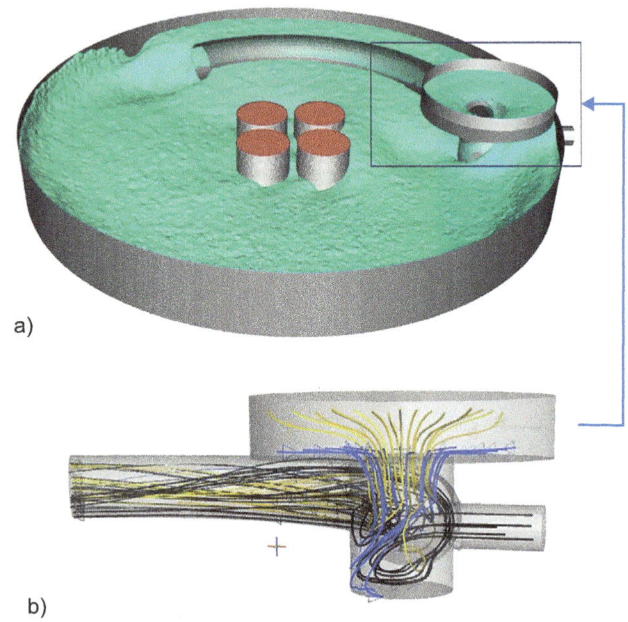

Figure 19

a) Liquid guard shape in the quench box system. b) Gas-liquid streamlines in the quench injection system. Liquid (blue), gas (yellow), quench fluid (black).

4.2.2 Gas-Liquid Flow Distributor

Once the fluid phases have been cooled down and properly mixed, they must be redistributed to further react in the bed downstream, which is done *via* the use of distributing devices. Many distributors consist of a large number of chimneys or tubes through which vapour and liquid phases flow to further contact the catalyst below the distributor (Maiti and Nigam, 2007; Bazer-Bachi *et al.*, 2013). For a long period of time, distributors were developed with no special attention to the vapour phase, considering that the main issue was to distribute the liquid phase. This was justified from common experiments, where no gas influence was indeed observed. However, many experiments are performed in ambient conditions and gas inertia effects are usually not well reproduced. Indeed, in HDT conditions, the liquid density can be as low as 550 kg.m^{-3}, while vapour density can be as high 40 kg.m^{-3}. With according gas and liquid velocities only between air/water laboratory experiments and industrial conditions, the gas to liquid inertia ratio

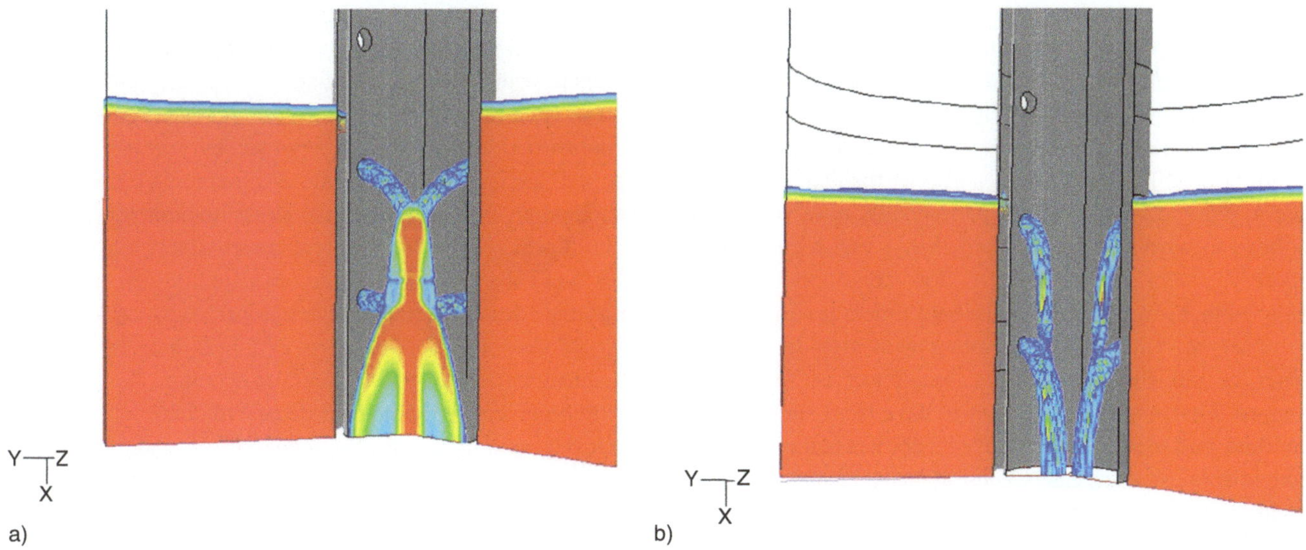

Figure 20

Liquid volume fraction contours of the flow through a HDT distribution tray for V_{SL} = 0.4 cm/s and V_{SG} = 10 cm/s; a) with air and water; b) with industrial conditions physical properties (but for surface tension kept constant).

Figure 21

Pictures of the flow through a HDT distribution tray for V_{SL} = 0.2 cm/s; a) V_{SG} = 0 cm/s; b) V_{SG} = 10 cm/s with N_2 in ambient conditions; c) V_{SG} = 10 cm/s with SF6 at 3 bar.

would be underestimated by a factor of 60 in experiments. This has been shown by Raynal and Harter (2001), who observed a dramatic change in the flow inside the downcomer when changing from standard experimental to industrial conditions, which could affect the initial wetting performance in the catalyst bed downstream. Indeed, as can be seen in Figure 20, when real industrial conditions are considered, the liquid jets coming out from the orifices along the downcomer are deviated and may not interact themselves to form fine droplets as initially expected from experiments (for details on the flow configuration and simulation assumptions, Raynal and Harter, 2001). This effect was so important, that a dedicated experimental set-up has been built with the possibility of operating the unit reaching moderate pressure with high gas densities and using sulphur hexafluoride (SF_6), a gas 5 times heavier than air. As can be seen in Figure 21, what was experimentally observed corresponded to what had been obtained by CFD. Indeed, from Figure 21, which correspond to flow conditions with no gas or with air in ambient conditions, almost no differences are observed, while when operating with high-density gas closer to industrial conditions, one observes two main changes. First, the liquid jet shapes have been significantly modified, and second, the liquid level above the tray is found to be lower. The gas density being higher, the pressure drop across the tray is higher; then for a given pressure at the catalyst bed inlet, the pressure above the tray and on the liquid holdup is higher, resulting in a liquid level reduction for a constant liquid flow rate. Since the size, number and location of the orifices have to be determined to result in adequate liquid holdup on the tray, this gas influence clearly has to be taken into account in distributor design rules.

These examples show how CFD can be used to develop reactor internals, but one can easily understand how, from such results, one can develop intellectual property, which is of key importance for process licensors or technology vendors.

CONCLUSION, PERSPECTIVES AND CHALLENGES

This paper shows that CFD is a very useful tool to be considered during all the steps of process development. It can be used to prepare the development phase by helping design tools or by identifying critical parameters that should be considered during the development. It clearly completes experiments for developing a better knowledge of all complex phenomena that occur in chemical reactors, in particular for sensitivity analysis or extrapolation purposes, which are very difficult to address experimentally. CFD can also help in the design of equipment which is essential to ensure that industrial units match the design expectations. Many types of flows can be considered coupled with mass and heat transfer and with chemical reactions. Depending on their complexity, different levels of answers can be obtained, from qualitative trends to design rules. In all cases, CFD can provide solutions, which, even in cases where significant simplifications are required, should not be neglected, in particular when considering the low cost of CFD when compared with experiments, both in time and budget. Despite all these advantages, there are still challenges to address, as discussed below.

As mentioned in Section 1, 3D simulations are challenging and problems are often too complex with the present models. Although simulations for single-phase flows or local simulations of two-phase flows with interface tracking can directly compete with experiments, this is not yet the case for multiphase systems with a Euler/Euler approach. Indeed, both experimental and numerical approaches complement each other. Experiments can provide information for closure law development, which can be used in simulations for extrapolation purposes. In that sense, there are still many types of complex flows for which interaction laws are not well understood and which deserve further research from the academic community. Moreover, too often mass and heat transfer, and also chemical reactions are not taken into account. This should be the main objective for the next decade to come.

Along with model developments, multi-scale approaches (Charpentier, 2009) and parallelisation will support the possibility of addressing more and more complex flows with less and less assumptions to make. Finally, it is believed that the use of simulation tools is still too complex and can only be handled by highly skilled people. In order to develop chemical reactor simulation skills, chemical engineering schools could either further integrate CFD in their educational programmes or develop closer relationships with mechanical engineering schools.

Combining all this, the authors believe that the use of CFD could be as popular in the oil and gas sector as it is in aeronautics and the automotive industry, and more importantly, it will help bring innovations to the market and make chemical reactors more and more efficient.

REFERENCES

Alvarez A., Ramirez S., Ancheyta J., Rodriguez L.M. (2007) Key role of reactor internal in hydroprocessing of oil fraction, *Energy Fuels* **21**, 1731-1740.

Antos G.J., Aitani A.M. (2004) *Catalytic Naphtha Reforming*, Marcel Dekker, New York.

Attou A., Ferschneider G. (2000) A two-fluid hydrodynamic model for the transition between trickle an pulse flow in a cocurrent gas-liquid packed-bed reactor, *Chem. Eng. Sci.* **55**, 491-511.

Augier F., Laroche C., Brehon E. (2008) Application of computational fluid dynamics to fixed bed adsorption calculations: Effect of hydrodynamics at laboratory and industrial scale, *Separation and Purification Technology* **63**, 466-474.

Augier F., Idoux F., Delenne J.Y. (2010a) Numerical simulations of transfer and transport properties inside packed beds of spherical particles, *Chem. Eng. Sci.* **65**, 3, 1055-1064.

Augier F., Koudil A., Royon-Lebeaud A., Muszynski L., Yanouri Q. (2010b) Numerical approach to predict wetting and catalyst efficiencies inside trickle bed reactors, *Chem. Eng. Sci.* **65**, 1, 255-260.

Bazer-Bachi F., Augier F., Santos B. (2011) 1D and 2D simulations of partially wetted catalyst particles: A focus on heat transfer limitations, *Chem. Eng. Sci.* **66**, 1953-1961.

Bazer-Bachi F., Haroun Y., Augier F., Boyer C. (2013) Experimental evaluation of distributor technologies for trickle-bed reactors, *Ind. Eng. Chem. Res.* **52**, 11189-11197.

Boisson N., Malin M.R. (1996) The numerical prediction of two-phase flow in bubble columns, *Int. J. Numerical Methods Fluids* **23**, 12, 1289-1310.

Boivin M., Simonin O., Squires K.D. (2000) On the prediction of gas-solid flows with two-way coupling using large eddy simulation, *Phys. Fluids* **12**, 8, 2080-2090.

Boyer C., Koudil C., Chen P., Dudukovic M.P. (2005) Study of liquid spreading from a point source in a trickle-bed via gamma-ray tomography and CFD simulation, *Chem. Eng. Sci.* **60**, 22, 6279-6288.

Braga M. (2013) Étude des phénomènes de transfert et de l'hydrodynamique dans des réacteurs agités à panier catalytique, *PhD Dissertation*, Université Claude Bernard, Lyon, France.

Capelle M., Deves J.-M., Hoffmann F., Thery M. (2000) Process for regenerating a catalyst for the production of aromatic compounds or for reforming, with improved oxychlorination, US patent **6**, 133, 183.

Chalermsinsuwan B., Piumsomboon P., Gidaspow D. (2009a) Kinetic theory based computation of PSRI riser: Part I-Estimate of mass transfer coefficient, *Chem. Eng. Sci.* **64**, 6, 1195-1211.

Chalermsinsuwan B., Piumsomboon P., Gidaspow D. (2009b) Kinetic theory based computation of PSRI riser: Part II- Computation of mass transfer coefficient with chemical reaction, *Chem. Eng. Sci.* **64**, 6, 1212-1222.

Charpentier J.-C. (2009) Perspective on multiscale methodology for product design and engineering, *Comput. Chem. Eng.* **33**, 936-946.

Chen P., Dudukovic M.P., Sanyal J. (2005) Three-dimensional simulation of bubble column flows with coalescence and breakup, *AIChE J.* **51**, 696-712.

Chen X., Closmann F., Rochelle G.T. (2011) Accurate screening of amines by the Wetted Wall Column, *Energy Procedia* **4**, 101-108.

Cooper B.H., Donnis B.B.L., Moyse B. (1986) *Oil Gas J.* **8**, December, 39-44.

Decoodt X., Plais C., Bazer-Bachi F. (2014) Continuous catalyst regeneration reactor with a chamber for mixing gas and distributing gas in the oxychlorination zone, US patent 2014/0027348.

Deen N.G., van Sint Annaland M., Van der Hoef M.A., Kuipers J. A.M. (2007) Review of discrete particle modeling of fluidized beds, *Chem. Eng. Sci.* **62**, 1-2, 28-44.

Derouin C., Nevicato D., Forissier M., Wild G., Bernard J.R. (1997) Hydrodynamics of riser units and their impact on FCC operation, *Ind. Eng. Chem. Res.* **36**, 4504-4515.

Dry M.E. (2002) The Fischer–Tropsch process: 1950–2000, *Catal. Today* **71**, 227-241.

Forni L. (1997) Laboratory reactors, *Catal. Today* **34**, 353-367.

Forret A., Schweitzer J.-M., Gauthier T., Krishna R., Schweich D. (2003) Influence of scale on the hydrodynamics of bubble column reactors: An experimental study in columns of 0.1, 0.4 and 1 m diameters, *Chem. Eng. Sci.* **58**, 719-724.

Fourati M., Roig V., Raynal L. (2013) Liquid dispersion in packed columns: experiments and numerical modeling, *Chem. Eng. Sci.* **100**, 266-278.

Gavi E., Marchisio D., Barresi A.A., Olsen M.G., Fox R.O. (2010) Turbulent precipitation in micromixers: CFD simulation and flow field validation, *Chem. Eng. Res. Design* **88**, 9, 1182-1193.

Gidaspow D. (1994) *Multiphase Flow and Fluidization: Continuum and Kinetic Theory Description*, Academic Press, Boston.

Ginestra J.C., Jackson R. (1985) Pinning of a bed of particles in a vertical channel by a cross flow of gas, *Ind. Eng. Chem. Fundam.* **24**, 121-128.

Greenwood A.R. (1987) Catalyst regeneration apparatus, US patent **4**, 687, 637.

Guedes de Carvalho J.R.F., Delgado J.M.P.Q. (2001) Radial dispersion in liquid flow through packed beds for $50<Sc<750$ and $103<Pem<105$, *Fifth World Conference on Experimental Heat Transfer, Fluid Mechanics Thermodynamics*.

Gupta R.K., Kumara V., Srivastava V.K. (2007) A new generic approach for the modeling of fluid catalytic cracking (FCC) riser reactor, *Chem. Eng. Sci.* **62**, 4510-4528.

Haroun Y., Legendre D., Raynal L. (2010) Volume of fluid method for interfacial reactive mass transfer: Application to stable liquid film, *Chem. Eng. Sci.* **65**, 10, 2896-2909.

Haroun Y., Augier F. (2014) Device for injecting and mixing fluids in a downward-flow reactor, US patent 2014/0224707.

Haroun Y., Raynal L. (2015) Use of Computational Fluid Dynamics for absorption packed column design, *Oil Gas Sci. Technol.* (this issue). DOI: 10.2516/ogst/2015027.

Heidari A., Hashemabadi S.H. (2013) Numerical evaluation of the gas-liquid interfacial heat transfer in the trickle flow regime of packed beds at the micro and meso-scale, *Chem. Eng. Sci.* **104**, 18, 674-689.

Hills J.H. (1974) Radial, non-uniformity of velocity and voidage in bubble column, *Trans. Inst. Chem. Eng.* **52**, 1-9.

Hong H.-S., Cai Z.-J., Li J.-Q., Shi D.-S., Wan W.-Q., Li L. (2014) Simulation of gas-inducing reactor couples gas-liquid mass transfer and biochemical reaction, *Biochemical Eng. J.* **91**, 1-9.

Horgue P., Augier F., Duru P., Quintard M., Prat M. (2013) Arrays of cylinders as an experimental and numerical laboratory for the study of trickling and pulsing flows, *Chem. Eng. Sci.* **102**, 335-345.

Jiang J., Wu J., Zhang J., Poncin S., Li H.Z. (2014) Multiscale hydrodynamic investigation to intensify the biogas production in upflow anaerobic reactors, *Bioresource Tech.* **155**, 1-7.

Jiradilok V., Gidaspow D., Damronglerd S., Koves W.J., Mostofi R. (2006) Kinetic theory based CFD simulation of turbulent fluidization of FCC particles in a riser, *Chem. Eng. Sci.* **61**, 17, 5544-5559.

Kataoka H., Takeuchi H., Nakao K., Yagi H., Tadaki T., Otake T. (1979) Mass transfer in a large bubble column, *J. Chem. Eng. Japan* **12**, 2, 105-110.

Kirillov V.A., Mikhailova I.A., Fadeyev S.I., Korolev V.K. (2002) Critical phenomena of an exothermic reaction proceeding on a partially wetted porous catalyst grain, *Combustion, Explosion and Shock Waves* **38**, 508-517.

Koide K., Morooka S., Ueyama K., Matsuura A., Yamashita F., Iwamoto S., Kato Y., Inoue H., Shigeta M., Suzuki S., Akehata T. (1979) Behavior of bubbles in large scale bubble column, *J. Chem. Eng. Japan* **12**, 2, 98-104.

Kojima E., Unno H., Sato Y., Chida T., Imai H., Endo K., Inoue I., Kobayashi J., Kaji H., Nakanishi H., Yamamoto K. (1980) Liquid phase velocity in a 5.5 m diameter bubble column, *J. Chem. Eng. Japan* **13**, 1, 16-21.

Krishna R., van Baten J.M., Ellenberger J., Higler A.P., Taylor R. (1999a) CFD simulations of sieve tray hydrodynamics, *Chem. Eng. Res. Design* **77**, A7, 639-646.

Krishna R., Urseanu M.I., van Baten J.M., Ellenberger J. (1999b) Influence of scale on the hydrodynamics of bubble columns operating in churn-turbulent regime: experiments *vs.* Eulerian simulations, *Chem. Eng. Sci.* **54**, 4903-4911.

Krishna R., van Baten J.M., Urseanu M.I. (2000) Three-phase Eulerian simulations of bubble column reactors operating in the churn-turbulent regime: a scale up strategy, *Chem. Eng. Sci.* **55**, 16, 3275-3286.

Krishna R., van Baten J.M. (2001) Eulerian simulations of bubble columns operating at elevated pressures in the churn turbulent flow regime, *Chem. Eng. Sci.* **56**, 21-22, 6249-6258.

Krishna R., van Baten J.M. (2002) Scaling up bubble column reactors with highly viscous liquid phase, *Chem. Eng. Tech.* **25**, 10, 1015-1020.

Krishna R., van Baten J.M. (2003) Modelling sieve tray hydraulics using computational fluid dynamics, *Chem. Eng. Res. Design* **81**, A1, 27-38.

Kuipers J.A.M., van Swaaij W.P.M. (1997) Application of computational fluid dynamics to chemical reaction engineering, *Rev. Chem. Eng.* **13**, 3, 1-118.

Lappalainen K., Manninen M., Alopaeus V. (2009) CFD modeling of radial spreading of flow in trickle-bed reactors due to mechanical and capillary dispersion, *Chem. Eng. Sci.* **64**, 207-218.

Lappin A., Lübbert A. (1994) Numerical simulation of the dynamics of two-phase gas-liquid flows in bubble columns, *Chem. Eng. Sci.* **49**, 21, 3661-3674.

Larachi F., Desvigne D., Donnat L., Schweich D. (2006) Simulating the effects of liquid circulation in bubble columns with internals, *Chem. Eng. Sci.* **61**, 4195-4206.

Larachi F., Hannaoui R., Horgue P., Augier F., Haroun Y., Youssef S., Rosenberg E., Prat M., Quintard M. (2014) X-ray microtomography and pore network modeling of single-phase fixed-bed reactors, *Chem. Eng. J.* **240**, 290-306.

Li X., Yang N., Sun Y., Zhang L., Li X., Jiang B. (2014) Computational Fluid Dynamics Modeling of Hydrodynamics of a New Type of Fixed Valve Tray, *Ind. Eng. Res.* **53**, 1, 379-389.

Lopes G.C., Rosa L.M., Mori M., Nunhez J.R., Martignoni W.P. (2011) Three-dimensional modeling of fluid catalytic cracking industrial riser flow and reactions, *Comput. Chem. Eng.* **35**, 2159-2168.

Luo H., Svendsen H.F. (1996) Theoretical model for drop and bubble breakup in turbulent dispersions, *AICHE J.* **42**, 5, 1225-1233.

Luo S.J., Fei W.Y., Song X.Y., Li H.Z. (2008) Effect of channel opening angle on the performance of structured packings, *Chem. Eng. J.* **144**, 2, 227-234.

Luo X., Hartono A., Svendsen H.F. (2012) Comparative kinetics of carbon dioxide absorption in unloaded aqueous monoethanolamine solutions using wetted wall and string of discs columns, *Chem. Eng. Sci.* **82**, 31-43.

Magnico P., Fongarland P. (2006) CFD simulations of two stirred tank reactors with stationary catalytic basket, *Chem. Eng. Sci.* **61**, 1217-1236.

Mahoney J.A., Robinson K.K., Myers E.C. (1978) Catalyst evaluation with the gradientless reactor, *Chemtech.* **8**, N12, 758-763.

Maiti R.N., Nigam K.D.P. (2007) Gas-Liquid distributors for trickle-bed reactors: A review, *Ind. Eng. Chem. Res.* **46**, 6164-6182.

Marcandelli C., Lamine A.S., Bernard J.R., Wild G. (2000) Liquid distribution in trickle-bed reactor, *Oil Gas Sci. Technol.* **55**, 407-415.

Mehmood N., Olmos E., Marchal P., Goergen J.-L., Delaunay S. (2010) Relation between pristinamycins production by Streptomyces pristinaespiralis, power dissipation and volumetric gas-liquid mass transfer coefficient, k_La, *Process Biochemistry* **45**, 11, 1779-1786.

Mitrovic M., Pitault I., Forissier M., Simoens S., Ronze D. (2005) Liquid-solid mass transfer in a three-phase stationary catalytic basket reactor, *AIChE J.* **51**, 1747-1757.

Moula G., Nastoll W., Simonin O., Andreux R. (2013) Multiscale Study of Reactive Dense Fluidized Bed for FCC Regenerator, *Oil Gas Sci. Technol.* **68**, 6, 1073-1092.

Neri A., Gidaspow D. (2000) Riser hydrodynamics: Simulation using kinetic theory, *AIChE J.* **46**, 1, 52-67.

Ng K.M., Chu C.F. (1987) Trickle-Bed Reactors, *Chem. Eng. Prog.* **83**, 11, 55-63.

Nijemeisland M., Dixon A.G. (2004) CFD study of fluid flow and wall heat transfer in a fixed bed of spheres, *AIChE J.* **50**, 5, 906-921.

Olmos E., Gentric C., Vial C., Wild G., Midoux N. (2001) Numerical simulation of multiphase flow in bubble column reactors. Influence of bubble coalescence and break-up, *Chem. Eng. Sci.* **56**, 21-22, 6359-6365.

Oryx GTL (2015) http://www.oryxgtl.com.qa/.

Perego C., Peratello S. (1999) Experimental methods in catalytic kinetics, *Catal. Today* **52**, 133-145.

Petre C.F., Larachi F., Illiuta I., Grandjean B.P.A. (2003) Pressure drop through structured packings: breakdown into the contributing mechanisms by CFD modeling, *Chem. Eng. Sci.* **58**, 163-177.

Pitault I., Fongarland P., Mitrovic M., Ronze D., Forissier M. (2004) Choice of laboratory scale reactors for HDT kinetic studies or catalyst tests, *Catal. Today* **98**, 31-42.

Pitault I., Fongarland P., Koepke D., Mitrovic M., Ronze D., Forissier M. (2005) Gas–liquid and liquid–solid mass transfers in two types of stationary catalytic basket laboratory reactor, *Chem. Eng. Sci.* **60**, 6240-6253.

Raynal L. (2001) Simulation numérique des colonnes à bulles, *IFP Internal Report* No. 56049.

Raynal L., Harter I. (2001) Studies of Gas-Liquid flow through distributing devices using VOF-CFD simulations, *Chem. Eng. Sci.* **56**, 6385-6391.

Raynal L., Forret A., Schweitzer J.-M. (2001) Simulation numérique Euler/Euler de l'hydrodynamique de colonnes à bulles, in *Récents progrès en Génie des Procédés*, Lavoisier Tech. et Doc. Ed. **15**, 227-234.

Raynal L. (2005) Use of CFD for applications in the oil and gas industry, *Revue de la Houille Blanche* **5**, 75-78.

Raynal L., Gomez A., Caillat B., Haroun Y. (2013) CO_2 Capture Cost Reduction: use of a Multiscale Simulations Strategy for a Multiscale Issue, *Oil Gas Sci. Technol.* **68**, 6, 1093-1108.

Santos-Moreau V., Brunet-Errard L., Rolland M. (2012) Numerical CFD simulation of a batch stirred tank reactor with stationary catalytic basket, *Chem. Eng. J.* **207-208**, 596-606.

Sanyal J., Marchisio D.L., Fox R.O., Dhanasekharan K. (2005) On the comparison between population balance models for CFD simulation of bubble columns, *Ind. Eng. Chem. Res.* **44**, 14, 5063-5072.

Schweitzer J.-M. (2003) Personal communication.

Schwidder S., Schnitzlein K. (2012) A new model for the design and analysis of trickle bed reactors, *Chem. Eng. J.* **207-208**, 758-765.

Sederman A.J., Gladden L.F. (2001) Magnetic resonance imaging as a quantitative probe of gas-liquid distribution and wetting efficiency in trickle-bed reactors, *Chem. Eng. Sci.* **56**, 8, 2615-2628.

Servia A., Laloue N., Grandjean J., Rode S., Roizard C. (2014) Modeling of the CO_2 Absorption in a Wetted Wall Column by Piperazine Solutions, *Oil Gas Sci. Technol.* **69**, 5, 885-902.

Solomenko Z., Haroun Y., Fourati M., Larachi F., Boyer C., Augier F. (2015) Liquid spreading in trickle-bed reactors: Experiments and numerical simulations using Eulerian-Eulerian two-fluid approach, *Chem. Eng. Sci.* **126**, 698-710.

Spogis N., Nunhez J.R. (2009) Design of a High-Efficiency Hydrofoil Through the Use of Computational Fluid Dynamics and Multiobjective Optimization, *AIChE J.* **55**, 7, 1723-1735.

Tosun G. (1984) A study of cocurrent downflow of nonfoaming gas-liquid systems in a packed bed. 2. Pressure drop: Search for a correlation, *Ind. Eng. Chem. Process Des. Dev.* **23**, 35-39.

Trambouze P. (1993) Computational Fluid Dynamics applied to Chemical Reaction Engineering, *Revue de l'Institut Français du Pétrole* **48**, 6, 595-613.

van Baten J.M., Krishna R. (2004) Eulerian simulation strategy for scaling up a bubble column slurry reactor for Fischer-Tropsch synthesis, *Ind. Eng. Chem. Res.* **43**, 16, 4483-4493.

van der Hoef M.A., Annaland M.V., Kuipers J.A.M. (2004) Computational fluid dynamics for dense gas-solid fluidized beds: a multi-scale modeling strategy, *Chem. Eng. Sci.* **59**, 22-23, 5157-5165.

van Houwelingen A.J., Kok S., Nicol W. (2010) Effectiveness Factors for Partially Wetted Catalysts, *Ind. Eng. Chem. Res.* **49**, 17, 8114-8124.

Wang Y., Jinwen C., Larachi F. (2013) Modelling and simulation of trickle-bed reactors using computational fluid dynamics: a state-of-the-art review, *The Canadian Journal of Chemical Engineering* **91**, 136-180.

Whitaker S. (1999) *The method of volume averaging, theory and applications of transport in porous media*, Kluwer Academic Publishers, Dordrecht, The Netherlands.

Zehner P. (1986) Momentum, mass and heat transfer in bubble columns. Part 1. Flow model of the bubble column and liquid velocities, *Int. Chem. Eng.* **26**, 22.

Zhenyuan W., Zhenmin C., Zibin H., Kun Y. (2013) Intensified Gas–Liquid Mixing in a Quench Box under the Driving of Super gravitational Swirling Flow, *Ind. Eng. Chem. Res.* **52**, 12802-12811.

Zou R. (1995) The packing of spheres in a cylindrical container: The thickness effect, *Chem. Eng. Sci.* **50**, 9, 1504-1507.

8

Spatially Resolved Experimental and Numerical Investigation of the Flow through the Intake Port of an Internal Combustion Engine

Frank Hartmann[1]*, Stefan Buhl[1], Florian Gleiss[1], Philipp Barth[2], Martin Schild[2], Sebastian A. Kaiser[2] and Christian Hasse[1]

[1] *Chair of Numerical Thermo-Fluid Dynamics, TU Bergakademie Freiberg, Fuchsmühlenweg 9, 09599 Freiberg - Germany*
[2] *Institute for Combustion and Gas Dynamics, University of Duisburg-Essen, Lotharstraße 1, 47057 Duisburg - Germany*
e-mail: frank.hartmann@iec.tu-freiberg.de - sebastian.kaiser@uni-due.de

* Corresponding author

Abstract *— Modern spark-ignited internal combustion engines have intake ports designed to introduce high levels of so-called "tumble" charge motion. Correspondingly high shear rates can lead to high fluctuations and turbulence within the combustion chamber. A suitable test case to characterize the intake flow is a steady-state flow bench. Although routinely used in the engine development process to determine the global discharge coefficients, only a few detailed numerical and experimental studies use this test case to analyze the flow in the vicinity of the valve with high spatial and temporal resolution. In this paper, we combined highly resolved two-dimensional, two-component Particle Image Velocimetry (PIV) measurements and numerical simulations using a Detached-Eddy Simulation (DES) model to characterize engine-relevant flow features on a flow bench. The spatial resolution of numerical simulations on two different grids is assessed and compared to that of the PIV measurement. The results of simulations and experiment are then compared in terms of their mean and fluctuation velocity fields and the jet orientation. A detailed study of the area around the valve seats investigates the validity of wall functions in this region. Finally, we examine structures induced by vortex-shedding at the valve stem and if they are transported into the combustion chamber.*

Résumé — Étude numérique et expérimentale résolue dans l'espace de l'écoulement à travers le conduit d'admission d'un moteur à combustion interne — Les moteurs à combustion interne modernes à allumage commandé disposent de conduits d'admission conçus pour générer des niveaux importants de mouvements de charge dits « *tumble* ». Les niveaux importants de taux de cisaillement qui en résultent peuvent conduire à de fortes fluctuations et turbulences dans la chambre de combustion. Un cas test adapté pour caractériser l'écoulement d'admission est un banc volute. Bien que son utilisation durant les phases de conception moteur pour déterminer les coefficients de perte de charge globaux soit très répandue, seules quelques études numériques et expérimentales détaillées utilisent ce test pour étudier l'écoulement au voisinage de la soupape avec des résolutions spatiales et temporelles élevées. Dans le présent article, nous avons combiné des mesures PIV bi-composants, bidimensionnelles hautement résolues et des simulations numériques utilisant une approche de type « *Detached-Eddy Simulation* » pour caractériser des structures d'écoulement d'importance pour le moteur sur un banc volute. La résolution spatiale des simulations numériques est évaluée sur deux maillages et comparée à celle de la mesure « *Particle Image Velocimetry* », PIV. Les résultats numériques et expérimentaux sont ensuite comparés en termes de champ de vitesse moyen et fluctuant et d'orientation des jets. Une étude détaillée du siège de soupape est réalisée pour vérifier la validité de lois de paroi dans cette zone. Enfin, nous examinons les structures induites par les détachements tourbillonnaires autour de la tige de soupape et si elles sont transportées dans la chambre de combustion.

INTRODUCTION

The intake jet is the main contributor to the kinetic energy of the internal combustion engine's in-cylinder flow. Thus, it is primarily responsible for the development of the flow pattern in the compression stroke and during combustion. During the intake stroke, the intake valves throttle the incoming gas flow and induce high velocities in the valve gap. Towards the combustion chamber, the resulting jet separates from the valve seat and lip creating shear layers with large gradients generating turbulence [1]. During an engine cycle, the penetrating jet transports momentum through the valve gap into the combustion chamber, which initiates large-scale motions at the beginning of the compression stroke [2].

Modern spark-ignited internal combustion engines are characterized by high levels of intake-induced in-cylinder flow, organized into a large-scale motion usually called "tumble". Such high levels of charge motion are achieved by flat intake ports and machined edges upstream of the valve seat. These measures yield a non-uniform distribution of the mass flow over the intake valve, creating the distinctive tumble motion.

In the past, there have been several investigations of the global in-cylinder flow and studies of the local flow-field near the intake port. On the one hand, steady-state experiments on flow benches were performed to investigate the flow physics in a simplified configuration [3]; on the other hand, optically accessible engines were used for measurements with more realistic boundary conditions [4]. A small number of studies investigate the flow field in or near the valve gap in detail. Schlieren cinematography, which is based on density gradients, was applied to qualitatively image the intake jet's motion within the cylinder of a transparent research engine [5]. Weclas *et al.* [6] used Laser Doppler Anemometry (LDA) to acquire quantitative pointwise information on the velocity and its fluctuations in the valve gap. In a second publication, the same authors [7] discussed the flow separation in the intake valve gap in steady and transient flows under engine-like conditions. Measurements were performed using LDA and hot-wire anemometry. In later studies, Particle Image Velocimetry (PIV) was applied in engine geometries to measure two velocity components instantaneously, spatially resolved within a plane defined by the laser sheet [8]. In principle, with stereoscopic PIV, the out-of-plane velocity component is also accessible [9]. "High-speed" PIV allows time-resolved investigations of in-cylinder flow structures and their development within a single cycle, while tomographic PIV enables volumetric measurements [10]. The mean three-dimensional flow in the intake port was fully quantified using Magnetic Resonance Velocimetry (MRV) by Freudenhammer *et al.* [11].

On the other hand, a very limited number of numerical investigations of the intake flow in such flow-bench scenarios were performed. Mclandress *et al.* [12] compared PIV measurements and simulations on a flow bench using the KIVA-3 code. Similar studies were performed by Inagaki *et al.* [13] using Large-Eddy Simulations (LES) with a mixed-time-scale subgrid-scale model, which were compared to experiments and RANS simulations.

Despite the relevance of the intake flow on the charge motion and the subsequent combustion process, no combined experimental-numerical studies using state-of-the-art approaches in both areas, namely highly resolved PIV and scale-resolving flow simulations, respectively, have been reported.

This work aims to start filling this gap by applying a combination of PIV and LES, both at relatively high spatial resolution, to the flow through the intake port on a flow bench. In the following section, the experiment will be described including flow bench, camera, laser system and image acquisition. The quality and reliability of the PIV data is assessed based on the primary peak ratio of the underlying cross-correlation. The simulation's general approach and grids are discussed next. In the results, we first compare the mean velocity fields in simulation and experiment. Further investigations on the boundary layer on the valve seat evaluate the widely-used assumption of a logarithmic wall function. Finally, large structures close to the intake valve and the jet-flapping are examined in order to investigate whether such instationary structures survive the acceleration through the valve gap, which in an operating engine could have an effect on the stability of the in-cylinder flow.

1 EXPERIMENT

1.1 Flow Bench

The head of an optically accessible single-cylinder engine [14] is mounted on a steady-state flow bench. In previous investigations, this flow bench was used to determine the

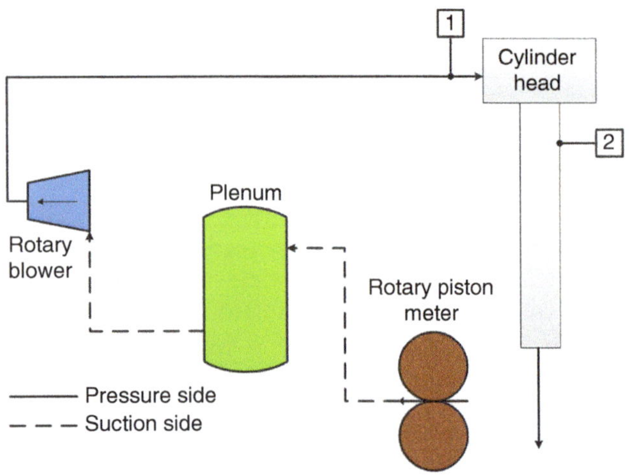

Figure 1

Flow diagram of the flow-bench with the two measurement points for pressure.

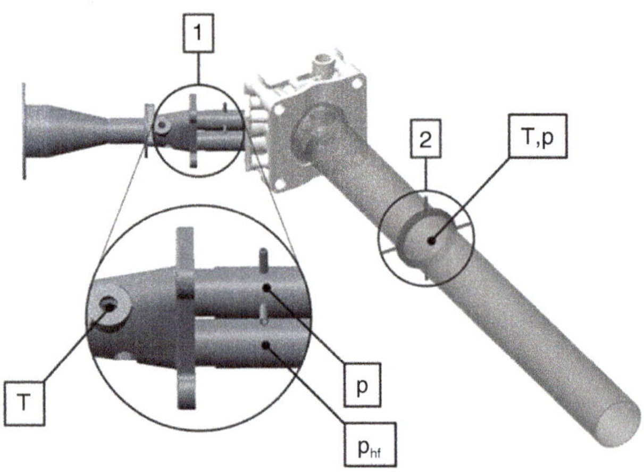

Figure 2

CAD rendering of intake system, cylinder head, and measurement tube.

cylinder head's discharge coefficients to be used in one-dimensional engine simulations [15]. The arrangement is shown in Figure 1. A rotary blower, powered by a speed-controlled Alternating Current (AC) motor, provides the pressure difference which drives the flow. On the suction side of the blower, a rotary piston meter measures the volume flow. The temperature measured at the intake of this meter and the atmospheric pressure are used to calculate the mass flow. A plenum of approx. 500 L downstream of the piston meter eliminates flow fluctuations caused by the blower.

The cylinder head is located on the pressure side of the blower. The PVC piping upstream of the head is straight, smooth-walled, and with no change in its inner diameter ($d = 76.6$ mm) for 2 840 mm (37 d). A conical adapter connects the cylinder head with its intake system, which is different in diameter, to the piping of the flow bench. Figure 2 shows the geometry of the head with the components immediately up- and downstream. In the laboratory frame, the whole assembly lies on its side on an optical table. Nevertheless, throughout this paper we will refer to the orientation of the cylinder head, valve, and flow as if the head were mounted upright with the valves above the combustion chamber, as in a typical automotive application.

In this work, a valve lift of 5 mm is fixed by a micrometer screw pushing onto the valve cups. Only the intake port is examined and the exhaust valves are closed. An acrylic tube with an inner diameter of 84 mm corresponding to the bore of the engine is placed below the cylinder head as optically accessible "cylinder walls". With an overall length of 950 mm from the pent-roof to the outlet, it provides an oriented outflow into the ambient air. At two locations, the temperature T and relative pressure p are measured: in the intake runners (reference point 1) and in the measurement tube (reference point 2, Fig. 2). Additionally, at point 1, the pressure is measured with a fast pressure transducer (*Kistler* 4049A5) to monitor for potential high-frequency fluctuations in the flow. For the measurements presented here, only reference point 2 (and the upstream volume flow rate) is required to define the boundary conditions, but point 1 enables data cross-check.

1.2 Velocity Measurements

In the vertical intake-valve plane, the two in-plane components of the instantaneous velocity field were measured by high-resolution PIV (2D-2C-PIV). Figure 3 illustrates the geometry of the optical access on the flow bench with the engine head, the location of the laser light sheet, and the imaged Region Of Interest (ROI).

For PIV, the intake air was seeded with Di-Ethyl-Hexyl-Sebacate droplets (DEHS, nominal diameter: 0.2-0.3 μm). The aerosol droplets were generated by a nebulizer with a Laskin nozzle design [16] (*LaVision*, Göttingen). To ensure that the particle were homogeneously distributed, the droplets were injected through four nozzles located 500 mm upstream of the intake valves. The field of view was illuminated by a dual-cavity, double-pulsed Nd:YAG laser (*Litron*) at 532 nm with an energy of 30 mJ/pulse. A telescopic lens arrangement formed a light sheet of 0.5 mm-0.9 mm thickness, entering the ROI diagonally from below the head.

a)

b)

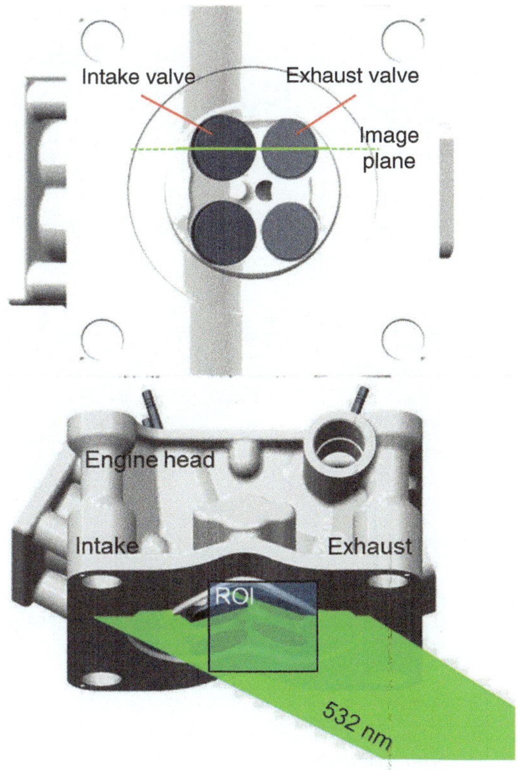

Figure 3

a) Side view and b) bottom view of engine head with ROI in the laser sheet.

a)

b)

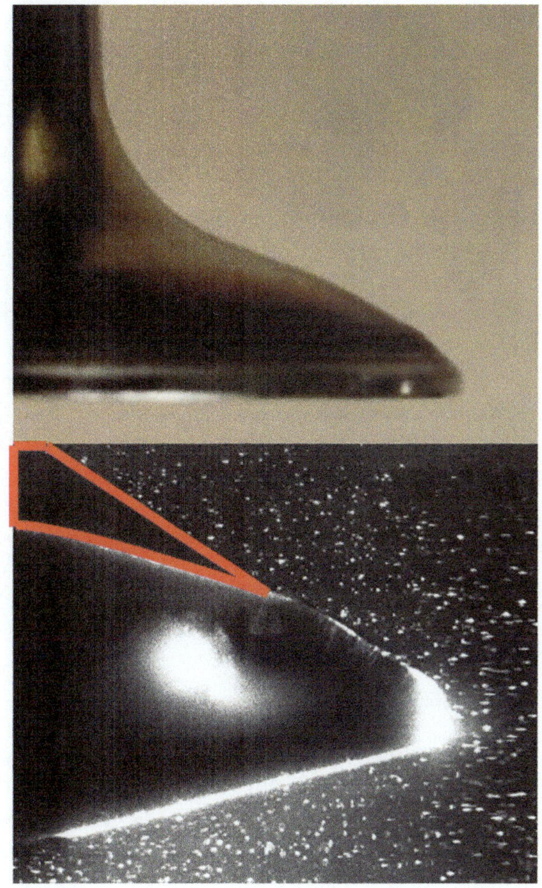

Figure 4

a) Intake valve and b) raw PIV data near the intake valve. The red triangle marks the region shadowed by the valve.

Complete optical access to the valve gap is difficult. Figure 4 illustrates some of the issues. With the laser sheet arranged as shown in Figure 3b, a region above the valve disc is shadowed and scattered light cannot be detected in this area, marked red in Figure 4b. Thus, no experimental data is reported for this area.

Laser-light reflections were suppressed by black surfaces in the cylinder head and on the back of the measurement tube. A 532 nm band pass filter was mounted in front of the camera lens, suppressing ambient room light. The light scattered by the seed particles was detected through the acrylic cylinder by a CMOS camera (*PCO Edge* 5.5, $2\,560 \times 2\,160\,px^2$) digitized to 16 bit. A commercial camera lens with a focal length of 50 mm (*Nikon* AF-D) was combined with a close-up lens (*Canon* 52 mm, 500*D*) to yield a magnification of 0.29. This projects an image section of 58 mm × 48.9 mm onto the detector. The lens aperture was set to $f/5.6$. PIV was performed with a repetition rate of 15 Hz, which is much longer than the integral time scale of the intake flow. For the current work, 1200 image pairs from 12 data sets with 100 shots each were recorded. After 300 images the acrylic glass tube was cleaned to limit refractive image distortion caused by the build-up of a liquid film of seed material on the inner surfaces. This effect leads to defocused particle images and a lower signal-to-background ratio. The apparent particle "size" (the area covered by scattering from a single particle on the sensor) and seeding density were also cross-checked by the imaging software, with the particle size estimated from the width of the auto-correlation peak of the pixels inside the ROI. To avoid peak locking effects resulting in inaccuracies in the velocity calculation, the lens aperture was set to achieve an apparent size of 2-4 pixels. Particle density was determined by means of simple thresholding segmentation with a signal threshold three times that of the background signal [17].

The image acquisition, scaling, and vector map computation were carried out using DaVis 8.1.4 (*LaVision*). Image distortion caused by the round measurement tube was corrected with a 3rd order polynomial fit based on a target placed in the field of view. The cross-correlation algorithm is described in Section 1.3.

Figure 5 illustrates the ROI with a final spatial resolution of 725 μm. Figure 5a shows an example of PIV raw data with background light scattering due to laser reflections on the surfaces of the acrylic glass cylinder. The resulting vector field is shown in Figure 5b. The ROI was chosen to investigate in detail the instationary behavior of the intake jet flow (jet flapping), its penetration depth, and potential interactions between the jet and the cylinder wall. Figures 5c and 5d show close-ups of the vector field. The jet break-up position and transition to turbulence are clearly visible. The high image resolution allows particles to be detected quite close to the valve. The detrimental effect of the shadowing illustrated in Figure 4 can be seen in Figure 5b. Additional limitations to flow measurements very close to the top of the valve disk, especially towards the valve stem, are imposed by seeding material collecting on the valve, thereby increasing reflections.

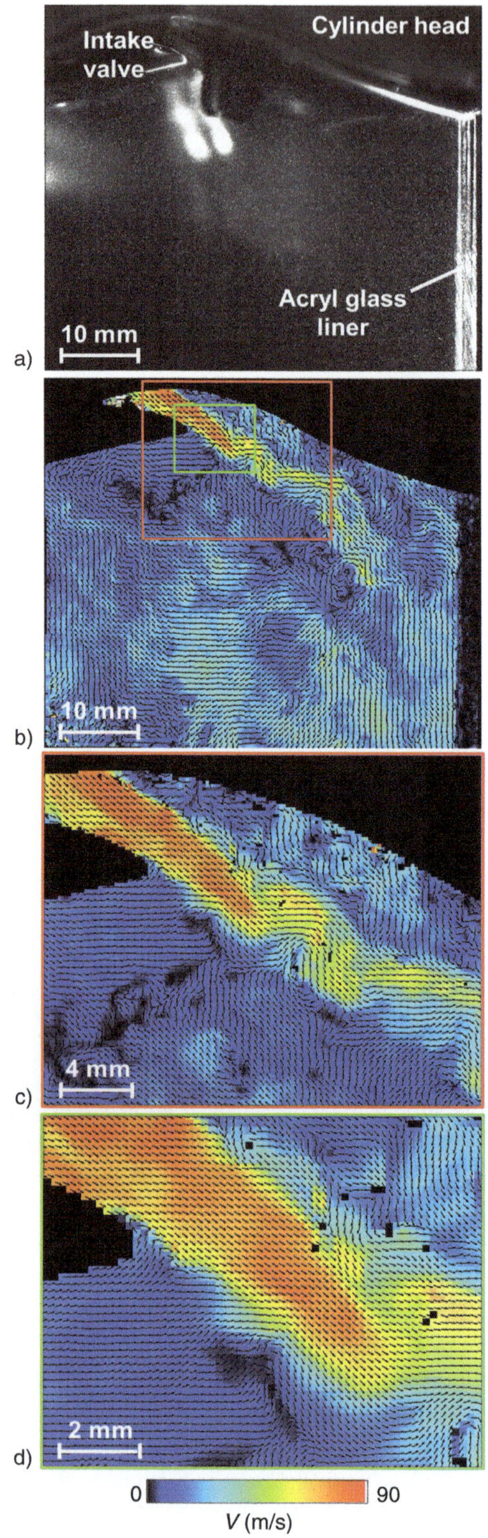

Figure 5

a) PIV raw image of the ROI in the center plane of the intake valve; b) velocity field determined from a single exposure in the ROI (every 4th vector is shown); c) and d) successively enlarged close-ups of this vector field, as indicated by the frames in b).

1.3 PIV Evaluation Procedure

Software Settings in DaVis

In the first step of processing the PIV image, the background light scatter is subtracted from the original PIV images. Next, objects that do not belong to the flow domain are masked out. Afterwards, the particle image intensity is normalized with a min/max-filter to obtain homogeneous particle intensities. The vector calculation is carried out with a cyclic FFT-based cross-correlation algorithm and multi-passes with decreasing grid size and the "adaptive PIV" option [18]. This option automatically optimizes the interrogation window shape according to the local seeding density and flow gradients from the previous pass. The first passes are calculated with round Gaussian-weighted interrogation windows. In the first two passes, an interrogation window of 64×64 pixel^2 with an overlap of 50% is set. In the following two passes, the interrogation window size is set to 32×32 pixel^2 with an overlap of 75%. Between each evaluation pass, a 4-pass regional median filter is applied, suppressing spurious vectors. Additionally, groups with less than 10 vectors are removed. Finally, a denoise filter (a fit with a polynomial of 2nd order) with 3×3 kernel size is applied in order to remove high-frequency noise from the vector field. Rejected vector data are interpolated using the *MATLAB* function "inpaint-nans". At the chosen magnification, the interrogation window size of 32×32 pixel^2 corresponds to 725×725 μm^2, and a 75% overlap yields 4×4 vectors in each of these interrogation windows.

Quality and Reliability of the PIV Data

The first quality criteria is the peak lock value, which error due to peak locking is negligible [17]. Another measure of

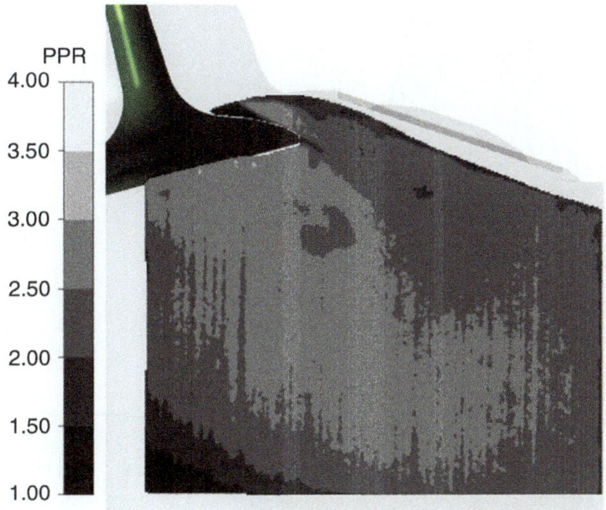

Figure 6
Mean PPR for 100 image pairs.

the quality of the PIV data is the Primary Peak Ratio (PPR), which is calculated according to [17] as:

$$PPR = \frac{P_1 - \min}{P_2 - \min} \tag{1}$$

with P_1 being the peak correlation value within the interogation window, P_2 the second-highest peak, and min the base value far from these peaks. It is a relative ratio that takes the background level into account and should be higher than 1 to ensure valid correlation peak detection. The mean PPR for 100 image pairs is illustrated in Figure 6.

The figure shows that the PPR is always larger than 1 except in the lower left corner, in the upper part of the valve gap, and in the shadowed region above the valve. This is caused by insufficient illumination and spurious vector data. However, in the remaining ROI, the PPR is larger than 2 and the total fraction of spurious vector data is below 5%.

2 SIMULATION

2.1 Numerical Method

All simulations are performed using *ANSYS CFX* Release 15.0. This implicit and node-centered code solves the governing equations based on a finite volume approach.

For constant-density flows, the following Reynolds-filtered equations for mass – Equation (2) – and momentum – Equation (3) – have to be solved:

$$\frac{\partial \bar{u}_i}{\partial x_i} = 0 \tag{2}$$

$$\rho\left(\frac{\partial \bar{u}_i}{\partial t} + \bar{u}_j\frac{\partial \bar{u}_i}{\partial x_j}\right) = -\frac{\partial \bar{p}}{\partial x_i} + \frac{\partial}{\partial x_j}\left(\tau_{ij} + \tau_{ij}^{mod}\right) \tag{3}$$

The quantity τ_{ij}^{mod} requires an adequate turbulence model, which is described in Section 2.2.

The numerical discretization schemes for scale-resolving engine simulations were chosen as discussed in [19-22]. The time step was chosen such that the Courant-Friedrichs-Lewy number (CFL), defined as:

$$\text{CFL} = \frac{\bar{u}\cdot\Delta t}{\Delta x} \tag{4}$$

is smaller than unity.

2.2 Turbulence Model

A closure is required for τ_{ij}^{mod} in the momentum equations (Eq. 3). The stationary flow bench is a challenging test case for a scale-resolving simulation because of the combination of large velocity gradients in the jet region, the wall boundary layer in the valve gap and flow separation at the valve. A standard LES approach, *e.g.* using the Smagorinsky model, requires very high spatial, nearly isotropic resolution, especially in the boundary layer. Since the time-step has to be adjusted accordingly, the numerical cost increases significantly. To overcome this problem, the simulations were performed based on the Detached-Eddy Simulation (DES) approach developed by Travin *et al.* [23]. This model offers a good compromise between accuracy, resolution of turbulent fluctuations, and computational effort, because it combines a URANS and an LES formulation. It is based on the Shear Stress Transport (SST) turbulence model [24] and solves a transport equation for turbulent kinetic energy k and turbulent frequency ω. The turbulent dissipation in the k equation of the SST model, written as:

$$D_k^{SST} \sim \rho k\omega \sim \rho\frac{k^{\frac{3}{2}}}{l_t} \tag{5}$$

is replaced by:

$$D_k^{DES} \sim \rho\frac{k^{\frac{3}{2}}}{l_{DES}} \text{with } l_{DES} = \min(l_t, C_{DES}\Delta) \tag{6}$$

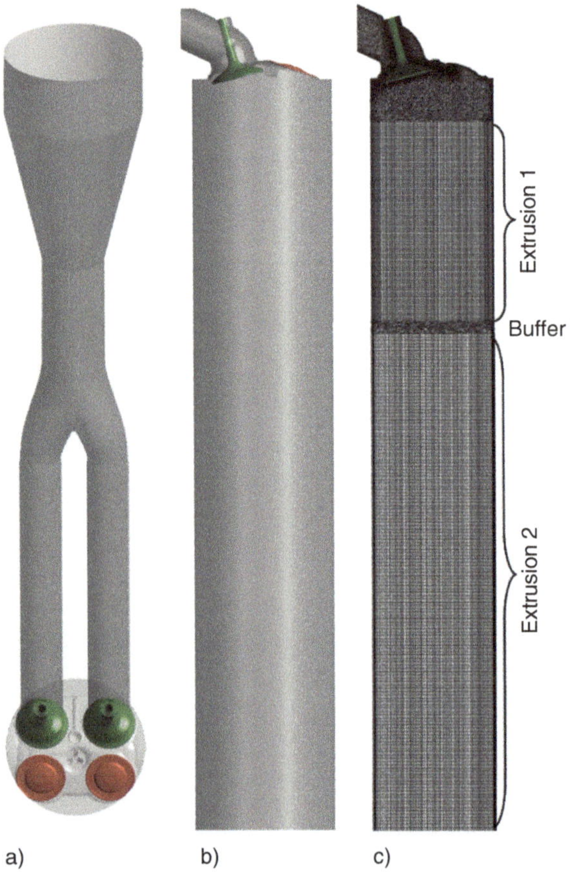

Figure 7

a) Top view; b) side view of the simulation domain and c) numerical mesh with 20 million grid points.

Figure 8

Detailed mesh view illustrating the volumetric refinement regions around the valve stem, in the valve gap and in the jet shear layers.

which enables local switching from URANS to scale-resolving simulations. The capability of the turbulence model to resolve turbulent fluctuations is directly linked to the local filter size Δ. For *ANSYS CFX* Menter and Kuntz [25] altered the pure grid-induced blending function and introduced an additional SST-based blending to overcome the problem of erroneously pressure-induced flow separation.

The DES-SST model guarantees at least URANS behavior in regions where grid resolution is not sufficient for LES. Here, such regions are found in particular close to the walls, where an order of magnitude more grid points would otherwise be required.

2.3 Numerical Model

Figure 7 shows the top view and the side view of the numerical domain for the flow bench test rig as well as the corresponding mesh for a valve lift of 5 mm. The simulation's boundary conditions were extracted from the experiments, see Section 1. For a realistic velocity distribution and a fully developed boundary layer at the end of the intake pipe almost the entire intake port illustrated in Figure 2 was included in the simulation. A mass flow of 3 kg/min was set at the inlet according to the experimental value and an outflow boundary condition with ambient pressure was used at the end of the cylinder tube.

Two different meshes, M_1 and M_2, are investigated in the following with a focus on the high resolution mesh M_2. In both cases, the numerical domain is discretized using a hybrid mesh consisting of tetrahedral, hexahedral, and prism elements. In the upper part of the domain, which has a complex geometry, the mesh consists of tetrahedral elements with mesh refinements at the separation edge at the intake port, at the valve shaft, in the valve gap, and in the shear layer region of the intake jet (Fig. 8). Prism elements at the wall are used to resolve the boundary layer flow.

Based on this upper part, the cylinder volume was generated by extruding the surface mesh, and it consists of prism and hexahedral elements, which yields a reduced number of elements compared to a pure tetrahedral-prism grid. The extrusion mesh was split into two parts (Extrusion 1 and Extrusion 2; Fig. 7), which have lengths of 2 and 3 times the bore diameter, respectively. To avoid hanging nodes, a buffer volume with a tetrahedral-based transition zone is

TABLE 1

Grid information and estimated time step requirements. BL: boundary layer

Grid type	M_1	M_2
Elements	10×10^6	105×10^6
Nodes	1.5×10^6	20×10^6
Domain volume	3 416 cm^3	3 416 cm^3
Δ_{max}	2 mm	2 mm
Δ_{min}	0.5 mm	0.25 mm
BL resolution	10 nodes	10 nodes
u_{max}	80 m/s	80 m/s
Δt	6 µs	3 µs
t_{tot}	30 s	3 s

placed between these two extrusion volumes. This methodology allows the maximum grid size to increase from 1 mm in the Extrusion 1 volume to 2 mm in the Extrusion 2 volume. To reduce numerical errors, a small cell growth ratio of about 1.1 is used over the entire domain. Overall, the larger mesh M_2 consists of 105 million elements. Further mesh information is given in Table 1.

The minimum cell width Δ_{min} occurs in the volumetric refinement for the jet shear layer. This cell width, together with the maximum jet velocity and the CFL condition of unity, yields the time step Δt in Table 1. The total simulation time t_{tot} is chosen to obtain sufficient statistical information on large-scale motions and to generate a sufficient number of snapshots which can be compared to the experimental results.

3 RESULTS

In this section, a comparison of the numerical and experimental results is presented and discussed. First, we consider the spatial resolution of the PIV measurements and compare it to the simulations, followed by comparisons of the instantaneous and mean flow fields. The "relevance index" and other statistical tools are used to evaluate the general agreement between simulation and two-dimensional, two-component PIV (2D-2C PIV) measurements. Additionally, an investigation of the boundary layer at the intake valve seat examines the flow separation and the accuracy of the standard wall-function formulation. Finally, large-scale structures induced by flow separation behind the valve stem will be examined using Fourier analysis in order to determine how far their influence reaches into the combustion chamber.

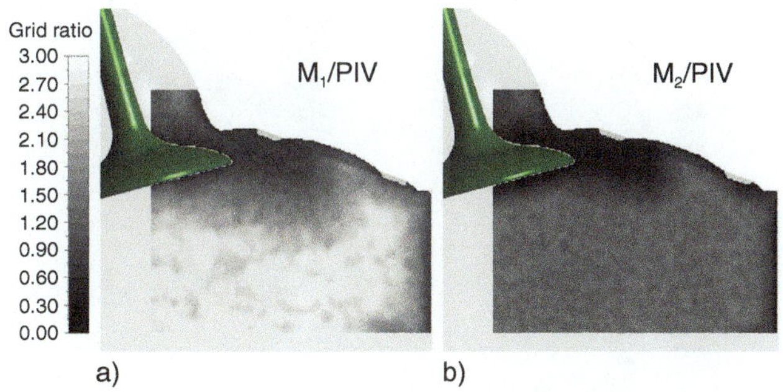

Figure 9

Cell width ratio (grid ratio) of a) M_1/PIV and b) M_2/PIV illustrating the spatial resolution of the meshes for the simulation compared to that of the PIV.

3.1 Comparison between Simulation and Experiment

First, the spatial resolution of each simulation mesh is considered and compared to the PIV grid. The local cell size for M_1 and M_2 is approximated as the cubic root of the finite volumes. The equidistant cell width from PIV is restricted by the interrogation window size of 32×32 pixel^2. The cell width divided by the equidistant cell width from PIV yields the GridRatio, which then reads

$$GridRatio = \frac{V_{\text{Sim}}^{1/3}}{\Delta_{\text{PIV}}} \tag{7}$$

Figure 9 compares this ratio between the two different computational grids. A grid ratio of unity means that the grid spacing in the experiment Δ_{PIV} and the mean cell width in the simulation $V_{\text{Sim}}^{1/3}$ are identical. Values smaller than unity indicate higher resolution in the simulation and *vice versa*. Figure 9 shows locally refined areas for M_1 as well as for M_2 that are of similar shape, whereas M_2 has much higher resolution in the combustion chamber and jet area corresponding to a smaller cell size (Tab. 1). The local refinement in the vicinity of the jet shear layer can be clearly identified. Here, the numerical grid has even higher resolution than the PIV. However, the resolution in regions far from shear layers is not as good as in the PIV.

While the grid ratio in Figure 9 was based on the size of the PIV interrogation area, further evaluation of the experiment will use all 4×4 vectors calculated in each interrogation area. A qualitative impression of the spatial resolution of the velocity field of the simulation and the experiment is given in Figure 10. 2D-2C PIV measurements provide the two velocity components (u, v) only. Accordingly, the instantaneous resolved velocity magnitude:

$$\text{vel}_{\text{mag}} = \sqrt{\bar{u}^2 + \bar{v}^2} \tag{8}$$

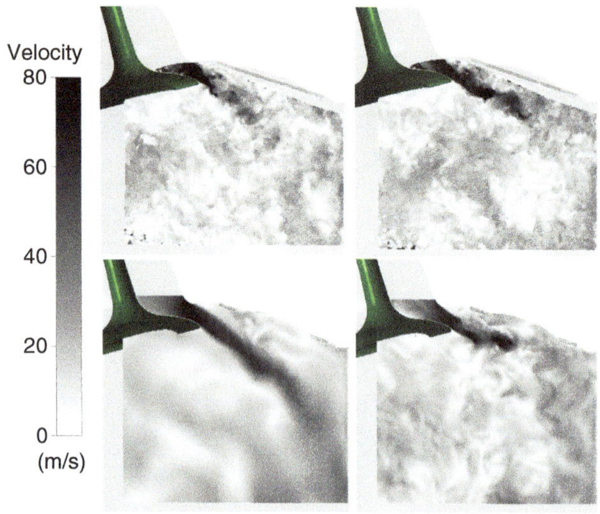

Figure 10

Velocity magnitude: two time steps from PIV (top) and a single simulation time step on mesh M_1 (left, bottom) and M_2 (right bottom).

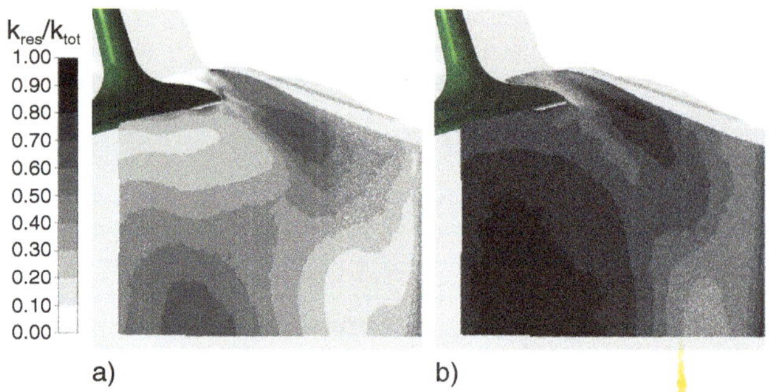

Figure 11

Time-averaged ratio of resolved and total TKE on meshes a) M_1 and b) M_2.

and the time-averaged resolved velocity magnitude:

$$\langle \text{vel}_{\text{mag}} \rangle = \sqrt{\langle \bar{u}^2 \rangle + \langle \bar{v}^2 \rangle} \tag{9}$$

from the simulations is based also on two velocity components only.

As expected, PIV shows more small-scale structures compared to the simulation on both simulation meshes. The flow field on mesh M_1 shows the least resolved turbulent scales. The result of mesh M_2 shows significantly more fluctuations, compared to mesh M_1. For mesh M_2, the spatial resolution is sufficiently fine to resolve large-scale fluctuations.

An approximation for the quality of scale-resolving simulations is the ratio of the resolved to the total turbulent kinetic energy k_{res}/k_{tot}. Pope [26] proposed that an LES should resolve at least 80% of the Turbulent Kinetic Energy (TKE). However, this criterion might not be suitable for very coarse grids. In view of the relatively fine mesh, we adopt Pope's criterion here. Further, analysis of grid quality indicators for scale-resolving simulations can be found in Celik *et al.* [27] and Kempf *et al.* [28].

The total TKE energy is calculated by:

$$k_{tot} = k_{res} + k_{sgs} \tag{10}$$

While the calculation of the kinetic energy of the resolved fluctuations $u'_i = \bar{u}_i - \langle \bar{u}_i \rangle$ is straightforward:

$$k_{res} = \frac{1}{2} \langle u'_i u'_i \rangle \tag{11}$$

the modeled velocity fluctuations have to be estimated based on the eddy viscosity ν_t and the turbulent dissipation rate ε:

$$\nu_t = C_M \frac{k_{\text{sgs}}^2}{\varepsilon} \tag{12}$$

The turbulent dissipation can be approximated using the resolved shear strain S_{ij} and the turbulent viscosity [29] and reads:

$$\varepsilon = 2\, \nu_t S_{ij} S_{ij} \tag{13}$$

Finally, the modeled TKE can be approximated as:

$$k_{sgs} = \left(\frac{2\, \nu_t^2 S_{ij} S_{ij}}{C_M} \right)^{0.5} \tag{14}$$

Based on the underlying SST turbulence model, the constant C_M is set to 0.09.

Figure 11 shows the locally estimated ratio of resolved to total for meshes M_1 and M_2 averaged over all snapshots. As expected, mesh M_1 shows significantly lower resolution compared to mesh M_2 in terms of velocity fluctuations, while neither M_1 nor M_2 resolve the entire section investigated here with a ratio greater than 80%. In particular, close to the intake valve, the ratio in both simulations is very small, which indicates insufficient resolution. In contrast, the modeled velocity fluctuations in the combustion chamber are low, due to low velocity gradients, which yields TKE ratios close to unity.

Figure 12 shows the time-averaged velocity magnitude (Eq. 8) and velocity fluctuation of the resolved flow field for the simulation on mesh M_2 compared to PIV. The lines L_1 and JL will be consider below in further investigation of velocity magnitude, velocity fluctuation, and jet orientation. The two-dimensional resolved fluctuating

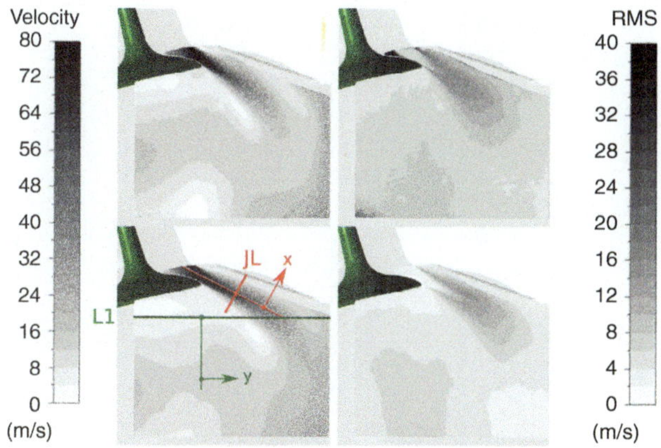

Figure 12

Time-averaged velocity magnitude (left) and RMS velocity (right) from PIV (top) and simulation (bottom) on mesh M_2.

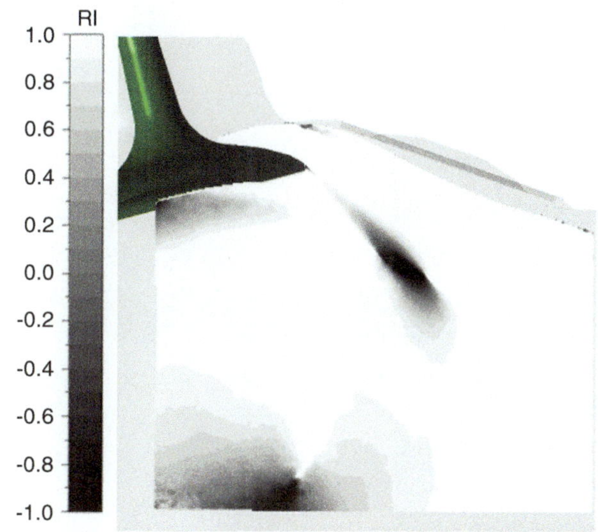

Figure 13

RI between the time-averaged velocity fields from simulation and PIV measurements.

velocity magnitude in simulation and PIV are quantified by Equation (15):

$$u_{rms} = \sqrt{\frac{1}{2}(\langle u'u' \rangle + \langle v'v' \rangle)} \tag{15}$$

Comparing the velocity magnitude from the simulation (Fig. 12, bottom) and experiment (Fig. 12, top), all characteristics of the mean flow field from the experiment can be found in the simulation. Since the jet is the dominating flow structure in this case, its orientation is important. Figure 12 shows that the simulated jet is slightly tilted upwards towards the cylinder head compared to the PIV experiment. One explanation is the manufacturing tolerance of the valve seat, which has a well-defined edge according to the CFD model, while the real edge has a finite radius. That means the point of flow separation might be slightly shifted, which influences the mean orientation of the jet. Nevertheless, with respect to the velocity magnitude and resolved velocity fluctuations, in general we find good agreement between simulation and experiment. Concerning the fluctuations, the PIV experiment shows areas of high Root Mean Square (RMS) close to the valve stem and in the bottom left corner of the ROI. In these areas, the primary peak ratio in the PIV is low, indicating that vector noise may contribute significantly to the RMS. In contrast, the high values of RMS velocity close to the valve jet are characteristic of turbulent jets. These high-RMS-zones may also partially be due to the jet "flapping" up and down. This non-stationary character of the jet can be identified in both PIV image series as well as in the time-resolved simulation.

In addition to the magnitudes, the orientation of the velocity vectors can be investigated by the Relevance Index (RI), defined as:

$$RI = \frac{\vec{u} \cdot \vec{v}}{\|\vec{u}\| \, \|\vec{v}\|} \tag{16}$$

This index was introduced by Liu and Haworth [30] and it is well suited for finding correlations between different flow fields. It compares vector orientations by projecting the velocity vectors of a flow field $\vec{u}$ onto the velocity vectors of another flow field $\vec{v}$. The range of RI lies between -1 (parallel, opposite direction) and 1 (parallel, same direction). A value of zero means that the vectors are orthogonal. The vector $\vec{u}$ is represented by the two time-averaged velocity components of the PIV and the vector $\vec{v}$ is represented by the corresponding two time-averaged velocity components of the simulation.

Figure 13 shows the local deviations in orientation between the experiment and the simulation using the RI. Averaged over the ROI, the RI is 0.85. Regarding the local RI, the largest deviations are found in the lower jet shear layer. There, the penetrating jet establishes a local recirculation area. Since the downward inclination of the jet differs between the experiment and the simulation, the recirculation areas from the simulation and the jet from the experiment overlap in the area with a relevance index of -1. The low RI in the bottom left corner of the section again is an indicator of uncertainties in the PIV measurements, caused by insufficient illumination, as discussed in Section 1.3.

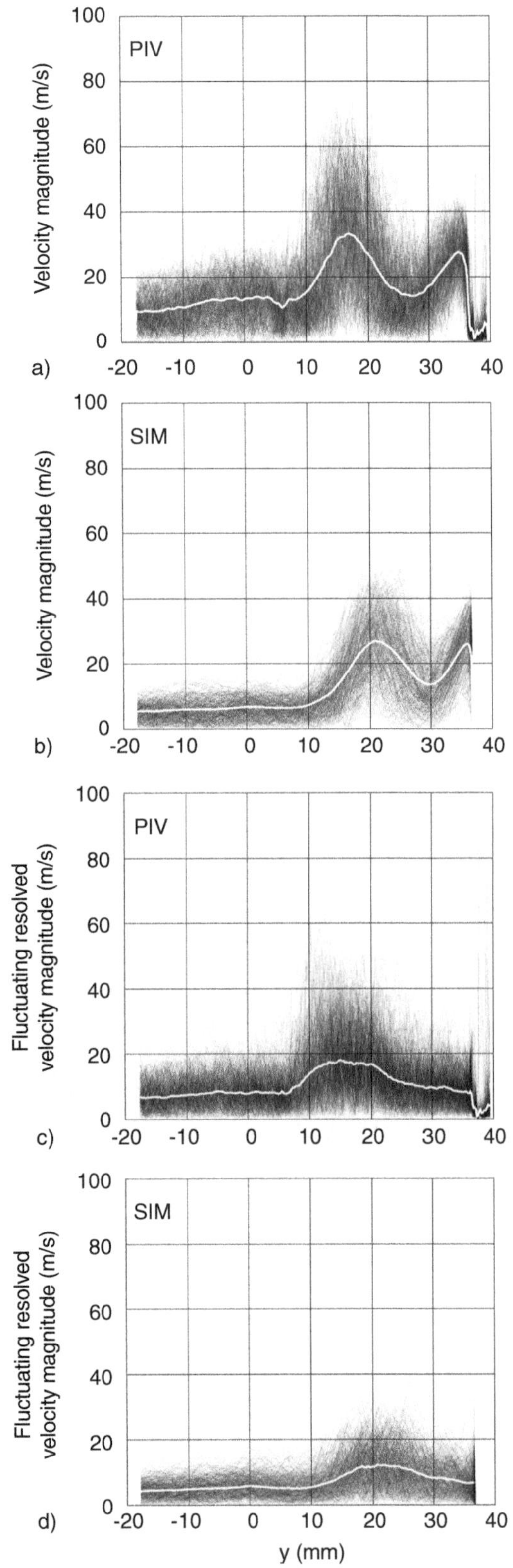

Figure 14

Velocity magnitude from simulation a) and experiment b) at line L_1; c) and d) show the fluctuating velocity from experiment and simulation. The white line represents the time-average, while the black lines are from individual snap-shots.

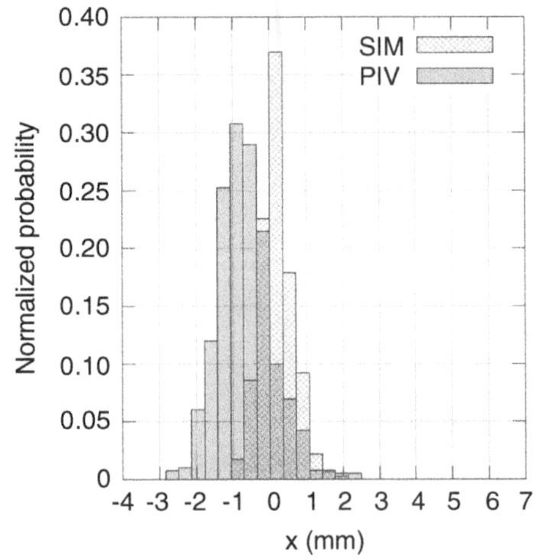

Figure 15

Probability of the jet-centerline being located at some value x along line JL, accumulated over the ensemble of snap-shots.

A more detailed analysis is presented in Figure 14 based on the velocity magnitude and fluctuating velocity at line L_1. Figure 14 plots values from the individual snap-shots in black, while the white line represents the time-average. Comparing the time-averaged velocities between the simulation and the experiment, the peak velocities and the velocities near the jet are in good agreement, but the location of the maximum velocities is different. The maximum velocity is located at $y = 22$ mm in the simulation and $y = 17$ mm in the PIV. As discussed in the previous section, the deviation of the jet orientation between experiment and simulation influences the entire flow field. Because of that, the locations of the peak velocities and the shape of the velocity profile are different. For example, at $y = 5$ mm, the simulations show the velocity magnitude slightly increasing to higher y-coordinate, while the PIV result shows a sharp local minimum. The time-averaged velocity fluctuations from the simulation is similarly shifted to a higher y-value with respect to the PIV results. In both simulation and experiment, the snapshots have large scatter, highlighting again the strongly fluctuating character of the flow. The maximum velocity magnitude values are about 50 m/s and 70 m/s for the simulation and the experiment, respectively. Although a rigorous comparison is not possible with the number of snapshots depicted here, the scatter of the velocity fluctuations in the PIV exceeds that in the simulation by about a factor of two.

An analysis at line JL provides some more detail on the deviation in jet orientation between simulation and experiment. Figure 15 presents the probability of the jet centerline being located at some value x along line JL, accumulated over the ensemble of snapshots. To calculate this probability, in each snapshot the instantaneous velocity magnitude sampled along line JL was thresholded at 95% of its maximum value. The jet centerline was then determined by averaging the x-values of the samples above the threshold. Line JL is orthogonal to the intake-valve seat, with the origin of the coordinate x being located at the intersection between JL and a second line, which is parallel to the valve seat with a distance of 2.5 mm (half valve lift) from the valve seat (Fig. 12). Positive values of x are towards the cylinder head and negative values towards the combustion chamber. The highest probability of finding the jet centerline can be found at $x = -1$ mm for PIV and $x = 0$ mm for the simulation, *i.e.*, a shift of the jet of about 1 mm. Additionally, the range of jet fluctuations is higher in the PIV compared to the simulation. This illustrates that the jet is more "stable" in the simulation, another indicator that the resolution should be increased.

3.2 Boundary Layer Profiles Upstream of the Intake Valve Seat

The following section analyses the time-averaged boundary layer profiles upstream of the flow separation and specifically looks at the general agreement with the standard wall function. Here, the flow separation at the intake valve is geometrically fixed at the intake valve lip, which is marked by a yellow line in Figure 16. Following Schlichting *et al.* [31], the standard wall boundary layer can be classified into three zones. A viscous-sublayer is given for $y^+ < 5$, while the Log-Law is taken to be valid for $y^+ > 70$. The transition zone lies in between and neither the viscous-sublayer nor the Log-Law formulation is valid.

Figure 16 shows the non-dimensional boundary layer along the lines L_3, L_4, L_9, L_{10}, L_{15}, L_{16} and the analytical formulation for the viscous sublayer and for the Log-Law region. The first line pair (L_{15} and L_{16}) is located upstream of the valve seat. The second line pair (L_3 and L_4) is located at the smallest gap between intake port and intake valve. The lines L_9 and L_{10} are located at the valve seat. The distances between each line pair are identical.

All of the simulated boundary layer profiles reach into the viscous sublayer. The boundary layer profiles at L_3, L_4, L_{15} and L_{16} match pair-wise, while the lines L_9 and L_{10} show distinctively different results. This shows that the velocity profile upstream of the valve seat is fully developed, while the lines downstream of the valve seat do not have a fully developed profile, potentially because of the upstream edge of the valve seat. At none of the cross-sections are the simulated boundary layer profiles in good agreement with the analytical formulation. Similar numerical investigations by Jainski *et al.* [32] concentrate on the area around the cylinder head of an operating engine at various crank angles and confirm the potential inaccuracies when using the logarithmic wall functions.

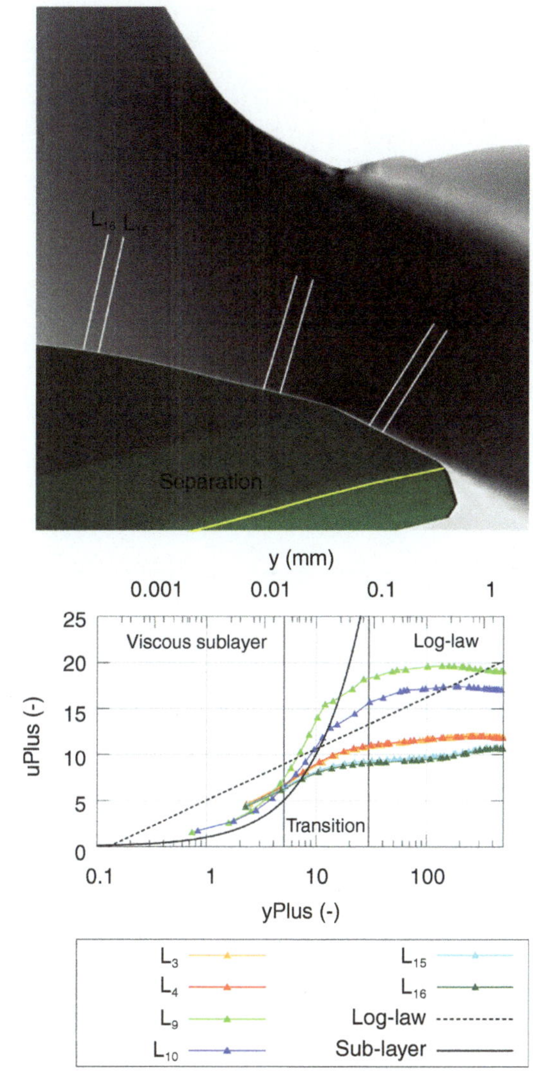

Figure 16

Dimensionless boundary layer profile above the intake valve including the analytical formulation for a fully turbulent flat-plate boundary layer (bottom) and the lines of investigation upstream of the flow separation (top). The second x-axis of the plot presents the wall distance in mm.

3.3 Frequency Analysis in the Vicinity of the Intake Valve Stem

The vicinity of the intake valve is illustrated in Figure 17. Additionally, the points of interest for a detailed frequency analysis are presented. A recirculation area with vortex-shedding develops in the wake. Based on these observations,

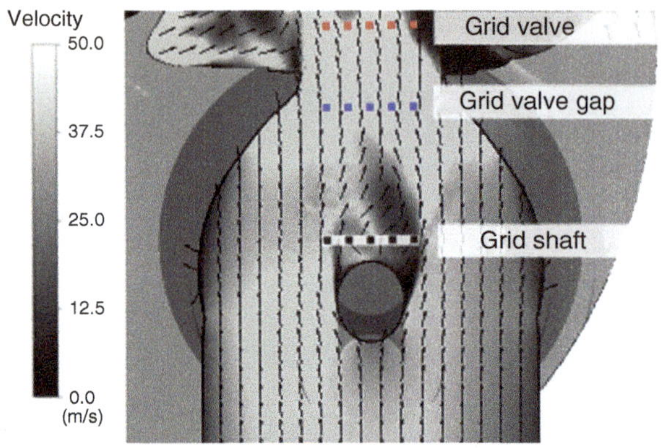

Figure 17

Snap-shot of the flow field around intake valve stem on a plane at the intake-pipe center line.

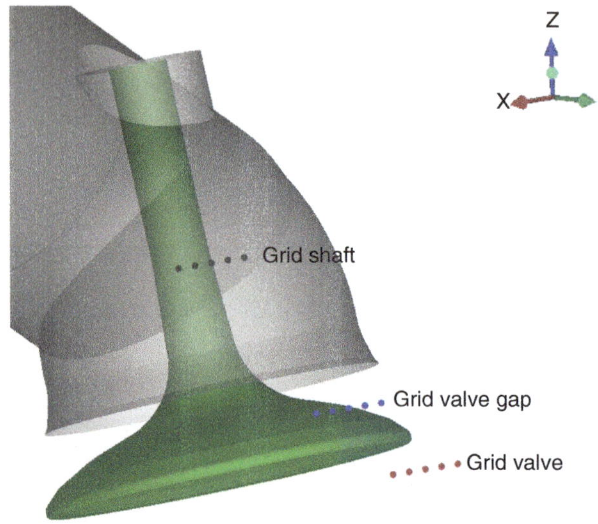

Figure 18

Locations of investigation Grid Shaft, Grid Valve Gap and Grid Valve within the flow bench geometry. The points along the x-axis show the locations for the detailed frequency analysis.

further analysis to identify dominant flow separation frequencies in the wake of the intake valve stem is carried out. We first calculate the Strouhal number, which with the frequency f, the characteristic diameter d_{ch} and the velocity u_∞ is defined as:

$$Sr = \frac{f\, d_{ch}}{u_\infty} \tag{17}$$

The characteristic diameter (d_{ch}) is taken to be the hydraulic diameter for the elliptical cross-section ($d_h = 4A/U = 3.53$ mm) of the valve stem in the cut plane presented in Figure 17. The velocity u_∞ is taken to be 45 m/s, which is the time-averaged velocity magnitude on the intake pipe centerline upstream of the valve stem. The Strouhal number of a flow past a cylinder was determined experimentally by Roshko [33]. In their work, the Strouhal number is plotted as a function of the Reynolds number ($Re = d_{ch}\, u_\infty/v$). Here, the Reynolds number based on the intake valve stem is 8 680, calculated using the hydraulic diameter of the ellipse and the molecular viscosity of air (1.83×10^{-5} m^2/s). This yields a Strouhal number of 0.196, which corresponds to a frequency of about 2 500 Hz.

To investigate the dominant frequencies of the velocity field behind the valve stem, we use a discrete Fourier transformation. If a dominant frequency can be identified, the question is whether it can "survive" the acceleration in the valve gap and thus can be a potential source of fluctuations in the jet. Probe points are placed within the numerical domain as illustrated in Figures 17 and 18. At each point and each time-step, all three velocity components are captured at three locations. The first one, Grid Shaft (GS), is placed directly behind the valve stem. The second one, Grid Valve Gap (GVG), is located within the valve gap.

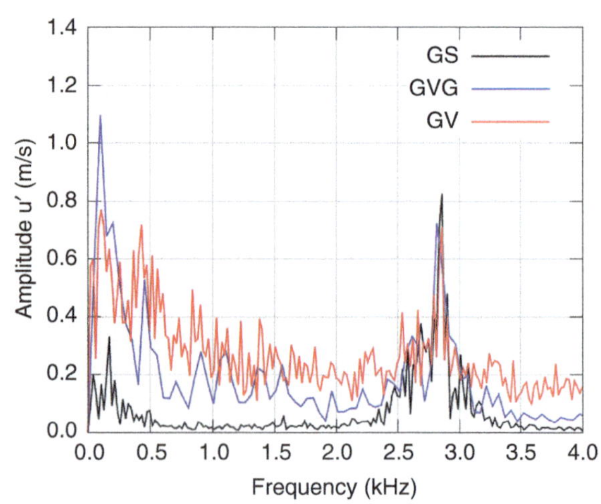

Figure 19

Fourier-transformed time series of the velocity fluctuation u' of all locations (Fig. 18).

The third one, Grid Valve (GV), lies downstream of the valve lip in the combustion chamber in the free-jet region.

First, all sampled data at the illustrated locations are time-averaged to calculate the mean velocity. Subtracting the resulting mean velocity from instantaneous resolved velocity yields the resolved fluctuation. Averaging the velocity fluctuation along the points at GS, GVG and GV at each sampled time step yields a time series, which is used for the Fourier analysis. The Fourier analysis of each velocity component shows that indeed there is a dominant frequency in the fluctuating velocity component u' in x-direction. The amplitudes of the velocity fluctuations u' are presented in Figure 19.

The location GS shows a significant fluctuation peak at 2 860 Hz, which corresponds approximately to that calculated from the Strouhal number. The locations GVG and GV show a broader range of relevant frequencies, but a distinctive peak at 2 860 Hz can still be detected. The importance of other frequencies increase with increasing distance from the valve stem. Overall, the analysis indicates that the vortex structure behind the valve stem survives the acceleration through the valve gap into the combustion chamber.

SUMMARY

This work investigated the unsteady flow through the intake valve of a typical spark-ignited engine's cylinder head and how it influences the jet entering the combustion chamber. With the cylinder head on a flow bench, the valve lift was fixed at 5 mm. The study utilized two-dimensional, two-component highly-resolved PIV and scale-resolving simulations on two different grids. Instantaneous flow fields illustrate the high resolution of the experimental data and locally even finer structures could be resolved than in the simulations on the finest grid. Generally, good agreement between the numerical and experimental velocity fields was observed. Statistical analysis of field-wide quantities as well as more detailed examination along selected line cuts through the data reveal some differences between experiment and simulation, in particular, in the orientation of the intake jet. Based on this, we suggest that even higher resolutions for the CFD simulation close to the jet should be the scope of future studies.

An analysis of the boundary layer profiles above the valve in the vicinity of the separation point revealed that the commonly-used wall formulation for $y^+ > 70$ was not applicable and the simulation grid should be fine enough to place the first grid point in the viscous sublayer as it is done here.

Finally, we investigated the vortex shedding at the valve stem. A Fourier analysis to identify dominant frequencies and a Strouhal-number-based approximation were compared. The Fourier analysis identified a dominant frequency at 2 860 Hz in the wake of the stem, which is close to the Strouhal approximation of 2 500 Hz. This frequency was also found in the fluctuating velocity in the combustion chamber, downstream of the acceleration between the valve and its seat. This indicates that such vortices generated upstream of the valve gap can influence the flow field as far downstream as the combustion chamber and thus in an operating engine could be a potential trigger of cyclic variations.

ACKNOWLEDGMENTS

The authors kindly acknowledge the financial support by the Saxon Ministry of Science and Fine Arts, the SAB and the European Union in the project „DynMo" (project number 100113147), the Forschungsvereinigung Verbrennungsmotoren (FVV) and the Deutsche Forschungsgemeinschaft (DFG) in the project "BSZ II" (FVV project number 6011333, DFG project ID Schu 1369/11-2) and the Rückkehrerprogramm of the NRW Ministry for Innovation, Science, and Research. The simulations were performed on the national supercomputer Cray XC40 (Hornet) at the High Performance Computing Center Stuttgart (HLRS) under the grant number ICECCV/44054.

REFERENCES

1 Bicen A.F., Vafidis C., Whitelaw J.H. (1985) Steady and unsteady airflow through the intake valve of reciprocating engine, *Journal of Fluids Engineering, Transactions of the ASME* **107**, 3, 413-420.

2 Heywood J.B. (1988) *Internal combustion engine fundamentals*, Mcgraw-Hill, New York.

3 Imberdis O., Hartmann M., Bensler H., Kapitza L., Thevenin D. (2007) A numerical and experimental investigation of a DISI-engine intake port generated turbulent flow, *SAE Technical Paper* 2007-01-4047.

4 Stansfield P., Wigley G., Justham T., Catto J., Pitcher G. (2007) PIV analysis of in-cylinder flow structures over a range of realistic engine speeds, *Experiments in Fluids* **43**, 1, 135-146.

5 Namazian M., Hansen S., Lyford-Pike E., Sanchez-Barsse J., Heywood J., Rife J. (1980) Schlieren visualization of the flow and density fields in the cylinder of a spark-ignition engine, *SAE Technical Paper* 800044.

6 Weclas M., Melling A., Durst F. (1995) Unsteady intake valve gap flows, *SAE Technical Paper* 952477.

7 Weclas M., Melling A., Durst F. (1998) Flow separation in the inlet valve gap of piston engines, *Progress in Energy and Combustion Science* **24**, 3, 165-195.

8 Valentino G., Kaufman D., Farrell P. (1993) Intake valve flow measurements using PIV, *SAE Technical Paper* 932700.

9 Bücker I., Karhoff D.C., Klaas M., Schröder W. (2012) Stereoscopic multi-planar PIV measurements of in-cylinder tumbling flow, *Experiments in Fluids* **53**, 6, 1993-2009.

10 Baum E., Peterson B., Surmann C., Michaelis D., Böhm B., Dreizler A. (2013) Investigation of the 3D flow field in an IC engine using tomographic PIV, *Proceedings of the Combustion Institute* **34**, 2, 2903-2910.

11 Freudenhammer D., Baum E., Peterson B., Böhm B., Jung B., Grundmann S. (2014) Volumetric intake flow measurements of an IC engine using magnetic resonance velocimetry, *Experiments in Fluids* **55**, 5.

12 McLandress A., Emerson R., McDowell P., Rutland C. (1996) Intake and in-cylinder flow modeling characterization of mixing and comparison with flow bench results, *SAE Technical Paper* 960635.

13 Inagaki M., Nagaoka M., Horinouchi N., Suga K. (2010) Large eddy simulation analysis of engine steady intake flows using a mixed-time-scale subgrid-scale model, *International Journal of Engine Research* **11**, 3, 229-241.

14 Kaiser S.A., Schild M., Schulz C. (2013) Thermal stratification in an internal combustion engine due to wall heat transfer measured by laser-induced fluorescence, *Proceedings of the Combustion Institute* **34**, 2, 2911-2919.

15 Gessenhardt C. (2013) Endoskopische Bestimmung des Temperaturfeldes im Brennraum von Ottomotoren mittels laserinduzierter Fluoreszenz, *PhD Thesis,* Universität Duisburg-Essen.

16 Melling A. (1997) Tracer particles and seeding for particle image velocimetry, *Measurement Science and Technology* **8**, 12, 1406-416.

17 LaVision (2013) *Product-Manual Flow Master for DaVis 8.1.*

18 Wieneke B., Pfeiffer K. (2010) Adaptive PIV with variable interrogation window size and shape, *15th International Symposium on Applications of Laser Techniques to Fluid Mechanics.*

19 Hasse C., Sohm V., Durst B. (2009) Detached eddy simulation of cyclic large scale fluctuations in a simplified engine setup, *International Journal of Heat and Fluid Flow* **30**, 1, 32-43.

20 Hasse C., Sohm V., Wetzel M., Durst B. (2009) Hybrid URANS/LES turbulence simulation of vortex shedding behind a triangular flameholder, *Flow, Turbulence and Combustion* **83**, 1, 1-20.

21 Hasse C., Sohm V., Durst B. (2010) Numerical investigation of cyclic variations in gasoline engines using a hybrid URANS/LES modeling approach, *Computers and Fluids* **39**, 1, 25-48.

22 Sohm V. (2007) Hybrid turbulence simulation to predict cyclic variations in internal combustion engines, *PhD Thesis*, RWTH Aachen.

23 Travin A., Shur M., Strelets M., Spalart P.R. (2004) Physical and numerical upgrades in the detached-eddy simulation of complex turbulent flows, *Fluid Mechanics and its Applications* **65**, 239-254.

24 Menter F.R. (1994) Two-equation eddy-viscosity turbulence models for engineering applications, *AIAA journal* **32**, 8, 1598-1605.

25 Menter F.R., Kuntz M. (2003) Development and application of a zonal DES turbulence model for CFX-5, ANYSYS CFX validation report, *Technical report*, ANSYS.

26 Pope S.B. (2000) *Turbulent Flows*, Cornell University, 1st ed,

27 Celik I., Klein M., Janicka J. (2009) Assessment measures for engineering LES applications, *Journal of Fluids Engineering* **131**, 3, 031102.

28 Kempf A., Geurts B., Ma T., Pettit M., Stein O. (2011) Quality issues in combustion LES, *Journal of Scientific Computing* **49**, 1, 51-64.

29 Tennekes H., Lumley J.L. (1972) *A first course in turbulence*, MIT press.

30 Liu K., Haworth D. (2011) Development and assessment of POD for analysis of turbulent flow in piston engines, *SAE Technical Paper* 2011-01-0830.

31 Schlichting H., Gersten K., Gersten K. (2000) *Boundary-layer theory*, Springer Science & Business Media.

32 Jainski C., Lu L., Dreizler A., Sick V. (2013) High-speed micro particle image velocimetry studies of boundary-layer flows in a direct-injection engine, *International Journal of Engine Research* **14**, 3, 247-259.

33 Roshko Anatol (1961) Experiments on the flow past a circular cylinder at very high reynolds number, *Journal of Fluid Mechanics* **10**, 03, 345-356.

9

Combustion, Performance and Emission Analysis of an Oxygen-Controlling Downsized SI Engine

J.X. Zhou[1], B. Moreau[2], C. Mounaïm-Rousselle[2] and F. Foucher[2]*

[1] College of Mechanical and Electrical Engineering, Wenzhou University, Wenzhou 325035 - PR China
[2] Université d'Orléans, laboratoire PRISME, 8 Rue Léonard de Vinci, 45072 Orléans - France
e-mail: zhoujxi@wzu.edu.cn - bruno.moreau@univ-orleans.fr - christine.rousselle@univ-orleans.fr - fabrice.foucher@univ-orleans.fr

* Corresponding author

Abstract — *In the present study, experiments were carried out in a single-cylinder downsized SI engine with different rates of oxygen (15% to 27% by volume in the total mixture of intake gases except fuel) and equivalence ratios (from 0.45 to 1). Therefore, the oxygen volume fraction is due to oxygen enrichment or nitrogen dilution. The study of the impact of controlling the oxygen concentration on the combustion characteristics and emissions was performed at 1 400 rpm, at several loads (Indicated Mean Effective Pressure (IMEP) from 400 to 1 000 kPa). For each operation point, the spark advance and the intake pressure were adjusted simultaneously in order to maintain the load and obtain a minimum value of the indicated Specific Fuel Consumption (SFC). The effect of the oxygen concentration on the engine combustion characteristics was simulated by using the commercial software AMESim, with the combustion model developed by* IFP Energies nouvelles, *and an adapted algorithm was used to avoid residual gas calibration. By implementing a correlation for the laminar burning velocity, the in-cylinder pressures were perfectly predicted with a maximum pressure relative error of less than 2% for almost all the operating points. The classification of engine combustion according to the Peters-Borghi diagram, gives a deeper insight into the interaction between turbulence and the flame front.*

Résumé — Combustion, performance et analyse des émissions d'un moteur essence suralimenté avec contrôle de l'oxygène admis — Dans cette étude, des expériences sur monocylindre essence suralimenté ont été réalisées avec comme variable le taux d'oxygène (15 % à 27 % en volume dans le mélange total de gaz d'admission à l'exception du carburant) admis et la richesse (0,45 à 1). La fraction volumique d'oxygène représente en fait soit un enrichissement en oxygène de l'admission moteur ou de l'azote de dilution. L'étude de l'impact du contrôle de la concentration d'oxygène sur les caractéristiques de combustion et les émissions a été effectuée à 1 400 tr/mn, pour plusieurs charges, de 400 à 1 000 kPa. Pour chaque point de fonctionnement, l'avance à l'allumage et la pression d'admission ont été ajustées simultanément afin d'obtenir la charge maximale et la valeur minimale de consommation spécifique indiquée de carburant. L'effet de la concentration en oxygène dans la combustion du moteur a été simulé à partir du logiciel AMESim, avec le modèle de combustion développé par l'*IFP Energies nouvelles*, et un algorithme adapté a été utilisé pour éviter l'étalonnage de gaz résiduels. À partir d'une corrélation pour la vitesse de combustion laminaire, les valeurs de pression cylindre ont été prédites avec une erreur relative inférieure à 2 %, pour tous les points de fonctionnement. Le classement de la combustion selon le diagramme de Peters-Borghi a permis d'avoir une idée plus précise de l'interaction entre la turbulence et le front de flamme dans ces conditions.

LIST OF SYMBOLS

CAD	Crank Angle Degree
CFD	Computational Fluid Dynamics
CFM	Coherent Flame Model
EGR	Exhaust Gas Recirculation
IMEP	Indicated Mean Effective Pressure
k	Turbulent kinetic energy
l_t	Integral length scale
m	Mass
P	Pressure
Q_{comb}	Heat release by combustion
r_{bg}	Mean radius of burnt gases
SFC	Specific Fuel Consumption
S_L^0	Laminar burning velocity
T	Temperature
UHC	Unburned HydroCarbon
u'	Instantaneous velocity fluctuation
V	Volume
V_{cyl}	Displacement volume of cylinder
W	Molar weight
Y	Mass fraction
ff	Flame front
pf	Post-flame
α	Correlation constant
β	Correlation constant
γ	Correlation constant
δ_l	Laminar flame thickness
ϕ	Equivalence ratio
ρ	Density
τ	Thermal expansion rate
σ_m	Mean flame surface
σ_t	Turbulent flame surface
χ	Volume fraction
υ	Stoichiometric coefficient
v_{O_2}	O_2 percentage in the air
Ξ	Flame front wrinkling factor

INTRODUCTION

The objective of this paper is to study the combustion of a downsized SI engine in the case of controlled oxygen concentration at the intake. On the one hand, oxygen-enriched combustion can improve engine power density with the same intake pressure level. Thus, oxygen-enriched combustion can be used either as a booster to increase engine output or as a combustion enhancer when the engine operates at low loads or in cold-start conditions. Several studies performed with oxygen enrichment in SI engines showed positive effects such as reduction of Unburned HydroCarbon (UHC) fuel and CO [1-4], optimization of the downsized operating mode due to the direct effect on the combustion process and overall engine thermodynamics [5], and a decrease in the combustion phase [6-8]. On the other hand, a low oxygen concentration in the air (or N_2 dilution) can be considered as an alternative to Exhaust Gas Recirculation (EGR), which has been widely investigated in recent years [9, 10].

The oxygen concentration in the air can be controlled by using elective permeation through a non-porous polymeric membrane. This kind of polymeric membrane is now extensively studied and was developed recently to solve a part of the CO_2 sequestration problem for fixed-site power generation, for example [6, 11, 12]. It would deplete or enrich oxygen in the intake air while being compact, lightweight, embeddable in the vehicle and inexpensive by overcoming the current problems of a loop EGR. Different from the current EGR dilution system, air is pre-selected to meet the N_2 dilution or O_2 enrichment level as desired for the oxygen controlled by the membrane. Then, the air is recomposed with oxygen enrichment or nitrogen dilution for engine applications [13].

The first purpose of this paper is to investigate experimentally the impact of controlling the oxygen concentration on engine performance and pollutant emissions, and to give a global overview of the optimized engine operating conditions with the lowest Specific Fuel Consumption (SFC) and exhaust gas emissions. In order to provide more details about the specificity of combustion processes with oxygen control, such as turbulent flame velocity evolution during the engine cycle or flame wrinkling, the AMESim commercial software with the combustion model developed by Richard *et al.* [14] was used. The AMESim software provides the following advantages: first, it enables the various physical domains involved in vehicle and engine system simulation to be addressed with a high level of detail [15]. Thanks to this advantage, an AMESim model can be set up with reference to the experimental configuration; for example, intake and exhaust pipes, engine valves, turbo-compressors, and so on. Secondly, the combustion model developed by Richard *et al.* [14] gives an accurate description of the physical processes. Although this combustion model is based on a weak assumption for the mean flame geometry, the combustion parameters (*i.e.* flame surface, turbulent intensity, laminar and turbulent burning velocities) provided by the model give a realistic physical meaning for the in-cylinder combustion process.

1 EXPERIMENTAL SETUP

1.1 Engine Characteristics

Experiments were carried out in a single-cylinder SI engine (*PSA* EP6) characterized by a four-valve pent-roof chamber

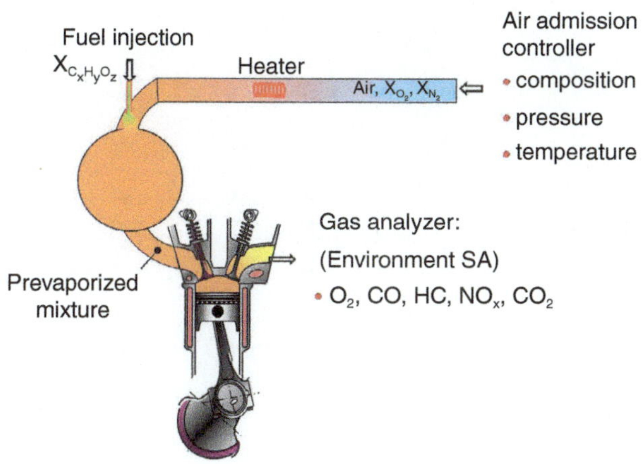

Figure 1

Scheme of air-fuel mixture system for the single-cylinder engine.

with a displacement volume (V_{cyl}) of 399.5 cm^3 and a compression ratio of 10.5. The bore, stroke and connecting rod lengths were 77 mm, 85.8 mm and 138.5 mm, respectively. The engine was driven by an electric motor at a fixed engine speed, and equipped with an optical encoder mounted on the main shaft, giving a 0.1 Crank Angle Degree (CAD) as resolution. A conventional spark plug with an electrode space of 1 mm was used. A timer card ensured synchronization of the various trigger signals and data acquisition systems. The engine sucks in the air through a thermal mass flowmeter and all other flows were evaluated from this reference flow. Oxygen, nitrogen and air flows were controlled by using thermal mass flow meters with an accuracy of ±0.7% for the instantaneous flow. Before the intake pipe, all the gases passed through a plenum volume, to avoid pressure oscillations inside the intake port. The isooctane quantity was regulated by using a 0-8 kg/h *Bronkhorst* Coriolis mass flow controller with a maximum combined standard uncertainty of ±1% for the minimum flow rate considered in the present work. To provide a premixed air-fuel mixture inside the intake pipe, the fuel was vaporized and mixed inside the intake pipe just after the plenum to obtain a homogeneous air-fuel mixture. The air-fuel mixture temperature was kept constant at 70°C by an electric heater. A scheme of the air-fuel injection system in the engine environment is presented in Figure 1.

Cylinder pressure was recorded with a water-cooled *AVL* quartz pressure transducer connected to a charge amplifier at 0.1 CAD resolutions. The linearity (<±0.6%) of the transducer was verified starting from a maximum pressure of 10 MPa (> maximum pressure in the engine). In this work, the absolute cylinder pressure was deduced by equalizing the in-cylinder pressure at 20 CAD after inlet valve opening timing to the intake mean absolute pressure. Thus, 100 consecutive cycles of cylinder pressure data were acquired by a PC equipped with a National Instruments acquisition board.

The exhaust emissions of gases (NO_x, HC, CH_4, CO, CO_2 and O_2) were measured by a classical emission analyzer (from *Environment S.A.*) with an accuracy <2% of the measured value and 1% of full scale.

1.2 Operating Conditions

Oxygen control was obtained by adding oxygen or nitrogen to ambient air. For simplification, we assume that the air consists of 21% O_2 and 79% N_2. To provide an O_2 percentage higher than 21%, O_2 was supplied by a compressed O_2 high-pressure tank. For an O_2 percentage less than 21%, N_2 was produced by an N_2 generator. Here, the percentage is defined as the volume fraction:

$$\chi_{O_2} = \frac{V_{O_2}}{V_{O_2} + V_{N_2}}.100\%$$

Thus, for $\chi_{O_2} > 21\%$, we have:

$$\chi_{O_2} = \frac{0.21V_{air} + V_{O_2}}{V_{air} + V_{O_2}}$$

with: $V_{air} + V_{O_2} = 1$.

For $\chi_{O_2} < 21\%$,

$$\chi_{O_2} = \frac{0.21V_{air}}{V_{air} + V_{N_2}}$$

with: $V_{air} + V_{N_2} = 1$.

The volume fraction of air, O_2 and N_2 can then be determined for a given O_2 percentage as shown in Table 1. The equivalence ratio φ was also predefined during experiments based on the quantity of oxygen and not on the quantity of air:

$$\phi = \frac{\text{stoichiometric } O_2/\text{fuel ratio}}{\text{actual } O_2/\text{fuel ratio}}$$

TABLE 1

Determination of the volume fraction of air, O_2 and N_2 for different cases

For $\chi_{O_2} > 21\%$	$V_{O_2} = \frac{\chi_{O_2} - 0.21}{0.79}$
	$V_{air} = \frac{1.21 - \chi_{O_2}}{0.79}$
For $\chi_{O_2} < 21\%$	$V_{N_2} = \frac{0.21 - \chi_{O_2}}{0.21}$
	$V_{air} = \frac{\chi_{O_2}}{0.21}$

TABLE 2

Correspondence between the O_2 rate and EGR percentage at the stoichiometric equivalence ratio

O_2 (%)	15	17	19	21	23	25	27
N_2-dilutio (%)	26.9	17.7	8.6	−0.3	−9.2	−17.9	−26.6

The global reaction of the isooctane/N_2/O_2 mixture can be defined as follows:

$$C_8H_{18} + \frac{12.5}{\phi}\left(O_2 + \frac{\chi_{N_2}}{\chi_{O_2}}N_2\right)$$
$$\rightarrow \frac{8}{\phi}CO_2 + \frac{9}{\phi}H_2O + \frac{12.5}{\phi}\frac{\chi_{N_2}}{\chi_{O_2}}N_2 + M, \text{ for rich mixture}$$
$$\rightarrow 8CO_2 + 9H_2O + 12.5\frac{\chi_{N_2}}{\chi_{O_2}}N_2 + M, \text{ for lean mixture}$$

where M is the quantity of excess air or unburned fuel for lean or rich mixtures, respectively. Thus, for a given equivalence ratio ϕ, the fuel mass flow rate is directly linked to the O_2 mass flow rate.

In order to compare the O_2 rate with the EGR percentage for a fixed equivalence ratio, the equivalence between them was calculated by replacing the inert gases CO_2 and H_2O with N_2. For $O_2 < 20.9\%$, combustion takes place with N_2 dilution, and for $O_2 > 20.9\%$, with O_2 enrichment (or 'negative' dilution). An example of equivalence between the O_2 and N_2 dilution rates for a fixed equivalence ratio is presented in Table 2. Negative N_2 dilution corresponds to O_2 enrichment, which also means that the N_2 is 'removed' from the air. As many studies have been carried out with EGR, represented by N_2 dilution in this study, Table 2 facilitates comparison between our data and the data reported by other studies.

For the test, the IMEP was adjusted to keep a quasi-constant value, and the spark ignition was set at the optimum spark ignition timing, to allow lower SFC. The intake pressure was also adjusted during the operation, so that for a given equivalence ratio and O_2 percentage, the IMEP is adjusted by advancing the spark timing and intake pressure. The covariance value of the IMEP was used as a criterion in this study to determine the lean and N_2 dilution operating limits: a maximum of 5% was considered to be acceptable.

1.3 Studied Cases

The different operating points studied experimentally and numerically are presented in Figure 2a with various conditions of the IMEP and oxygen percentage for an air-fuel mixture at the stoichiometric equivalence ratio. In Figure 2b, for a fixed condition of 8 bar IMEP, the effect of different equivalence ratios was improved.

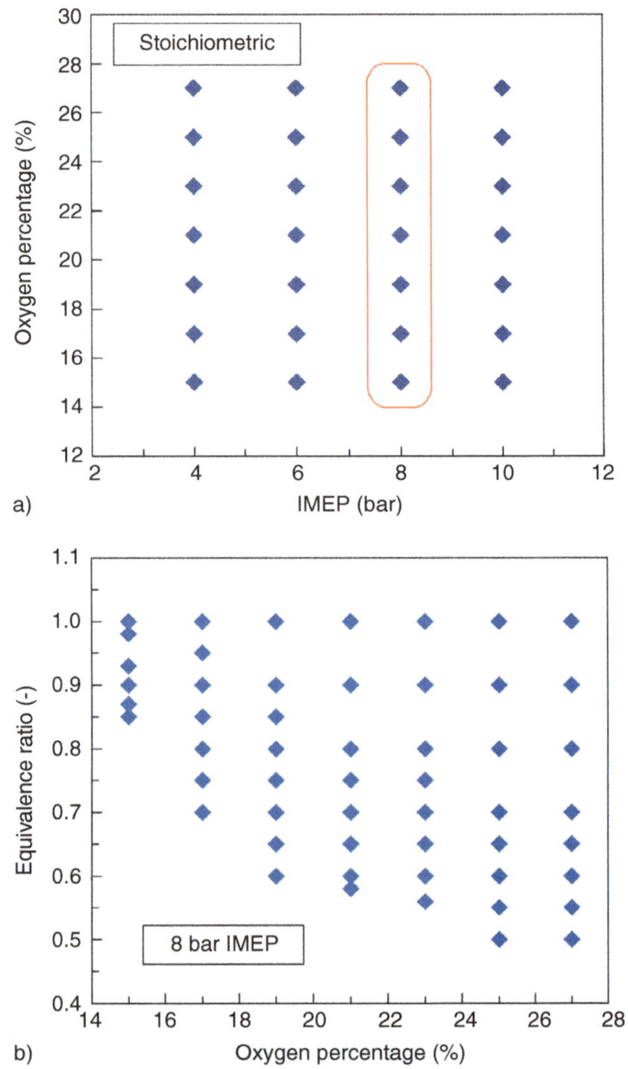

Figure 2

All different operating conditions were studied at fixed 1 400 rpm. a) In fixed stoichiometric conditions, b) in fixed 8 bar IMEP conditions.

2 DESCRIPTION OF THE MODEL

The combustion model used in this work is a physical 0D model, obtained by reduction of the 3D Coherent Flame Model (3D CFM) [14]. CFM formalism distinguishes two zones: fresh and burned gases, which are separated by a flame front propagating from the burned gases toward the fresh gas mixture. Chemical reactions of fuel oxidation occur in the flame front, a very thin layer compared with all scales of the turbulent flow, and post-flame chemistry takes place in the burned gases. The main assumptions made to derive the 0D model are the following [14, 16]:

- in each zone, the mixture composition is assumed to be homogeneous and hence to have uniform properties;
- the pressure is uniform throughout the cylinder in both zones;
- the gaseous mixture consists of 15 species (Fuel, N_2, O_2, H_2, H_2O, CO, CO_2, NO, NO_2, HC, NH_3, soot, O, H and OH);
- the gases considered are ideal gases;
- the turbulent kinetic energy field is assumed to be uniform in the cylinder.

2.1 Heat Release Calculation

The enthalpy balance equations were described in detail in the previous study of Richard *et al.* [14]. The heat release by combustion is expressed as:

$$\frac{dQ_{comb}}{dt} = \sum_i h_{fi} \left(\left.\frac{dm_i}{dt}\right|_{ff} + \left.\frac{dm_i}{dt}\right|_{pf} \right) \tag{1}$$

where h_{fi} is the formation enthalpy of species i, and $dm_i|_{ff}$ and $dm_i|_{pf}$ are the mass variations of this species, respectively, in the flame front (ff) and in the burned gases due to post-flame (pf) chemistry reactions. $dm_i|_{ff}$ can be expressed as a function of the fuel consumption rate:

$$\left.\frac{dm_i}{dt}\right|_{ff} = \upsilon_i \frac{W_i}{W_f} \left.\frac{dm_{fuel}}{dt}\right|_{ff} \tag{2}$$

where υ_i is the stoichiometric coefficient of gas i in the reaction of fuel oxidation through the flame, and W_i denotes the molar weight of this gas. The fuel consumption rate is then given by a flame propagation model described in the next paragraph. Finally, $dm_i|_{pf}$ is computed using a reduced chemistry approach [14] accounting for CO and NO formation in the burned gas zone.

2.2 Flame Propagation Model

The consumption rate of fuel throughout the flame front depends on the turbulent flame surface σ_t and the fresh gas properties according to:

$$\left.\frac{dm^{fg}_{fuel}}{dt}\right|_{ff} = -\rho_{fg} Y^{fg}_{fuel} S^0_L \sigma_t \tag{3}$$

where ρ_{fg} is the fresh gas density, $Y^{fg}_{fuel} = m^{fg}_{fuel}/m_{fg}$ is the fuel mass fraction in the fresh gases and S^0_L is the laminar burning velocity computed from a correlation, a function of the pressure and the fresh gas composition and temperature, described in the subsection below. The turbulent flame surface σ_t is written as the product of a mean surface σ_m and the flame front wrinkling factor, Ξ ($\sigma_t = \Xi\sigma_m$). The mean surface is computed by considering a spherical expansion of the flame from the spark plug to the cylinder walls in a realistic geometry. For this purpose, σ_m is *a priori* tabulated as a function of the burned gas volume and the piston position as described [14, 17]. The flame wrinkling factor evolution is described by a physical equation describing the progressive transition from the laminar kernel to the fully turbulent flame. This equation was obtained by reduction of the 3D equation for the flame surface density [18, 19] and includes the unsteady effect of the turbulent flow on the flame stretch through the efficiency function Γ proposed in [20]:

$$\frac{1}{\Xi}\frac{d\Xi}{dt} = \Gamma\left(u'/S^0_L,\, l_t/\delta_l\right)\frac{u'}{l_t}\left(\frac{\Xi_{equ} - \Xi}{\Xi_{equ} - 1}\right) - \frac{2}{r_{bg}}(1+\tau)(\Xi - 1)S^0_L \tag{4}$$

u' is the instantaneous velocity fluctuation, l_t is the integral length scale, δ_l is the laminar flame thickness estimated using Blint's correlation [21], $\tau = \rho_{fg}/\rho_{bg}$ is the thermal expansion rate, $r_{bg} = (3V_{bg}/4\pi)^{1/3}$ is the current mean radius of burnt gases and Ξ_{equ} corresponds to the Ξ value when an equilibrium is reached between turbulence and flame wrinkling Ξ [18]. The main unknowns of the problem finally correspond to the turbulent flow field properties u' and l_t.

2.3 Turbulence Model

The integral length scale is assumed to be close to the piston-cylinder head distance D, as proposed by many authors [17, 22], while $u' = \sqrt{2/3k}$ is obtained from a classical two- equation model describing the evolution of the mean and turbulent kinetic energy (resp. K and k) [14]. This model accounts for energy transfer from the mean flow into turbulence during the compression stroke and for dissipation of k on the Kolmogorov scale. Such a 0D model cannot be entirely predictive as the complex flow field evolution highly depends on the engine geometry and cannot directly be resolved as in 3D CFD calculations. Classically, the parameters of the turbulence model such as the integral length scale and the tumble number are then usually adjusted.

2.4 Adaptation of the AMESim Model for the Case of an 'Oxygen-Controlled Engine'

Figure 3 presents the sketch of the AMESim model. On the left is the intake part, in the middle the main engine block,

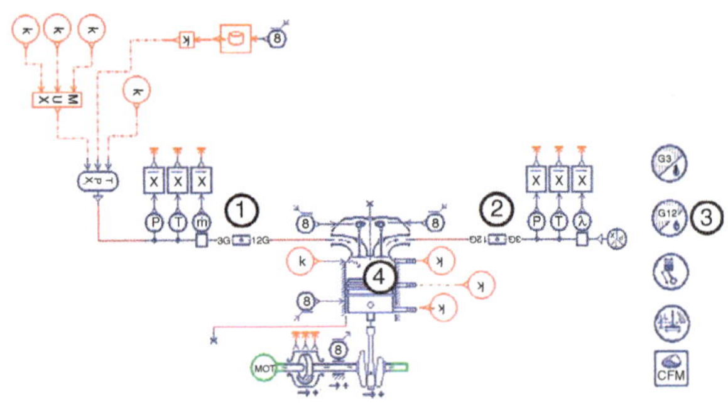

Figure 3

Scheme of AMESim modeling for an oxygen ratio-controlled engine.

containing the model of the intake and exhaust valves, the model of the 12 gases, the engine model, and on the right the exhaust part.

In this AMESim model, four parts of the subroutine (marked as ① to ④) were modified to adapt the oxygen control calculation using an advanced AMESim tool, called 'AMESet':

- ① and ② are pipe models which transform the mixture of 3 gases (air, fuel and burned gases) into 12 species (fuel, N_2, O_2, H_2, H_2O, CO, CO_2, NO, NO_2, HC, NH_3 and soot) when gases pass though the intake or exhaust pipes to the engine block. Inversely, from the engine block to the intake or exhaust side, 12 gases are transformed into 3 gases. In the pipe model, an integer parameter for the O_2 mole percentage was defined; thus, by changing this parameter to above or less than 21%, the air can be considered as oxygen-enriched or nitrogen-diluted, respectively;
- ③ is the model of the 12 gas engine fluid data, where instead of a stoichiometric air-fuel ratio, a stoichiometric oxygen-fuel ratio is calculated to determine the equivalence ratio based on oxygen, as defined in the previous section;
- ④ the engine combustion model, was changed by using the integer parameter for the O_2 mole percentage to control the oxygen percentage in the air and as a variable for laminar burning velocity correlation and auto-ignition delay time correlation.

2.5 Laminar Burning Velocity Correlation

The laminar burning velocity for different quantities of O_2 was experimentally investigated in previous studies [23, 24]. In order to take into account the effect of oxygen control on the engine combustion simulation, a correlation based on an experimental estimation [24] was implemented in the AMESim model:

$$S_L^0 = S_{Lref}^0 \left(\frac{T_i}{T_{ref}}\right)^{\alpha_s} \left(\frac{P_i}{P_{ref}}\right)^{\beta_s}$$

In the above relationship, S_{Lref}^0 is a function of the equivalence ratio:

$$S_{Lref}^0 = A + B(\phi - \phi_m) + C(\phi - \phi_m)^2 + D(\phi - \phi_m)^3 + E(\phi - \phi_m)^4$$

and,

$$\alpha_s = \alpha_1 + \alpha_2(\phi - \phi_m)$$

$$\beta_s = \beta_1 + \beta_2(\phi - \phi_m)$$

T_{ref} and P_{ref} are the reference temperature and pressure, respectively and S_{Lref}^0 the laminar burning velocity of these conditions. In this correlation, pressure, temperature and laminar burning velocity are respectively expressed in bar, K and cm/s.

In the study of Galmiche *et al.* [24], an empirical term was added to take into account both the effect of N_2 dilution and O_2 enrichment:

$$S_L^0(\nu_{O_2}) = S_L^0(\nu_{O_2ref}) \left(\frac{\nu_{O_2}}{\nu_{O_2ref}}\right)^{\gamma}$$

where $\gamma = \gamma_1 + \gamma_2(\phi - \phi_m)$, with two constant values of γ_1 and γ_2. ν_{O_2} is the percentage in the air defined by:

- in the case of N_2 dilution:

$$\nu_{O_2} = \frac{n_{O_2(synthetic\ air)}}{n_{O_2(synthetic\ air)} + n_{N_2(synthetic\ air)} + n_{N_2(diluent)}}.100$$

- in the case of O_2 enrichment:

$$\nu_{O_2} = \frac{n_{O_2(synthetic\ air)} + n_{O_2(enrichment)}}{n_{O_2(synthetic\ air)} + n_{N_2(synthetic\ air)}}.100$$

ν_{O_2ref} is the reference percentage of O_2 in the air, *i.e.* the percentage of O_2 in the synthetic air.

The parameters used in the correlation are listed in Table 3.

In order to apply this correlation to the engine combustion calculation, the effect of residual gas can be defined by:

$$S_L^0(\nu_{O_2})\ \text{with residual gas} = S_L^0(\nu_{O_2})(1 - 2.1\chi_{residual})$$

where $\chi_{residual}$ is the residual gas proportion inside the cylinder.

TABLE 3
Correlation parameters used to estimate laminar burning velocity

Parameter	Optimized value
A	56.21
B	−14.44
C	−214.08
D	43.47
E	267.17
α_1	1.88
α_2	−0.19
β_1	−0.24
β_2	0.27
γ_1	2.67
γ_2	0.17
ϕ_m	1.07

2.6 Calibration Process

The calibrations were accomplished by coupling AMESim with a Matlab program, to input experimental results in the requisite format.

Figure 4 illustrates the simulation process for all the engine operating points. First, experimental data (such as intake pressure, intake temperature, exhaust pressure, exhaust gas temperature, equivalence ratio, advanced ignition crank angle and in-cylinder pressure) are read by a Matlab program. In this program, Matlab-AMESim coupling was used to optimize the in-cylinder data.

The second part of this algorithm focuses on two-zone AMESim model simulation with oxygen control. In this part, the AMESim development tool, AMESet, was applied to induce laminar burning velocity as a sub-routine inside the main combustion model. In addition, the main two-zone combustion model was modified to adapt it to the oxygen control.

Multi-cycle simulations were performed with the initial 'guess value' of the integral scale and tumble value, and then the simulation results of in-cylinder pressure of the last simulation cycle were compared with the experimental data for a range of crank angles from 250 to 450 CAD. The discrepancy between the two values can be calculated by:

$$\text{error} = \sum_{i=250°CA}^{450°CA} (P_{cyl}(\text{simu}) - P_{cyl}(\text{expe}))^2 + \alpha * (P_{max}(\text{simu}) - P_{max}(\text{expe}))^2$$

The second term of the above expression gives a more precise P_{max} calibration with regard to the importance of P_{max} for heat release and exhaust gas emission. The parameter α gives a flexible value for considering the greater or lesser effect of P_{max}. Once the error between simulation and experimental in-cylinder pressure data is minimized, the value of l_t and N_{tumble} can be obtained by loop calculation. Finally, the above l_t and N_{tumble} maps were used to analyze the simulation results.

The present algorithm differs from the calibration process methodology employed by Richard *et al.* [14] using 'IFP-Optilab'. In the present study, the initial conditions that

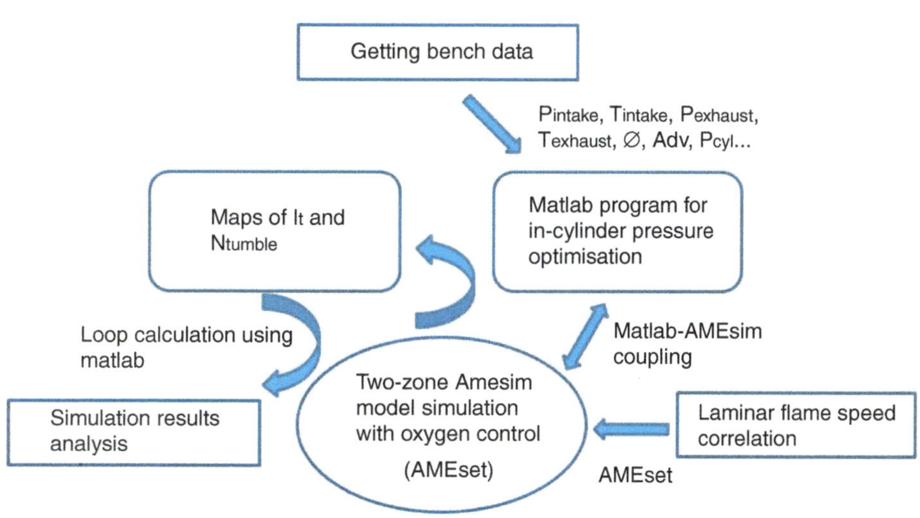

Figure 4
Global scheme of different simulation and model coupling and looping.

concern the mass fraction of air and of isooctane fuel, intake pressure and temperature are taken directly from experimental data. Residual burned gases, which have a significant impact on the combustion process (heat release, knock, pollutants, etc.), were estimated by a multi-cycle simulation with five real engine cycles. Only the results of the last cycle were used to calculate the error between in-cylinder simulation and experimental data and to analyze the results.

3 RESULTS AND DISCUSSION

3.1 Experimental Results

3.1.1 Fuel Consumption

The evolution of indicated SFC *versus* the equivalence ratio is presented in Figure 5 for different IMEP (400 kPa, 600 kPa, 800 kPa and 1 000 kPa). In these figures (Fig. 5a-d), the O_2 percentage in the air varies from 15 to 27%. For each O_2 percentage, the equivalence ratio is decreased to reach the lean operating limits. With the decrease in the equivalence ratio, SFC was decreased for a fixed O_2 percentage, although an SFC increase is usually observed at the operating limits.

For IMEP 400 kPa and 600 kPa (Fig. 5a, b), by increasing the O_2 percentage in the air and decreasing the equivalence ratios, a lower SFC can be obtained. Compared with the conventional operating point (with 21% O_2, and a stoichiometric case), SFC can be reduced by 15%. The lowest SFC is found at an O_2 percentage of around 25% to 27% with an equivalence ratio of 0.5 to 0.55, which means that oxygen-enriched combustion with a low equivalence ratio can be a potential method for improving fuel efficiency.

For IMEP 800 kPa (Fig. 5c), and an O_2 percentage from 15% to 19%, a lower SFC can be obtained by increasing the O_2 percentage while reducing the equivalence ratio, which is similar to the cases of IMEP 400 kPa and 600 kPa. However, for an O_2 percentage from 21% to 27%, this tendency is reversed. The lowest SFC for IMEP 800 kPa is found at 19% O_2 with an equivalence ratio of 0.7.

For IMEP 1 000 kPa (Fig. 5d), the lowest SFC is obtained at 15% O_2 with an equivalence ratio of 0.9. In this case, the dilution of N_2 will be a better way than oxygen enrichment to limit fuel consumption. Compared with the regular operating point (with 21% O_2, and a stoichiometric equivalence ratio), N_2 dilution has the potential to achieve about 10% fuel economy.

It can therefore be concluded that for low IMEP (400 kPa, 600 kPa), oxygen enrichment is a potential way to improve fuel efficiency, whereas for high IMEP (800 kPa, 1 000 kPa), N_2 dilution should be used to have a better fuel economy.

3.1.2 HC Emissions

Figure 6 represents the evolution of indicated specific HC emissions *versus* the equivalence ratio for two different IMEP (400 kPa, 1 000 kPa). The effect of oxygen enrichment on HC emissions is obvious. For each fixed equivalence ratio, the HC emission decreases with the increase in the O_2 percentage and the lowest values are always found at 27% O_2. Oxygen enrichment can decrease the quenching distance of the mixture. In the study by Friedman and Johnston [25], the quenching distance was found to be a decreasing function of the flame temperature, and it is well known that the flame temperature increases in the case of oxygen enrichment [23]. Mazas *et al.* [26] also showed that quenching effects at the flame front are reduced due to the increase in the flame temperature when the oxygen-enrichment ratio is increased. Thus, oxygen enrichment allows the flame to propagate much closer to the cylinder walls than the N_2 dilution mixture.

The specific HC emission increases when the equivalence ratio varies from 0.9 to 1; however, the HC emission decreases from the lean limits to an equivalence ratio of 0.9. The lowest value can be found around an equivalence ratio of 0.9.

3.1.3 NO_x Emissions

The evolution of NO_x emissions *versus* the equivalence ratio is presented in Figure 7. Contrary to HC emissions, at a fixed equivalence ratio value, NO_x emissions increase when the oxygen percentage is increased, with the exception of a few points which may be due to measurement error. As the formation of NO_x depends mainly on the burned gas temperature, due to the thermal formation process, the higher the burned gas temperature, the higher the rate of NO_x formation. Thus, the higher emission of NO_x for oxygen enrichment can be explained by the high temperature of the burned gases. The stoichiometric engine operating conditions show a slight decrease in NO_x emissions for the highest O_2 volume fraction, contrary to the literature. This is probably due to different reasons: as the later spark timing changes the thermodynamic conditions, it may have a significant effect on NO_x, especially at 10 bar IMEP, for 27% O_2, where the spark timing is retarded to prevent the knock. Moreover, although the O_2 enrichment produces a higher burned gas temperature and then causes higher NO_x emission, a lower concentration of N_2 is not conducive to NO_x formation. In the case of a highly lean mixture, the leaner the mixture is, the less NO_x emission it gets, even though the SFC is greatly increased (the lean limit points in Fig. 5). In the case of a lean mixture, the leaner the mixture is, the less NO_x emission it gets; the interval of the operation points for the equivalence ratio is normally set to 0.05, which may be too high, and induces the fluctuation of NO_x emissions. As shown in the

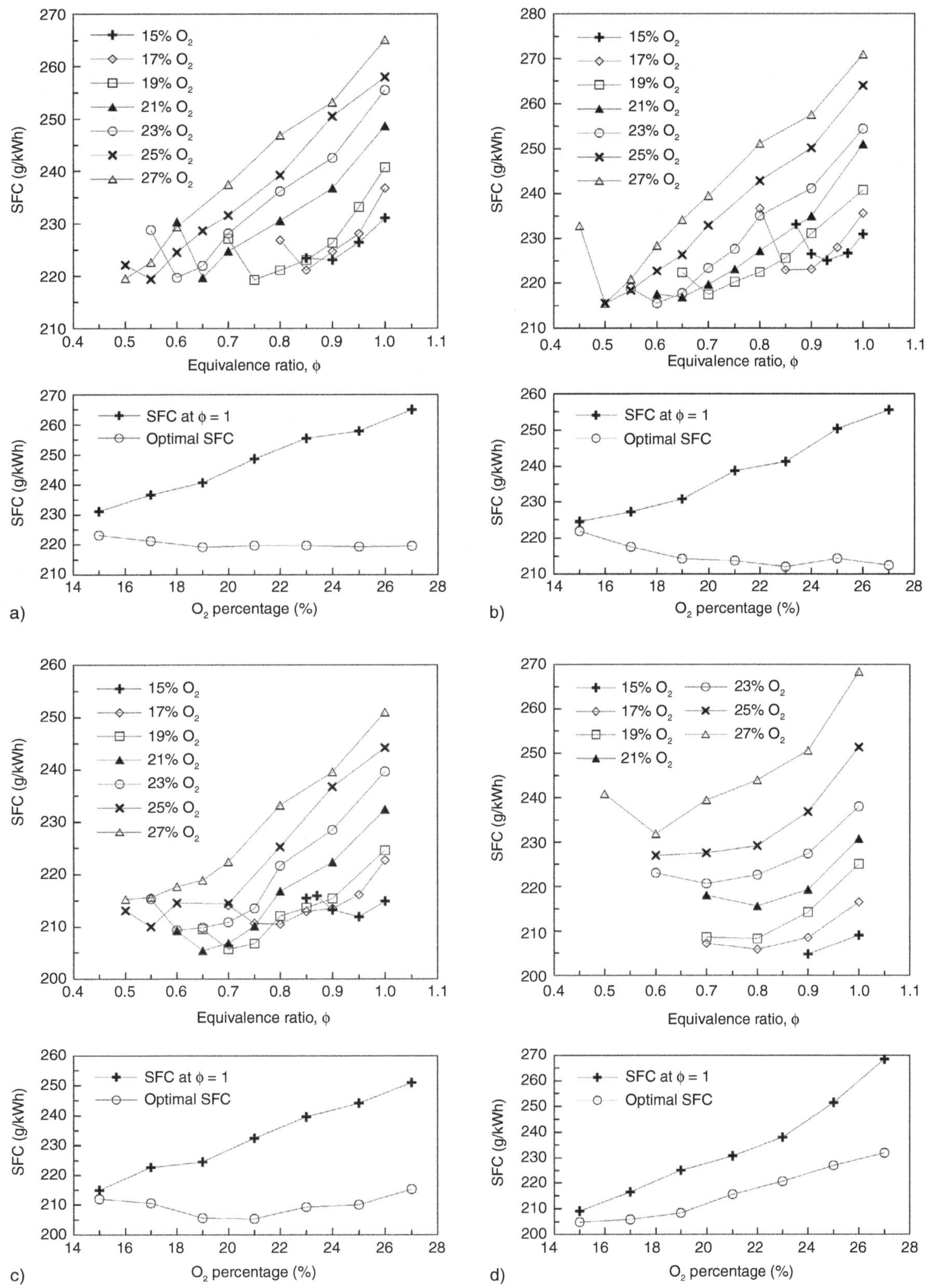

Figure 5

Evolution of indicated SFC *versus* the equivalence ratio for different IMEP; a) 400 kPa, b) 600 kPa, c) 800 kPa, d) 1 000 kPa.

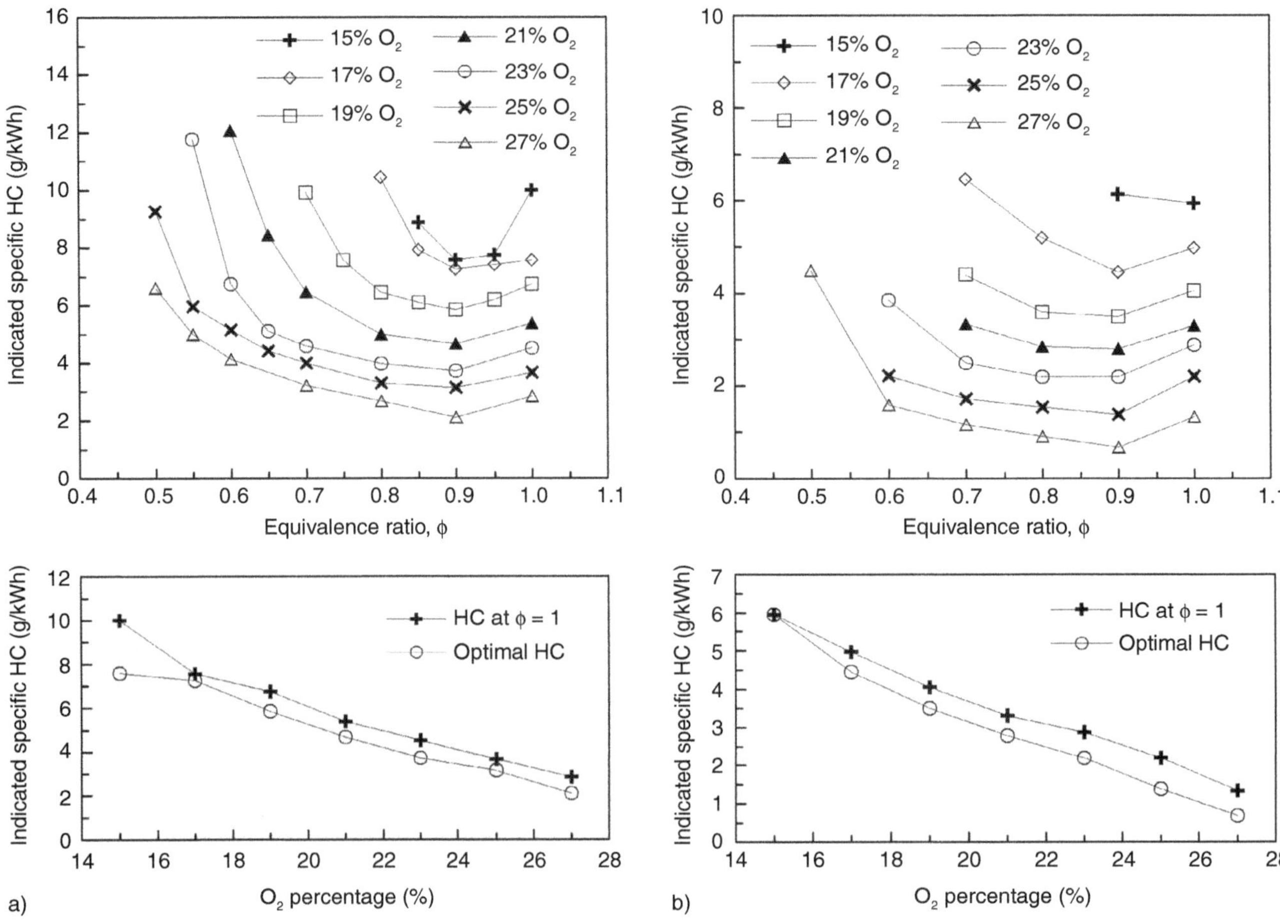

Figure 6

Evolution of indicated specific HC emissions *versus* the equivalence ratio for different IMEP; a) 400 kPa, b) 1 000 kPa.

lower part of Figure 7a, the filled circles represent the NO_x with high SFC (the lean limit points in Fig. 5a). The closer the lean limit is touched, the lower the NO_x emission.

3.1.4 CO Emissions

Figure 8 represents the evolution of indicated specific CO emissions *versus* the equivalence ratio for different IMEP (400 kPa, 1 000 kPa). CO emissions are mainly affected by the equivalence ratio ϕ; especially for ϕ near unit value, they decrease in lean conditions due to the increase in the O_2 concentration, which helps the oxidation of CO into CO_2. Compared with the equivalence ratio, the O_2 percentage in the air does not present a clear tendency. In Figure 8a, the local slope of the equivalence ratio between 0.95 and 1 for 17% O_2 can reach 240 (the closer it is to the unit ϕ value, the greater it is), which means that with one percent variation in the equivalence ratio, the CO emission will have more than 2.4 g/kWh discrepancy. Conversely, the biggest variation from 15% to 27% O_2 is about 7 g/kWh. Thus, to ensure precise CO measurement, the equivalence ratio should be controlled and the precision of the mass flow meters of fuel and gases should be checked.

3.2 ENGINE MODEL RESULTS

3.2.1 Engine in-Cylinder Pressure

Experimental and simulated in-cylinder pressure curves for different engine loads at 1 400 rpm are presented in Figure 9. The in-cylinder pressure evolution is plotted for different cases of the oxygen percentage at a stoichiometric equivalence ratio, 15% O_2, 21% O_2 and 27% O_2, respectively. The discrepancies can be found in Figure 9a, especially for

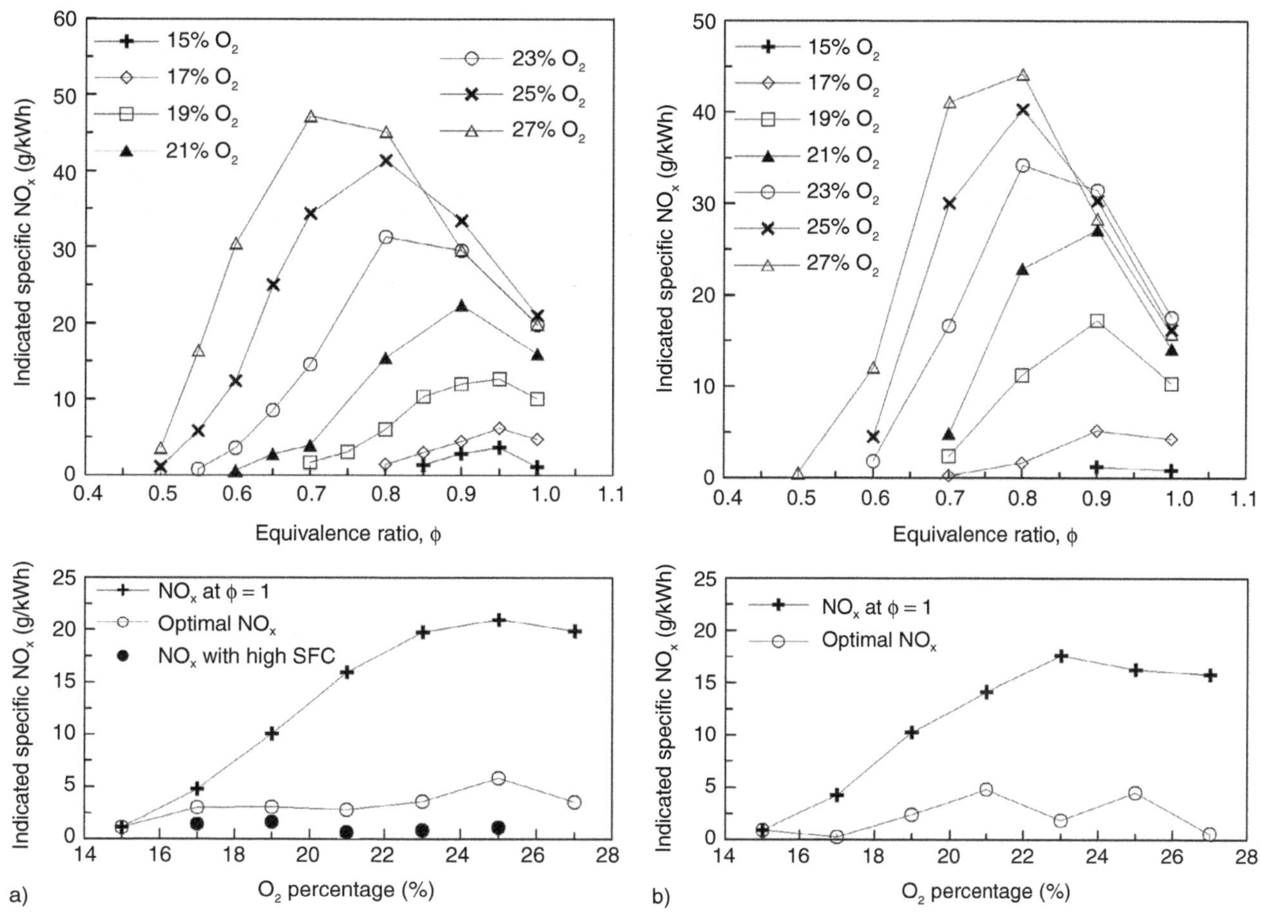

Figure 7

Evolution of indicated specific NO_x emissions *versus* the equivalence ratio with different IMEP; a) 400 kPa, b) 1 000 kPa.

the cases of IMEP 8 and 10 bar, which may be due to the assumption of spherical flame propagation inside the combustion volume. For a high nitrogen dilution mixture, the flame evolution in the cylinder becomes irregular, as previously observed by Mounaïm-Rousselle *et al.* [9]. However, good calibration results can be found in Figure 9b, c for 21% and 27% of O_2, respectively. The simulated results show good agreement with our experimental data, which indicates that the evolution of the burning velocity and therefore the heat release rate is well described for all cases [16].

In Figure 10, the IMEP absolute error and maximum pressure relative error *versus* the O_2 percentage for different cases of IMEP are presented. For all cases, IMEP relative errors do not exceed 3.5% and maximum pressure relative errors are less than 2%. The largest values of these errors are found at 15% O_2 and the smallest at an O_2 percentage above 21%. The maximum pressure relative error roughly decreases with increasing O_2 percentage.

In Figure 11, the integral length and the tumble number *versus* the O_2 percentage are presented for IMEP from 400 to 1 000 kPa. As these two parameters are used to adjust the simulation, their values evolve: with the increase in the O_2 percentage, they increase, except in the case of 15% O_2 for the tumble number.

Although some operating points are beyond the general tendency, the evolution of the integral length scale and the tumble number is smooth when the IMEP, equivalence ratio and O_2 percentage are varied. Thus, reasonable values of l_t and N_{tumble} can be obtained by interpolation of the data in look-up tables, in order to calculate transient engine operations.

3.2.2 Combustion Characteristics

The evolution of the calculated in-cylinder pressure at the stoichiometric equivalence ratio and IMEP = 1 000 kPa for

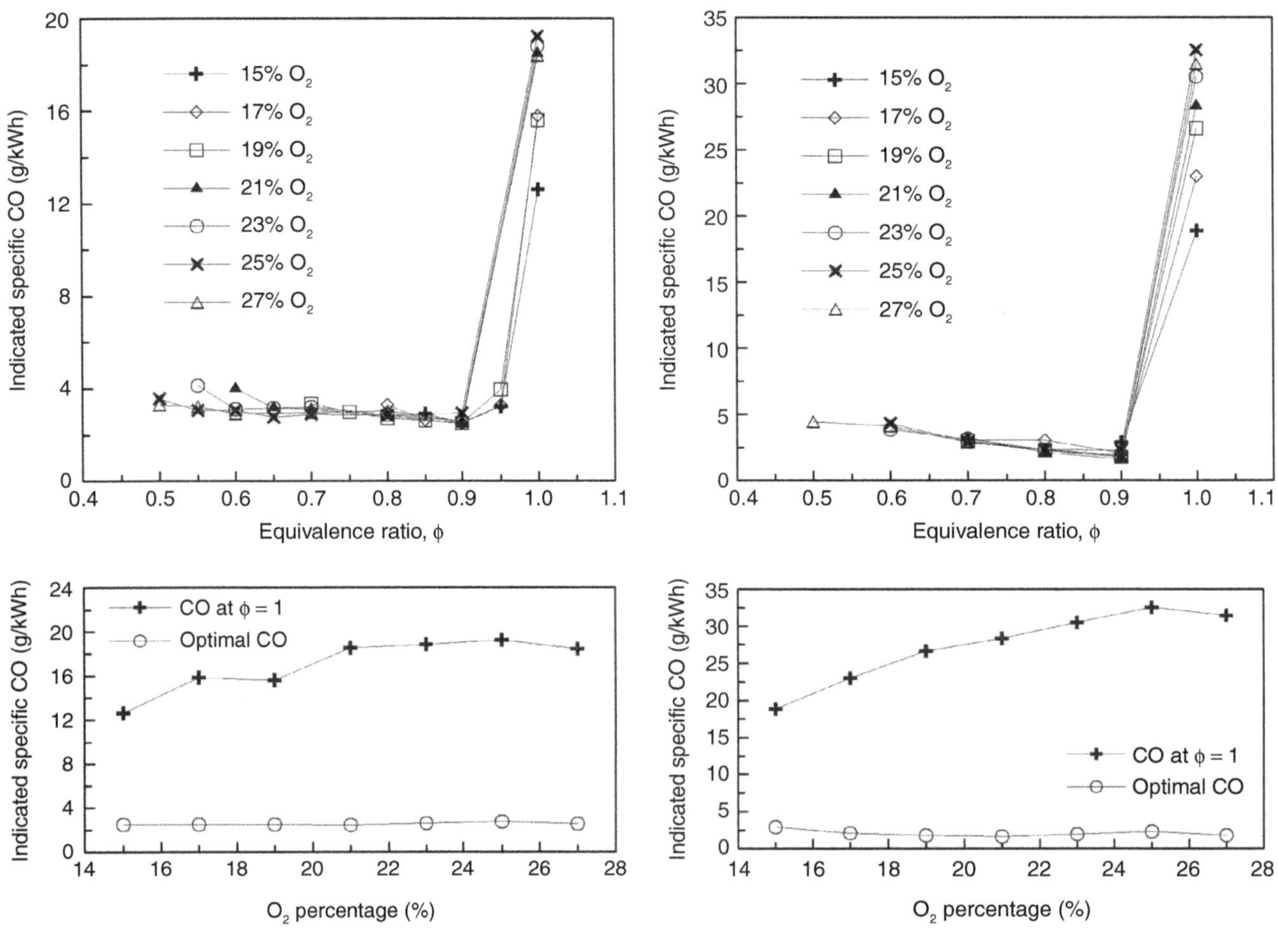

Figure 8

Evolution of indicated specific CO emissions *versus* the equivalence ratio for different IMEP; a) 400 kPa, b) 1 000 kPa).

different cases of the O_2 percentage is presented in Figure 12. The in-cylinder pressure is started at the spark ignition degree with the operation conditions listed in Table 4. The increase in the O_2 percentage results in a decrease in intake pressure while maintaining the same IMEP. However, for 27% O_2, in order to maintain IMEP at 10 bar, the intake pressure was kept at the same value as for 25% O_2 while increasing the spark timing to prevent knock. For relatively high O_2 percentages, engine operation was limited by knock phenomena, which can be seen in the in-cylinder pressure trace in Figure 12. To avoid engine knock, the higher spark timing was set for higher oxygen enrichment.

Figure 13 presents the laminar burning velocity evolution *versus* CAD with different O_2 percentages. The laminar burning velocity increases as a function of the O_2 percentage. Compared with 15% O_2, the laminar burning velocity of 27% O_2 is more than 8 times higher. The reasonable results of laminar burning velocity under engine operation conditions were obtained thanks to the reliability of the experimental correlation proposed by Galmiche *et al.* [24].

The evolution of the flame surface and flame wrinkling *versus* CAD for different O_2 percentages at the stoichiometric equivalence ratio and IMEP = 1 000 kPa are plotted in Figure 14. A previous experimental study (Mounaïm-Rousselle *et al.* [9]) showed that dilution enhances flame-turbulence interactions and that the corrugation generated through dilution occurs on smaller scales than the integral length scales. This phenomenon is well reproduced in the present simulation, as shown in Figure 14a, for 15 and 17% O_2, which correspond to 26.87% and 17.65% of EGR, respectively, as presented in Table 2. The flame surface is much higher than in the case of lower dilution

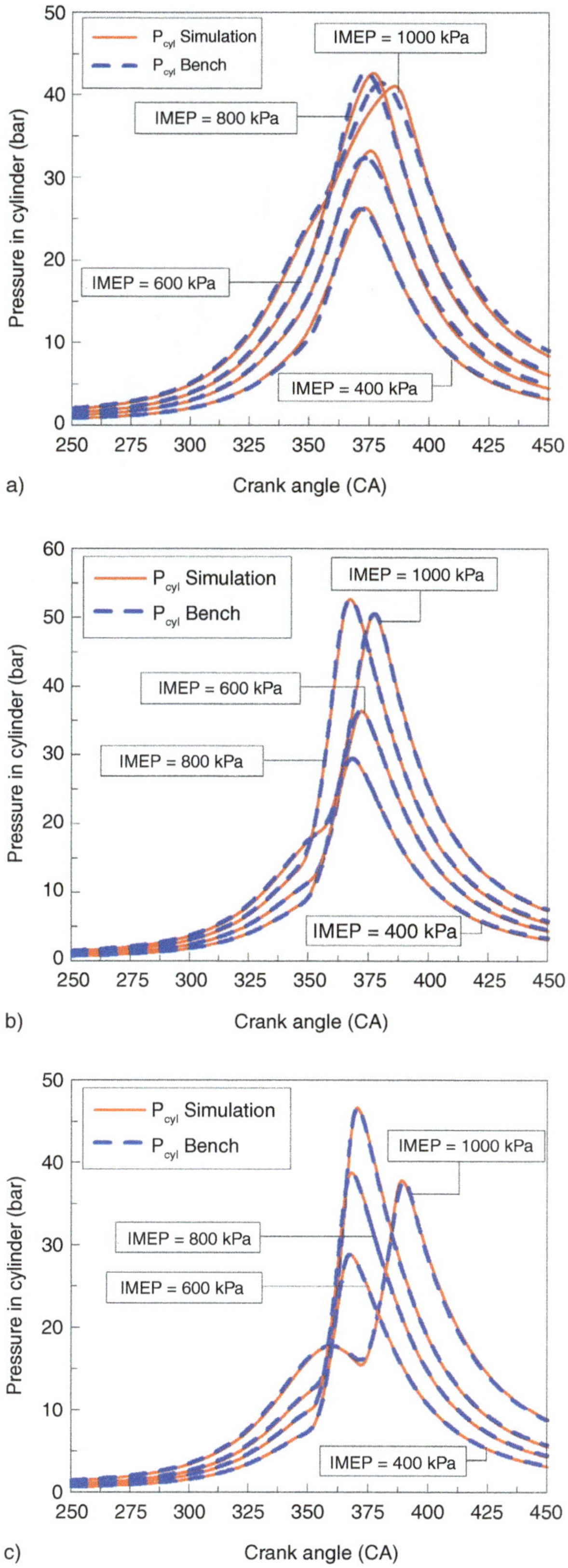

Figure 9

Comparison between experimental and simulated in-cylinder pressure evolution for different engine loads. $\varphi = 1$, a) 15% O_2, b) 21% O_2, c) 27% O_2.

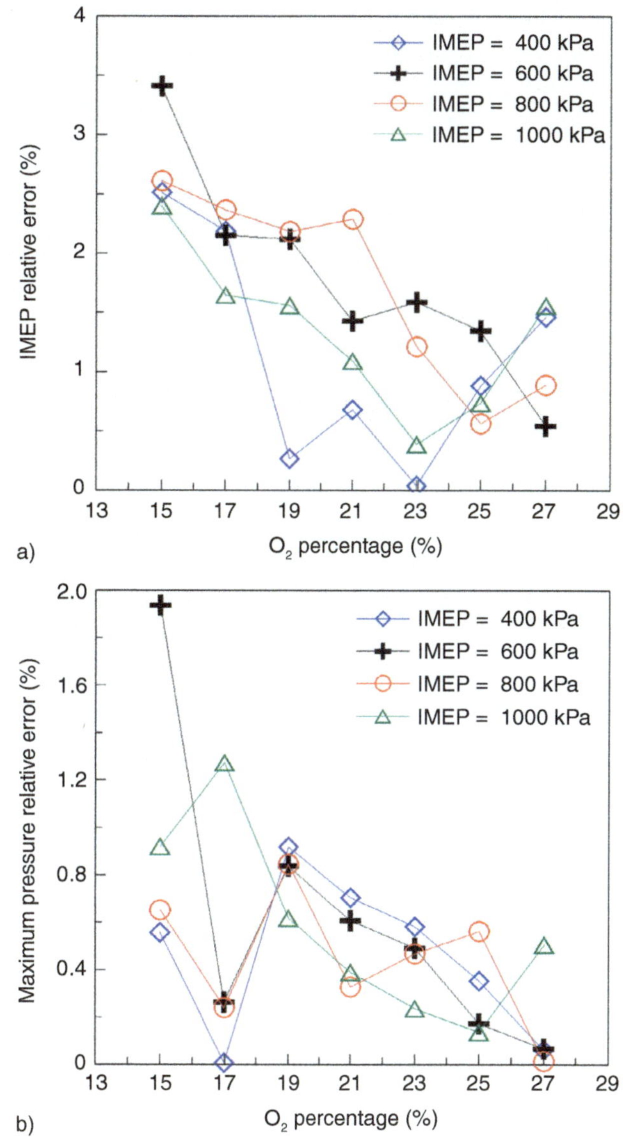

Figure 10

a) Relative error of IMEP and b) maximum in-cylinder pressure.

or oxygen enrichment. This tendency is mainly due to the high flame wrinkling in the case of high dilution, as presented in Figure 14b.

Finally, Figure 15 shows the evolution of the turbulent flame velocity *versus* CAD. The flame kernel growth is initially laminar-like as described in [27], and once the flame has grown to more than 5 mm, it can become turbulent [28].

Thus, based on the above findings, as the laminar burning velocity is much lower in the case of dilution, the laminar flame state lasts longer than with oxygen enrichment: about 7.5 CAD for 15% O_2, 5 CAD for 21% O_2 and 2.5 CAD for

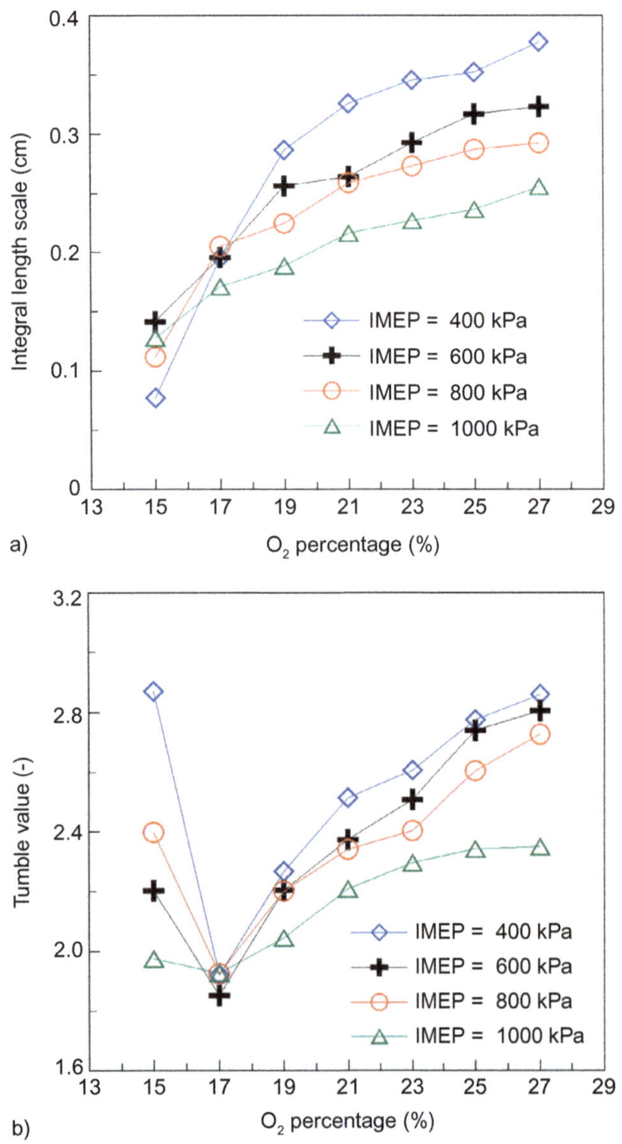

Figure 11

a) Integral length scale and b) tumble values obtained to optimize the in-cylinder pressure.

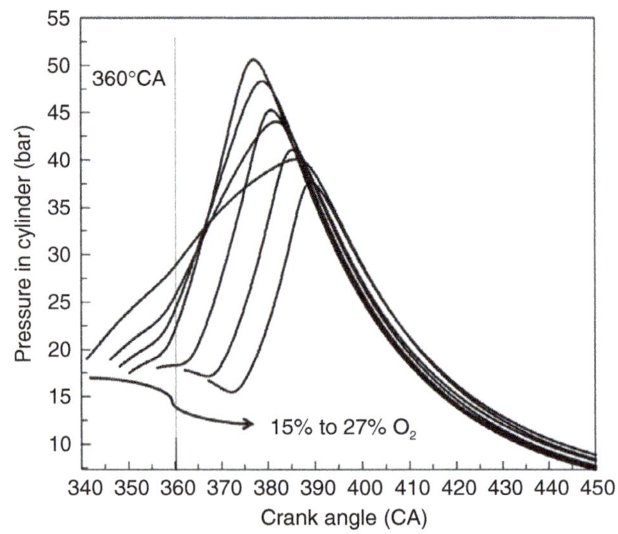

Figure 12

Evolution of the mean in-cylinder pressure at the stoichiometric equivalence ratio and IMEP = 1 000 kPa.

TABLE 4

Operating conditions for combustion analysis

O_2 (%)	Intake pressure (bar)	Spark timing advanced (°CA)	Fuel mass flow rate (kg/h)
15	1.60	−19	0.98
17	1.39	−14	1.01
19	1.27	−12	1.05
21	1.19	−10	1.08
23	1.13	−4	1.12
25	1.12	2	1.19
27	1.12	7	1.28

27% O_2. Although substantial differences are found for laminar burning velocities (8 times higher with 27% O_2 compared with 15% O_2), the turbulent burning velocities show comparatively little difference (less than 2 times higher with 27% O_2 compared with 15% O_2). This is mainly due to the reverse evolution of flame wrinkling: the lower the O_2 percentage, the higher the flame wrinkling.

In order to characterize the combustion regime in the Peters-Borghi diagram [9], the evolution of the characteristic length and velocity scale ratio *versus* CAD for different O_2 percentages is plotted in Figure 16. As previously described, after the spark, the flame is initially 'laminar-like', and the duration of this phase is from about 2.5 to 7.5 CAD. Here, the curve starts at 10 CAD after spark timing (360 + ST + 10 CAD) for all the cases. A higher the value of l_t/δ_L is found for a higher value of the O_2 percentage; the reverse is true for u'/S_L. Two CAD cases of these ratios (10 CAD and 20 CAD after spark timing) are shown in Figure 17.

Finally, in Figure 18, the combustion is classified according the Peters-Borghi diagram, which gives a deeper insight into the interaction between turbulence and the flame front. According to the study of Linse *et al.* [29], engine

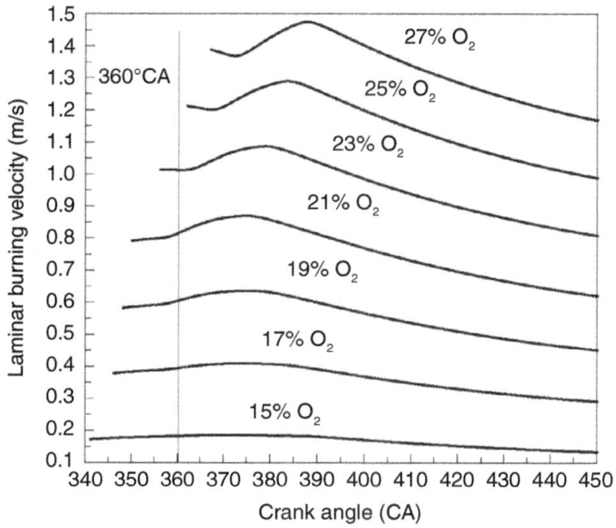

Figure 13

Laminar burning velocity *versus* CAD with different O_2 percentages at the stoichiometric equivalence ratio and IMEP = 1 000 kPa.

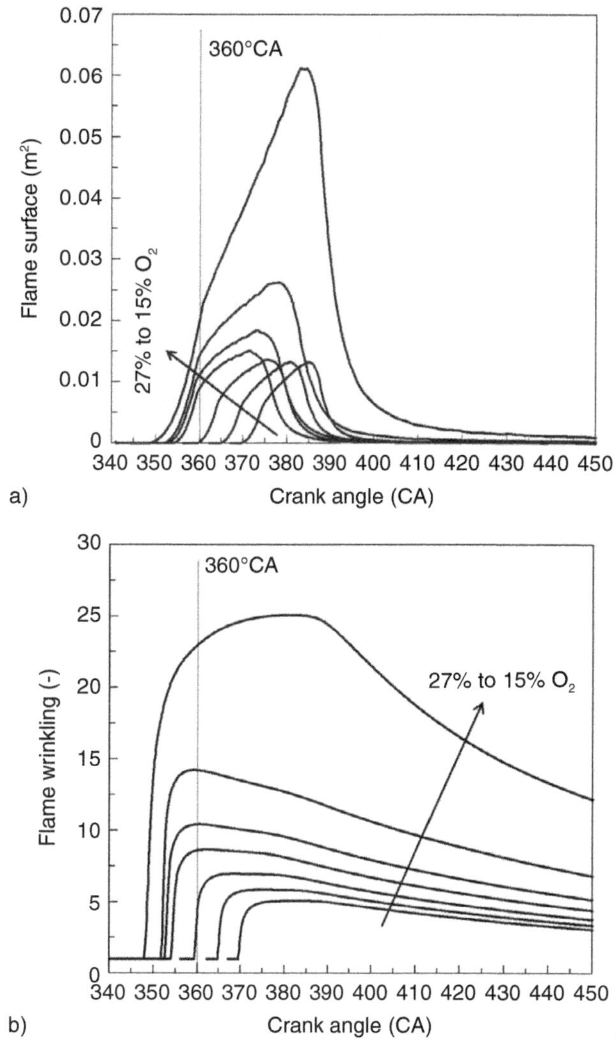

Figure 14

a) Flame surface and b) flame wrinkling *versus* the crank angle with different O_2 percentages at the unit equivalence ratio and IMEP = 1 000 kPa.

combustion is normally located in the thin reaction zone and the corrugated flamelet zone. Generally, the combustion of the earlier generation of SI engines takes place in the corrugated flamelet regime, while the combustion of the new generation of engines takes place in the thin reaction zone [30]. A higher turbulent level and charge motion are created when turbocharging and/or direct injection are used in new generation engines, which is the main reason for the different regime between new and earlier generation engines [29].

In Figure 18, from 15% O_2 to 27% O_2, the combustion trace passes from the broken reaction zone to the thin reaction zone, and finally to the corrugated flamelet zone. For the cases of 17% to 23% O_2, combustion takes place in the thin reaction zone, which may occur when the tested engine is supercharged and the high turbulence level is produced. For the cases of 25% and 27% O_2, the combustion traces are mainly in the corrugated flamelet zone, and for the case of 15% O_2, the combustion occurs in the broken reaction zone. Mainly, flamelet theory is used for engine combustion simulation, as validated by Poinsot *et al.* [31] based on direct simulation. Their results show that the resistance of the flame front to vortices is underestimated by classical diagrams which neglect viscous, transient and curvature effects. The limit line of the flamelet regime determined by Poinsot *et al.* [31] is drawn as the *v*-formed green line in Figure 18. Therefore, for the case of 15% O_2, the spherical propagation of the flame might not be well adapted, which certainly contributes to the imperfect agreement between simulated and experimental in-cylinder pressure (Fig. 9a).

The effect of N_2 dilution on the combustion trace corresponds well with the previous experimental study of Mounaïm-Rousselle *et al.* [9]. Dilution moves the combustion trace toward the upper left side in the Peters-Borghi diagram, and the effect of O_2 enrichment shows a reverse tendency compared with N_2 dilution. For all the operating cases, the combustion trace has a 'turning point'. Compared with the simulation results of Linse *et al.* [29], the 'turning angle' is smaller, which is caused by the simplification of the l_t definition in the AMESim model. l_t is supposed to be a constant value and equal to its value at spark timing

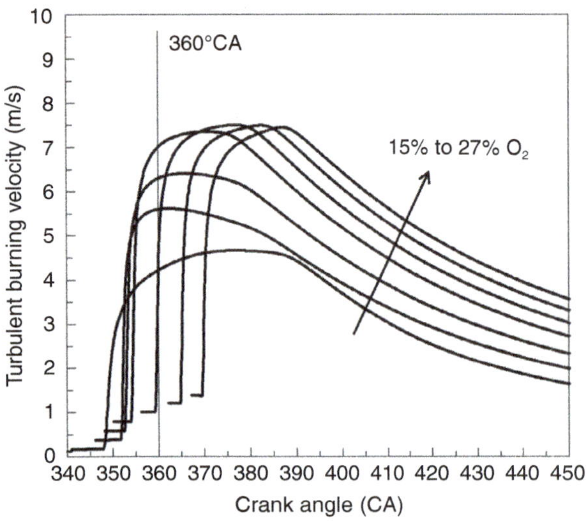

Figure 15

Evolution of turbulent flame velocity with different O_2 percentages at the stoichiometric equivalence ratio and IMEP = 1 000 kPa.

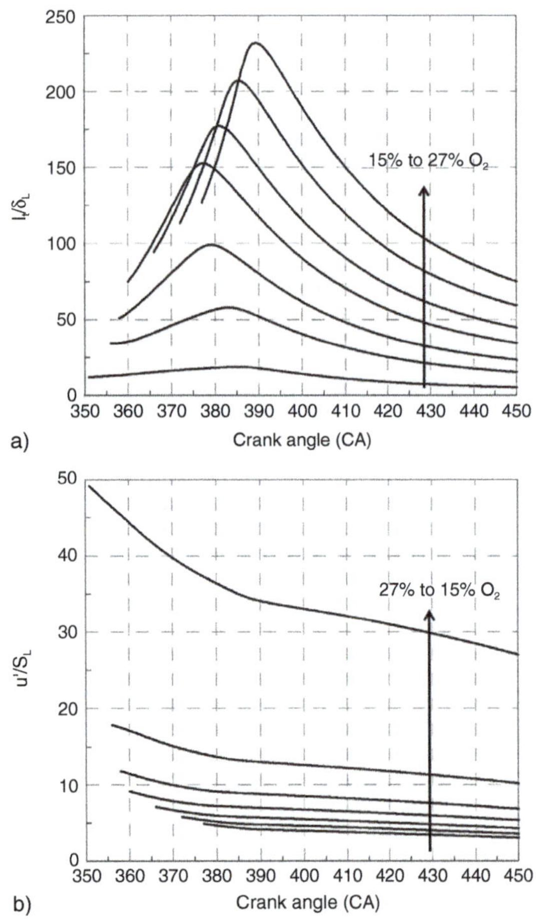

Figure 16

Ratio of the characteristic length a) and velocity scales b) *versus* CAD for different O_2 percentages.

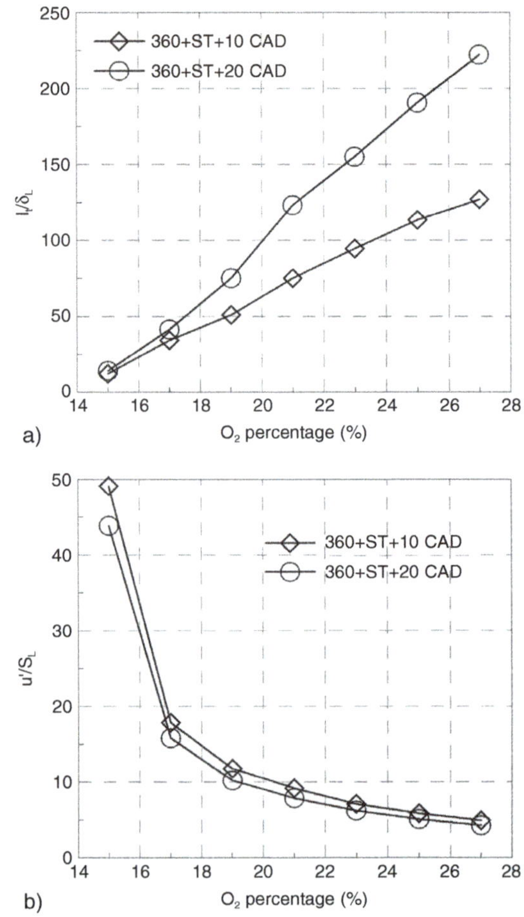

Figure 17

Ratio of the characteristic length a) and velocity scales b) *versus* the O_2 percentage for two CAD cases.

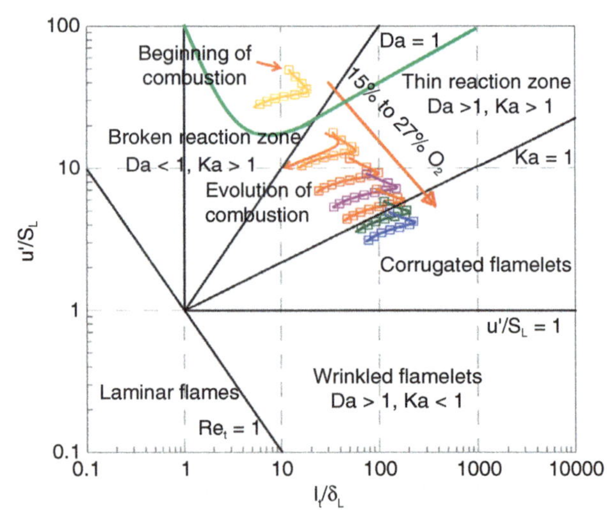

Figure 18

Comparison of the combustion traces in the Peters-Borghi diagram for different O_2 percentages.

during the whole combustion stroke in the combustion model [14], while l_t normally depends on the distance between the piston and cylinder head. Nevertheless, the general tendency is in good agreement with the previous studies of Linse *et al.* [29] and Mounaïm-Rousselle *et al.* [9].

CONCLUSION

In this study, the effect of controlling the oxygen concentration in the air on the performance and emissions in a downsized engine is presented and discussed, using both experimental and simulation methods.

The experimental results indicate that, by increasing the O_2 percentage in the air and decreasing the equivalence ratios, lower fuel consumption can be achieved for IMEP of 400 and 600 kPa. However, for 800 and 1 000 kPa, the lowest fuel consumption is obtained for 19% O_2 at 0.7 of the equivalence ratio and 15% O_2 at 0.9, respectively. It can be concluded that for low loads, oxygen enrichment has a real potential to improve fuel efficiency, whereas for relatively high IMEP, N_2 dilution still remains the best way. For all equivalence ratios, the HC emission decreases with the increase in the O_2 percentage and the lowest value can always be found at 27% O_2 but in this case, NO_x emissions increase.

The numerical simulation was conducted with the AMESim software by taking into account the effect of O_2 control. In-cylinder pressure was well reproduced, although discrepancies can be found for high levels of dilution. The increase in the O_2 percentage results in a decrease in intake pressure while maintaining the same IMEP. The laminar burning velocity increases as a function of the O_2 percentage. Compared with 15% O_2, the laminar burning velocity of 27% O_2 is more than 8 times higher. The flame surface of high dilution cases is much greater than that with lower dilution or oxygen-enriched cases. Therefore, turbulent flame velocities show few differences compared with laminar ones. The combustion trace passes from the broken reaction zone to the thin reaction zone, and finally to the corrugated flamelet zone by varying the O_2 percentage from 15% to 27%. Only for the case of 15% O_2 does the combustion take place in the broken reaction zone, in which the spherical propagation of the flame may not be well adapted.

ACKNOWLEDGMENTS

The authors acknowledge Stephane Richard for discussions to improve the AMESim software and the new procedure for residual gases. This study was financially supported by *PSA Peugeot Citroën* and Region Centre. This work was also financed by the MACDOC project of the French National Research Agency, TDM program, ANR-12-VPTT-0008. The authors are also grateful for the support from the Foundation of Zhejiang Educational Committee (No. Y201430810) and Natural Science Foundation of Zhejiang Province (No. LQ15E060001), China.

REFERENCES

1 Poola R.B., Ng H.K., Sekar R.R., Baudino J.H., Colucci C.P. (1995) *Utilizing Intake-Air Oxygen-Enrichment Technology to Reduce Cold-Phase Emissions*, Vol. **104**, New York, NY, Society of Automotive Engineers. 15.

2 Kajitani S., Clasen E., Campbell S., Rhee K. (1993) Partial-load and Start-up Operations of Spark-ignition Engine with Oxygen Enriched Air, *SAE Technical Paper* 932802.

3 Poola R.B., Sekar R., Ng H.K., Baudino J.H., Colucci C.P. (1996) *The effects of oxygen-enriched intake air on FFV exhaust emissions using M85*, Argonne National Lab., IL, United States.

4 Ng H.K., Sekar R.R., Kraft S.W., Stamper K. (1993) *The potential benefits of intake air oxygen enrichment in spark ignition engine powered vehicle*, Society of Automotive Engineers.

5 Caton J.A. (2005) Use of a cycle simulation incorporating the second law of thermodynamics: results for spark-ignition engines using oxygen enriched combustion air, *SAE Technical Paper* 2005-01-1130.

6 Poola R.B., Stork K.C., Sekar R., Callaghan K., Nemser S. (1998) Variable Air Composition with Polymer Membrane–A New Low Emissions Tool, *SAE Paper* 980178.

7 Kajitani S., Sava N., McComiskey T., Rhee K. (1992) *A spark ignition engine operated by oxygen enriched air*, Society of Automotive Engineers.

8 Quader A. (1978) Exhaust Emissions and Performance of a Spark Ignition Engine Using Oxygen Enriched Intake Air, *Combustion Science and Technology* **19**, 1-2, 81-86.

9 Mounaïm-Rousselle C., Landry L., Halter F., Foucher F. (2013) Experimental characteristics of turbulent premixed flame in a boosted Spark-Ignition engine, *Proceedings of the Combustion Institute* **34**, 2, 2941-2949.

10 Van Blarigan A., Seiser R., Chen J., Cattolica R., Dibble R. (2012) Working fluid composition effects on methane oxycombustion in an SI-engine: EGR *vs.* CO_2, *Proceedings of the Combustion Institute* **34**, 2, 2951-2958.

11 Favre E., Bounaceur R., Roizard D. (2009) A hybrid process combining oxygen enriched air combustion and membrane separation for post-combustion carbon dioxide capture, *Separation and Purification Technology* **68**, 1, 30-36.

12 Koros W.J., Chern R.T. (1987) *Separation of gaseous mixtures using polymer membranes*, Handbook of separation process technology, pp. 862-953.

13 Coombe H.S., Nieh S. (2007) Polymer membrane air separation performance for portable oxygen enriched combustion applications, *Energy Conversion and Management* **48**, 5, 1499-1505.

14 Richard S., Bougrine S., Font G., Lafossas F.-A., Le Berr F. (2009) On the reduction of 3D CFD combustion model to build a physical 0D model for simulating heat release, knock and pollutants in SI engines, *Oil & Gas Science and Technology - Rev. IFP* **64**, 3, 223-242.

15 *IFP Energies nouvelles* (2010) AMESim IFP Engine library user manual, LMS IMAGINE S.A.

16 Bougrine S., Richard S., Veynante D. (2011) On the combination of complex chemistry with a 0-D coherent flame model to account for the fuel properties in spark ignition engines simulations: Application to methane-air-diluents mixtures, *Proceedings of the Combustion Institute* **33**, 2, 3123-3130.

17 Bozza F., Gimelli A. (2004) A Comprehensive 1D Model for the Simulation of a Small-Size Two-Stroke SI Engine, *SAE Technical Paper* 2004-01-0999.

18 Richard S., Colin O., Vermorel O., Benkenida A., Angelberger C., Veynante D. (2007) Towards large eddy simulation of combustion in spark ignition engines, *Proceedings of the Combustion Institute* **31**, 2, 3059-3066.

19 Colin O., Benkenida A., Angelberger C. (2003) 3d Modeling of Mixing, Ignition and Combustion Phenomena in Highly Stratified Gasoline Engines, *Oil & Gas Science and Technology - Rev. IFP* **58**, 1, 47-62.

20 Charlette F., Meneveau C., Veynante D. (2002) A power-law flame wrinkling model for LES of premixed turbulent combustion Part I: non-dynamic formulation and initial tests, *Combustion and Flame* **131**, 1-2, 159-180.

21 Blint R.J. (1986) The Relationship of the Laminar Flame Width to Flame Speed, *Combustion Science and Technology* **49**, 1-2, 79-92.

22 Poulos S.G., Heywood J.B. (1983) The effect of chamber geometry on spark-ignition engine combustion, *SAE Technical Paper* 830334.

23 Zhou J.X., Cordier M., Mounaïm-Rousselle C., Foucher F. (2011) Experimental estimate of the laminar burning velocity of iso-octane in oxygen-enriched and CO_2-diluted air, *Combustion and Flame* **158**, 12, 2375-2383.

24 Galmiche B., Halter F., Foucher F. (2012) Effects of high pressure, high temperature and dilution on laminar burning velocities and Markstein lengths of iso-octane/air mixtures, *Combustion and Flame* **159**, 11, 3286-3299.

25 Friedman R., Johnston W.C. (1950) The Wall-Quenching of Laminar Propane Flames as a Function of Pressure, Temperature, and Air Fuel Ratio, *Journal of Applied Physics* **21**, 8, 791-795.

26 Mazas A.N., Fiorina B., Lacoste D.A., Schuller T. (2011) Effects of water vapor addition on the laminar burning velocity of oxygen-enriched methane flames, *Combustion and Flame* **158**, 12, 2428-2440.

27 Heywood J. (1994) *Combustion and its modeling in spark-ignition engines*, in *International symposium COMODIA*.

28 Boulouchos K., Steiner T., Dimopoulos P. (1994) *Investigation of flame speed models for the flame growth period during premixed engine combustion*, Society of Automotive Engineers.

29 Linse D., Hasse C., Durst B. (2009) An experimental and numerical investigation of turbulent flame propagation and flame structure in a turbo-charged direct injection gasoline engine, *Combustion Theory and Modelling* **13**, 1, 167-188.

30 Wirth M. (1993) *Die turbulente flammenausbreitung im ottomotor und ihre charakteristischen längenskalen*, in *RWTH Aachen*.

31 Poinsot T., Veynante D., Candel S. (1991) Quenching processes and premixed turbulent combustion diagrams, *Journal of Fluid Mechanics* **228**, 561-606, 230.

10

Identification of Large-Scale Structure Fluctuations in IC Engines using POD-Based Conditional Averaging

Stefan Buhl*, Frank Hartmann and Christian Hasse

Chair of Numerical Thermo-Fluid Dynamics, TU Bergakademie Freiberg, Fuchsmühlenweg 9, 09599 Freiberg - Germany
e-mail: stefan.buhl@iec.tu-freiberg.de

* Corresponding author

Abstract — *Cycle-to-Cycle Variations (CCV) in IC engines is a well-known phenomenon and the definition and quantification is well-established for global quantities such as the mean pressure. On the other hand, the definition of CCV for local quantities,* e.g. *the velocity or the mixture distribution, is less straightforward. This paper proposes a new method to identify and calculate cyclic variations of the flow field in IC engines emphasizing the different contributions from large-scale energetic (coherent) structures, identified by a combination of Proper Orthogonal Decomposition (POD) and conditional averaging, and small-scale fluctuations. Suitable subsets required for the conditional averaging are derived from combinations of the the POD coefficients of the second and third mode. Within each subset, the velocity is averaged and these averages are compared to the ensemble-averaged velocity field, which is based on all cycles. The resulting difference of the subset-average and the global-average is identified as a cyclic fluctuation of the coherent structures. Then, within each subset, remaining fluctuations are obtained from the difference between the instantaneous fields and the corresponding subset average. The proposed methodology is tested for two data sets obtained from scale resolving engine simulations. For the first test case, the numerical database consists of 208 independent samples of a simplified engine geometry. For the second case, 120 cycles for the well-established Transparent Combustion Chamber (TCC) benchmark engine are considered. For both applications, the suitability of the method to identify the two contributions to CCV is discussed and the results are directly linked to the observed flow field structures.*

Résumé — **Identification de fluctuations de structure aux grandes échelles dans les moteurs à combustion interne utilisant des moyennes conditionnelles basées sur la POD** — Les variations cycle à cycle (VCC) dans les moteurs à combustion interne sont un phénomène bien connu, et leur définition et quantification sont bien établies pour des quantités globales telles que la pression moyenne. Toutefois, la définition des VCC pour des quantités locales, par exemple la vitesse ou le mélange, est moins simple. Le présent article propose une nouvelle méthode visant à identifier et à calculer des variations cycliques de l'écoulement dans les moteurs à combustion interne, en mettant l'accent sur les différentes contributions provenant de structures énergétiques aux grandes échelles (cohérentes), identifiées en combinant la détermination d'une moyenne conditionnelle et par décomposition orthogonale aux valeurs propres (*Proper Orthogonal Decomposition*, POD) et de fluctuations aux petites échelles. Des sous-ensembles adaptés requis pour l'établissement d'une moyenne conditionnelle sont tirés de la combinaison de coefficients POD des second et troisième modes. Dans chaque sous-ensemble, la vitesse est moyennée et ces moyennes sont comparées à la moyenne d'ensemble du champ de vitesse basée sur l'ensemble des cycles. La différence résultante entre la moyenne du sous-ensemble et la moyenne globale est identifiée comme une fluctuation cyclique des structures cohérentes. Ensuite, dans chaque sous-ensemble, les fluctuations résiduelles sont obtenues à partir de la différence entre les champs instantanés et la moyenne correspondante de sous-ensemble. La méthodologie proposée est testée pour deux ensembles de données obtenues à partir de simulations de moteur à résolution d'échelle. Pour le premier cas, la base de données numérique est constituée de 208 réalisations indépendantes pour une géométrie moteur simplifiée. Pour le second cas, 120 cycles pour le moteur de référence à chambre de combustion transparente (*Transparent Combustion Chamber*, TCC) sont considérés. Pour les deux applications, l'adéquation de la méthodologie permettant d'identifier les deux contributions aux VCC est discutée, et les résultats sont directement reliés aux structures de l'écoulement observé.

INTRODUCTION

The development of Internal Combustion (IC) engines is characterized by a large number of conflicting targets such as reducing the fuel consumption and the pollutant emissions while at the same time, the engine performance and the combustion stability has to be increased. Cycle-to-Cycle Variations (CCV) have a strong impact on these targets and their reduction has become a major issue in engine development. However, up to now, the complex non-linear cause-and-effect chain starting from the intake flow and finally leading to cyclic variations of the combustion process is not yet fully understood.

Numerical simulations using URANS, often in combination with experiments, is an established method in the engine development process. However, URANS is not capable to describe CCV and only Scale Resolving Simulations (SRS), *e.g.* Large-Eddy Simulations (LES), have the potential to capture cyclic fluctuations. One major drawback for the application of SRS are the significant computational demands of such investigations. The high spatial and temporal resolution for each individual cycle and the necessity to calculate a statistically relevant number of cycles pushes the computational requirements of SRS orders of magnitudes beyond what is needed for URANS. Thus, SRS for IC engine flows is still a topic of research rather than a standard method in the engine design and development process.

One important issue, especially in the context of numerical simulations, is the characterization and sometimes even the definition of CCV. While in experimental investigations CCV can be identified clearly for example by global quantities such as the cyclic variations of the peak and/or the mean pressure, such definitions are less straightforward when looking at local numerical results based on a limited number of cycles (being usually smaller than the number of measured cycles). In the following, a heuristic approach for the identification of CCV based on the velocity field in IC engines is proposed, which allows to identify fluctuations of large-scale (coherent) and smaller structures. It is important to note that there is an active discussion concerning the separation between large-scale variability and turbulence already mentioned in [1]. This study is not meant to enter into this debate or to provide a clear definition of the separation or the associated length or time scales. As already mentioned by Lumley [2], there is no scale separation in engine flows. Rather it is our aim to use the well-known capability of POD to identify large scale energetic structures to quantify the cyclic fluctuations of the coherent motions with respect to a ensemble average. Furthermore, the remaining fluctuations are expected not to be associated with a variation of a coherent motion but rather being smaller scale turbulent fluctuations. To achieve this, our method combines POD, which has been used previously for the identification and investigation of velocity field fluctuations based both on experimental and numerical data [3-9], and conditional averaging [10].

In the next sections, an overview of the general workflow to identify CCV using the SRS data and the modeling approach is given. Afterwards, the presented workflow will be tested for two engine applications, a simplified engine configuration [11] and the well-established benchmark Transparent Combustion Chamber (TCC) engine [12], respectively.

1 IDENTIFICATION OF CYCLIC VARIATIONS

In the following, an introduction of the proposed workflow to identify CCV will be given (Sect. 1.4). The workflow is based on conditional averaging (presented in Sect. 1.3), with the cycle-dependent POD coefficients (Sect. 1.2) as conditioning variables. The subsequent analysis uses the velocity field.

1.1 Definition of Cyclic Variations

CCV in IC engines are a well-known but not fully understood phenomenon. Compared to the effects of CCV, like the variation of the peak pressure during combustion being directly accessible both in experiments and simulations, the relevant underlying physics are not yet fully identified despite a significant number of investigations, [13-17] and references therein as well review articles [18, 19].

Although the term CCV is used frequently, to the authors' knowledge there is no generally accepted and unique definition. Here, we define CCV of the velocity field based on the following considerations:

- each cycle is influenced by intrinsically (*e.g.* turbulence) and extrinsically (*e.g.* acoustic effects) induced fluctuations and represents an individual process;
- at any instant in time within a cycle, the specific flow field directly results from these fluctuations;
- each of these specific flow fields is based on a less complex fundamental flow topology, which is unique and should be clearly identifiable. The fundamental flow topology is associated with a large-scale motion;
- cyclic variations of these fundamental flow topologies are considered to be CCV;
- additional fluctuations, *e.g.* from non-coherent structures, do exist but are not considered as CCV here;

In summary, this means that for IC engines, CCV of the velocity field are characterized here by the cyclic variation of large-scale coherent structures such as *e.g.* the intake jet or the tumbling motion. Of course, the smaller scales do fluctuate from cycle to cycle as well but our focus in this investigation is on the coherent structures. The aim is to extract these underlying flow topologies using a suitable mathematical procedure and quantify their variations denoting this as large-scale fluctuation. Unfortunately, there is no straightforward and unique way to extract an underlying flow topology from a single flow field realization. To identify these flow topologies, a new method based on POD and conditional averaging is proposed and evaluated. It should be mentioned that recently a cluster-based analysis [20] was proposed for IC engine flows, which seems to open up a very promising route but is not the scope of this study.

1.2 Proper Orthogonal Decomposition

Proper Orthogonal Decomposition (POD) is a powerful and well-established method to analyse complex processes. By means of a linearisation, a low-dimensional approximation of the problem can be obtained, which is often easier to handle. POD has been successfully applied in model reduction and data analysis in several scientific fields. In fluid dynamics, POD was introduced by Lumley [21] to identify coherent structures in turbulent flows followed by a number of investigations of complex flow fields in several applications [22, 23].

Here, we focus on the snapshot POD method [6, 7, 24]. A detailed description of the underlying theory and the application for experimental and simulation data can be found in [21, 24, 25]. Starting point for the POD is an approximation of an arbitrary function f by a sum of weighted linear basic functions:

$$f(\mathbf{x},t) \approx \sum_{k=1}^{N} a_k(t)\Psi_k(\mathbf{x}) \tag{1}$$

To obtain such an approximation, several approaches such as the Fourier decomposition exist. For POD, the base functions $\Psi_k(\mathbf{x})$ have to fulfill two requirements. In addition to orthogonality of the basis function $\Psi_k(\mathbf{x})$, the approximation of f has to be as good as possible in a least square sense for any arbitrary k. The singular value decomposition corresponding to Equation (1) reads as follows:

$$A = U\Sigma V^T \tag{2}$$

Matrix A represents the discrete form of function $f(\mathbf{x},t)$ and includes the data set under investigation. In general, there is no explicit rule how the data in matrix A in terms of rows and columns should be arranged. However, depending on the data set structure, a particular arrangement can reduce the computational costs to solve Equation (2). For a typical numerical investigation, the number of points in space is usually large compared to the number of points in time and this also applies for the data sets discussed later. Thus, the following structure is used here:

- each row of A represents a single variable over the considered time period;
- each column of A represents a single time step and contains all variables of interest.

This results in an $M \times N$ matrix A, while the number of rows directly results from the absolute number of considered values in the complete domain of interest and N represents the number of time steps ("snapshots"). The columns of the orthogonal $M \times M$ matrix U, which represents $\Psi_k(\mathbf{x})$ in Equation (1), are the proper orthogonal modes. The diagonal elements Σ_{ii}, also called singular values, represent the energy content of the individual modes and contain only non-negative values. All off-diagonal elements of the $M \times N$ matrix Σ are zero. Each row of the orthogonal $N \times N$ matrix V corresponds to a POD mode and acts as a time-dependent weighting factor describing the dynamics of the mode. The factor $a_k(t)$ in Equation (1) is obtained from the product of the row and column of Σ and V^T.

In this work, an eigenvalue decomposition of a temporal correlation matrix using the data set under investigation is used to solve the singular value decomposition. For this reason, Equation (2) is multiplied with its transpose, see Equation (3):

$$A^T A = V\Sigma^T U^T U\Sigma V^T \tag{3}$$

For a data set only consisting of real values, as it is the case here, it follows for the orthogonal matrix $U^T = U^{-1}$, leading to:

$$A^T A = V\Sigma^2 V^T \tag{4}$$

Equation (4) can be solved by an eigenvalue decomposition and the eigenvalues λ are the squared singular values Σ_{ii}. Afterwards, the matrix U, consisting of the proper orthogonal modes, can be calculated from Equation (2). Finally, the rows and columns of the matrices are sorted in descending order (in terms of the energy content *i.e.* singular values). This means that the first mode is the highest energetic and the last mode is the lowest energetic proper orthogonal mode.

It is interesting to note that POD has been applied regularly in the analysis of IC engine flows to obtain a better understanding of cyclic variations [3, 5, 7, 8, 14, 17, 26]. Based on these investigations, there is a broad consensus that the first (highest energetic) mode represents the ensemble-averaged flow field. Furthermore, it has become common practice to associate the other modes to fluctuations, while the modes with more energy are often associated with large-scale fluctuations. The scope of this study is to extend this idea by finding within these higher modes a number of flow topologies, which can be used for conditional averaging.

1.3 Principles of Conditional Averaging

Statistical analysis is a fundamental part for the investigation of turbulent and complex fluid flows. The use of this methodology is discussed in the context of IC engine analysis. Instead of averaging all available flow fields, the calculation of mean values and statistics is intuitively limited to flow fields, which are identical in terms of the crank angle. The advantage of this methodology is that the resulting average is based on instantaneous flow fields, which are similar in terms of the geometry and its boundary conditions (piston speed) and accordingly to the fundamental flow topology. The ensemble average is calculated as:

$$\bar{\mathbf{u}}(\mathbf{x},\varphi) = \frac{1}{N}\sum_{n=1}^{N} \mathbf{u}(\mathbf{x},\varphi,n) \tag{5}$$

based on N available flow realisations for the velocity field $\mathbf{u}(\mathbf{x},\varphi,n)$ at a specified crank angle φ within an individual cycle n. The Reynolds decomposition separates the instantaneous velocity field into an ensemble averaged and a fluctuating part:

$$\mathbf{u}(\mathbf{x},\varphi,n) = \bar{\mathbf{u}}(\mathbf{x},\varphi) + \mathbf{u}'(\mathbf{x},\varphi,n) \tag{6}$$

Typically, turbulent fluctuations are quantified by the root mean square:

$$u'_{rms}(\mathbf{x},\varphi) = \sqrt{\frac{1}{N}\sum_{n=1}^{N}\frac{1}{3}\sum_{i=1}^{3} u_i'^2(\mathbf{x},\varphi,n)} \tag{7}$$

while the components of velocity fluctuation $u_i'(\mathbf{x},\varphi,n)$ are determined by the difference of the instantaneous and the ensemble averaged flow field. For simplicity, in the following the dependency on space and crank angle is dropped unless necessary. However, a straightforward use of this averaging procedure and decomposition can lead to less meaningful results in terms of large-scale fluctuations, which can be linked to CCV as described in Section 1.1.

To remedy this problem, the average can be calculated based on a reduced set of instantaneous flow fields, which are identical in terms of a conditioning vector ϕ, whereby this conditioning vector must be able to detect the cycle-dependent dynamics of the large-scale structures. This method of conditional averaging [10] is defined according to:

$$\hat{\mathbf{u}}(\Phi) = \frac{1}{M(\Phi)}\sum_{n=1}^{M(\Phi)} (\mathbf{u}(n,\phi)|\phi = \Phi) \tag{8}$$

As can be seen, Equation (8) is very similar to Equation (5), while the value $M(\Phi)$ denotes the number of instantaneous flow fields for which $\phi = \Phi$ is valid.

The high-energetic POD modes are linked to large-scale structures [3, 7, 26], therefore the cycle-dependent coefficients of these modes (matrix V in Sect. 1.2) offer a starting point to define the conditioning vector ϕ. For this specific application here (and this will be shown later), it is possible to reduce the conditioning vector to a single conditioning variable. This conditioning variable can be represented by a set of four combinations of POD coefficients. These combinations, called subsets below, are denoted by the simple variable $s(n)$, whereby the dependency of the individual cycle n is not explicitly shown in the following for notational convenience. Furthermore, the conditioning variable s can take the discrete values 1, 2, 3 and 4 (see details below). Accordingly, the subset-dependent averaged flow field is denoted as $\hat{\mathbf{u}}(s)$.

Based on this conditional average, the following decomposition of the instantaneous flow field becomes feasible:

$$\mathbf{u}(n) = \hat{\mathbf{u}}(s) + \mathbf{u}^*(n) \tag{9}$$

Compared to a Reynolds decomposition (Eq. 6), which results into a cycle-independent averaged flow field $\bar{\mathbf{u}}$ and cycle-dependent fluctuations $\mathbf{u}'(n)$, a subset-dependent averaged flow field $\hat{\mathbf{u}}(s)$ and cycle-dependent fluctuations

$\mathbf{u}^*(n)$ appear in Equation (9). Averaging the resulting subset-dependent flow fields yields the ensemble average:

$$\bar{\mathbf{u}} = \sum_{s=1}^{S} \alpha(s)\,\hat{\mathbf{u}}(s) \qquad (10)$$

where $\alpha(s)$, defined as:

$$\alpha(s) = \frac{M(s)}{N} \qquad (11)$$

is the ratio of independent samples within each subset and the overall number of independent samples. The quantification of the subset specific small-scale fluctuations is performed similarly to Equation (7):

$$u^*_{rms}(s) = \sqrt{\frac{1}{M(s)}\sum_{n=1}^{M(s)}\frac{1}{3}\sum_{i=1}^{3}u_i^{*2}(n)} \qquad (12)$$

whereby the subset specific small-scale fluctuations $u_i^*(n)$ are calculated by Equation (9). Please note that each cycle n is associated with one subset s. The root mean square of the coherent structure fluctuations is:

$$u^\dagger_{rms} = \sqrt{\sum_{s=1}^{S}\alpha(s)\frac{1}{3}\sum_{i=1}^{3}\left(\hat{u}_i(s) - \bar{u}_i\right)^2} \qquad (13)$$

and this can be seen as an indicator for large-scale CCV. The fluctuations are defined by the difference of the globally averaged flow field (Eq. 5) and the subset average (Eq. 8) with an overall number of S subset averages (for the applications here, $S = 4$).

1.4 Workflow to Identify Cyclic Variations

In this section, based on the above ideas, the procedure for the identification of cyclic variations in IC engines is described. As mentioned before the starting point of this method is the established capability of POD to identify coherent structures. A significant portion of an engine cycle is characterized by high-energetic flow structures (*e.g.* intake jet and large vortex structures), which are associated with high energetic POD modes. Cyclic variations of these characteristic high-energetic flow structures are reflected by a variation of the high-energetic POD coefficients from cycle to cycle. For that reason, these coefficients, which act as a "cycle-dependent weighting factor" for the proper orthogonal modes, provide a good base to cluster cycles into subsets. Following the discussion above and the definitions introduced in Section 1.1, each subset is defined in such a way that it includes all independent samples with an almost identical fundamental flow structure (*i.e.* the structure of the most energetic flow patterns). Averaging within each of these subsets yields the fundamental flow topology of each subset. The deviation of the subset average $\hat{u}(s)$ to the ensemble average $\bar{u}$ does represent large-scale cyclic variations. The general workflow is shown in Figure 1.

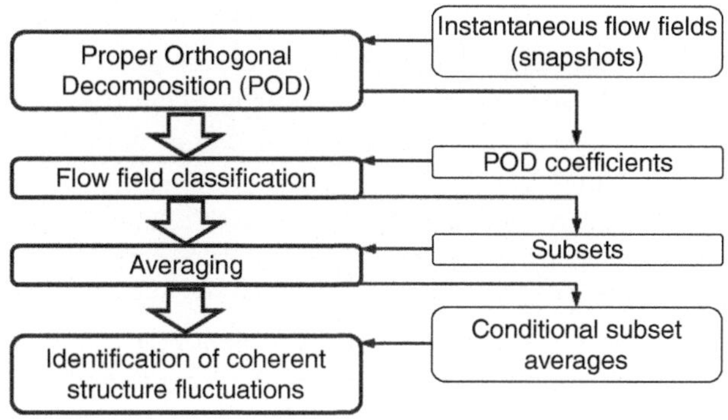

Figure 1

General workflow to calculate conditional averages and fluctuations in context of IC engines at a specified crank angle.

It is interesting to note that this approach, in contrast to previously published methods, does not require an interpretation of the resulting POD modes. This is an important point because POD is a purely mathematical method and the resulting modes do not necessarily allow a physical interpretation. However, the conditionally averaged flow fields are intuitive in terms of the flow physics and always allow an interpretation.

In a previous investigation [27], we looked at the vortex shedding behind a triangular obstacle. We revisited this configuration (not shown here) as a first step to evaluate and test our workflow according to Figure 1 for the identification of coherent structures representative for the vortex formation at the edge of intake valve during the intake stroke, which we focus on below. Starting with two subsets (based on the first POD mode) the number of subsets was increased by adding two subsets (based on the highest energetic mode not used up to this point). The conclusions from this simple test case were:

- the coefficient of the first POD mode (RANS-like field) is almost constant with respect to the different samples. Thus, it is not suitable in terms of a subset classification;
- the coefficients of the other POD modes vary in time and can be used for subset classification;
- vortex shedding can be captured with even a small number of conditional averages/subsets based on the high energetic POD modes;
- although the use of more subsets leads to an improved and more detailed information about the large-scale

structures, 4 subsets based on mode 2 and 3 provide a reasonable representation of the vortex shedding.

It has to be mentioned that this finding might differ for other, especially more complex, flow situations. Of course, in practical applications, there is only a limited number of independent samples available. This leads to a limited number of subsets, which is directly linked to the number of considered POD modes. Thus, every subset can only be an approximation of the underlying flow topology of an individual cycle. For this reason, a compromise has to be found between:

- the statistical convergence within each subset;
- the quality of the fundamental flow topology approximation.

Based on the discussion above, for the cases considered in the following, we restrict ourselves to four subsets as a compromise between statistical convergence and representation of the underlying flow topology of the subsets.

2 NUMERICAL METHOD AND TURBULENCE MODELING

In this section, an overview of the CFD solver, the numerical method and the used turbulence model will be given.

2.1 Numerical Method

All simulations are performed using *ANSYS CFX* Release 15.0, which is an implicit and node centered code. The governing equations are solved based on a finite volume method. For fully compressible flow (*e.g.* Sect. 3.2), the governing Equations (14) to (16) for mass, momentum and enthalpy are Favre-filtered:

$$\frac{\partial \bar{\rho}}{\partial t} + \frac{\partial(\bar{\rho}\tilde{u}_i)}{\partial x_i} = 0 \tag{14}$$

$$\frac{\partial(\bar{\rho}\tilde{u}_j)}{\partial t} + \frac{\partial(\bar{\rho}\tilde{u}_i\tilde{u}_j)}{\partial x_i} = -\frac{\partial \bar{p}}{\partial x_j} + \frac{\partial}{\partial x_i}\left(\tau_{ij} + \tau_{ij}^{mod}\right) \tag{15}$$

$$\frac{\partial(\bar{\rho}\tilde{H})}{\partial t} + \frac{\partial(\bar{\rho}\tilde{u}_j\tilde{H})}{\partial x_j} = \frac{\partial \bar{p}}{\partial t} + \frac{\partial}{\partial x_j}\left(\bar{q}_j + q_j^{mod}\right) + \frac{\partial}{\partial x_j}\left[\tilde{u}_i\left(\tau_{ij} + \tau_{ij}^{mod}\right)\right] \tag{16}$$

where q_i is the molecular heat flux. The quantity H, defined as:

$$\tilde{H} = \tilde{h} + \frac{1}{2}\tilde{u}_j\tilde{u}_j + k \tag{17}$$

describes the total enthalpy, while τ_{ij} is the molecular stress tensor. The quantities q_j^{mod} and τ_{ij}^{mod} are associated to turbulent fluctuations and have to be modelled. Furthermore, the system is closed by the equation of state for perfect gases:

$$p = \rho\frac{R}{W}T \tag{18}$$

For an incompressible fluid, which is the case for the simplified engine setup described in Section 3.1, only a reduced set (no enthalpy equation) is considered and no Favre-filtering is applied. For all simulations, we use a central differencing scheme in space and a second order backward Euler scheme in time.

A crucial topic for the simulation of IC engines is the rather complex mesh generation taking into account the moving piston and valves, which require special treatment. Here we use the so-called key grid approach. Starting from an initial mesh, the grid is deformed according to the moving boundaries until a critical mesh quality is reached. At this point, a new mesh is generated and the previous result is interpolated onto it. Both, the implications of the mesh morphing and the interpolation in *ANSYS CFX* was investigated in detail by Hasse *et al.* [16].

2.2 SAS-SST Turbulence Model

The Scale Adaptive Simulation Shear Stress Transport (SAS-SST) turbulence model was introduced by Menter and Egorov [28]. The fundamental idea is derived from the exact transport equation of the two-point correlation developed by Rotta [29]. This transport equation is linked to the turbulent kinetic energy k and the integral length scale L and includes the von Kármán length scale. This length scale, including the second derivative of the velocity field, was finally introduced in the ω-equation of the well known SST turbulence model [30]. The modified ω-equation for the SAS-SST model reads:

$$\frac{D(\rho\omega)}{Dt} = P_\omega + Q_{SAS} + \mathrm{Diff}_\omega - D_\omega + (1 - F_1)\frac{2\rho}{C_\omega \omega}\frac{\partial k}{\partial x_j}\frac{\partial \omega}{\partial x_j} \tag{19}$$

The last term on the right-hand side defines the transformation of ε to ω and is a fundamental part of the SST model. The additional source term:

$$Q_{SAS} = \max\left[C_{SAS_1}\rho\kappa S^2\left(\frac{L}{L_{vK}}\right)^2 - C_{SAS_2}\rho k \max\left(\frac{1}{\omega^2}\left(\frac{\partial\omega}{\partial x_j}\right)^2, \frac{1}{k^2}\left(\frac{\partial k}{\partial x_j}\right)^2\right), 0\right] \tag{20}$$

includes the von Kármán length scale, which is defined as:

$$L_{vK} = \kappa\frac{\sqrt{2S_{ij}S_{ij}}}{|U''|} \text{ with } |U''| = \sqrt{\left(\sum_i \frac{\partial^2 U_i}{\partial x_j^2}\right)^2} \tag{21}$$

Due to this additional source term, the SAS-SST turbulence model has the capability, unlike URANS models, to resolve fluctuations smaller than the integral length scale [31, 32] in case of sufficient spatial and temporal resolution. For an insufficient resolution in space and time the model reverts back to standard SST behavior.

3 ENGINE APPLICATIONS

In this section, the methodology presented above to identify and quantify CCV will be tested on two different engine setups. The first case, based on a simplified engine geometry, is meant to investigate the workflow in general in a well-defined and geometrically simple setup, which however still retains a significant number of engine specific characteristics. Numerical studies of this engine have been published by several authors [33-35].

In a second step, the method will be evaluated using data from a realistic engine case. The analysis will be performed for two specified crank angles.

Both engines are motored and therefore no injection, ignition and combustion are considered. It is important to note that we focus on fluctuations of the intake jet, sometimes referred to as "jet flapping", characterized by the flow separation at the intake valve and large-scale vortical flow structures. As explained below, this has a significant impact on the choice of the conditioning variables.

3.1 Simplified Two-Stroke Engine

This case is based on the engine setup experimentally investigated by Morse *et al.* [11]. Due to the simplicity of the geometry, this setup is suitable for the first analysis. The direct numerical simulations by Schmitt *et al.* [34] provide a very detailed insight into the complex flow structure inside the combustion chamber and confirmed an axisymmetric average flow field as well as axial symmetry for the coherent large-scale and small-scale structures. The data sets are generated from slices rotated around the cylinder axis with a non-dimensional radius of 0.2 to 1 (Fig. 2). Although the slices and thus the POD snapshots are only 2D, for a sufficiently large number of samples, are relevant flow structures should be captured since the azimuthal direction is homogeneous. The average distance between the adjacent samples is chosen such that it is more than one integral length scale, see Figure 2. This yields 26 samples for each cycle. A total number of nine consecutive cycles for the analysis, leaving out the first cycle, leads to a total number of 208 samples considered below.

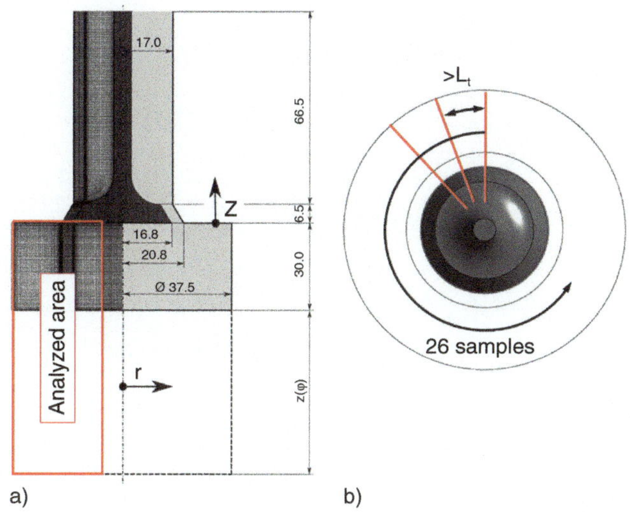

Figure 2

a) Side and b) top view of hexahedral engine mesh and engine geometry (measurements in millimeter). Red lines illustrates the location of slices for data analysis.

3.1.1 Numerical Setup

The geometry features a straight intake pipe, a flat dome and piston and a single non-moving valve. Although the experiments were performed in an horizontal arrangement, Figure 2 and all subsequent figures use the more common vertical arrangement.

All parts are concentric with respect to the cylinder liner with a bore diameter of 75 mm. The engine runs at 200 rpm with a sinus-shaped piston motion $z(\varphi)$, which results in a mean and maximum piston speed of 0.40 m/s and 0.63 m/s, respectively. A compression ratio of 3 results from a clearance height of 30 mm and a piston stroke of 60 mm. Because of the non-moving valve, there is a constant valve gap size of 4 mm and a full engine cycle corresponds to 360 °CA with an intake and exhaust stroke of 180 °CA each. Corresponding to the experimental setup, the simulations were performed with air at an ambient pressure and temperature of 1.03 bar and 300 K. Resolved turbulent fluctuations were superimposed on the filtered inlet boundary condition.

The simulations are carried out on a highly-resolved hexahedral mesh with about 8 million grid points and a mesh refinement at the wall leading to a minimum edge length of 40 μm. Due to the non-moving valve and the simple geometry, the complete cycle can be simulated on a single mesh preventing interpolation errors. The necessary mesh morphing is limited to the combustion chamber. The time step width was set to 0.025 °CA, which leads to a CFL number smaller than 1 for the largest part of the simulated domain.

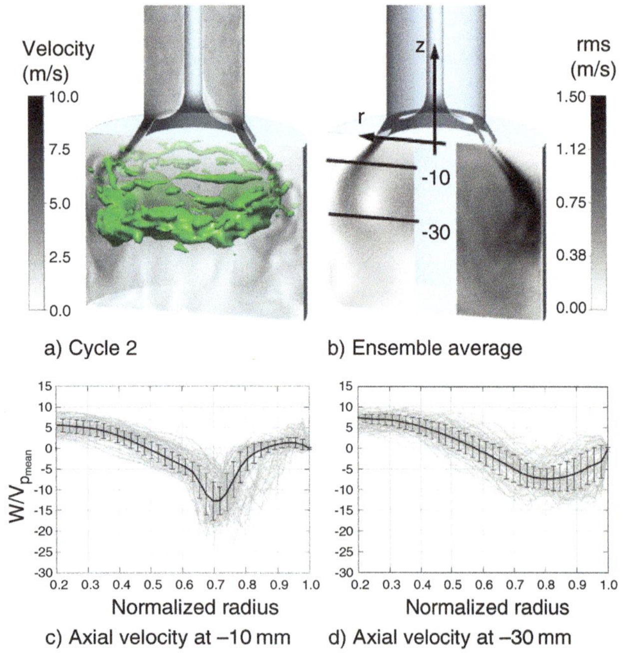

Figure 3

a) Instantaneous velocity magnitude in the mid plane and isosurface at relative pressure of −1 855 Pa in cycle 2; b) shows the ensemble-averaged velocity field (left half plane; legend as a)) and the root mean square of velocity fluctuations (right half plane), c,d) show the averaged (black line) and instantaneous (grey lines) axial velocities (normalized by average piston speed) as well as the standard deviation of velocity (bars) at −10 mm and −30 mm. All results are shown for 90 °CA.

3.1.2 Results

Figure 3a shows the instantaneous velocity field on the mid plane at 90 °CA with a maximum velocity magnitude of about 12 m/s in the intake jet region. In addition an isosurface for a relative pressure of $p_{\mathrm{rel}} = -1\,855$ Pa is shown.

Figure 3b shows the ensemble-averaged velocity field (left half plane) as well as the ensemble-averaged rms values of the velocity fluctuations (right half plane) extracted from slices as shown in Figure 2. The averaged velocity field shows a well-defined intake jet as well as a vortex structure corresponding to the isosurface illustrated in Figure 3a. As expected, the velocity fluctuations show a maximum in the shear layer region of the intake jet. It is interesting to note that the rms values also have a local maximum close to the vortex core region of the averaged velocity field (Fig. 3b) although no significantly increased level of fluctuations can be recognized in the independent samples (not shown here). This is a first indication that this local maximum, which occurs for wide ranges of the intake stroke, does not result from turbulent small-scale fluctuations but rather from a large-scale structure fluctuation, which is represented by the deformation and displacement of the large vortex structure with respect to the cylinder axis. Thus, the high rms values directly result from the ensemble-averaging procedure. Borée *et al.* [5] identified experimentally such a large scale jet fluctuation in another simplified engine setup. Hasse *et al.* [15] performed scale-resolving simulations of this setup and concluded in agreement with the experimental observations, that fluctuations of the large scale structure are falsely identified as turbulent kinetic energy when standard ensemble-averaging is used. These observations are another indicator that the standard averaging procedure might lead to results difficult to interpret and in fact, our method aims to reduce this issue. Figures 3c and d show the averaged and the instantaneous axial velocities (normalized by mean piston speed) on lines (normalized by cylinder radius) placed at 10 mm and 30 mm below the cylinder head. Furthermore, the standard deviation of the instantaneous velocities (related to the averaged velocity) is shown as error bars. The largest deviations for the instantaneous velocity profiles at −10 mm can be identified at a nondimensional radius of 0.6 to 0.8, which represents the unstable jet position during the intake stroke. At –30 mm a broad region between 0.45 and 0.95 with large fluctuations can be identified.

3.1.3 Identification of Cyclic Variations

To identify large-scale fluctuations, the workflow described in Section 1.4 is applied at 50 °CA. The first step is a POD analysis using all 208 snapshots. Figure 4 shows the POD modes (each normalized by its maximum value) resulting from the velocity field at 50 °CA.

As mentioned before, there is an obvious similarity between the first (*i.e.* highest energetic) POD mode in Figure 4a and the ensemble-averaged velocity field in Figure 5a. The interpretation of the 2nd, 3rd and 4th mode is less clear at this point due to the more complex flow topology. It seems that especially the 2nd and 3rd mode are strongly connected to the intake jet and the vortex structure due to the dominating influence of the jet shear layer and the vortex core close to the cylinder axis. In connection with the POD coefficients, it seems that these modes are linked to the jet direction and its width as well as the position and strength of the vortex structure of the flow field in physical space. An interpretation of the 4th mode appears more difficult at this crank angle due to missing dominant structures and associated field topologies. Most likely, this mode can be associated to the vortex core.

As a next step, the instantaneous flow fields are grouped into different subsets followed by conditional averaging. As described in Section 1.4, this classification is based on the POD coefficients. The relevance of the individual POD

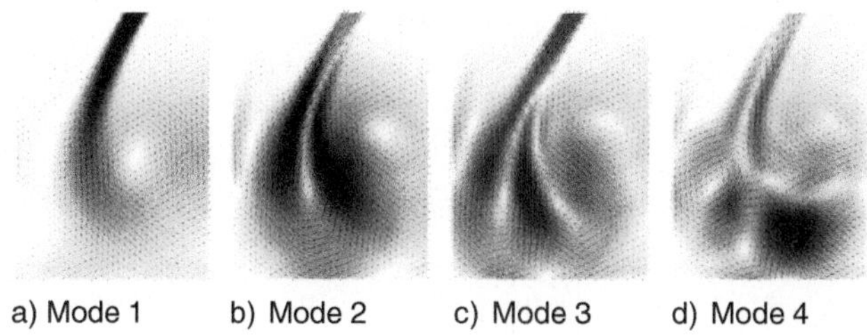

Figure 4

Proper orthogonal modes 1 to 4 (each normalized by its maximum value) resulting from the velocity field at 50 °CA in the simplified engine setup.

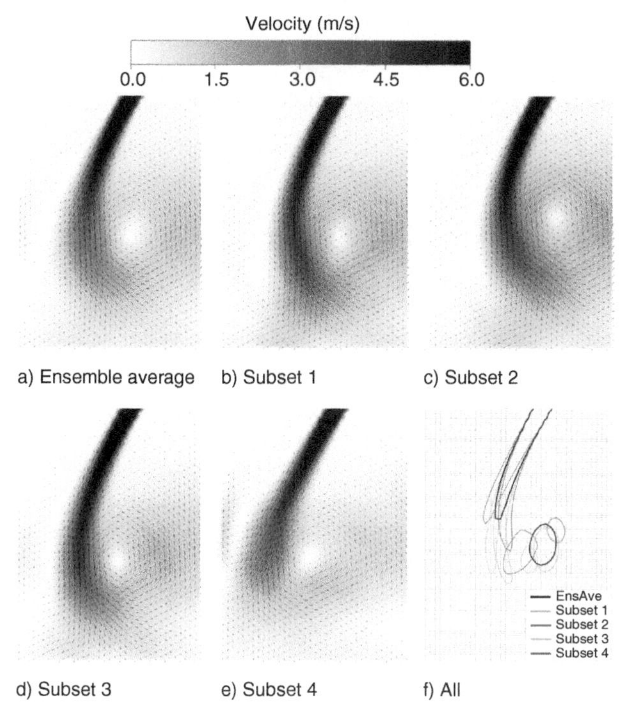

Figure 5

a) Ensemble-averaged and b-e) conditionally-averaged velocity fields at 50 °CA in the simplified engine setup. f) Comparison of the conditionally-averaged (thin lines) and ensemble-averaged (thick line) using velocity isolines (6 m/s for intake jet and 1 m/s for vortex core). Corresponding structures are marked by the same color.

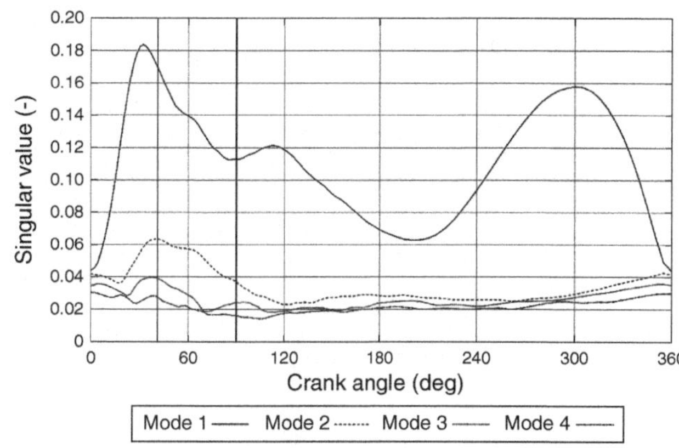

Figure 6

Singular values of the first 4 proper orthogonal modes as a function of crank angle. 0 °CA to 180 °CA for intake and 180 °CA to 360 °CA for exhaust stroke. The two crank angles, 50 °CA and 90 °CA, respectively, discussed in Section 3.1 are marked.

modes can quantified by their singular values, which are shown in Figure 6.

As expected the singular value of the first mode is high compared to the other modes. For wide ranges over the intake stroke, the singular value of the second mode is also comparatively large, which is an indicator for a large-scale structure in physical space and the higher modes seem to contain only a smaller fraction of the energy. This suggests that these modes are connected to less dominant flow structures in physical space.

For the further discussion, we use the following observations, strictly valid only for these crank angle:

- due to the structure of mode 1, 2 and 3, see Figure 4, these modes can be associated with the intake jet;
- for most of the intake stroke, the singular values of the first and second mode are high compared to the singular values of higher modes (Fig. 6);
- the POD coeficient of the first mode is almost constant from sample to sample and thus less useful for a classification, see Figure 7. This is similar to the findings described in Section 1.4;
- the POD coeficients of the other modes show strong variations from sample to sample (Fig. 7; only coefficients of 1st, 2nd and 3rd POD modes are shown for readability).

Based on this observation and the necessity of a compromise between the number of subsets and the samples within each subset, as described in Section 1.4, in the following we focus on a small number of high energetic modes for the analysis of large-scale fluctuations. This results into four subsets, defined by the coefficient of 2nd and 3rd POD mode, which ensures a significant number of individual samples necessary for statistical convergence within each subset, Table 1. It is important to note that this is a restriction of the current analysis and we cannot claim that the general applicability reaches beyond this jet intake flow configuration. Certainly further investigations are necessary to reach more generally applicable procedures. Nevertheless, as will

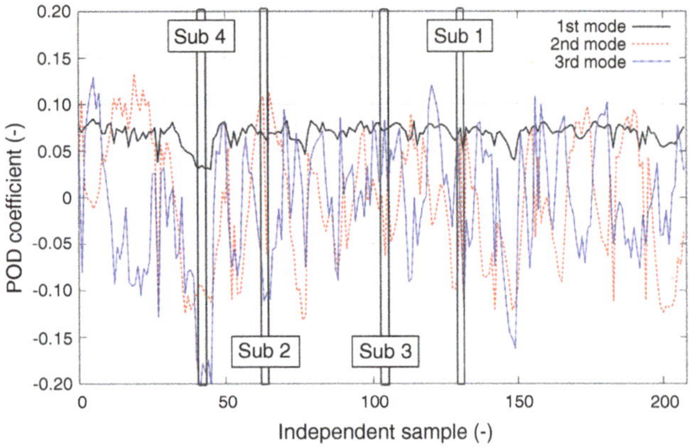

Figure 7

Sample dependent POD coefficients of 1st, 2nd and 3rd mode at 50 °CA. Furthermore, exemplarily a subset classification (corresponding to Tab. 1 is shown).

TABLE 1

Subset classification based on POD coefficients of the 2nd and 3rd mode at 50 °CA. Number of snapshots in each subset are shown in the last column

Subset	M2 coefficient	M3 coefficient	Snapshots
1	≥ 0	≥ 0	36
2	≥ 0	< 0	65
3	< 0	≥ 0	64
4	< 0	< 0	43

be shown below, this very straightforward approach (shown in Tab. 1 and Fig. 7) chosen here can provide an interesting physical insight and offers a different way of looking at large scale fluctuations.

Beside the classification criterion, Table 1 also shows the number of instantaneous velocity fields clustered into each subset. In a final step, an averaging within the subsets leads to the subset averages conditioned by the POD coefficients. Figure 5 shows the ensemble-average as well as the subset averages at 50 °CA.

As expected, the general flow topology is characterized by an intake jet and a vortex structure, which can be identified for the ensemble average as well as for the subset averages. An explicit comparison of the averages is illustrated in Figure 5f by means of velocity isolines of 6 m/s (intake jet) and 1 m/s (vortex structure), respectively. The intake jet is identical for all averages in the region close to the valve gap. This is the direct result of the strong wall influence on the flow. However, a variation of the jet direction and the penetration depth can be clearly seen in the combustion chamber. The intake jet of the 1st and 3rd subset average in Figures 5b and d show a increased penetration depth compared to the ensemble average in Figure 5a. In the direct comparison the averaged velocity fields of the 2nd and 4th subset in Figures 5c and e show a larger deviation from the ensemble-averaged velocity field with respect to the jet direction.

A variation can also be identified for the vortex structure. With the exception of the 3rd subset, each subset average shows a clear variation of the vortex core position compared to the ensemble average. It can be clearly seen in Figure 5f that the shift of the vortex core position is directly linked to the position of the intake jet. For example, in the second subset the most significant bending of the intake jet towards the cylinder axis can be identified. The corresponding vortex core is shifted significantly towards the cylinder axis. Since the vortex core and the jet are strongly coupled, their observed fluctuations are also well correlated.

Figure 8a shows the rms of the velocity fluctuations calculated by the ensemble-averaged and the total number of instantaneous velocity fields according to Equation (7). Close to the valve gap, the velocity fluctuations are observed in a well-defined and thin shear layer generated by the intake jet.

Figures 8b-e show the rms values calculated from the subset averages and the corresponding instantaneous velocity fields as described in Equation (12). Most notable is that the velocity fluctuations are very well-defined with respect to the subset averages in Figures 5b-e. As in Figure 8a, each subset shows very thin regions of velocity fluctuations in the vicinity the valve gap, corresponding to the thin shear layer. In contrast to the ensemble average, the fluctuations of the subset averages are constrained to the shear layer region near to the intake jet. An increased level of velocity fluctuations can be seen near the piston or the cylinder liner, which causes the final jet breakup. The well-defined structure of velocity fluctuations indicates that the subset averages are a good approximation for the fundamental flow topology representative for the corresponding instantaneous velocity fields. Figure 8f shows the large-scale fluctuations, calculated by Equation (13), which represents the as defined in Section 1.1. The structures of the large-scale fluctuations look very similar to the 2nd POD mode although the conditional averaging was done based on the POD coefficients of the 2nd the 3rd mode. This quite remarkable effect can be explained by the energy contents of the modes. As can be seen in Figure 6 the second POD mode shows an increased energy level of about 75% compared to the third POD mode and represents a more dominant flow field topology (*i.e.* large-scale structures). Another interesting observation is the occurrence of the thin region, bounded by the high

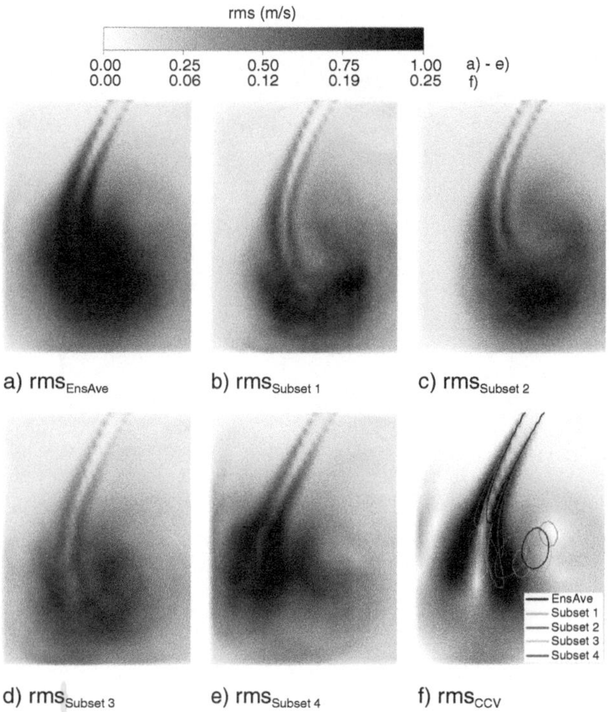

Figure 8

Root mean square of velocity fluctuations calculated from ensemble-averaged and the total number of instantaneous velocity fields are shown in a), conditionally-averaged and subset velocity fields in b-e) and the conditionally-averaged and subset averaged velocity fields in f) at 50 °CA, respectively. Isolines are defined as in Figure 5. Corresponding structures are indicated by the same color.

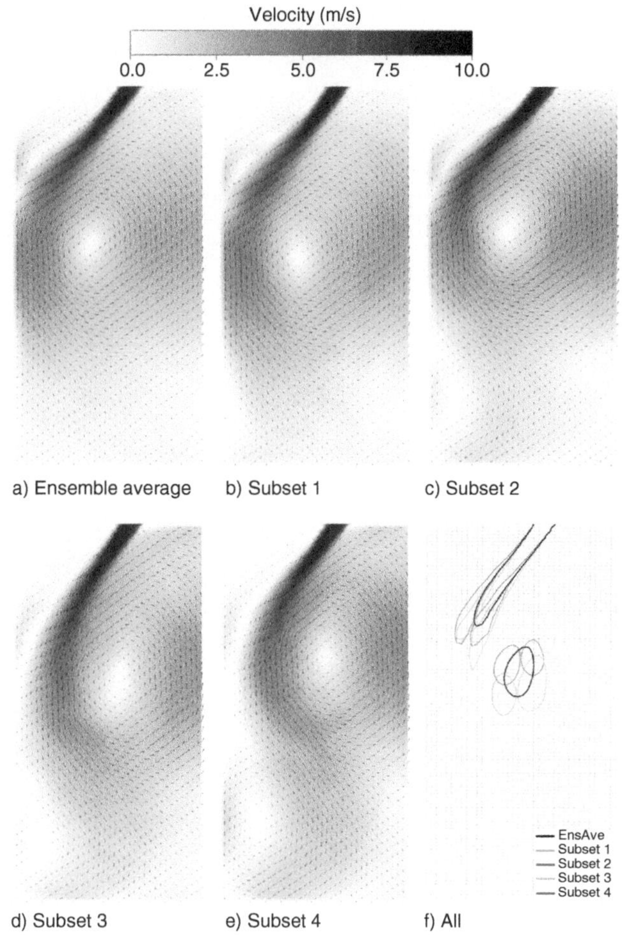

Figure 9

a) Ensemble-averaged and b-e) conditionally-averaged velocity fields at 90 °CA in the simplified engine setup. f) Comparison of the conditionally-averaged (thin lines) and ensemble-averaged (thick line) using velocity isolines (6 m/s for intake jet and 1 m/s for vortex core). Corresponding structures are indicated by the same color.

rms values, which marks an overlap between all the subset and the ensemble average.

The ensemble and subset-averaged velocity fields at 90 °CA, which corresponds to the crank angle of maximum piston speed, can be seen in Figures 9a-e, respectively. In general, the same trends as for 50 °CA can be observed. The downwards moving piston induces an intake jet. Due to the constant height of the valve gap, the velocity of the intake jet and the piston speed are directly coupled and also the maximum jet velocity is found at this crank angle. The jet direction remains almost unchanged until it is finally deflected by the cylinder liner. The direct comparison to 50 °CA shows that this jet bending is pushed further away from the axis. It is interesting to note that the influence of the liner seems to reduce the cyclic variations of the intake jet, see Figure 9f illustrating the conditionally and ensemble-averaged positions of the jet and the vortex, respectively. This liner influence will also be observed in the second example below.

The vortex structure appears more regular when compared to 50 °CA. The subset averages of the vortex core position seem to indicate a precession around the ensemble-averaged vortex core. Looking at the ensemble-averaged velocity field, this precessing vortex core results in an increased level of turbulent fluctuations depicted in Figure 10a and this finding agrees well with the investigations of Borée *et al.* [5]. In contrast to the ensemble-averaged data field, the subset averages yield more homogeneous fluctuating fields within the combustion chamber and no local maximum at the vortex core, Figure 10b, is found. Thus, the potentially false identification of a turbulent fluctuation due to the precessing vortex core can be avoided using the proposed subset averaging.

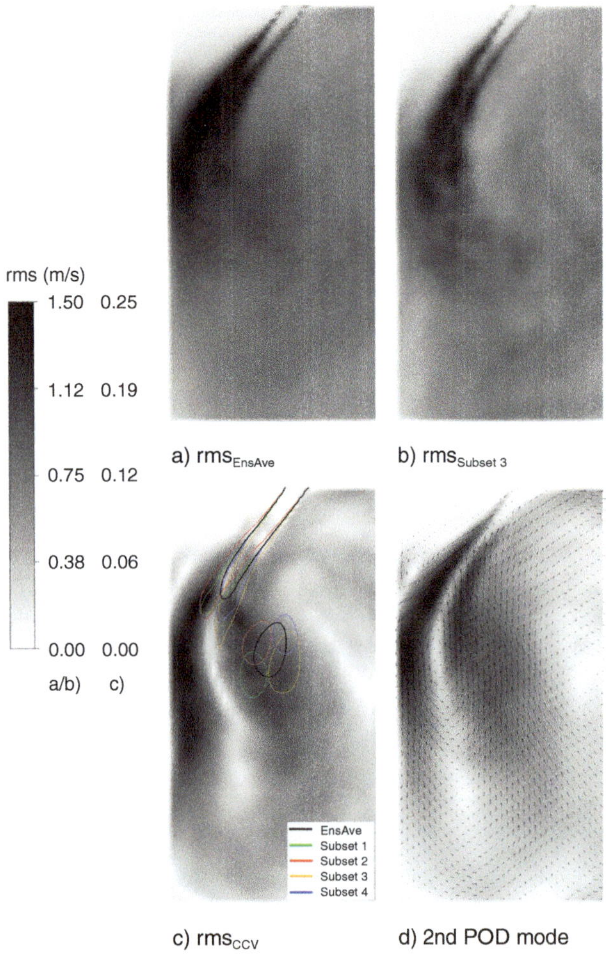

Figure 10

a) Root mean square of velocity fluctuations calculated from ensemble-averaged with the total number of instantaneous velocity fields at 90 °CA. Root mean square of velocity fluctuations in b) subset 3 and c) large-scale fluctuations. d) 2nd normalized proper orthogonal mode. Isolines defined as in Figure 9. Corresponding structures are indicated by the same color.

Figure 10c shows the large-scale fluctuations calculated by Equation (13) and Figure 10d shows the 2nd POD mode. Interestingly, the structures of the large-scale fluctuations and the 2nd POD mode appear to be almost identical, which can be explained by the higher energy content (more than 50% compared to mode 3) of the 2nd mode.

3.2 TCC Engine

The second application considered here is a realistic engine setup. The considered case is motored and a full cycle takes 720 °CA, while the firing TDC is at 0 °CA. Due to the fact that this engine has an straight vertical intake port without machined tumble edge, the resulting tumble motion is relatively small. For this reason, we focus on the intake phase (358 °CA-593 °CA) in the following. 120 cycles/ ensembles of the TCC engine [12], being an internationally well-established benchmark case, were simulated. The optically accessible combustion chamber provides a high-quality experimental data base for comparison to numerical investigations.

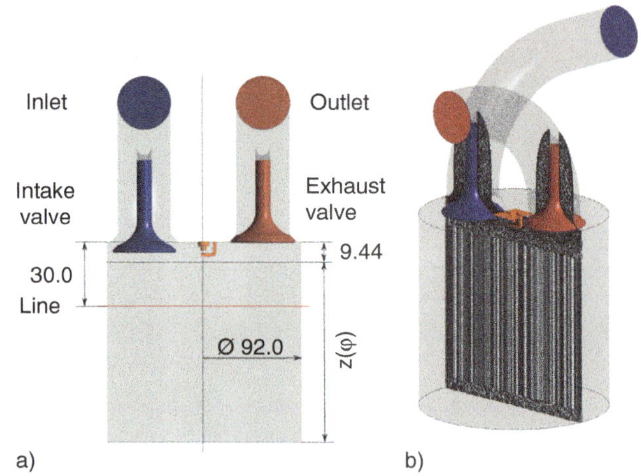

Figure 11

a) Geometry and b) numerical mesh of the TCC.

3.3 Numerical Setup

The domain for the numerical simulation is shown in Figure 11.

The setup features a flat piston and dome and a single intake and exhaust valve. The bore and the stroke are 92 mm and 86 mm, respectively. An effective compression ratio of 8 is based on the clearance height of 9.44 mm. All simulations were performed at the same atmospheric conditions as in the experiment. The engine runs at 800 rpm in motored operation.

As can be seen in Figure 11, the simulations use a hybrid tetrahedra/prism mesh with a maximum number of 2.5 million grid points during the phase when both valves are open. The time step varies during the cycle and reaches a minimum value of 0.25 °CA during the early phase of the intake stroke. Resolved turbulent fluctuations (similar to Sect. 3.1.1) were imposed additionally as inlet boundary condition in combination with the SAS-SST model (Sect. 2.2).

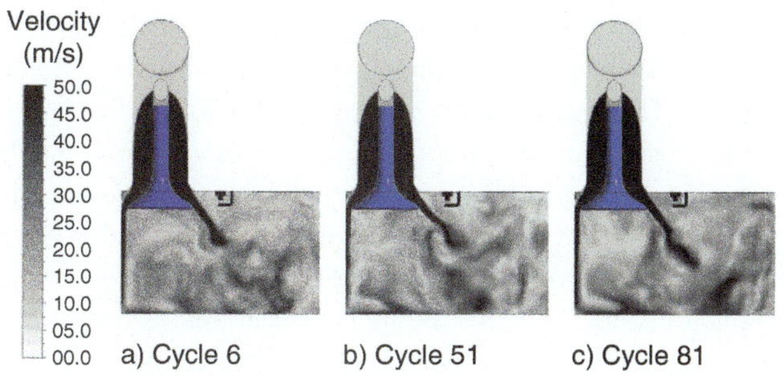

Figure 12

Instantaneous velocity magnitude in the TCC engine in the middle plane at 450 °CA (*i.e.* intake stroke).

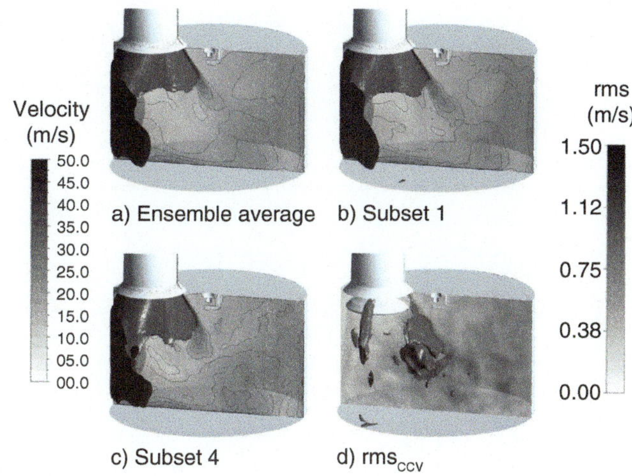

Figure 13

a) Ensemble-averaged velocity fields and subset-averaged velocity fields b,c) in the TCC engine at 450 °CA; d) shows the root mean square of the large-scale fluctuation according to Equation (13). In addition two isosurfaces for the velocity magnitude (50 m/s) and for $\tilde{u}_{rms}$ (1.5 m/s) are included in all plots.

3.4 Results

Figures 12 shows the instantaneous velocity fields in the middle plane of cycle 6, 51 and 81 at 450 °CA.

Due to the orientation and shape of the intake port, the engine does not produce a significant tumbling motion. Indeed, the fluid motion shows no large-scale structures other than the intake jet with a velocity magnitude of about 50 m/s. After deflection, the intake jet is mainly following the direction of the cylinder liner and only small variations from cycle-to-cycle can be seen, which is consistent with the finding above for the simplified engine at 90 °CA. Furthermore, no other dominant structures are found in the rest of the flow field, which mostly features small-scale structures. Following up on the discussion above for the first test case, it is interesting to see whether the proposed method is also capable of describing this different situation with mostly small-scale and only limited large-scale fluctuations.

3.4.1 Identification of Cyclic Variations

The following investigation is carried out at 450 °CA during the intake stroke. The conditioning variables for this analysis are chosen in a fully analogous manner as in Section 3.1.3. A POD analysis based on the full 3D velocity field in the combustion chamber is performed. Identically to Section 3.1.3 the POD coefficient of the first mode is almost constant with respect to the cycle count. Thus, the 2nd and 3rd mode are used for a flow field classification. Due to the relatively small number of 120 independent samples/cycles, the averages generated by the four subsets are potentially less reliable in a statistical sense than in Section 3.1.3. After classification, the averaging procedure, see Equation (8), within the different subsets is applied and the rms of the large-scale fluctuations according to Equation (13) is calculated.

Figure 13 compares the ensemble-averaged velocity field with the rms of the large scale fluctuations and the conditional averages of subsets 1 and 4. At a first glance, the ensemble average and the subset averages 1 and 4 appear almost identical. However, there is a notable deviation from subset to subset, which is found in close proximity of the intake jet region, which is not directly influenced by the cylinder liner and correspondingly the largest values for the rms of the large-scale fluctuations are found in this specific region. In the wall-guided part of the intake jet, the differences are very small.

In Figure 14, the velocity magnitude of selected averaged and instantaneous velocity fields along a line in the middle plane 30 mm below the cylinder head (Fig. 11) is illustrated. A direct comparison of the ensemble and subset-averaged velocity fields is shown in Figure 14a. In the region of the wall-guided intake jet, all averages are almost identical. A larger deviation can be identified for the position of the maximum velocity magnitude and subset 4 shows the largest deviation with respect to the ensemble average.

Figures 14b and c show the averaged as well as the instantaneous velocity profiles of subset 1 and 4. In general, both subsets confirm that the instantaneous velocity fields are classified correctly in terms of the dominant structure and thus the instantaneous profiles can be associated with the corresponding subset average, *e.g.* looking in detail at the difference in velocity magnitude near the cylinder axis or the shift of the maximum velocity in terms of the dimensionless radius.

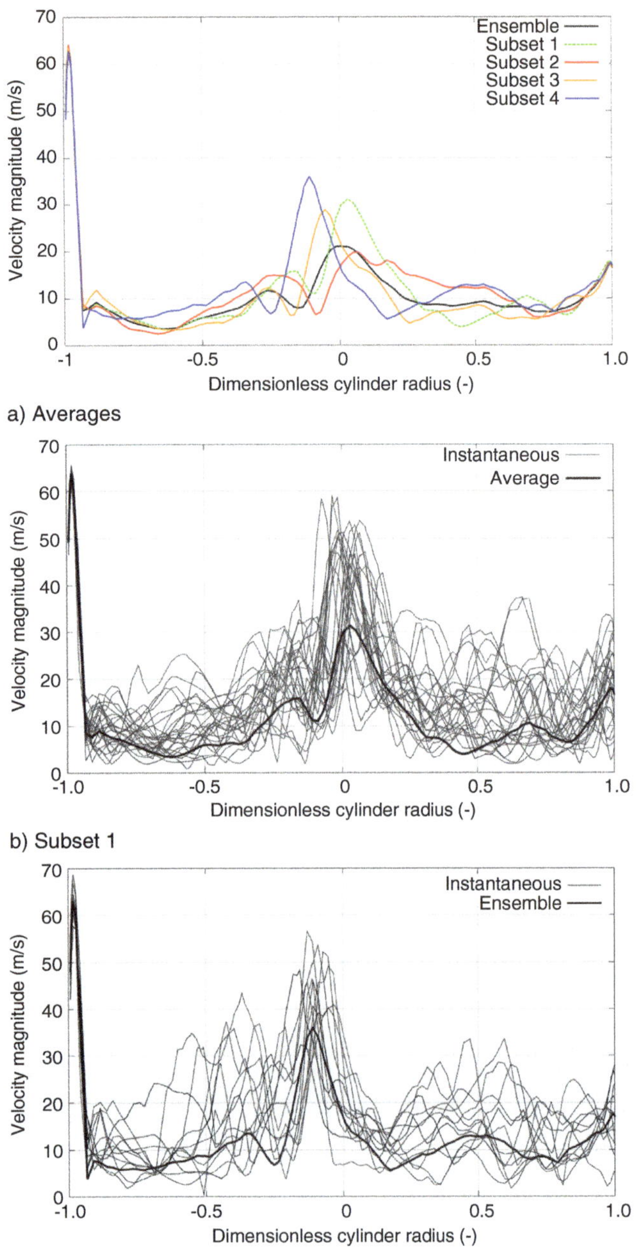

Figure 14

a) The velocity magnitude of the ensemble and subset-averaged flow fields on line $z = 30$ mm below cylinder head at 450 °CA; b) and c) the averaged as well as the instantaneous velocity magnitudes of subset 1 and subset 4.

SUMMARY

A new methodology to identify and quantify large-scale CCV in internal combustion engines is proposed. The method is applicable both for experimental and numerical data. Here, it is used and evaluated for 2D and 3D velocity data sets from multi-cycle scale resolving simulations of two different engines. The general idea behind the method is to identify the large-scale coherent motion and its deviation from the ensemble average. This is achieved by first decomposing the instantaneous velocity field into a mean and a large (energetic)- as well as a smaller-scale contribution to the fluctuations, respectively. While the mean field is determined by a standard ensemble average, the identification of the two fluctuating quantities is based on a combination of POD and conditional averaging. POD is known to have the capability to identify coherent structures, and we use combinations of the cycle-dependent POD coefficients of the first modes to identify suitable subsets, which are then the conditioning variable.

The instantaneous flow fields are then grouped into different subsets with the aim of averaging and calculating fluctuations within these subsets. Since the number of realizations within each subset is smaller than the overall number of realizations, this averaging potentially requires a larger number of engine cycles. The difference of the subset average to the ensemble average is then denoted as large-scale fluctuation. On the other hand, the difference between the instantaneous velocity field with respect to its specific subset average is denoted a small-scale fluctuation.

This method was tested using two numerical data sets from scale resolving engine simulations. The focus was to assess the ability to detect large scale flow topologies and their cyclic fluctuations in addition to small-scale fluctuations. For the first engine test case, 208 independent samples were available and 4 subsets could be identified based on the POD coefficients of the first modes. The results show a clear difference of the position and penetration depth of the intake jet for each of the 4 subsets and this is directly identified by the method. Furthermore, a variation of the vortex core position, similar to a precessing vortex core, can be detected. Small-scale structures are found in regions, where high turbulent fluctuations are expected, *e.g.* in the high shear region of the jet.

The second case is the well-established TCC engine benchmark, for which previously computed 120 cycles were used for the analysis. Again, 4 different subsets could be identified and the same averaging procedure was performed. Similar conclusions concerning the decomposition of the velocity field as for the first case could be drawn and detailed results were shown.

Finally, it can be concluded that the newly proposed method allows to separate fluctuations of the velocity field into a large-scale coherent and a small-scale turbulent contribution. Thus, using this method allows to distinguish the large-scale fluctuation, such as a precessing vortex core, from the classical turbulent contributions.

ACKNOWLEDGMENTS

The simulations were performed on the national supercomputer Cray XC40 (Hornet) at the High Performance Computing Center Stuttgart (HLRS) under the grant number ICECCV/44054.

The authors kindly acknowledge the financial support by the Saxon Ministry of Science and Fine Arts, the SAB (Sächsische Aufbaubank) and the European Union in the project "DynMo" (project number 100113147), as well as the FVV in the project "BSZII" (project number 6011333).

Furthermore the authors would like to thank Dave Reuss, Volker Sick (University of Michigan), Jens Neumann (*BMW AG*) and Mark Sastuba (TU Bergakademie Freiberg) for the helpful and fruitful discussions.

REFERENCES

1 Heywood J.B. (1988) *Internal Combustion Engines Fundamentals*, McGraw-Hill.

2 Lumley J.L. (1999) *Engines. An Introduction*, Cambridge University Press.

3 Abraham P., Liu K., Haworth D., Reuss D., Sick V. (2013) Evaluating large-eddy simulation (LES) and high-speed particle image velocimetry (PIV) with phase-invariant proper orthogonal decomposition (POD), *Oil & Gas Science and Technology.*

4 Arányi P., Janiga G., Zähringer K., Thévenin D. (2013) Analysis of different pod methods for piv-measurements in complex unsteady flows, *International Journal of Heat and Fluid Flow* 204-211.

5 Borée J., Maurel S., Bazile R. (2002) Disruption of a compressed vortex. *Physics of Fluids (1994-present)* **14**, 2543-2556.

6 Chen H., Reuss D., Hung D., Sick V. (2013) A practical guide for using proper orthogonal decomposition in engine research, *International Journal of Engine Research* **14**, 4, 307-319.

7 Chen H., Reuss D., Sick V. (2012) On the use and interpretation of proper orthogonal decomposition of in-cylinder engine flows, *Measurement Science and Technology* **23**, 085302.

8 di Mare F., Knappstein R., (2014) Statistical analysis of the flow characteristics and cyclic variability using proper orthogonal decomposition of highly resolved les in internal combustion engines, *Computers & Fluids* **105**, 101-112.

9 Juergens W., Kaltenbach H. (2003) Eigenmode decomposition of turbulent velocity fields behind a swept, backward-facing step, *Journal of Turbulence* **4**, 18.

10 Block D., Teliban I., Greiner F., Piel A. (2006) Prospects and limitations of conditional averaging, *Physica Scripta* **2006**, 25.

11 Morse A., Whitelaw J., Yianneskis M. (1979) Turbulent flow measurements by laser-doppler anemometry in motored piston-cylinder assemblies, *J. Fluids Eng.* **101**, 2, 208-216.

12 Sick V., Reuss D., Rutland C., Howarth D., Oefelein J., Janicka J., Kuo T.-W., Yang X., Freitag M. (2010) A common engine platform for engine les development and validation, *International Conference on Large-Eddy Simulation for Internal Combustion Engine Flows (LES4ICE)*, Rueil-Maimaison, France, 18-19 Nov.

13 Baby X., Dupont A., Ahmed A., Deslandes G., Charnay W., Michard M. (2002) A new methodology to analyze cycle-to-cycle aerodynamic variations. Technical report, *SAE Technical Paper* 2002-01-2837.

14 Böhm B., di Mare F., Dreizler A. (2010) Characterisation of cyclic variability in an optically accessible IC engine by means of phase-independent POD, *les Rencontres Scientifiques de l'IFP, LES for Internal Combustion Engine Flows (LES4ICE)*, 18-19 Nov.

15 Hasse C., Sohm V., Durst B. (2009) Detached eddy simulation of cyclic large scale fluctuations in a simplified engine setup, *International Journal of Heat and Fluid Flow* **30**, 1, 32-43.

16 Hasse C., Sohm V., Durst B. (2010) Numerical investigation of cyclic variations in gasoline engines using a hybrid URANS/LES modeling approach, *Computers and Fluids* **39**, 1, 25-48.

17 Voisine M., Thomas L., Borée J., Rey P. (2011) Spatio-temporal structure and cycle to cycle variations of an in-cylinder tumbling flow, *Experiments in Fluids* **50**, 1393-1407.

18 Haworth D. (1999) Large-eddy simulation of in-cylinder flows, *Oil & Gas Science and Technology* **54**, 2, 175-185.

19 Rutland C.J. (2011) Large-eddy simulations for internal combustion engines - a review, *International Journal of Engine Research* **12**, 5, 421-451.

20 Cao Y., Kaiser E., Borée J., Noack B.R., Thomas L., Guilain S. (2014) Cluster-based analysis of cycle-to-cycle variations: application to internal combustion engines, *Experiments in Fluids* **55**, 11.

21 Lumley J. (1967) The structure of inhomogeneous turbulent flows, *Atmospheric Turbulence and Radio Wave Propagation*, Yaglom A.M., Tatarski V.I. (eds.), pp. 166-178.

22 Mureithi N., Huynh K., Rodriguez M., Pham A. (2010) A simple low order model of the forced karman wake, *International Journal of Mechanical Sciences* **52**, 11, 1522-1534.

23 Orellano A., Wengle H. (2001) POD analysis of coherent structures in forced turbulent flow over a fence, *Journal of Turbulence* **2**, 8.

24 Chatterjee A. (2000) An introduction to the proper orthogonal decomposition, *Current Science* **78**, 7, 808-817.

25 Holmes P., Lumley J., Berkooz G., Rowley C. (2012) *Turbulence, Coherent Structures, Dynamical Systems and Symmetry*, Cambridge University Press.

26 Liu K., Haworth D.C., Yang X., Gopalakrishnan V. (2013) Largeeddy simulation of motored flow in a two-valve piston engine: Pod analysis and cycle-to-cycle variations, *Flow, Turbulence and Combustion* **91**, 373-403.

27 Hasse C., Sohm V., Wetzel M., Durst B. (2009) Hybrid URANS/LES turbulence simulation of vortex shedding behind a triangular flameholder, *Flow, Turbulence and Combustion* **83**, 1, 1-20.

28 Menter F.R., Egorov Y. (2010) The scale-adaptive simulation method for unsteady turbulent flow predictions. part 1: Theory and model description, *Flow, Turbulence and Combustion* **85**, 113-138.

29 Rotta J. (2010) *Turbulente Strömungen*, Göttinger Klassiker der Stroemungsmechanik, Universitaetsverlag Goettingen.

30 Menter F.R. (1994) Two-equation eddy-viscosity turbulence models for engineering applications, *AIAA Journal* **32**, 8, 1598-1605.

31 Egorov Y., Menter F., Lechner R., Cokljat D. (2010) The scale-adaptive simulation method for unsteady turbulent flow predictions. part 2: Application to complex flows, *Flow, Turbulence and Combustion* **85**, 139-165.

32 Lucius A., Brenner G. (2010) Unsteady CFD simulations of a pump in part load conditions using scale-adaptive simulation, *International Journal of Heat and Fluid Flow* **31**, 6, 1113-1118.

33 Haworth D., Jansen K. (2000) Large-eddy simulation on unstructured deforming meshes: towards reciprocating IC engines, *Computers & Fluids* **29**, 493-524.

34 Schmitt M., Frouzakis C.E., Tomboulides A.G., Wright Y.M. Boulouchos K. (2014) Direct numerical simulation of multiple cycles in a valve/piston assembly, *Physics of Fluids (1994-present)* **26**, 035105.

35 Schmitt M., Frouzakis C.E., Tomboulides H.A.G., Wright A., Boulouchos K. (2014) Direct numerical simulation of the effect of compression on the flow, temperature and composition under engine-like conditions, *Proceedings of the Combustion Institute* **35**, 3069-3077.

11

Risk Assessment-Led Characterisation of the SiteChar UK North Sea Site for the Geological Storage of CO_2

Maxine Akhurst[1]*, Sarah D. Hannis[1], Martyn F. Quinn[1], Ji-Quan Shi[2], Marielle Koenen[3], Florence Delprat-Jannaud[4], Jean-Claude Lecomte[4], Daniel Bossie-Codreanu[4], Stanislaw Nagy[5], Łukas Klimkowski[5], Davide Gei[6], Maarten Pluymaekers[3] and David Long[1]

[1] British Geological Survey, West Mains Road, Edinburgh, EH9 3LA - UK
[2] Imperial College London, South Kensington Campus, London SW7 2AZ - UK
[3] TNO, Princetonlaan 6, 3584 CB Utrecht - The Netherlands
[4] IFP Energies nouvelles, 1-4 avenue de Bois-Préau, 92852 Rueil-Malmaison Cedex - France
[5] AGH University of Science and Technology, 30 Mickiewicza Avenue, 30-059 Krakow - Poland
[6] Istituto Nazionale Di Oceanografia E Geofisica Sperimentale, Borgo Grotta Gigante, 42/C - 34010 Sgonico (TS) - Italy
e-mail: mcak@bgs.ac.uk - sann@bgs.ac.uk - mfq@bgs.ac.uk - j.q.shi@imperial.ac.uk - marielle.koenen@tno.nl - florence.delprat-jannaud@ifpen.fr
jean-claude.lecomte@ifpen.fr - dan.bossie-codreanu@ifpen.fr - stanislaw.nagy@agh.edu.pl - lklimkow@agh.edu.pl - dgei@ogs.trieste.it
maarten.pluymaekers@tno.nl - dal@bgs.ac.uk

* Corresponding author

Abstract — *Risk assessment-led characterisation of a site for the geological storage of CO_2 in the UK northern North Sea was performed for the EU SiteChar research project as one of a portfolio of sites. Implementation and testing of the SiteChar project site characterisation workflow has produced a 'dry-run' storage permit application that is compliant with regulatory requirements. A site suitable for commercial-scale storage was characterised, compatible with current and future industrial carbon dioxide (CO_2) sources in the northern UK. Pre-characterisation of the site, based on existing information acquired during hydrocarbon exploration and production, has been achieved from publicly available data. The project concept is to store captured CO_2 at a rate of 5 Mt per year for 20 years in the Blake Oil Field and surrounding Captain Sandstone saline aquifer. This commercial-scale storage of 100 Mt CO_2 can be achieved through a storage scenario combining injection of CO_2 into the oil field and concurrent water production down-dip of the field. There would be no encroachment of supercritical phase CO_2 for more than two kilometres beyond the field boundary and no adverse influence on operating hydrocarbon fields provided there is pressure management.*

Components of a storage permit application for the site are presented, developed as far as possible within a research project. Characterisation and technical investigations were guided by an initial assessment of perceived risks to the prospective site and a need to provide the information required for the storage permit application. The emphasis throughout was to reduce risks and uncertainty on the subsurface containment of stored CO_2, particularly with respect to site technical performance, monitoring and regulatory issues, and effects on other resources. The results of selected risk assessment-led site characterisation investigations and the subsequent risk reassessments are described together with their implications for the understanding of the site. Additional investigations are identified that could further reduce risks and uncertainties, and enable progress toward a full storage permit application. Permit performance conditions are presented as SiteChar-recommended useful tools for discussion between the competent authority and operator.

Résumé — Caractérisation d'un site de stockage géologique de CO_2 situé en Mer du Nord (Royaume-Uni) sur la base d'une analyse de risques — Dans le cadre du projet de recherche EU SiteChar, la caractérisation d'un site pour le stockage géologique du CO_2 situé au Royaume-Uni au nord de la mer du Nord a été réalisée sur la base d'une analyse de risques. Cette évaluation a permis de déployer et de tester la méthodologie préconisée dans SiteChar jusqu'au développement d'un dossier de permis de stockage en blanc conforme aux exigences réglementaires. Il s'agit d'un site approprié pour un stockage de CO_2 à une échelle commerciale, compatible avec les sources industrielles, actuelles et futures, émettrices de dioxyde de carbone dans le nord du Royaume-Uni. Une pré-caractérisation du site à partir d'informations existantes acquises au cours de l'exploration et de la production d'hydrocarbures, a été réalisée sur base de données accessibles au public. Le projet vise à stocker 5 millions de tonnes (Mt) de CO_2 par an pendant 20 ans dans le champ de pétrole Blake et l'aquifère salin Captain Sandstone dans lequel il se trouve. Un stockage à l'échelle commerciale de 100 Mt de CO_2 pourrait être réalisé grâce à un scénario combinant injection du CO_2 dans le champ pétrolier et production de saumure simultanée dans l'aquifère en aval. Le CO_2 à l'état supercritique ne s'étendrait pas sur plus de deux kilomètres au-delà de la limite du champ pétrolier et aucun effet défavorable n'affecterait la production des champs d'hydrocarbures en cours d'exploitation dans la mesure où des approches de gestion de la pression seraient mises en œuvre.
Les éléments d'une demande de permis de stockage pour ce site sont présentés ; ils ont été développés autant qu'il était possible dans le cadre d'un projet de recherche. La caractérisation du site et les investigations techniques ont été initiées par une évaluation des risques et menées de façon à fournir les informations requises pour le dossier de permis de stockage. Tout au long de l'étude, l'objectif a été de réduire les risques et les incertitudes sur le confinement souterrain du CO_2 stocké, en particulier ceux relatifs à la performance technique du projet de stockage, la surveillance et l'adéquation aux exigences réglementaires, ainsi que l'interaction avec les autres ressources du sous-sol. Les résultats de l'analyse de risque, de la caractérisation du site évaluation menée ainsi que la mise à jour des risques après analyse des résultats sont décrits et les implications pour la compréhension du site exposées. Les besoins d'investigations supplémentaires sont identifiés de façon à réduire encore davantage les risques et les incertitudes et permettre de compléter le permis de stockage. Des critères portant sur les conditions de performance du site sont présentés ; ce sont des éléments que SiteChar recommande comme base de discussion entre l'autorité compétente et l'opérateur.

INTRODUCTION

Depleted hydrocarbon fields and brine-saturated (saline aquifer) sandstones are the CO_2 storage options for proposed demonstrator sites or operational Carbon Capture and Storage (CCS) projects in the North Sea (Chadwick *et al.*, 2008; DECC, 2013; ROAD, 2013). The potential capacity for storage of CO_2 offshore UK is sufficient to meet the national need (DTI, 2006; SCCS, 2009; Bentham *et al.*, 2014), as one of a suite of low-carbon technologies, to achieve UK national targets for greenhouse gas emissions reduction. A prospective site for the geological storage of CO_2 within the UK sector of the North Sea was investigated as one of a portfolio of realistic storage sites proposed by the EU FP7-funded SiteChar research project (Delprat-Jannaud *et al.*, 2015). Characterisation of an offshore storage site that comprises a depleted hydrocarbon field and adjacent saline aquifer sandstone in the northern North Sea, anticipates commercial-scale storage site development following on from initial demonstrator projects. Characterisation of the SiteChar UK North Sea site evaluated a storage complex to provide a credible storage injection scenario over a 25 to 50 year term compatible with likely current and future industrial CO_2 sources in the northern UK. Characterisation implemented and tested the workflow of the SiteChar project (Nepveu *et al.*, 2015) and was sufficient to present a 'dry-run' storage permit application to comply with the requirements of the EU storage directive (EU, 2009, 2011) as far as possible within a research project. The usage of the terms storage site and storage complex are based in Figure 3 of Guidance Document 2 for implementation of the EU CCS Directive (EU, 2011).

The objectives of this paper are to present the process of risk assessment-led site characterisation to meet the requirements of a storage permit application. Pre-characterisation of the storage site used publicly available existing data acquired during hydrocarbon exploration and production. A summary of the technical aspects of characterisation of the prospective site and implications to risk reduction are presented. Full and detailed technical characterisation is reported and available from the project web site (www.sitechar-co2.eu). The outcomes of previous assessments of the prospective storage site are reviewed in Section 1. The prospective storage site is described in Section 2, including selection of the hydrocarbon field component, the site geology and the storage scenario. Section 3 presents risk assessment-led characterisation and the required outputs, the risk assessment process in SiteChar and a summary of risk reduction investigations for selected risks. Outputs derived from risk assessment-led characterisation for an application for a CO_2 storage permit are summarised in Section 4. Permit performance conditions are presented in Section 5.

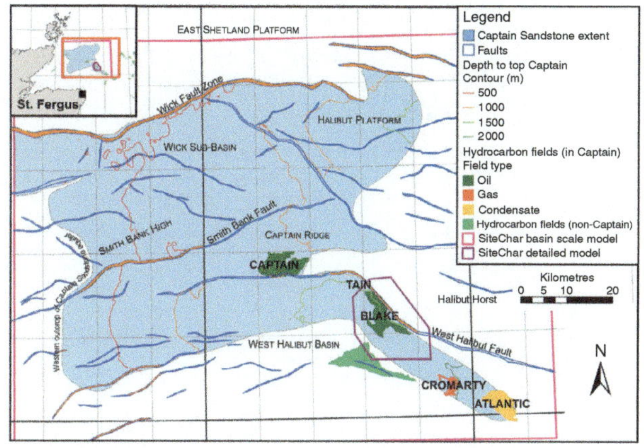

Figure 1

Map of the UK North Sea SiteChar storage site, structural contours (depth in metres True Vertical Depth Sub Sea) and faults at top Captain Sandstone surface (after SCCS, 2011) showing the SiteChar detailed model area and hydrocarbon fields hosted within the Captain Sandstone and older strata (Quinn *et al.*, 2012). Contains public sector information licensed under the Open Government Licence V.1.0. Field outlines and wells from DECC 2011. Data available from: https://www.og.decc.gov.uk/pprs/full_production.htm.

1 PREVIOUS ASSESSMENTS OF THE PROSPECTIVE SITE

The site lies offshore in the UK sector of the North Sea approximately 75 km north-east of the St Fergus gas terminal on the north-east coast of Scotland (Fig. 1). The project concept is for commercial-scale storage of CO_2 sourced from industrial plant along the eastern coast of Scotland and transported offshore *via* existing oil and gas pipelines from St Fergus. Opportunities for carbon capture and storage offshore Scotland have been assessed and developed by previous studies (SCCS, 2009, 2011). Output of CO_2 from existing industrial sources has been quantified, the future CO_2 output forecast, storage sites assessed, and cost for pipeline and shipping for the transportation of CO_2 estimated (SCCS, 2009).

The Captain Sandstone was amongst those saline aquifers shortlisted as prospective storage assets by Scottish Centre for Carbon Storage (SCCS, 2009). It meets acceptable geological characteristics as indicated by Chadwick *et al.* (2008) and has sufficient potential storage capacity. The Captain Sandstone has an estimated static CO_2 storage capacity ranging from 36 Mt (applying a storage efficiency of 0.2%) to 363 Mt (applying a storage efficiency of 2%) (SCCS, 2009). Characterisation and calculation of storage capacity by dynamic modeling of CO_2 injection endorses a potential capacity of more than 360 Mt for the Captain Sandstone saline aquifer with a storage efficiency of more than 0.6% by Scottish Carbon Capture and Storage (SCCS, 2011).

The Captain Sandstone was selected for research assessment as an example of a commercial-scale CO_2 storage asset for Scotland following an examination of three sandstones. Each was appraised on its proximity to existing onshore CO_2 sources, estimated storage capacity, access to existing oil and gas pipelines, whether known from oil and gas exploration data and having sufficient data accessible to a research project (SCCS, 2011). Selected 2D seismic survey and well data were interpreted to map and characterise the site and a basin-scale model was constructed and attributed with petrophysical characteristics (SCCS, 2011). Dynamic simulation of CO_2 calculated the potential storage capacity to be at least as much as the previously estimated upper value and possibly much greater (Jin *et al.*, 2012). Further assessment of the Captain Sandstone as a potential CO_2 store was considered as justified (SCCS, 2011). The basin-scale model of Scottish Carbon Capture and Storage (SCCS, 2011) was also used for the SiteChar UK northern North Sea storage site (Fig. 1).

The SiteChar research benefited from publicly available assessments of the Goldeneye Gas Field for CO_2 storage (Shell, 2011a, b). The Goldeneye Field is also within the Captain Sandstone and lies 30 kilometres to the east of the SiteChar UK site study area. During 2013 the Captain Sandstone was proposed as the storage reservoir for two project entrants to the UK CCS demonstrator competition, one preferred and one reserve project (DECC, 2013). The site characterisation research presented here assesses a site that is in addition to others proposed within the Captain Sandstone as a prospective multi-user storage asset.

2 SITECHAR UK NORTH SEA STORAGE SITE

2.1 Geological Setting

The SiteChar UK North Sea storage site is within the Captain Sandstone Member of the Lower Cretaceous Wick

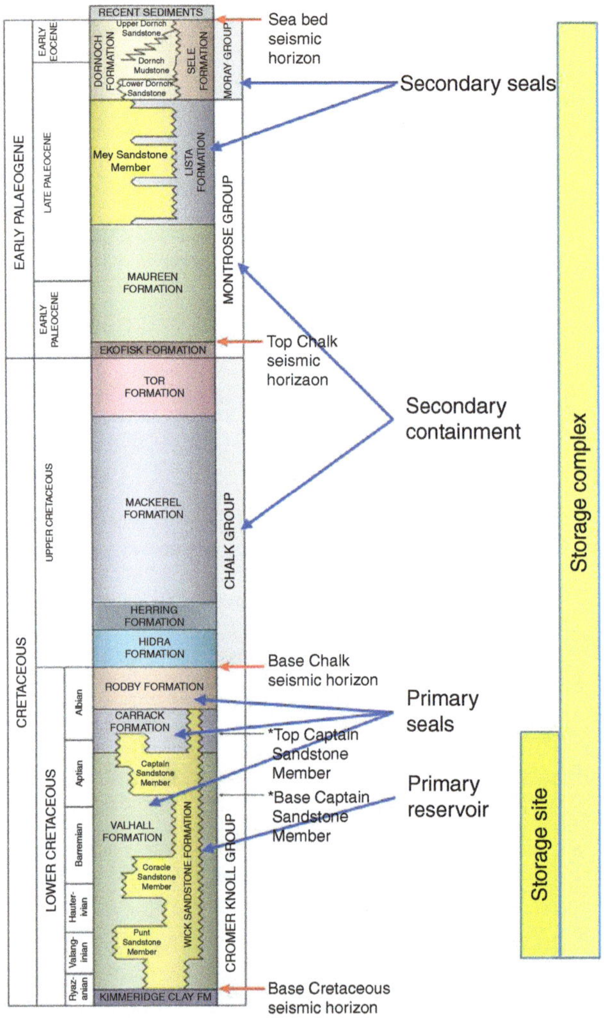

Figure 2

Stratigraphy of the SiteChar UK North Sea storage site and storage complex components (Quinn *et al.*, 2012).

Sandstone Formation (Johnson and Lott, 1993) (Fig. 1, 2) (Quinn *et al.*, 2012; Akhurst *et al.*, 2014). The Captain Sandstone member extends over an area of at least 3 400 km^2 in the Outer Moray Firth, offshore Scotland in the UK northern North Sea (Fig. 1). The sandstone was delineated by Johnson and Lott (1993) and remapped by SCCS (2011). The Captain Sandstone is underlain and laterally equivalent to the argillaceous Valhall Formation, laterally equivalent to and overlain by the argillaceous Carrack Formation. Both of these formations are in turn overlain by the Rodby Formation that also has sealing potential (Johnson and Lott, 1993) (Fig. 2). Rocks of the Upper Cretaceous Chalk Group may directly overlie the Captain Sandstone where the Carrack and Rodby formations are absent (Fig. 2). The Palaeogene overburden sequence comprises sandstone, mudstone, claystone, siltstone and limestone of the Montrose and Moray groups (Fig. 2) (Knox and Holloway, 1992; Johnson and Lott, 1993).

2.2 Hydrocarbon Field Site Selection

The Captain Sandstone contains four oil, gas and gas condensate fields within the extent of the SiteChar study area (Fig. 1). The Blake Oil Field was selected because it meets the storage criteria of Chadwick *et al.* (2008), has sufficient estimated static storage capacity (28 Mt CO_2) for a component of a commercial-scale storage site and data from the field is publicly available. The Blake Oil Field has the largest potential storage capacity of the three fields that lie at a depth greater than 800 metres, to ensure CO_2 injection in supercritical phase. It lies at a minimum depth of 1 350 metres (true vertical sub-sea depth) in approximately 100 m water depth. The Blake Field has produced oil and gas since 2002 and is now supported by water injection. There are publicly available 3D seismic surveys across the Blake Field and data from 23 well penetrations within the field to constrain the seismic interpretation and attribute the site geological model.

The Blake Oil Field is located in the north-western end of the Captain Sandstone 'pan-handle' (Fig. 1). The SiteChar UK site detailed model area lies where the 'pan-handle' begins to widen out into the main extent of the Captain Sandstone (Fig. 1). Dynamic simulation of CO_2 injection by SCCS (2011) has shown that predicted pressure increase due to CO_2 injection is greatest in the south-eastern part of the 'pan handle' where the width of the Captain Sandstone is thought to be at its narrowest. Thus the Blake Oil Field is in an appropriate position, compared to fields further south-east within the study area, to minimise the impact of the increase in pressure modeled by Jin *et al.* (2012).

2.3 Geology and Geological Models of the UK North Sea Storage Site

The Captain Sandstone was deposited in a deep-water submarine environment, sediment was derived from a source to the North and West and deposited by gravity-flow processes including high- and low-density turbidites, mud flows, debris flows, slides and slumps and settling of hemi-pelagic sediment (Law *et al.*, 2000).

Within the Blake Oil Field, the Captain Sandstone has been divided into two distinct areas (Hilton, 1999) based on facies associations that reflect deep marine depositional processes: a 'Channel' facies area in the down-dip, south-western part of the field; a 'Flank' facies area up-dip to the North-East. The Channel facies area is likely to comprise sandstone with consistently high net sandstone to gross

TABLE 1

Summary of net sandstone to gross thickness (NTG) and porosity from well log data, and permeability from core measurement in the Channel and Flank facies areas of the Blake Oil Field (derived from well log and well report data) from Quinn *et al.* (2012)

Blake Oil Field	Average NTG	Average effective porosity	Average total porosity	Average permeability
Channel facies area (15 wells)	0.985	24.6%	29.2%	2 324 mD
Flank facies area (5 wells)	0.334	15.6%	20.7%	153.6 mD

thickness (NTG) values and with good permeability and porosity. However, the Flank area is expected to have much more variable petrophysical characteristics. Consequently, the juxtaposition of the Channel and Flank facies associations will result in a wide range of contact relationships, for example, between sandstone and sandstone, sandstone and shale and a high number of variations between these two end members. The Channel and Flank facies areas of the Blake Oil Field have been developed separately with the Flank area exhibiting complex reservoir architecture with possible compartmentalisation of the reservoir sandstone.

Interpretation of geophysical logs of 23 wells within the Blake Oil Field enabled porosity and NTG values to be derived for both Channel and Flank areas. These log interpretations along with permeability values taken from core analyses enabled a porosity-permeability relationship to be established and attribution of the model to be facilitated. These values are summarised in Table 1. The Channel area of the Blake Oil Field is interpreted to comprise a series of stacked and amalgamated channels that together form part of a major north-west-trending submarine channel system approximately 700 metres wide and at least 6 kilometres in length (mapping from seismic data). The sandstone contains very few shale intercalations, these being either remnant layers that were deposited in shallow depressions at times between the sand depositional episodes, and/or modified or caused by erosion as indicated by the presence of mudstone clast-rich sands. From their interpreted derivation these layers are assumed to have limited areal extent, and not expected to form significant baffles to flow within the Channel facies sandstone.

The Flank area of the Blake Oil Field is interpreted to comprise deep-marine slope strata, dominated by mud deposition, but with minor amounts of thin-bedded sandstone and moderately thick-bedded sandstone deposited by low-density and high-density turbidity currents, respectively. Evidence of sedimentary slumping, syn-sedimentary faulting and minor injection of sand have all been observed in core and Formation Micro Imagery (FMI) (Hilton and Morris, 2002). The reservoir architecture in the Flank area is complex. Hilton and Morris (2002) note that sandstone bodies recognised in flank wells are not laterally continuous. This is borne out by Du (2002) whose interpretation of modular formation dynamics tester tool (MDT) analyses suggests that the reservoir is compartmentalised and/or layered. This interpretation is based on well pressure differences both between the underlying Coracle Sandstone (Fig. 2) and the Captain Sandstone and also the presence of oil-bearing strata and water-bearing strata over the same depth interval in two different wells, suggesting faulting or structural heterogeneity. This compartmentalisation is fully in accordance with the interpretation for a mud-rich deep marine depositional environment. The episodic sedimentation processes deposit sandstone bodies of lensoid shape, some possibly amalgamated, but generally with some lateral and vertical separation.

2.4 Storage Scenario

The concept for the storage site is for commercial-scale storage within the depleted Blake Oil Field and surrounding Captain Sandstone saline aquifer, that is, assuming storage of a total of 100 Mt CO_2. It is also assumed that the storage capacity would be maximised by active management of the reservoir pressure. Containment within the site is beneath the sealing formation that is a proven stratigraphical trap for the Blake Field and four other hydrocarbon fields in the vicinity, the latter also with reservoirs within the Captain Sandstone. The injection scenario for the site is CO_2 injection into the Blake Oil Field and up-dip migration into the adjacent Captain Sandstone saline aquifer beyond the extent of the field. The scenario assumes commercial-scale injection at a rate of 5 Mt per year, to permanently store output of CO_2 captured at industrial point sources along the eastern coast of Scotland and northern England. The duration of injection modeled is for 20 years with a total of 100 Mt CO_2 stored. It is assumed that the site must continuously receive, accept and store CO_2 at an unvarying annual rate. For the purposes of SiteChar research, injection would start in 2016 when the Blake Oil Field is estimated to have ceased production and change of use to geological storage of CO_2. Two constraints for the injection scenario are that the integrity of the site will not be compromised and also that other users of the subsurface will not be adversely affected (outside of the inter-operator agreements) by increased reservoir pressure due to the injection of CO_2 at the storage site.

3 RISK ASSESSMENT-LED SITE CHARACTERISATION

The objective of risk assessment-led site characterisation is to provide the information required by and in agreement with the competent authority for award of a permit for geological storage of CO_2. A storage licence application is undertaken in two parts, an application for an exploration permit and an application for a storage permit. For the SiteChar UK North Sea site, comprising in part a depleted hydrocarbon field, it is assumed that an exploration permit has previously been awarded. For award of a storage permit the applicant (SiteChar research project) must demonstrate sufficient understanding of the site and proposed site operation to securely and permanently contain CO_2. The resources for a research project are necessarily less than those that would be expected for a real storage site permit application. The objectives of SiteChar research, to demonstrate geological characterisation and assess long-term storage complex behaviour, are addressed by a 'dry-run' application for a storage permit. The storage permit application template completed for the UK northern North Sea site was developed within SiteChar from stipulations and requirements in the EU Directive on the geological storage of CO_2 regulations and guidelines (EU, 2009, 2011).

Geological site characterisation activities were targeted to reduce the perceived risks, decrease uncertainty and increase understanding of the storage site (Tab. 2). Geological (static) 3D model construction and model attribution (Quinn *et al.*, 2012) was followed by coupled dynamic simulation and geomechanical modeling (Shi *et al.*, 2013) and dynamic modeling of CO_2 injection (Delprat-Jannaud *et al.*, 2012). Regional migration pathway analysis and wellbore integrity modeling was also undertaken (Delprat-Jannaud *et al.*, 2012). Shallow geohazards were assessed and the geochemical response to CO_2 injection was evaluated (Shi *et al.*, 2013). Feasibility of monitoring was assessed by rock physics studies and synthetic seismic modeling investigations (Hannis *et al.*, 2013c).

3.1 Information Required from Site Characterisation

Information and components required for a storage permit application are determined by an assessment of risks to the CO_2 storage site. Information required includes a description of the storage project and a description of the storage site. Required storage permit components include details of measures to prevent significant irregularities. Monitoring, corrective measures and post-closure plans are also required. The preventative measures component is based on a register of risks and includes a plan of risk mitigation. The risk register is a key document as the starting point for site characterisation (Hannis *et al.*, 2013b). Characterisation activities seek to reduce the perceived risks to subsurface containment of CO_2, and uncertainties due to lack of information, to a level as low as reasonably possible.

3.2 Assessment of Risks to the CO_2 Storage Site

At a very early stage in the site characterisation process an assessment of risks to the storage site was undertaken. It is the first technical activity in the characterisation process as the results determine all other site characterisation investigations and is required by the EU CCS Directive (EU, 2009, 2010). A risk register of 79 perceived risks to the prospective CO_2 storage site were identified by SiteChar research experts. They estimated the probability of each risk occurring and the severity of impact if each risk did actually occur. This list was presented as a summary risk register (Nepveu *et al.*, 2015). The initial assessment was qualitative rather than quantitative and focused on injection- and storage-related risks. Other risks relating to capture and transport of CO_2, CO_2 stream composition, offshore health and safety operations and financial risks, were considered outside the scope of the SiteChar research project. For a real CCS project, investigation of all risks would be required for a storage permit application.

The risks were ranked according to the initial estimates of probability and severity and the ranked list used to guide site characterisation, working toward reduction of risks to an acceptable 'as low as reasonably possible' level. The resources of the research allowed investigation of selected technical risks identified within the project. Those risks that were investigated followed the methodology as far as possible. Where mitigation of a risk could not be achieved within research project resources further risk- and uncertainty-reduction activities were suggested, such as laboratory analysis, re-logging of wells, drilling new wells and baseline monitoring surveys (Hannis *et al.*, 2013a). For a storage permit application submitted by a prospective operator it is assumed that most of the risks would be reduced to an acceptable level by storage design; relatively few residual risks would remain to be monitored (Hannis *et al.*, 2013c). A list was prepared of those risks that were not mitigated by the SiteChar project investigations. This list of residual risks was used to guide the monitoring planning to ensure that any remaining risks are monitored (Hannis *et al.*, 2013c).

3.3 Risks Addressed

The technical investigations in SiteChar were targeted to inform reduction of risks to the subsurface containment of stored CO_2, where site technical performance might be less than expected, associated with monitoring and regulatory issues or have a potentially adverse effect on other resources (Tab. 2). Investigation of risks to the prospective storage site is,

TABLE 2

Risk assessment-led site characterisation investigations for selected risks identified for the SiteChar storage site in the UK North Sea discussed in this paper (for risk ratings see Tab. 3)

Risk type	Risk description	Risk reduction investigations	SiteChar technical reports
Subsurface containment of stored CO_2	Connection of storage site reservoir sandstone to an adjacent fault	Geological site characterisation	Quinn *et al.*, 2012
	Primary cap rock thin or absent	Geological site characterisation and 3D modeling	Quinn *et al.*, 2012
	Fluid migration pathway north-westwards out of the storage site	Dynamic modeling of CO_2 injection, regional migration analysis	Delprat-Jannaud *et al.*, 2012
	Secondary reservoirs not present, laterally restricted or poor quality	Geological site characterisation	Quinn *et al.*, 2012
	Fracture pressure threshold of the primary cap rock exceeded	Coupled dynamic simulation and geomechanical modeling	Shi *et al.*, 2013
	Cap rock fracture pressure threshold lower than predicted	Dynamic modeling	Delprat-Jannaud *et al.*, 2012
	Cap rock capillary entry pressure threshold exceeded		
	Injection-induced faults or fractures in cap rock	Geomechanical modeling	Shi *et al.*, 2013
	Fluid escape pathways up abandoned wells	Wellbore integrity modeling. Shallow geohazards assessment	Delprat-Jannaud *et al.*, 2012 Shi *et al.*, 2013
Technical performance less than expected	Unpredicted reservoir permeability heterogeneities	Geological modeling. Dynamic modeling of CO_2 injection	Quinn *et al.*, 2012 Delprat-Jannaud *et al.*, 2012
	Hydrocarbon saturation to accurately inform dynamic modeling not known	Geological site characterisation	Quinn *et al.*, 2012
	Short-term CO_2 injection-induced permeability reduction near wellbore	Geochemical evaluation	Shi *et al.*, 2013
	Long-term CO_2-induced permeability reduction in the storage reservoir due to geochemical changes	Geochemical evaluation	Shi *et al.*, 2013
Monitoring or regulatory issues	Seismic monitoring ineffective due to the presence of a strong reflector in overlying strata	Synthetic seismic modeling	Hannis *et al.*, 2013c
	Seismic monitoring ineffective at detecting CO_2	Rock physics studies	Hannis *et al.*, 2013c
Adverse effects on other resources	Interference with hydrocarbon fields from migration of CO_2 or increased pressure due to CO_2 injection	Dynamic modeling of CO_2 injection. Storage site performance forecast	Delprat-Jannaud *et al.*, 2012 Hannis *et al.*, 2013a
	Induced seismicity	Geomechanical modeling	Shi *et al.*, 2013

of necessity, confined to those risks associated with technical matters as appropriate for a research project. Risks identified by SiteChar experts associated with economic, financial and environmental issues were not addressed. Technical investigations were directed to reduce risks and provide information needed for a storage permit application.

During the course of the investigations additional data were acquired or reviewed. The increased information, either on or relevant to the storage site, imparted a greater degree of certainty and informed the subsequent re-assessment of risks. During the progress of the investigations additional risks were identified, which is to be expected.

3.4 Risk Reduction Investigations

Site characterisation investigations were targeted to address the risks identified in the risk register and also to provide the information required for a storage permit application. Selected risk reduction investigations presented here are summarised in Table 2. The list of all perceived risks to the site and reassessed values of probability of occurrence, severity of impact and revised ranking of risks after completion of the investigations are discussed by Hannis *et al.* (2013a). SiteChar investigations completed a single iteration of risk reduction activities but could not address all risks or continue with additional site characterisation to reduce risks to a level 'as low as reasonably possible'. The full register of risks, explanation and designation of both initial and revised values for probability, severity and ranking, preventative and corrective measures, also further work that could be undertaken to more accurately define the risk are presented by Hannis *et al.* (2013b).

The results of the risk reduction investigations for the risks presented in Table 2 and implications to the understanding of each perceived risk are discussed in the following text sections.

3.4.1 Risk: Connection of the Storage Site Reservoir Sandstone to an Adjacent Fault

The Captain Sandstone is inferred to extend to the adjacent West Halibut Fault from interpretation of 2D seismic survey data by SCCS (2011) (Fig. 1). The possibility of a flow pathway from the storage site reservoir sandstone to the fault bounding the Halibut Horst to the north presented a risk to the subsurface containment of CO_2.

3D seismic survey data was interpreted for the SiteChar detailed model area (Fig. 1). In the revised interpretation of more finely resolved data, the principal sandstone of the storage site, the Captain Sandstone, is interpreted to pinch out in the hanging wall succession of the West Halibut Fault within the study area (Fig. 3) (Quinn *et al.*, 2012). Dynamic modeling of CO_2 injection into the Captain Sandstone Channel facies (Tab. 1) was calculated for well locations either within the extent of the Blake Oil Field or down-dip within the brine-saturated Captain Sandstone (Delprat-Jannaud *et al.*, 2012). Twenty years of injection and thirty years after the end of injection were simulated. In all of the injection scenarios modeled, the CO_2 (in supercritical

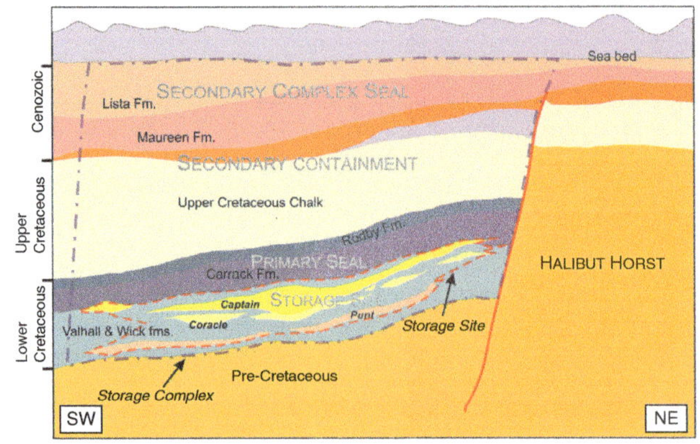

Figure 3

Diagrammatic geological cross-section across the SiteChar UK northern North Sea storage site, showing storage site boundary (red dashed line) and storage complex boundary (blue dashed line) from Quinn *et al.* (2012). Primary seal (cap rock), secondary containment reservoir and secondary seal (cap rock) to the storage complex are labelled.

phase and dissolved in water) was predominantly retained within the extent of the Channel facies. The injected CO_2 is predicted to migrate very slowly beyond the Channel into the Flank facies only after injection has ceased in the injection scenario selected for the prospective storage site.

The likelihood of migration of CO_2 to the West Halibut Fault for the injection scenario proposed is therefore very low. Reassessment of the initial risk from the results of remapping of the Captain Sandstone and dynamic modeling of CO_2 injection has reduced the probability of the risk occurring and so also the risk rating (Tab. 3).

3.4.2 Risk: Primary Cap Rock Thin or Absent

The top of the Captain Sandstone is generally sealed by overlying low permeability mudstone but in some places the mudstone is thin and it may be directly overlain by the unconformable Base Chalk surface. Some wells, *e.g.* Well 13/22a-28 in the Captain Oil Field, prove Chalk resting directly on the Captain Sandstone. Chalk cap rock may not be as efficient a seal as the mudstone.

The 3D seismic survey data interpreted for the SiteChar detailed study area does not resolve where the mudstone cap rock is absent but indicates where it is most likely to be thin (less than 10 metres). Although geological mapping shows the Captain Sandstone is expected to be overlain by Lower Cretaceous mudstone across the extent of the storage complex, an area where the cap rock is interpreted to be thin (<10 m) lies in the north-west part of the Blake Field within the Flank facies area (Quinn *et al.*, 2012).

TABLE 3

Risk ratings before and after the first iteration of risk reduction, based on site characterisation investigations (shown in Tab. 2) for the SiteChar UK North Sea storage site. Probability and severity ratings: 5 is high; 1 is low. NR, risk not reassessed

Risk type	Risk description	Initial risk assessment			Revised assessment		
		Probability	Severity	Risk rating	Probability	Severity	Risk rating
Subsurface containment of stored CO_2	Connection of storage site reservoir sandstone to an adjacent fault	3	3	**9**	1	3	**3**
	Primary cap rock thin or absent	4	3	**12**	2	3	**6**
	Fluid migration pathway north-westwards out of the storage site	2	3	**6**	1	3	**3**
	Secondary reservoirs not present, laterally restricted or poor quality	2	2	**4**	1	2	**2**
	Fracture pressure threshold of the primary cap rock exceeded	2	4	**8**	1	4	**4**
	Cap rock fracture pressure threshold lower than predicted	3	3	**9**	1	1	**3**
	Cap rock capillary entry pressure threshold exceeded	3	3	**9**	3	1	**3**
	Injection-induced faults or fractures in cap rock	2	4	**8**	1	4	**4**
	Fluid escape pathways up abandoned wells	4	4	**16**	2	2	**4**
Technical performance less than expected	Unpredicted reservoir permeability heterogeneities	4	4	**8**	2	2	**4**
	Hydrocarbon saturation to accurately inform dynamic modeling not known	4	3	**12**	2	2	**4**
	Short-term CO_2 injection-induced permeability reduction near wellbore	3	3	**9**	1	2	**2**
	Long-term CO_2 injection-induced permeability reduction in the storage reservoir due to geochemical changes	3	3	**9**	1	2	**2**
Monitoring or regulatory issues	Seismic monitoring ineffective due to the presence of a strong reflector in overlying strata	3	3	**9**	2	3	**6**
	Seismic monitoring ineffective at detecting CO_2	3	3	**9**	2	3	**6**
Adverse effects on other resources	Interference with hydrocarbon fields from: migration of CO_2,	2	4	**8**	2	2	**4**
	increased pressure due to CO_2 injection	4	3	**12**	NR	NR	NR
	Induced seismicity	2	3	**6**	1	3	**3**

However, dynamic modeling of the SiteChar storage site (Delprat-Jannaud *et al.*, 2012) predicts the distribution of the plume of dissolved and supercritical phase CO_2 to remain within the area where cap rock is mapped as being continuous and having acceptable thickness (approximately 40 metres thick).

The risk of leakage due to thin or absent primary cap rock is therefore thought to be low. Mapping of primary cap rock

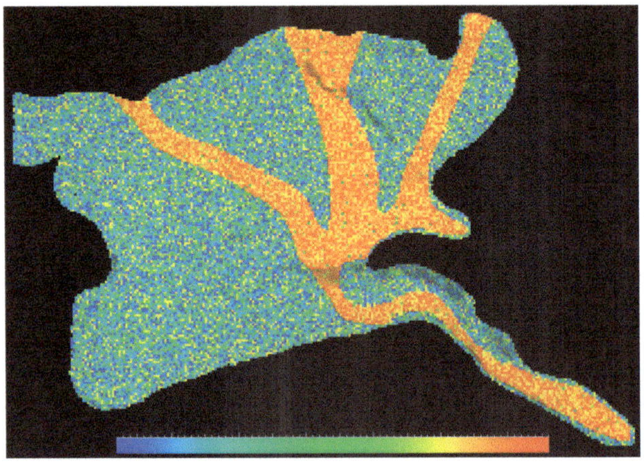

Figure 4

Modelled porosity within the SiteChar basin-scale study area illustrating the distribution of Channel facies (values mostly 0.25 and above) and Flank facies (values mostly 0.25 and below) within the Captain Sandstone, mapped from well data (Quinn *et al.*, 2012).

and dynamic modeling of the selected injection scenario (Tab. 2) has reduced the probability of the risk occurring and so also the risk rating (Tab. 3).

3.4.3 Risk: Fluid Migration Pathway North-Westwards out of the Storage Site

The Blake Oil Field forms a significant component to the storage site and is a combined structural and stratigraphical trap. The field reservoir dips to the west and south providing structural trapping to hydrocarbons filling the reservoir. The reservoir sandstone pinches out to the east providing stratigraphical trapping of hydrocarbons parallel to the Halibut Horst. However, the nature of the trap to hydrocarbons in the Blake Oil Field along its northern boundary is not clear as no structural closure to the field could be mapped within the limits of the seismic data available to the study (Quinn *et al.*, 2012). There is a significant hydrocarbon discovery, the Tain discovery, immediately to the north of the Blake Oil Field beyond the SiteChar detailed area of study (Fig. 1). It is possible that hydrocarbons have migrated beyond the northerly limits of the Blake Field, as defined by the Department of Energy and Climate Change (DECC), filling this small structure and presenting a risk to containment of CO_2 at the prospective storage site.

A regional migration analysis was conducted to identify possible secondary containment and migration paths by modeling substantial and excessive overfilling of the storage site. Prediction of CO_2 migration laterally within the Captain Sandstone (contained under the primary cap rock) if the site is overfilled, indicates flow to the north-west in the direction of the Tain discovery (Delprat-Jannaud *et al.*, 2012). However, the results of dynamic modeling of the chosen injection scenario suggest northward migration will not occur if injection is into the Channel facies of the Blake Oil Field (Delprat-Jannaud *et al.*, 2012). Although the Tain discovery lies north-west of the storage site along a potential regional migration pathway it is not thought to be connected directly to the main Channel area of the Blake Field and may have more in common with the Flank area.

The risk to containment from fluid migration northwestwards from the storage site can therefore be considered very low. The probability of a fluid migration pathway to the north-west is reduced (Tab. 3) by conforming with the injection scenario modeled and current understanding of poor connectivity from the Channel to the Flank facies within the Captain Sandstone.

3.4.4 Risk: Secondary Reservoirs not Present, Laterally Restricted or Poor Quality

In the unlikely event of migration of injected CO_2 through the primary cap rock and out of the storage site, the CO_2 would be expected to be contained within strata with available pore space within the storage complex. Definition of secondary containment systems in the overall storage system, *i.e.* porous strata within the storage complex, is described as a component of the site (EU, 2009). A risk to the storage site is that there are no suitable secondary reservoirs.

Review of published atlases for the stratigraphical sequences in the area of the storage site (Knox and Holloway, 1992; Johnson and Lott, 1993) indicates strata potentially suitable for both secondary containment and secondary sealing for the storage of injected CO_2. Migration of injected CO_2 beyond the storage site may be contained within formations that overlie, underlie or are laterally equivalent to the Captain Sandstone (Fig. 2, 3) (Quinn *et al.*, 2012). The overlying Upper Cretaceous Chalk Group succession has an average thickness of approximately 460 metres over the detailed study area. The Chalk Group could provide additional containment for CO_2 because although the Chalk has low rock porosity and permeability it may have fracture porosity and permeability sufficient to provide storage. The overlying Palaeogene succession comprises sandstone, mudstone, claystone, siltstone and limestone and has an average thickness of approximately 1 000 metres over the SiteChar detailed study area. The sandstone could provide very good secondary containment for migrating CO_2 while the mudstone and claystone successions present in the Palaeogene Lista and Dornoch formations (Fig. 2) could provide a secondary cap rock to the storage complex (Fig. 3). Further site-specific investigation would improve the characterisation of the secondary containment reservoir and seal rocks with an estimate of possible volumes of secondary containment (Hannis *et al.*, 2013a).

The risk associated with the possibility of inadequate secondary storage strata is much reduced based on the published information by reducing the probability of the risk happening (Tab. 3).

3.4.5 Risk: Fracture Pressure Threshold of the Primary Cap Rock Exceeded

The cap rock fracture pressure threshold defines the pressure at which local fracturing of the cap rock may occur. If the pressure threshold is exceeded the sealing properties of the cap rock could be compromised, leading to leakage from the primary storage reservoir. A value of around 75% of the lithostatic pressure is commonly assumed to approximate the fracture pressure threshold. Knowledge of the cap rock fracture pressure is vital to be able to derive realistic injection scenarios.

Coupled flow-geomechanical modeling of CO_2 injection and an evaluation of mechanical stability (Shi *et al.*, 2013) accompanied by dynamic simulations to define an appropriate CO_2 injection scenario (Delprat-Jannaud *et al.*, 2012) were investigated for the storage site. The geomechanical stability assessment used the pre-injection *in situ* stress state information (one of the main sources of uncertainties in the assessment of geomechanical stability) from the nearby Goldeneye Gas Field (Shell, 2011b). Based on these data, the overpressure value causing cap rock fracture has been estimated as between 75 and 80 bar for the depth range of the storage site (1 600-1 700 m) by Shi *et al.* (2013).

Coupled flow-geomechanical modeling of CO_2 injection into the Channel facies sandstone down-dip from the Blake Field at the commercial-scale rate without pressure management generated high overpressure values that would exceed the cap rock fracture pressure threshold after 10 years (Shi *et al.*, 2013). Even without pressure management no shear failure is predicted after 5 years of injection at a rate of 5 Mt CO_2 per year into the Captain Sandstone saline aquifer but continuation at this rate would exceed the fracture pressure threshold.

Different injection scenarios were simulated for the storage site all at a constant rate of 5 Mt CO_2 per year. The maximum reservoir pressure attained for CO_2 injection into the Blake Oil Field with pressure management does not exceed the cap rock fracture pressure (Delprat-Jannaud *et al.*, 2012). Pressure management by concurrent fluid (water) production from one well down-dip of the hydrocarbon field in the Captain Sandstone saline aquifer was modeled. The predicted maximum overpressure value for this scenario is 26 bar after 20 years of CO_2 injection. This value is substantially lower than the estimated threshold of 75-80 bar and decreases to 8 bar overpressure after cessation of injection (Delprat-Jannaud *et al.*, 2012). Further iterative dynamic modeling using a range of injection rates and well positions would allow the maximum rate and volume of injected CO_2 without exceeding the pressure thresholds to be determined. Pressure management by implementing the chosen injection and production scenario, although it may not be the optimum, would ensure the cap rock fracture pressure would not be exceeded.

The probability of exceeding the cap rock fracture pressure has been reduced by determining the cap rock fracture pressure threshold and including pressure management in the injection scenario (Tab. 3).

3.4.6 Risk: Cap Rock Fracture Pressure Threshold Lower than Predicted

The cap rock fracture pressure threshold estimated at 75-80 bar overpressure at the storage site depth by Shi *et al.* (2013) is very similar to the assumed values derived by calculating a proportion of the lithostatic or hydrostatic pressure. The maximum overpressure value predicted for the chosen injection scenario of 26 bar overpressure is approximately one third of the estimated fracture pressure threshold. Therefore, subject to pressure management, the probability and risk that the threshold is exceeded due to a lower than predicted cap rock fracture pressure has been reduced (Tab. 3) and the likelihood is very low.

3.4.7 Risk: Cap Rock Capillary Entry Pressure Threshold Exceeded

The cap rock capillary entry pressure threshold defines the pressure at which CO_2 will start to slowly permeate through the capillaries of the sealing rock. It is lower than the fracture pressure, and leakage of CO_2 through capillaries will be slow, at low volumes, and for long periods be contained within the cap rock itself.

The estimated cap rock capillary entry pressure is 36 bar (Delprat-Jannaud *et al.*, 2012), based on the theoretical relationship (Thomas *et al.*, 1967) with and corresponding to an absolute permeability of 10^{-5} mD. The theoretical entry pressure approximates the predicted maximum overpressure of 26 bar for the selected injection scenario. However, the small pressure difference between the two is not judged to be a significant concern in terms of CO_2 migration into the cap rock over the medium- to long-term due to the low permeability of the cap rock. In the short term, the driving overpressure will drop immediately once injection ceases and further dissipate over time. As the excess pressure will drop after injection ceases, the risk is restricted to the period of injection and unlikely to be a source of leakage.

No further information was available to change the assessment of the probability that the cap rock entry pressure threshold would be exceeded. However, the rates of leakage from this mechanism are likely to be low and occur only during the decades of CO_2 injection so the severity and overall risk rating are reduced (Tab. 3).

3.4.8 Risk: Injection-Induced Faults or Fractures in Cap Rock

Fault structures have not been identified in the Captain Sandstone reservoir and cap rock within the area of detailed study for the storage site from 3D seismic survey data (Quinn *et al.*, 2012). A risk remains that there are minor fault structures within the site that are not resolved in the seismic data and the increased pressure associated with CO_2 injection might induce fault reactivation to create flow pathways within the storage site reservoir and through the primary cap rock.

The outcomes of the geomechanical modeling and failure analysis by Shi *et al.* (2013) suggest that the likelihood of fault reactivation within the Captain Sandstone is low provided pressure is managed. An assessment of the likely presence of fractures and their connectivity within the cap rock and overburden sequence was undertaken for the UK offshore site by Shi *et al.* (2013). Fracture data for the cap rock and overburden from well log and core data sources from within and in the vicinity of the storage site were used to inform estimation of a percolating fracture network. The computed percolation probability is strongly influenced by cap rock thickness, proportion of open (conducting) fractures and the number of fractures over a specified vertical distance (line density). For the two scenarios with the lowest proportion of open fractures (0.1) and the thickest cap rock (227 m), the percolation probability is negligible.

Note here that natural gas and oil has been contained in the Blake Field and other hydrocarbon fields over the geological time scale which is evidence that should preclude the existence of a natural gas leakage pathway in the cap rock. It thus can be argued that it is unlikely that a percolating open fracture network exists in the cap rock.

The results of risk reduction investigations indicate the probability of the presence of minor fractures has been reduced and so also the risk that they might be reactivated by CO_2 injection (Tab. 3).

3.4.9 Risk: Fluid Escape Pathways Up Abandoned Wells

One of the major challenges associated with geological storage of CO_2 is the prediction of performance of the confining system and for the store to remain secure over long timescales. Abandoned exploration and production wells act as potential leakage pathways for CO_2 stored in depleted hydrocarbon fields and adjacent saline aquifer strata. Hence, it is essential to investigate well integrity during the process of storage site evaluation.

Migration path analysis for the SiteChar UK North Sea site has identified possible leaking wells as the most probable pathway to the surface, transmitting CO_2 directly to the atmosphere (Delprat-Jannaud *et al.*, 2012). Fluid-escape pathways and CO_2-induced fluid escape pathways up abandoned wells are rated as having the highest probability and severity and so ranked highest in the risk register (Hannis *et al.*, 2013a, b). The potential for leakage of injected CO_2 through the wellbore, and its sensitivity to cement bond condition and storage pressure, were investigated by numerical simulation of well integrity (Delprat-Jannaud *et al.*, 2012). The analysis is based on data from an appraisal well within the Channel facies of the Blake Oil Field (BG Exploration & Production, 1998). The scenarios simulated assume storage pressure and micro-annulus permeability values, 1 000 years of elevated pressure and predict the mass of leaked CO_2 at the top of the secondary containment strata. If reservoir pressures are maintained at 90% of the initial pressure no leakage of CO_2 *via* the wellbore is predicted, even if a pessimistic scenario for the micro-annulus permeability is assumed. At the initial reservoir pressure 1.3 tonnes of CO_2 is predicted to leak to secondary storage *via* the wellbore over a period of 1 000 years. Dynamic simulation (Delprat-Jannaud *et al.*, 2012) predicts up to 26 bar overpressure within the Captain Sandstone reservoir for the selected CO_2 injection and water production scenario. Well integrity modeling assuming CO_2 storage pressure equal to 130% of the initial reservoir pressure and pessimistic micro-annulus permeability values predicts up to 20 tonnes of CO_2 leakage to the secondary reservoir. All scenarios assume constant elevated pressure over a period of 1 000 years and the only way to reduce the pressure is *via* leakage through the wellbore. Thus the effect of relaxation after stopping injection after 20 years is not incorporated in the calculations. Such conditions may therefore be considered as very pessimistic as in reality after stopping injection the reservoir pressure will decrease with time and thus also the rate of CO_2 leakage.

The risk for the scenario investigated was reduced by the SiteChar investigations (Tab. 3). Uncertainty in the number of wells likely to fall within the predicted extent of the injected CO_2 plume (Fig. 6 in Sect. 3.4.16) and the mass of CO_2 leakage to the secondary reservoir was also reduced. Risk mitigation measures of additional well characterisation (Le Guen *et al.*, 2008) are suggested and included as part of the preventative measures plan (Hannis *et al.*, 2013c; Shi *et al.*, 2013) and 'dry-run' storage permit application for the site (Hannis *et al.*, 2013a). Baseline observation and resurvey of the sea bed during injection over existing well sites by sonar and echo sounder surveys (Shi *et al.*, 2013) are a component of the monitoring plan (Hannis *et al.*, 2013c).

3.4.10 Risk: Unpredicted Reservoir Permeability Heterogeneities

The risk of unpredicted heterogeneities in reservoir permeability, creating impediments to flow by creating low permeability 'baffles' (vertical, horizontal, or at a range of angles), could lead to a reduction in the rate of injected CO_2. The reservoir character is inferred from well geophysical logs and well core samples. The uncertainty lies in the extrapolation between wells, which may be very widely spaced, and beyond the extent of hydrocarbon fields.

Two end-members of a possible range of reservoir heterogeneity within the Captain Sandstone were examined to give an indication of the impact of heterogeneity on reservoir performance (Delprat-Jannaud *et al.*, 2012). CO_2 injection was simulated by assuming either homogeneous or heterogeneous Channel facies sandstone. The heterogeneous petrophysical parameters (porosity and permeability) were stochastically modeled assuming a porosity distribution derived from available core measurements (Fig. 4). The permeability was attributed using the relationship between porosity and permeability values measured from core samples (Quinn *et al.*, 2012). This method of attributing the model does not give the most realistic result in terms of the geology but gives petrophysical property distributions which are statistically representative of existing input data. Observation of the Captain Sandstone Channel facies in core recovered from within the basin-scale study area reveals it has an extremely uniform character. The Channel facies of the Captain Sandstone is interpreted to be amalgamated turbidite sandstone deposited from sand-rich turbidity currents (Law *et al.*, 2000). The homogeneous Channel facies Captain Sandstone was populated using average values of 0.25 for porosity from well data and 2 000 mD permeability from core measurements within the Blake Oil Field (Quinn *et al.*, 2012).

Comparison of the simulations of CO_2 injection into the Channel facies with either heterogeneous or homogeneous attribution reveals significant differences in the extent and magnitude of overpressure anomalies (Delprat-Jannaud *et al.*, 2012). Given the extremely permeable character of the Captain Sandstone, with the majority of samples exceeding the average value of 2 000 mD (Quinn *et al.*, 2012) the very high overpressure response predicted with heterogeneous attribution seemed unreasonable. The homogeneous attribution was selected as more representative of the Captain Sandstone Channel facies.

The risk to injectivity was reduced by the investigations of the storage site (Tab. 3) (Hannis *et al.*, 2013b). The relatively large range of the estimated overpressure values from the two end-members modeled reflects the sensitivity to the degree of heterogeneity represented in the model. Additional investigation of the attribution of permeability to the Channel and Flank facies is strongly advised and essential to impart greater confidence in the predicted performance of the storage site. However, an indication of connectivity between hydrocarbon fields hosted within the Captain Sandstone is suggested by pressure effects noted from one field to another (geological well report from the adjacent Cromarty Field). This suggests that if heterogeneities in reservoir permeability exist they do not significantly affect reservoir performance.

3.4.11 Risk: Hydrocarbon Saturation to Accurately Inform Dynamic Modeling not Known

A risk to prediction of the site performance by dynamic modeling of CO_2 injection into a depleted hydrocarbon field is the lack of input data for modeling. It is important to accurately determine residual hydrocarbon distribution and saturation. This is due to the greater compressibility of hydrocarbons than water and the lower solubility of CO_2 in water than in oil and resultant lower overpressure values when injecting CO_2 (Delprat-Jannaud *et al.*, 2012).

Uncertainty associated with this risk was reduced using water saturation information from well reports (Quinn *et al.*, 2012). Well data indicate that residual hydrocarbon saturation after water sweep within the Blake Oil Field is likely to be 30% oil, thus reducing the uncertainty on the probability of this risk occurring. Comparison of dynamic modeling results of injection into the hydrocarbon field and into the saline aquifer suggests the effects might be less severe than initially estimated (Tab. 3).

3.4.12 Risk: Short-Term CO_2 Injection-Induced Permeability Reduction Near Wellbore

The risk of CO_2-induced permeability reduction near the wellbore could restrict the rate of injection and hence the predicted technical performance of the storage site. Permeability can be reduced by salt or mineral precipitation or by microbial activity induced by CO_2 injection. An evaluation of the geochemical effects of CO_2 storage over geological time on both the Captain Sandstone aquifer and Blake Field gas cap components of the storage site was undertaken by geochemical modeling. The potential effect of the presence of oil was also evaluated (Shi *et al.*, 2013).

Predictions of geochemical changes, mineral precipitation and porosity changes were calculated for the storage site using an initial reservoir sandstone composition for the Captain Sandstone from Well 13/23b-05 up-dip from the Blake Field (Jappy, 2006). The mineralogical, pH and porosity change from mineral reactions calculated over a period of up to 100 000 years are summarised in Figure 5a, c and discussed in detail in Shi *et al.* (2013). In terms of the risk of CO_2 injection-induced permeability reduction over the geological short term, the model results show that the mineral reactions were predicted to be negligible during the injection phase. Porosity changes due to the reactions take place on

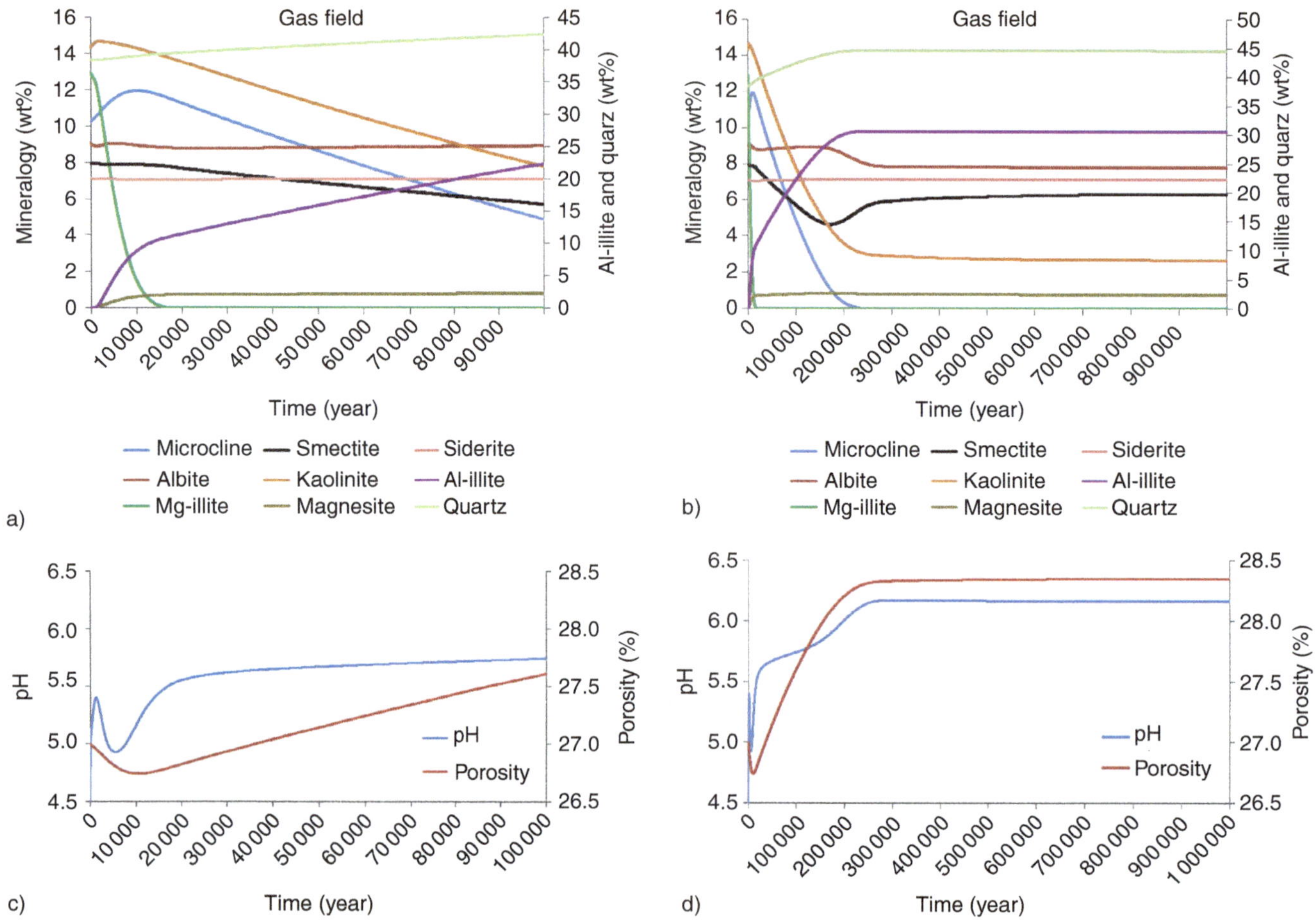

Figure 5

Geochemical modeling of changes due to CO_2 injection into the gas cap of the Blake Oil Field from Shi *et al.* (2013). a) Mineralogy changes and carbonate trapping up to 100 000 years. b) Mineralogy changes and carbonate trapping up to 1 million years. Note: % Al-illite and quartz are shown on the right *y*-axis. c) pH and porosity changes up to 100 000 years. d) pH and porosity changes up to 1 million years.

a timescale much longer than that of injection, as do the accompanying changes in permeability. The impact of CO_2 reacting with the reservoir rock on the injection process is therefore thought to be very small and the risk has been reduced (Tab. 3). The potential for halite precipitation causing salt clogging near the wellbore in a drying-out zone requires further investigation by transport modeling. The occurrence and impact of salt clogging could be prevented or reduced by regularly carrying out a water wash, which is common practice in the oil and gas industry. Future research could include a reactive transport model incorporating salt precipitation for further site characterisation. In addition, the effects of microbial activity should be assessed.

3.4.13 Risk: Long-Term CO_2-Induced Permeability Reduction in the Storage Reservoir Due to Geochemical Changes

The risk of long-term CO_2-induced permeability reduction in the Captain Sandstone reservoir resulting from geochemical changes as a result of the injection and storage scenario were evaluated by geochemical modeling. Mineralogical, pH and porosity changes from mineral reactions were calculated over a period of up to 1 million years (Shi *et al.*, 2013). The changes are modeled for the gas cap of the Blake Oil Field and summarised in Figure 5b, d and discussed in detail in Shi *et al.* (2013).

The geochemical effects of CO_2 injection in the Blake Field gas cap (Fig. 5) or Captain Sandstone saline aquifer (Shi *et al.*, 2013) are predicted by geochemical modeling. While the short term effect of storing CO_2 in the Blake Field is a significant reduction of the pH, subsequent mineral reactions are slow. A porosity decrease from 27% to 26.7% takes more than 10 000 years in both the gas cap and the saline aquifer sandstone. Reactions in the oil reservoir of the Blake Field are expected to be even slower than in the gas cap or the aquifer (Shi *et al.*, 2013) and the risk from long-term permeability changes has been reduced (Tab. 3). After 15 000 years the amount of mineral dissolution is enhanced compared to the amount of precipitation and porosity starts to increase slowly to a final value which is slightly above the initial value. However, the formation of large amounts of illite might significantly reduce the permeability. A permeability reduction would not necessarily pose any risks. In fact, it might enhance storage integrity by increasing capillary pressures which would hamper CO_2 migration out of the complex.

3.4.14 Risk: Seismic Monitoring Ineffective Due to the Presence of a Strong Reflector in Overlying Strata

3D seismic survey is the main method proposed at this site to monitor the CO_2 plume and surrounding environment (Hannis *et al.*, 2013c). A risk to the proposed monitoring plan is that the presence of a very strong reflector from the basal surface of the Hidra Formation, Chalk Group (Fig. 2), creates a seismic 'shadow'. Seismic reflections from the immediately underlying strata could be masked, including the primary cap rock and Captain Sandstone storage site (Fig. 2, 3), making storage site monitoring by time-lapse seismic surveys ineffective.

A detailed study of the feasibility of seismic surveys to monitor CO_2 stored in the Captain Sandstone was investigated for the SiteChar UK North Sea site (Hannis *et al.*, 2013c). The analysis was based on a 2D geological model cross-section across the site (Quinn *et al.*, 2012) attributed by seismic properties and a corresponding un-migrated seismic section (Hannis *et al.*, 2013c). A rock physics model of the reservoir was developed for a range of CO_2 saturations. A series of synthetic seismograms showing the seismic response of CO_2 in the subsurface were computed and the sensitivity to seismic noise levels was measured using standard metrics (Hannis *et al.*, 2013c).

The seismic numerical modeling concludes that stored CO_2 should be detected within the site, and even within the seismic shadow, by techniques to compare time-lapse offshore surface 3D seismic survey data (Hannis *et al.*, 2013c). The probability and so the risk associated with the presence of a strong reflector in the strata immediately overlying the storage site is reduced by the investigations (Tab. 3). However, the use of data acquisition techniques to guarantee a high signal-to-noise ratio and the best repeatability are recommended.

3.4.15 Risk: Seismic Monitoring Ineffective at Detecting CO_2

Time-lapse 3D seismic surveys are proposed to monitor the presence and migration of CO_2 injected into the storage site. A risk to monitoring is that seismic survey methods are ineffective if there is no significant change in acoustic impedance generated by the presence of CO_2 within the site or that the detection limit (resolution) of the seismic method is too low.

Rock physics studies, calculation of synthetic seismograms and a signal-to-noise sensitivity assessment were conducted for the site (Hannis *et al.*, 2013c). These were accompanied by calculation of the seismic amplitude-*versus*-angle response to test the sensitivity to CO_2 saturation (Hannis *et al.*, 2013c). These detailed investigations indicate that CO_2 injection results in a sufficient change in acoustic impedance to be detected using 'standard' offshore surface 3D seismic survey.

The probability, and so the risk, that 'standard' seismic monitoring methods would be ineffective were reduced by the site characterisation investigations (Tab. 3). However, an optimal alternative of seismic data acquisition from a permanent ocean bottom cables system is recommended to achieve higher signal-to-noise ratios, ensure repeatability and acquisition of additional seismic property data (Hannis *et al.*, 2013c). A cost-benefit analysis would be required to indicate if the increased certainty of monitoring is justified by the additional cost.

3.4.16 Risk: Interference with Hydrocarbon Fields from Migration of CO_2 or Increased Pressure Due to CO_2 Injection

The EC requires that CO_2 storage operations should not present a risk to groundwater, in the offshore setting of the SiteChar UK North Sea site this is not a concern. However, the UK Government requires that CO_2 storage operations should not interfere with hydrocarbon field operations. Interference might arise either from migration of injected CO_2 or brine into a field or reservoir pressure increases as a result of CO_2 injection.

The risk that the CO_2 storage operations proposed for the site might interfere with other hydrocarbon fields was investigated by dynamic modeling (Delprat-Jannaud *et al.*, 2012). Possible interference with respect to existing hydrocarbon fields was assessed in the 'dry run' storage permit application (Hannis *et al.*, 2013a). The predicted maximum extent of injected CO_2, in both supercritical phase and dissolved in water, was mapped from the results of dynamic modeling of CO_2 injection and concurrent water production (Hannis *et al.*, 2013a). Figure 6 shows the maximum extent of the CO_2 plume during injection, at completion of injection and 1 000 years after the start of storage site operations. During the 20-year injection period the CO_2 plume (supercritical phase and dissolved in water) gradually spreads within the Channel facies sandstone (Fig. 6). Initially, the CO_2 is retained within the Blake Field, subsequently extending one to two kilometres beyond the modeled field boundary (Fig. 6). After injection ceases the CO_2 in supercritical phase does not extend further although the dissolved CO_2 continues to migrate, mostly in an up-dip direction (Fig. 6). At cessation of injection, the mapped plume has

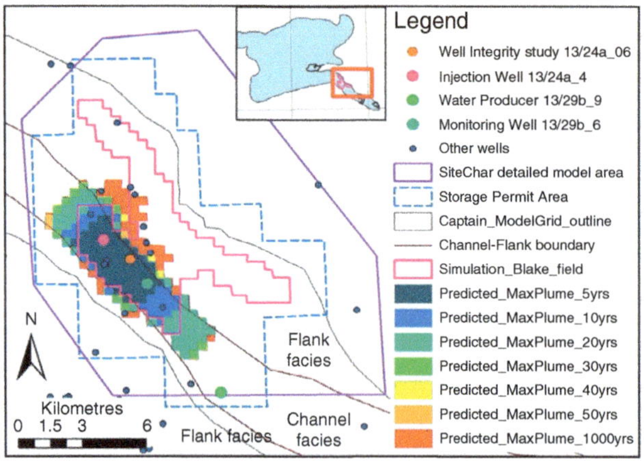

Figure 6

Maximum predicted extent of the CO_2 plume (supercritical phase and dissolved in water) mapped during and after the proposed storage site operations, showing the Channel and Flank facies Captain Sandstone, well positions and modeled extent of the Blake Oil Field from Hannis *et al.* (2013c).

a predicted area of 21 km^2, extending to 23 km^2 after a further 20 years which is the anticipated date of site closure (Hannis *et al.*, 2013a). Details of the distribution of the CO_2 in supercritical phase, dissolved CO_2 and hydrocarbon phases are discussed by Hannis *et al.* (2013a).

The predicted increase in reservoir pressure within the Blake Oil Field and adjacent sandstone, greater than the initial reservoir pressure of 161 bar, and hydrocarbon fields within the Captain Sandstone is illustrated in Figure 7 (Hannis *et al.*, 2013a). The position of the CO_2 injection and water production wells is shown in Figure 6. Little pressure change is noted at the position of the Captain Oil Field from the modeling of storage site operations. There is an initial pressure drop south-east of the site at the position of the Atlantic and Cromarty fields, due to water production. During CO_2 injection, there is a progressive increase to a maximum overpressure value of 27 bar (Fig. 7). The overpressure values drop immediately CO_2 injection ceases. The investigations did not include modeling of hydrocarbon fluids other than in the Blake Field.

The risk of migration of the injected CO_2 to other hydrocarbon fields in the vicinity of the storage site is reduced (Tab. 3) and negligible for the modeled CO_2 injection and water production (Fig. 6). There is little increase in reservoir pressure due to modeled storage operations predicted at the position of the Captain Oil Field, the only operating field hosted within the Captain Sandstone in the vicinity of the site (Fig. 7). The Atlantic and Cromarty fields have ceased production, although the modeled pressure changes might be beneficial to any future operations. The risk from interference with hydrocarbon fields due to increased reservoir pressure from CO_2 injection was not reassessed (Tab. 3). Further modeling is required, incorporating fluid properties of nearby fields, to quantify pressure changes from storage site operations and reassess the risk to nearby hydrocarbon fields.

3.4.17 Risk: Induced Seismicity

The possibility of increased seismicity associated with the operation of a prospective storage site is identified as a required component of the characterisation and assessment of a prospective storage site (EU, 2009, 2011). Subsurface pressure changes associated with CO_2 injection may reactivate faults or generate minor fractures which may be evident as increased measured seismicity.

The outcomes of the geomechanical modeling and failure analysis for the SiteChar UK North Sea site suggest that the probability of increased seismicity induced by CO_2 injection is low when the overpressure does not exceed the fracture pressure threshold (Shi *et al.*, 2103). The scenario of CO_2 injection with pressure management by concurrent water production, which is the case for the chosen injection scenario, is proposed to ensure fracturing of reservoir or cap rock should not occur. Given the low predicted magnitude of overpressure in the Captain Sandstone (Fig. 7) relative to the estimated cap rock fracture overpressure threshold of 75-80 (Shi *et al.*, 2013) there is a low probability of the occurrence of this risk.

The site characterisation investigations have reduced the probability and so the risk of induced seismicity due to storage site operations (Tab. 3). However, microseismic monitoring is included as a regular monitoring component of the monitoring plan to assure that site integrity is maintained or to identify if the operation is inducing adverse seismic events.

4 RESULTS OF RISK ASSESSMENT-LED SITE CHARACTERISATION

Risk assessment-led site characterisation focuses investigations at a prospective site to provide information required for a storage permit application. The outputs inform the definition of the site, determine storage site design and operation or form the components of an application for a CO_2 storage permit. An application for a storage permit submitted to the competent authority of a European member state should comply with the EU directive and guidance on the geological storage of carbon dioxide (EU, 2009, 2010, 2011) and is reviewed by the European Commission.

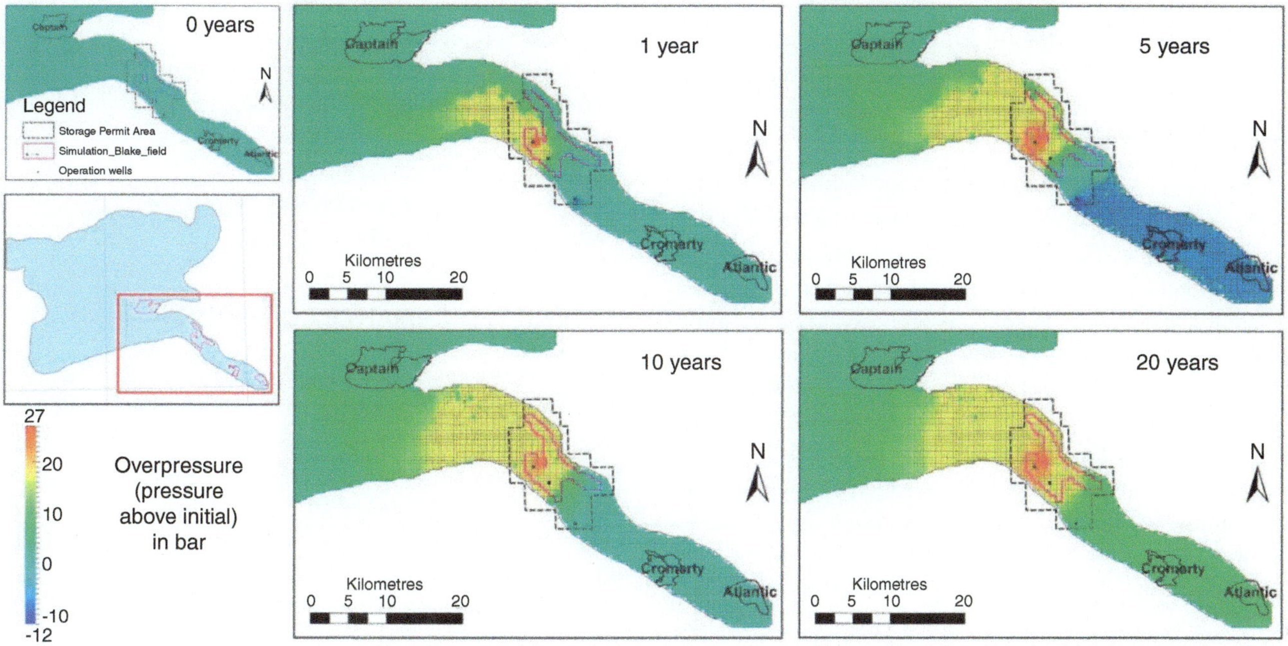

Figure 7

Reservoir pressure change in the 'pan handle' of the Captain Sandstone study area predicted from dynamic modeling of the proposed 20-year period of CO_2 injection and water production (Delprat-Jannaud *et al.*, 2012) showing the position of hydrocarbon fields hosted within the Captain Sandstone from Hannis *et al.* (2013a). Inset shows the position of the Blake Field (red outline) and other hydrocarbon fields (black outlines), see also Figure 1.

4.1 Definition of the Storage Site and Storage Complex

The SiteChar UK North Sea storage site (Fig. 2, 3) is located principally in the Captain Sandstone Member of the Lower Cretaceous Wick Sandstone Formation (Johnson and Lott, 1993). Vertically, the storage site is defined as the entire Wick Sandstone Formation because the Captain Sandstone may be hydraulically connected to the underlying Coracle and Punt sandstone members (Fig. 2, 3) (Quinn *et al.*, 2012; Akhurst *et al.*, 2014).

The storage complex is a defined volume that extends beyond the storage site. The SiteChar UK North Sea storage complex extends vertically to include the rocks directly above the storage site up to the sea bed (Fig. 3). The storage complex includes the primary cap rock to the storage site, comprising mudstone of the Valhall, Carrack and Rodby formations (Johnson and Lott, 1993) (Fig. 2). Rocks of the overlying Chalk Group may also act as seal if they are of sufficiently low permeability (Fig. 2). Secondary reservoirs for CO_2 are strata overlying and laterally continuous with the storage site that may be hydraulically connected. These include any rocks with available pore volume that overlie the primary cap rock, for example rocks of the Chalk Group and sandstones in the Lista Formation of the Montrose Group (Knox and Holloway, 1992; Johnson and Lott, 1993) (Fig. 2). Secondary seal rocks (cap rocks to the secondary reservoirs) are anticipated to be primarily non-calcareous mudstone of the Lista Formation of the Montrose Group and the mud-prone Moray Group (Fig. 2) (Quinn *et al.*, 2012; Akhurst *et al.*, 2014).

Laterally, the storage complex boundaries enclose a defined volume beyond the storage site. The lateral extent of the storage complex is the same as the storage permit area shown in Figure 6 (Hannis *et al.*, 2013a). The storage complex area encloses the maximum predicted plume extent (Fig. 6) and extends one kilometre beyond to reflect the single investigation of sensitivities conducted during site characterisation and recommendation for additional investigations (Hannis *et al.*, 2013a). The complex includes the lowest closing structural contour enclosing both the injection and water production well of the Captain Sandstone upper surface. It also encompasses the area up-dip (North-East) of the predicted plume extent to reflect the lack of knowledge of the property variations between the Channel and Flank facies sandstones in this direction (Delprat-Jannaud *et al.*, 2012). The overall recti-linear shape of the storage complex is defined by the existing hydrocarbon licence blocks for which agreement with existing licence holders would be expected to be sought for a storage site permit application (Hannis *et al.*, 2013a).

4.2 Injection Scenario

A component of the storage permit application is the project description. The project description includes an injection plan. For the SiteChar UK North Sea site, the injection plan is determined by the results of site characterisation summarised in this paper. CO_2 injection is proposed to be *via* multiple wells in the vicinity of Well 13/24a-4 (*i.e.* in the Blake Field) at a commercial-scale rate of 5 Mt of CO_2 per year for 20 years resulting in a total 100 Mt of CO_2 injected (Hannis *et al.*, 2013a). Increased reservoir pressure due to injection of CO_2 at the site will be managed by water production. Simultaneous water production through Well 13/29b-9 (*i.e.* down-dip in the saline aquifer) at a rate of 4.75 Mt per year is proposed (Delprat-Jannaud *et al.*, 2012) to manage reservoir pressure to a level one third of the estimated fracture pressure threshold (Shi *et al.*, 2013). This rate of water production will also ensure that CO_2 breakthrough does not occur at the production well. The pressure increase due to injection is expected to extend beyond the storage complex but not interfere with the Captain Oil Field, the only operating field within the Captain Sandstone in the vicinity of the site (Hannis *et al.*, 2013a). Once injection ceases after 20 years the site will be in the post-injection phase. Dynamic modeling results predict rapid pressure dissipation and little post-injection movement of the CO_2 plume (Hannis *et al.*, 2013c).

4.3 Storage Permit Application Components

The risk assessment-led characterisation process presents outputs that are required components of a storage permit application that complies with the EU directive for geological storage of CO_2 (EU, 2009, 2010, 2011). In SiteChar, all components presented for the UK North Sea site as a 'dry-run' storage permit application (Hannis *et al.*, 2013a) were completed as much as possible within the resources of a research project:

- register of risks (Hannis *et al.*, 2013b),
- preventative measures plan,
- monitoring plan,
- provisional corrective measures plan,
- provisional post-closure plan.

The provisional nature of some of the storage permit components reflects the early stage of site characterisation possible within a research project. Site characterisation was confined to those risks associated with technical matters. Some storage permit components that would be prepared for a real storage permit application are beyond the scope of the SiteChar research and are not included in the SiteChar 'dry-run' application:

- a full Environmental Impact Assessment (EIA),
- provisions relating to the acceptance and injection of CO_2,
- details of financial security,
- a fully developed post-closure plan,
- provisions for reporting.

5 PERMIT PERFORMANCE CONDITIONS

Permit Performance Conditions, criteria against which to measure storage site performance, are recommended as a finding from the SiteChar research. They are useful tools for discussion between the competent authority and operator throughout site characterisation to inform the monitoring plan, and to define and agree acceptance criteria to facilitate site closure.

Six Permit Performance Conditions (PPC) are proposed for the UK North Sea storage site and defined in Table 4. These aim to be unambiguous and achievable by defining ranges of acceptable expected performance. Specific justification for each condition and definition of evidence acceptable to support these statements is documented by Hannis *et al.* (2013a). These criteria were discussed and agreed with regulators from EC Member States and representatives of potential site operators (offshore hydrocarbon industry). PPC define limits to site behaviour which, if exceeded, indicate that a significant irregularity has occurred. This will trigger appropriate corrective measures. They were identified through risk assessment and the resulting risk register and inform the monitoring plan (Hannis *et al.*, 2013c). Although ordinarily the PPC would lie solely in the corrective measures and post-closure plans, they are included as they are necessary for the monitoring plan. PPC are not

TABLE 4

Permit Performance Conditions (PPC) for the SiteChar UK northern North Sea storage site (Hannis *et al.*, 2013a)

Number	Condition
PPC 1	Environmental or human health will not be adversely affected by the storage operation
PPC 2	CO_2 will not pass beyond the storage complex area boundaries
PPC 3	CO_2 plume shows migration within expected modeled behaviour
PPC 4	Pressure changes will remain within predefined/ predicted ranges
PPC 5	Geomechanical integrity of the site will be maintained
PPC 6	Cost per tonne of CO_2 will remain within a set limit

explicitly required by the EU storage directive (EU, 2009) but are SiteChar-recommended useful tools for discussion between the competent authority and operator and a way to define and agree acceptance criteria. This will ultimately also facilitate site hand-over and closure at the end of the term, if a storage permit is awarded.

CONCLUSIONS

Risk assessment-led site characterisation investigations targeted to meet the requirements of a storage permit application have been implemented for a prospective storage site in the UK northern North Sea. The process of risk assessment-led site characterisation has been followed, as specified in the EC directive for the geological storage of CO_2, to provide the required components as a 'dry-run' storage permit application. Pre-characterisation of a storage site comprising a hydrocarbon field and adjacent saline aquifer sandstone has been achieved based on publicly available data acquired during hydrocarbon exploration and production. Access to data available to the hydrocarbon field operator would greatly reduce uncertainties associated with site characterisation assessment.

Commercial-scale storage of 100 Mt CO_2, injected at an annual rate of 5 Mt for 20 years, can be achieved without threat to the integrity of the site by pressure management. The pressure is managed by injection of CO_2 into the Blake Oil Field and concurrent water production from the Captain Sandstone saline aquifer down-dip from the field. Optimisation of the CO_2 injection and water production well positions and the CO_2 injection rate is recommended to maximise storage capacity and minimise the cost of the storage operation.

Site characterisation and a first iteration of risk reduction have been conducted for the mostly highly rated technical risks. Site characterisation has reduced most of the risks investigated. However, further iterations of risk assessment-led site characterisation are required to reduce the risks to 'as low as reasonably possible' and to be sufficient to warrant submission of a storage permit application to the competent authorities.

Risk reduction and mitigation measures are proposed to further reduce risks, decrease uncertainties, and increase certainty in the prospective storage site. Abandoned wells remain the most likely fluid escape pathways. However, the mass of CO_2 migrating within the storage complex *via* a leaking wellbore is modeled as a maximum of 20 tonnes per 100 million tonnes injected, despite assuming very pessimistic conditions.

The impact of CO_2 reacting with the reservoir rock during the period of injection is predicted to be negligible by geochemical modeling with a modest net increase in porosity over a period of one million years. The potential for salt precipitation and the effects of microbial activity should also be assessed. Rock physics studies and seismic modeling indicate the injected CO_2 should be detectable using time-lapse surface 3D seismic survey data acquired using standard methods. However, a cost-benefit analysis for acquisition of sea bottom seismic data is recommended. Dynamic simulation of the Blake Oil Field and adjacent Captain Sandstone indicates that after 1 000 years the injected supercritical phase CO_2 would be retained within two kilometres beyond the field boundary. Provided the pressure of injection within the sandstone is maintained at acceptable levels there would be no adverse influence on operating hydrocarbon fields from the pressure of injection. The pressure changes on adjacent non-producing hydrocarbon fields are predicted for the proposed injection scenario and geology modeled.

The risk assessment-led characterisation process presents outputs that are required components of a storage permit application that complies with the EU directive for geological storage of CO_2 (EU, 2009, 2010, 2011). A prospective storage site, storage complex and storage scenario have been defined and storage permit components were presented, as much as possible within the resources of a research project. In addition, PPC, criteria against which to measure the site performance, are recommended as useful tools for discussion between the competent authority and operator throughout the site characterisation process to inform the monitoring plan, define and agree acceptance criteria to facilitate site closure.

ACKNOWLEDGMENTS

The research leading to these results has received funding from the European Union Seventh Framework Programme (FP7/2007-2013) under grant agreement No. 256705 as well as from *Enel, Statoil, Vattenfall, Véolia Environnement*, and the Scottish Government. The support and contribution of the Scottish Government in the conduct of the SiteChar research and funding of the underpinning, preceding Scottish CCS studies by SCCS is gratefully acknowledged. Two anonymous reviewers are thanked for their constructive reviews which greatly improved an earlier version of this paper.

REFERENCES

Akhurst M.C., Quinn M.Q., Delprat-Jannaud F., Bossie-Codreanu D., Hannis S.D., Lecomte J.-C., McCormac M., Frykman P. (2014) Risk assessment-led pre-characterisation of a site for commercial-scale storage of CO_2, EAGE CO_2 Geological Storage Workshop, *Demonstrating storage integrity and building confidence in CCS*, Stavanger, Norway, April.

Bentham M., Mallows T., Lowndes J., Green A. (2014) CO_2 Storage Evaluation Database (CO_2 Stored), the UK's online storage database, *Energy Procedia* **63**, 5103-5113.

BG Exploration & Production (1998) Well 13/24a-6 Geological Report, June.

Chadwick R.A., Arts R., Bernstone C., May F., Thibeau S., Zweigel P. (2008) Best practice for the storage of CO_2 in saline aquifers, *British Geological Survey Occasional Publication* No. 14, Keyworth, Nottingham.

DECC (2013) Preferred bidders announced in UK's £1bn CCS Competition. Press release 13/028, 20 March, UK Department of Energy & Climate Change, Available at: https://www.gov.uk/government/news/preferred-bidders-announced-in-uk-s-1bn-ccs-competition.

Delprat-Jannaud F., Pearce J., Akhurst M., Nielsen C., Neele F., Lothe A., Volpi V., Brunsting S., Vincké O. (2015) SiteChar-Methodology for a fit-for-purpose assessment of CO_2 storage sites in Europe, *Oil and Gas Science and Technology* **70**, 4, 531-554.

Delprat-Jannaud F., Maurand N., Roggero F., Bossie-Codreanu D., Pluymaekers M., Nagy S., Klimkowski L., Akhurst M.C., Kirby G. A. (2012) Storage design, UK site. SiteChar Characterisation of European CO_2 storage Deliverable No. D3.2, Available at: http://www.sitechar-CO_2.eu/.

DTI (2006) Industrial carbon dioxide emission and carbon dioxide storage potential in the UK, Report number COAL R308 DTI/Pub URN 06/2027, UK Government Department of Trade and Industry.

Du E.K. (2002) Blake Flank Well 13/24a- 8, Test Analysis Report, BG Group, Petroleum Engineering.

EU (2009) Storage Directive 2009/31/EC, Available at: http://eur-lex.europa.eu/LexUriServ/LexUriServ.do?uri=OJ:L:2009:140:0114:0135:EN:PDF.

EU (2010) Implementation of Directive 2009/31/EC on the geological storage of carbon dioxide – Guidance Document 1 – CO_2 storage life cycle risk management framework.

EU (2011) Implementation of Directive 2009/31/EC on the geological storage of carbon dioxide – Guidance Document 2 – Characterisation of the storage complex, CO_2 stream composition, monitoring and corrective measures.

Hannis S.D., Kirby G.A., Akhurst M.C. (2013a) Dry-run licence application, UK site SiteChar Characterisation of European CO_2 storage Deliverable No. D3.6, Available at: http://www.sitechar-CO_2.eu/.

Hannis S.D., Kirby G.A., Akhurst M.C. (2013b) Risk register for UK site, SiteChar Characterisation of European CO_2 storage Deliverable No. D3.7, Available at: http://www.sitechar-CO_2.eu/.

Hannis S.D., Kirby G.A., Gei D., Picotti S., Bohm G., Carcione J.M., Rossi G., Arts R.J. (2013c) Monitoring Plan, UK site SiteChar Characterisation of European CO_2 storage Deliverable No. D3.4, Available at: http://www.sitechar-CO_2.eu/.

Hilton V.C. (1999) Structural and sedimentological interpretation of STAR image data, core sedimentology and petrography Report for BG Exploration and Production, Blake Field, Well 13/24a- 6, Baker Atlas GEOScience.

Hilton V.C., Morris S. (2002) An integrated interpretation of core and borehole image data. Report for BG Group, Blake Flank, Wells 13/24a-8, 8z & 8y, Baker Atlas GEOScience.

Jappy I. (2006) Advance rock properties study for Talisman energy (UK) limited; well: 13/23b-5, Petroleum Services Division, Core Laboratories (UK) Limited, Advanced Technology Centre.

Jin M., Mackay E., Quinn M.F., Hitchen K., Akhurst M.C. (2012) Evaluation of the CO_2 Storage Capacity of the Captain Sandstone Formation, *SPE* 154539, Presented at *EAGE /SPE Europec Conference and Exhibition*, Copenhagen, Denmark.

Johnson H., Lott G.K. (1993) 2. Cretaceous of the Central and Northern North Sea, in *Lithostratigraphic nomenclature of the UK North Sea*, Knox R.W.O'B., Cordey W.G. (eds), British Geological Survey, Nottingham.

Knox R.W.O'B., Holloway S. (1992) 1. Palaeogene of the Central and Northern North Sea, in *Lithostratigraphic nomenclature of the UK North Sea*, Knox R.W.O'B., Cordey W.G. (eds), British Geological Survey, Nottingham.

Law A., Raymond R., White G., Atkinson A., Clifton M., Atherton T., Dawes I., Robertson E., Melvin A., Brayley S. (2000) The Kopervik fairway, Moray Firth, UK, *Petroleum Geoscience* **6**, 265-274.

Le Guen Y., Le Gouevec J., Chammas R., Gerard B., Poupard O. (2008) CO_2 Storage - Managing the Risk Associated with Well Leakage over Long Timescales, *Society of Petroleum Engineers Asia Pacific Oil & Gas Conference and Exhibition*, Perth, Australia, 20-22 Oct.

Nepveu M., Neele F., Delprat-Jannaud F., Akhurst M., Vincké O., Volpi V., Lothe A., Brunsting S., Pearce J., Battani A., Baroni A., Garcia B., Hofstee C., Wollenweber J. (2015) CO_2 storage feasibility: a workflow for site characterization, *Oil and Gas Science and Technology* **70**, 4, 555-566.

Quinn M.F., Akhurst M.C., Frykman P., Hannis S.D., Hovikoski J., Kearsey T., Lecomte J.-C., McCormac M. (2012) Geological report of the integrated Blake Oil Field and Captain Sandstone aquifer, UK multi-store site, SiteChar. SiteChar Report D3.1, http://www.sitechar-CO2.eu/.

ROAD (2013) The Rotterdam Capture and Storage Demonstration Project, http://road2020.nl/en/.

SCCS (2009) *Opportunities for CO_2 Storage around Scotland — an integrated strategic research study*, 56 pp., Available at: http://www.sccs.org.uk/images/expertise/reports/opportunities-for-co2/CO2-JointStudy-Full.pdf.

SCCS (2011) *Progressing Scotland's CO_2 Storage opportunities*, 60 pp., Available at: http://www.sccs.org.uk/images/expertise/reports/progressing-scotlands-co2/ProgressingScotlandCO2Opps.pdf.

Shell (2011a) Static Model Field Report (UKCCS - KT - S7.21 - Shell – 002), ScottishPower CCS Consortium, UK Carbon Capture and Storage Demonstration Competition, Available at: http://webarchive.nationalarchives.gov.uk/20121217150421/http://decc.gov.uk/assets/decc/11/ccs/sp-chapter3/ukccs-kt-s7.21-shell-002-smfr.pdf.

Shell (2011b) Geomechanics Summary Report (UKCCS - KT - S7.19 - Shell – 004), ScottishPower CCS Consortium, UK Carbon Capture and Storage Demonstration Competition, Available at: http://webarchive.nationalarchives.gov.uk/20121217150421/http://decc.gov.uk/assets/decc/11/ccs/sp-chapter3/ukccs-kt-s7.19-shell-004-gsr.pdf.

Shi J.-Q., Govindon R., Syed A., Tambach T., Koenen M., Neele F., Wildenborg A.F.B., Long D. (2013) Risk assessment, UK Site, SiteChar Characterisation of European CO_2 storage Deliverable No. D3.3, Available at: http://www.sitechar-CO_2.eu/.

Thomas L.K., Katz D.L, Tek M.R. (1967) Threshold Pressure Phenomena in Porous Media, *SPE* 1816, *42nd Annual Fall Meeting*, Oct.

12

Simultaneous Production of CH_4 and H_2 from Photocatalytic Reforming of Glucose Aqueous Solution on Sulfated Pd-TiO_2 Catalysts

Vincenzo Vaiano[1], Giuseppina Iervolino[1], Giuseppe Sarno[1], Diana Sannino[1*], Luigi Rizzo[2], Julie J. Murcia Mesa[3,4], Maria C. Hidalgo[4] and Jose A. Navío[4]

[1] *Department of Industrial Engineering, University of Salerno, via Giovanni Paolo II, 132, 84084 Fisciano (SA) - Italy*
[2] *Department of Civil Engineering, University of Salerno, via Giovanni Paolo II, 132, 84084 Fisciano (SA) - Italy*
[3] *Universidad Pedagógica y Tecnológica de Colombia, Avenida Central del Norte, Tunja, Boyacá - Colombia*
[4] *Instituto de Ciencia de Materiales de Sevilla (ICMS), Consejo Superior de Investigaciones Científicas (CSIC), Universidad de Sevilla, Américo Vespucio 49, 41092 Sevilla - Spain*
e-mail: dsannino@unisa.it

* Corresponding author

Abstract *— In this work, the simultaneous production of CH_4 and H_2 from photocatalytic reforming of glucose aqueous solution on Pd-TiO_2 catalysts under UV light irradiation by Light-Emitting Diodes (LED) was investigated. The Pd-TiO_2 catalysts were prepared by the photodeposition method. The Pd content was in the range 0.5-2 wt% and a photodeposition time in the range 15-120 min was used. Pd-TiO_2 powders were extensively characterized by X-Ray Diffraction (XRD), S_{BET}, X-Ray Fluorescence spectrometry (XRF), UV-Vis Diffuse Reflectance Spectra (UV-Vis DRS), TEM and X-Ray Photoelectron Spectroscopy (XPS). It was found that the lower Pd loading (0.5 wt%) and 120 min of photodeposition time allowed us to obtain homogeneously distributed metal nanoparticles of small size; it was also observed that the increase in the metal loading and deposition time led to increasing the Pd^0 species effectively deposited on the sulfated TiO_2 surface. Particle size and the oxidation state of the palladium were the main factors influencing the photocatalytic activity and selectivity. The presence of palladium on the sulfated titania surface enhanced the H_2 and CH_4 production. In fact, on the catalyst with 0.5 wt% Pd loading and 120 min of photodeposition time, H_2 production of about 26 µmol was obtained after 3 h of irradiation time, higher than that obtained with titania without Pd (about 8.5 µmol). The same result was obtained for the methane production. The initial pH of the solution strongly affected the selectivity of the system. In more acidic conditions, the production of H_2 was enhanced, while the CH_4 formation was higher under alkaline conditions.*

Résumé — Production simultanée de CH_4 et H_2 par réformage photocatalytique d'une solution aqueuse de glucose sur un catalyseur Pd-TiO_2 sulfaté — Dans cette recherche, la production simultanée de CH_4 et H_2 par le reformage photocatalytique de glucose sur les catalyseurs Pd-TiO_2 sous irradiation de lumière UV réalisée par des diodes électroluminescentes (LED) a été étudiée. Les échantillons de Pd-TiO_2 ont été préparés par la méthode de photodéposition. Le contenu de Pd était compris entre 0,5-2 % en poids et il a été utilisé un temps de photodéposition dans la gamme 15-20 minutes. Les poudres Pd-TiO_2 ont été caractérisées par diffraction des rayons X (XRD), surface spécifique BET (S_{BET}), fluorescence X (XRF), spectrométrie UV visible (UV vis DRS), microscopie électronique à transmission (TEM) et spectrométrie photo électronique X (XPS). Il a été constaté que le chargement inférieur de Pd (0,5 % en poids) et 120 minutes de temps de photodéposition conduit à une répartition homogène des nanoparticules métalliques avec une faible dimension ; il a aussi été observé que l'augmentation de la charge de métal et celle de durée du dépôt conduisent à augmenter les espèces de Pd^0 efficacement déposées sur la surface de TiO_2. La taille des particules et l'état d'oxydation du palladium ont été les principaux facteurs qui influent sur l'activité photocatalytique et sur la sélectivité. La présence de palladium sur la surface de l'oxyde de titane accroit la production de H_2 et CH_4. En fait, sur le catalyseur avec 0,5 % en poids du chargement de Pd et 120 min de temps de photodéposition, il a été obtenu une production de H_2 d'environ 26 μmol après 3 heures de temps d'irradiation, quantité supérieure à celle obtenue avec TiO_2 sans Pd (environ 8,5 μmol). Le même résultat a été obtenu pour la production de méthane. Le pH initial de la solution a fortement influencé la sélectivité du système. Dans des conditions plus acides, la production de H_2 a été améliorée, tandis que la formation de CH_4 a été plus élevée pour des conditions alcalines.

INTRODUCTION

In recent years, there have been intensive efforts toward the development of novel technologies for the production of hydrogen or methane from renewable resources, mainly water and biomass [1-3]. In particular, the food industry produces a large amount of solid and liquid wastes which could be valorized. In wastewater from the food industry, it is possible to find a high concentration of glucose. Glucose can be converted into hydrogen by several reactions, such as steam reforming [4, 5], wet oxidation of glucose [6], oxidative glucose reforming [7], pyrolysis [8], biophotolysis [9, 10], dark fermentation [11], electrolysis [12] and photocatalysis [13, 14]. Hydrogen has been identified as an ideal energy carrier for sustainable energy development [15-18] and it can be used in a fuel cell to generate electricity with high efficiency. In order to support a sustainable hydrogen economy, it is crucial to produce hydrogen cleanly and renewably. Typically, heterogeneous photocatalysis is studied with the aim of removing organic pollutants from wastewater [19, 20] through oxidation reactions which can finally result in the production of CO_2 and water. An interesting approach is to explore, in parallel to wastewater treatment, opportunities of mass recovery which can be sold as secondary raw material or used as energy resources, such as hydrogen and methane. Some wastewaters from food processes contain a high concentration of sugars, particularly glucose, which should be removed before effluent disposal or reuse, but if properly treated, hydrogen could be produced. In the past two decades, photocatalytic processes have been investigated as a possible method to obtain hydrogen glucose aqueous solutions [21]. The photocatalytic hydrogen production by decomposition of water containing glucose seems to have become a very powerful method for the practical and low-cost technologies in the hydrogen based energy system [22, 23]. Different catalysts containing noble metals on the surface (Au, Pt, Pd) or nanostructured Fe_2O_3 polymorphs [24] have been investigated [22, 25] because the presence of noble metals on the surface of the semiconductor improves photocatalytic performances. Linsebigler *et al.* (1995) [26] observed that by adding a small amount of noble metals, it is possible to suppress to some extent the charge recombination by forming a Schottky barrier. Rh or Pt, Pd, Cu or Ni supported on TiO_2 have been investigated in the conversion of glucose into H_2 [27]. However, the reactor configurations reported in the literature only include the use of mercury vapor lamps with high voltage as a light source (125-400 W) [28, 29]. Moreover, a few results are reported in the scientific literature about the production of methane through photocatalytic reforming of wastewater. Generally, the production of methane derives from processes such as anaerobic digestion [30, 31]. The production of hydrogen or methane from ethanol by photocatalytic reaction has also been reported [32].

In this work, the simultaneous production of CH_4 and H_2 from photocatalytic reforming of glucose on Pd-TiO_2 was investigated under UV light irradiation by Light-Emitting Diodes (LED) with low energy consumption (10 W). TiO_2 was prepared *via* a sol–gel technique, while Pd addition was carried out by photodeposition. The influence of some parameters, such as: photodeposition time and Pd loading on the final properties of the materials and their photocatalytic efficiency for glucose reforming was evaluated. The effect of the photocatalyst dosage and initial pH on photocatalytic performance was also investigated.

1 EXPERIMENTAL

1.1 Synthesis of Photocatalysts

TiO_2 was prepared by hydrolysis of titanium tetraisopropoxide (*Aldrich*, 97%) in isopropanol solution (1.6M) by the slow addition of distilled water (volume ratio isopropanol/water 1:1). The TiO_2 powder was recovered by filtration, dried at 110°C for 24 h and pre-treated by immersion in a 1M sulfuric acid solution under continuous stirring for 1 h. After pre-treatment, the solution was filtered, dried and calcinated at 650°C for 2 h; the resulting photocatalyst was used as starting material (S-TiO_2). The fresh TiO_2 powder was submitted to a sulfate treatment, taking into account that previous results have shown that sulfate pre-treatment stabilizes the anatase phase up to high temperatures and protects the surface area against sintering [33-36].

Photodeposition of palladium was performed over the sulfated TiO_2 powders using palladium chloride (II) ($PdCl_2$, *Aldrich* 99%) as the metal precursor. Under an inert atmosphere (N_2), a suspension of the corresponding TiO_2 sample in distilled water containing isopropanol (*Merck* 99.8%), which acts as the sacrificial donor, was prepared. Then, the appropriate amount of $PdCl_2$ solution to obtain the desired nominal palladium loading was added; in order to dissolve the metal precursor, the pH of this solution was adjusted to 2 by using a HCl 0.1M solution. Photodeposition of palladium was then performed by illuminating the suspension for 120 min with an *Osram* Ultra-Vitalux lamp (300 W) which possesses a sun-like radiation spectrum with a main emission line in the UVA range at 365 nm. The light intensity on the suspensions was 60 W/m^2 determined by a PMA 2200 UVA photometer (*Solar Light Co.*). After photodeposition, the powders were recovered by filtration and dried at 110°C overnight. Series of Pd-TiO_2 catalysts were prepared by using different metal contents (0.5% or 2% weight total to TiO_2) and photodeposition times (15 or 120 min). The photocatalysts prepared by TiO_2 metallization will be denoted in this document as 0.5Pd15, 0.5Pd120, 2Pd15 and 2Pd120.

1.2 Photocatalyst Characterization

All the materials were widely characterized using different techniques. The crystalline phase composition and degree of crystallinity of the samples were estimated by XRD. XRD patterns were obtained on a *Siemens* D-501 diffractometer with a Ni filter and graphite monochromator using Cu Kα radiation. Anatase crystallite sizes were calculated from the line broadening of the main anatase XRD peak (101) by using the Scherrer equation. Peaks were fitted by using a Voigt function.

Specific surface area (S_{BET}) measurements were carried out using low-temperature nitrogen adsorption in a *Micromeritics ASAP* 2010 instrument. Degasification of the samples was performed at 150°C.

The chemical composition and total palladium content in the samples were determined by XRF in a *PANalytical Axios* sequential spectrophotometer equipped with a rhodium tube as the source of radiation. XRF measurements were performed on pressed pellets (sample included in 10 wt% of wax).

The light absorption properties of the samples were studied by UV-Vis spectroscopy. The UV-Vis DRS were recorded on a *Varian* spectrometer (model Cary 100) equipped with an integrating sphere and using $BaSO_4$ as a reference. Band-gap values were calculated from the corresponding Kubelka-Munk functions, F(R∞), which are proportional to the absorption of radiation by plotting $(F(R\infty)\cdot h\nu)^{1/2}$ against hν.

Palladium particle sizes were evaluated by TEM, in a *Philips* CM 200 microscope. The samples were dispersed in ethanol using an ultrasonicator and dropped on a carbon grid.

XPS studies were carried out on a *Leybold–Heraeus* LHS-10 spectrometer, working with a constant pass energy of 50 eV. The spectrometer's main chamber, working at a pressure $< 2 \times 10^{-9}$ Torr, is equipped with an EA-200 MCD hemispherical electron analyzer with a dual X-ray source working with Al Kα ($h\nu$ = 1 486.6 eV) at 120 wand 30 mA. The C 1s signal (284.6 eV) was used as an internal energy reference in all the experiments. Samples were outgassed in the prechamber of the instrument at 150°C up to a pressure $< 2 \times 10^{-8}$ Torr to remove chemisorbed water.

1.3 Photocatalytic Tests

Photocatalytic experiments were carried out with a *Pyrex* cylindrical reactor (ID = 2.5 cm) equipped with a N_2 distributor device (Q = 0.122 NL/min). The continuous mixing of the glucose solution and the photocatalyst was carried out by external recirculation of water through the use of a peristaltic pump. A thermocouple was inserted inside the reactor to monitor the temperature during irradiation. The photoreactor was irradiated by a strip composed of 15 UV-LED (nominal power: 10 W) with wavelength emission in the range 375-380 nm. The LED strip was positioned around the external surface of the reactor so that the light source uniformly irradiated the reaction volume (light intensity: 1.5 W/m^2). Typically, 0.04 g of catalyst was suspended in an 80-mL aqueous solution containing 500 mg/L of glucose (D$^+$ Glucose VWR, *Sigma-Aldrich*). The suspension was left in dark conditions for 2 hours to reach the adsorption-desorption equilibrium of glucose on the photocatalysts' surface, and then a photocatalytic reaction was initiated under UV light for up to 3 hours. Hydrogen and methane yield during the irradiation time is evaluated according to the following relationship:

$$R_{H_2} = \frac{2 \cdot n_{H_2}}{12 \cdot n^0_{C_6H_{12}O_6}} \cdot 100$$

$$R_{CH_4} = \frac{4 \cdot n_{CH_4}}{12 \cdot n^0_{C_6H_{12}O_6}} \cdot 100$$

where: R = yield; n_{H_2} = moles of H_2 produced; n_{CH_4} = moles of CH_4 produced; $n^0_{C_6H_{12}O_6}$ = moles of glucose after dark adsorption.

To evaluate the influence of the initial pH on the production of H_2 and CH_4, the pH of the solution was changed by adding nitric acid (*Baker* Analyzed, 65%) or ammonium hydroxide (*Baker* Analyzed, 30%). About 2 mL of samples were taken from the photoreactor at different times and filtered in order to remove Pd-TiO_2 particles before analysis.

1.4 Chemical Analysis

The analysis of the gaseous phase from the photoreactor was performed by online gas chromatography (*MicroGC, Agilent*) with a TDX-01 column for the analysis of CO, CO_2, CH_4 and H_2 and equipped with a molecular sieve for the analysis of oxygen. The concentration of glucose was measured by a spectrophotometric method [37, 38]. Briefly, a 2-mL aliquot of a carbohydrate solution is mixed with 1 mL of 5 wt% aqueous solution of phenol (*Sigma-Aldrich*) in a test tube. Subsequently, 5 mL of concentrated sulfuric acid (*Sigma-Aldrich*) is added rapidly to the mixture. Then, light absorption at 490 nm is recorded on a UV-Vis spectrophotometer (*Perkin Elmer*).

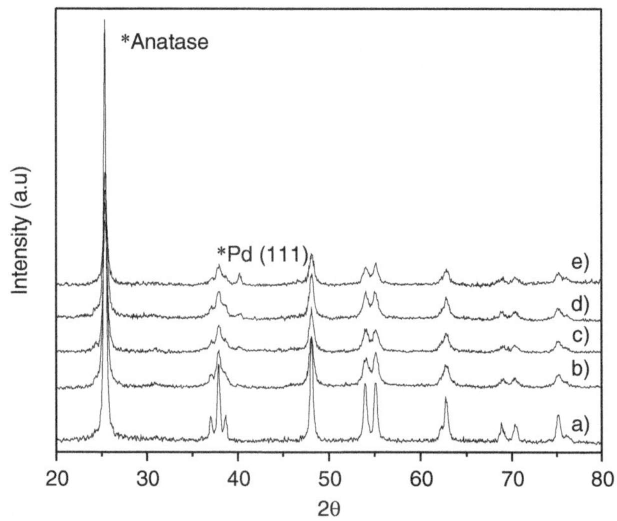

Figure 1

XRD patterns for TiO_2 and Pd-TiO_2 photocatalysts prepared with different Pd contents and deposition times. a) TiO_2; b) 0.5 wt% Pd-TiO_2, 15 min; c) 0.5 wt% Pd-TiO_2, 120 min; d) 2 wt% Pd-TiO_2, 15 min; e) 2 wt% Pd-TiO_2, 120 min.

2 RESULTS AND DISCUSSION

2.1 Photocatalyst Characterization

2.1.1 X-Ray Diffraction

Figure 1 shows the XRD patterns of the Pd-TiO_2 photocatalysts; the only crystalline phase present in all the samples was anatase, identified by the main XRD peak located at 25.25°, thus indicating that the sulfation pre-treatment inhibited the formation of the rutile phase of TiO_2 during the calcination process, as has been reported by different authors [33-36].

On the other hand, photodeposition of palladium did not affect the phase composition in any case. In the spectra of metallized samples, a diffraction peak for Pd was detected at 40°; this peak is assigned to Pd (111) [39]. It was observed that the intensity of this peak increased with the Pd loading, being highest in the catalysts prepared with 2 wt% of nominal content and 120 min of deposition time; this could be due to a higher metal particle size in these samples.

The anatase crystallite sizes calculated by the Scherrer equation from the (101) peak of the XRD pattern are listed in Table 1. As can be seen, the TiO_2 presents a value of 20 nm; the Pd addition did not significantly modify crystallite sizes beyond this value.

2.1.2 BET Surface Area

BET surface area values (S_{BET}) for all the samples are shown in Table 1. As can be seen, Pd addition did not induce significant modifications in the S_{BET} of the analyzed materials, and the observed slight decreases are probably due to pore blocking by metal deposits on the TiO_2 surface.

TABLE 1
Characterization of the investigated photocatalysts

Photocatalyst	Pd nominal loading (wt%)	Photodeposition time (min)	Real Pd content (wt%)*	Sulfate amount (wt%)	S_{BET} (m^2/g)	Band gap (ev)**
S-TiO_2	0	–	–	0.30	58	3.2
0.5Pd15	0.5	15	0.18	0.28	46	3.2
0.5Pd120	0.5	120	0.20	0.27	48	3.2
2Pd15	2	15	0.34	0.28	48	3.2
2Pd120	2	120	0.80	0.28	49	3.2

2.1.3 X-Ray Fluorescence

The real palladium content in the metallized samples was measured by XRF and the values are listed in Table 1. These values are less than the nominal metal content used to prepare these materials (0.5 or 2 wt%), indicating an incomplete reduction of the metal precursor ($PdCl_2$) on the TiO_2 surface during the photodeposition process. However, the amount of deposited Pd was highest in the samples prepared with the highest deposition time (120 min).

In all the samples, XRF analysis revealed that a certain amount of S and Cl^- species remained on the solid after preparation. The sulfur content was between 0.27 and 0.30%, and the chloride content was below 0.02%.

2.1.4 UV-Vis Diffuse Reflectance Spectra

The absorption spectra of all the samples are shown in Figure 2. The typical absorption band edge of the TiO_2 semiconductor was observed at around 400 nm for all the samples. As can be seen, metallization did not substantially alter the absorption properties of the samples; however, a slight increase in absorption throughout the visible range of the spectrum was observed due to the gray color of these materials; this increase is more evident in samples with the highest Pd loading.

From the UV-Vis DRS spectra, band-gap energies were calculated, being 3.20 eV for the TiO_2 corresponding to the anatase phase. Pd photodeposition did not induce any significant change in this value; estimated band-gap energies are between 3.2 and 3.3 eV.

2.1.5 Microscopic Analysis

The morphology and Pd particle size in the samples prepared was studied by TEM.

As was observed from the TEM images, in the samples prepared with 0.5 wt% Pd loading, metal nanoparticles appear to be homogenously distributed all over the surface, with sizes around 2-4 nm (images are not shown).

Figure 3 shows representative TEM images of TiO_2 after photodeposition of 2 wt% Pd with irradiation times of 15 and 120 min (Fig. 3a and 3b, respectively). In both samples, palladium particles can be seen as dark spherical spots placed on the larger anatase particles; in these samples Pd nanoparticles are heterogeneously distributed on the TiO_2 surface and particle sizes higher than 6 nm were observed.

2.1.6 XPS Analyses

XPS measurements were performed and a summary of these results is reported in Table 1.

Figure 4 shows spectra of Ti 2p and O 1s for the different samples analyzed. The Ti 2p core peaks exhibit a main component at around 458.5 ± 0.1 eV (Ti $2p_{3/2}$) in all the samples, representative of the Ti^{4+} ions in the TiO_2 lattice. The addition of Pd did not modify the oxidation state or the chemical environment of titanium atoms at the surface of the TiO_2. In the O 1s region, a peak located at 529.8 ± 0.2 eV can be observed for all the samples, assigned to oxygen atoms in the TiO_2 lattice. This peak is asymmetric, with a shoulder at higher binding energies assigned to surface OH groups. The shoulder is more prominent in the non-metallized TiO_2 sample, indicating a higher hydroxylation degree in

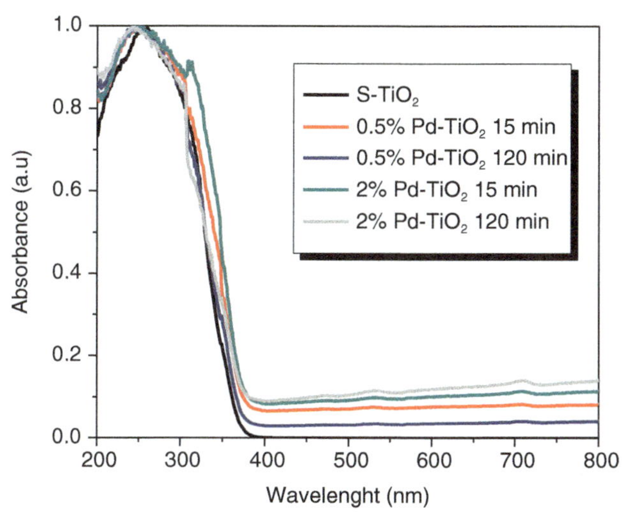

Figure 2

UV-Vis DRS spectra for the investigated photocatalysts.

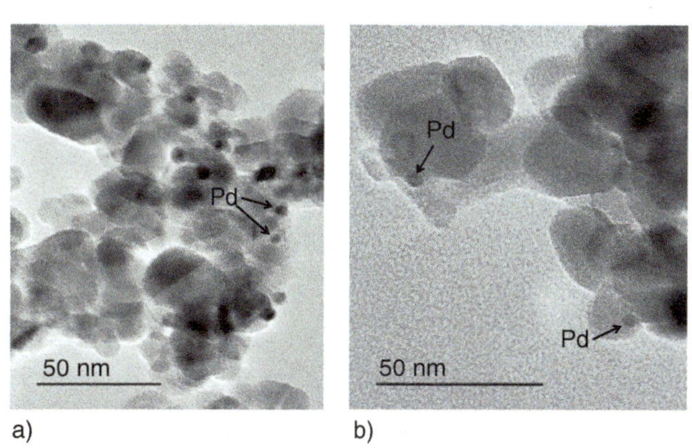

Figure 3

TEM images of 2 wt% Pd-TiO_2 samples. a) Pd-TiO_2, 15 min; b) Pd-TiO_2, 120 min.

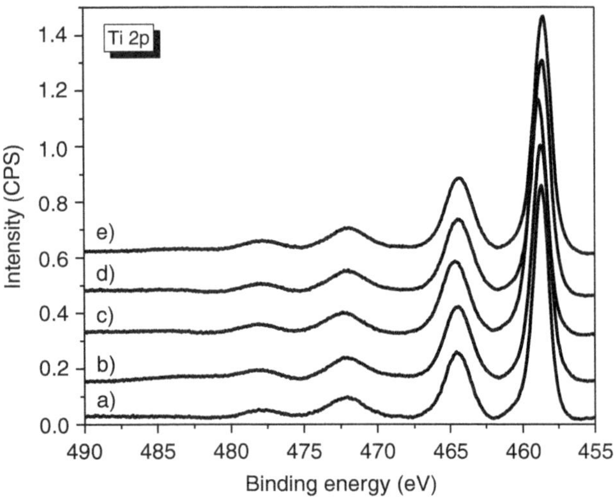

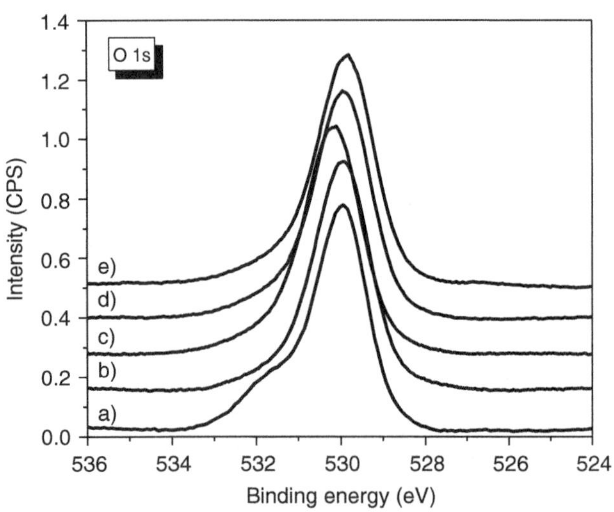

Figure 4

XPS core level spectra of Ti 2p and O 1s regions for the investigated photocatalysts. a) S-TiO_2; b) 0.5 wt% Pd-TiO_2, 15 min; c) 0.5 wt% Pd-TiO_2, 120 min; d) 2 wt% Pd-TiO_2, 15 min; e) 2 wt% Pd-TiO_2, 15 min.

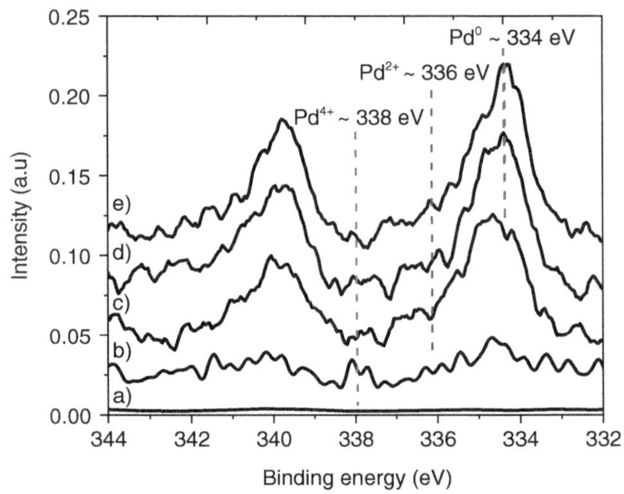

Figure 5

Comparative XPS spectra of Pd $3d_{7/2}$ and $3d_{7/2}$ core levels of Pd-TiO_2 catalysts prepared with different Pd contents and deposition times. a) S-TiO_2; b) 0.5 wt% Pd-TiO_2, 15 min; c) 0.5 wt% Pd-TiO_2, 120 min; d) 2 wt% Pd-TiO_2, 15 min; e) 2 wt% Pd-TiO_2, 120 min.

this sample. The intensity of this shoulder is lower in Pd-TiO_2 samples; this is mainly due to the presence of metallic nanoparticles partially covering the TiO_2 surface, as was observed by TEM (Fig. 3).

The study of XPS Pd 3d peak regions can provide information concerning the oxidation state of the Pd species adsorbed on the TiO_2 surface. A Pd 3d region is formed by a doublet corresponding to the signals for $3d_{7/2}$ and $3d_{5/2}$. The Pd $3d_{7/2}$ binding energy for metallic palladium (Pd^0) appears near to 334 eV, while for partially oxidized forms (Pd^{2+}/Pd^{4+}) it appears at higher binding energies with values of ca. 336 and 338 eV, respectively [40]. The XPS Pd 3d region for the Pd-TiO_2 samples prepared with 0.5 and 2 wt% metal content and deposition times of 15 and 120 min is shown in Figure 5. All the spectra were calibrated with the C 1s peak at 284.6 eV attributed to "adventitious" surface carbon. Pd^0 and $Pd^{\sigma+}$ species were detected in all the analyzed samples, thus indicating an incomplete reduction of the metal precursor on the TiO_2 surface; these observations are in agreement with the results obtained by XRF analysis (Sect. 2.1.3). However, the Pd^0 on TiO_2 increased with the deposition time and with the metal content. As can be seen qualitatively in Figure 5, the highest intensity of the peaks assigned to Pd^0 species content was observed in the catalyst prepared with 120 min of deposition time and 2 wt% of nominal Pd loading.

Pd $3d_{7/2}$ and $3d_{5/2}$ doublets were deconvoluted using the UNIFIT 2009 software assuming a doublet separation of 5.2 eV of the two components. The spectrum obtained for the 2 wt% Pd-TiO_2 catalysts prepared with 120 min of deposition time is shown in Figure 6. A Shirley-type background was subtracted from each spectrum. By the deconvolution of the doublets, it is possible to observe clearly the presence of $Pd^0/Pd^{\sigma+}$ species in the analyzed sample.

2.2 Photocatalytic Results

2.2.1 Influence of Photocatalyst Formulation

The preliminary test performed in dark conditions showed no reaction product in the gaseous phase.

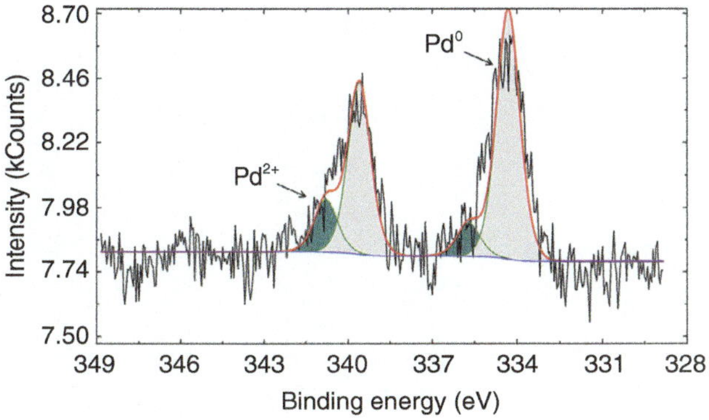

Figure 6

XPS Pd $3d_{7/2}$ and $3d_{7/2}$ regions for the 2 wt% Pd-TiO_2 catalyst prepared with 120 min of deposition time.

The final concentration of glucose after the dark period was 351 and 305 mg/L for S-TiO_2 and 0.5Pd120, respectively.

Control tests with UV as a standalone process and under photocatalytic conditions with UV/S-TiO_2 and UV/0.5Pd120 were carried out to evaluate the contribution of each process in the production of hydrogen and methane from glucose (Fig. 7). During the irradiation time, the production of H_2, CO_2 and CH_4 was observed, but no formation of CO and O_2 was observed. The presence of palladium photodeposited on the surface of the titania enhanced the production of hydrogen (Fig. 7a). In particular, a production of H_2 as high as 26 μmol (R_{H_2}= 3%) after 3 hours of irradiation with a 0.5Pd120 catalyst was observed. The hydrogen yield on 0.5Pd120 (3%) is significantly higher than that obtained on S-TiO_2 (0.9%). However, a large fraction of glucose is not converted into hydrogen and methane under the present conditions.

These results are in agreement with those reported in the literature. In particular, loading noble metals such as Pt [41] and Au [42] onto the photocatalyst surface can separate photogenerated electrons and holes more effectively and thus improve hydrogen production. In a similar way, the coexistence of Pd with TiO_2 in the Pd-TiO_2 composite photocatalyst may lower the recombination rate between photogenerated electrons and holes, thus improving the efficiency of the photocatalysis process. It is well known that there are two forms of electron-hole recombination in TiO_2, surface recombination and bulk recombination, and both can lower the efficiency of photocatalysis [43]. The Fermi energy level of Pd is lower than that of TiO_2, and since they are closer to each other, the electrons on the surface of TiO_2 can spontaneously transfer to the surface of Pd until their Fermi energy levels become equal. As a result, excessive negative charges accumulate on the surface of Pd and excessive positive charges accumulate on the surface of TiO_2. Therefore, the energy band of TiO_2 bends upwards, forming the Schottky barrier. Therefore, in the composite catalyst, Pd acts as an electron trap, which can effectively prevent the surface electron-hole recombination [43].

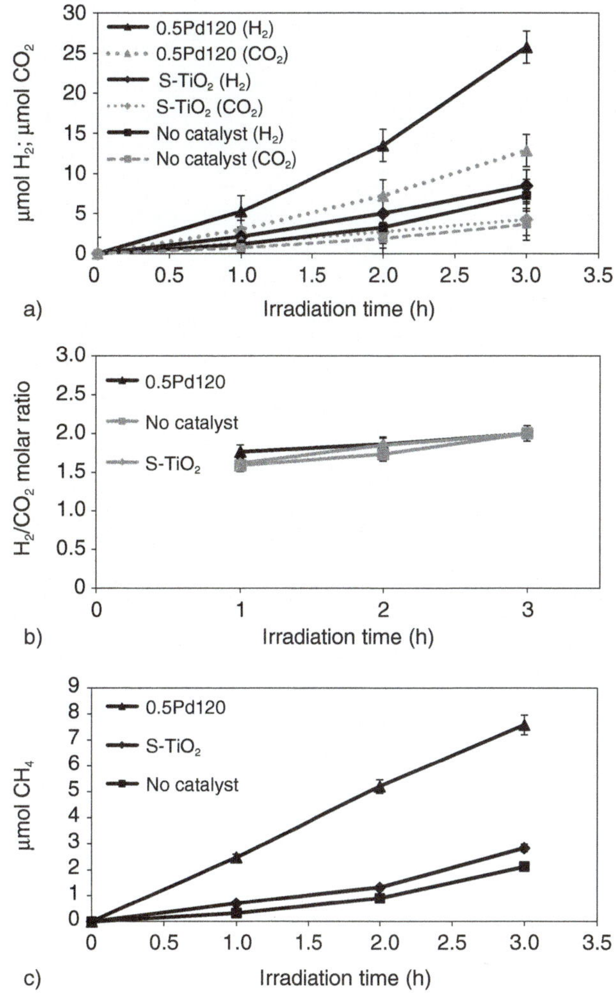

Figure 7

Control tests with UV as a standalone process and under photocatalytic conditions with UV/S-TiO_2 and UV/0.5Pd120: H_2 and CH_4 production a), H_2/CO_2 molar ratio b) and CH_4 production c).

The value of the H_2/CO_2 molar ratio did not change significantly as irradiation time increased, tending to a value approximately equal to 2 after three hours of irradiation (Fig. 7b). CO_2 and H_2 are mainly produced from the photocatalytic wet reforming of glucose, according to the following reaction (Eq. 1):

$$C_6H_{12}O_6(l) + 6\ H_2O(l) = 6\ CO_2(g) + 12\ H_2(g) \quad (1)$$

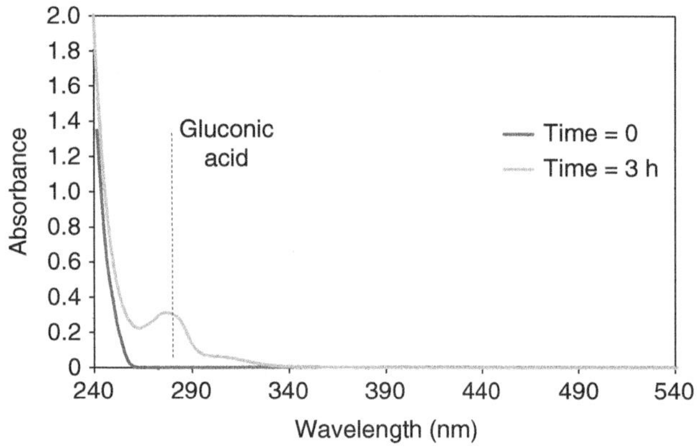

Figure 8

Spectrophotometric analysis of the aqueous solution before and after treatment with the 0.5Pd120 catalyst.

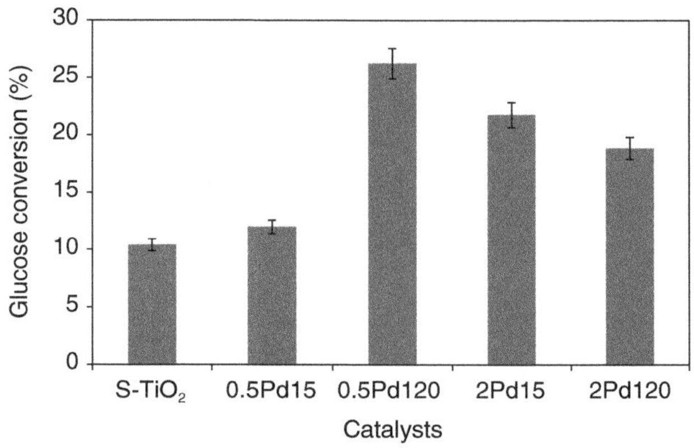

Figure 9

Glucose conversion after photocatalytic treatment by S-TiO_2 and Pd-TiO_2 catalysts (3 h irradiation time).

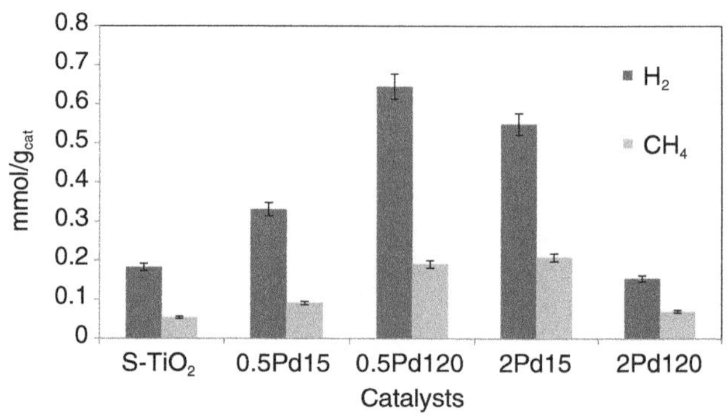

Figure 10

H_2 and CH_4 production after photocatalytic treatment by S-TiO_2 and Pd-TiO_2 catalysts (3 h irradiation time).

The deposition of Pd (0.5Pd120) resulted in an increased methane production (7.58 μmol) and yield (2%) compared with the S-TiO_2 catalyst, for which CH_4 production and yield were about 2.8 μmol and 0.6%, respectively (Fig. 7c).

In the absence of oxygen, methane could result from the degradation of glucose through the following reaction (Eq. 2) [44]:

$$C_6H_{12}O_6(l) = 3\ CH_4(g) + 3\ CO_2(g) \qquad (2)$$

Since CO_2 mainly comes from the photocatalytic wet reforming of glucose (Eq. 1), as confirmed by the molar ratio H_2/CO_2 being equal to 2, CH_4 could be formed through the following reaction (Eq. 3), similar to anaerobic digestion:

$$C_6H_{12}O_6(l) + 6\ H_2O(l) = 6\ CH_4(g) + 6\ O_2(g) \qquad (3)$$

This result is confirmed by the analysis of the liquid phase after 3 hours of irradiation, which showed the presence of gluconic acid (Fig. 8), a typical product of glucose oxidation [45]. Therefore, the oxygen produced according to Equation (3) possibly reacts with glucose, leading to the formation of gluconic acid. The presence of the latter was detected by spectrophotometric analysis of the liquid sample whose result is reported in Figure 8; gluconic acid is represented by a band with a peak at a wavelength of 254 nm [45]. The presence of any intermediate compounds that have been produced by the oxidation of glucose, such as gluconic acid, is important because they may possibly be recovered and used for different purposes (food, pharmaceutical and hygienic products) [45].

Glucose conversions observed during S-TiO_2 and Pd-TiO_2 photocatalytic treatment after 3 hours of irradiation are shown in Figure 9. The photocatalytic activity of Pd-TiO_2 was influenced by the photodeposition time and by the content of palladium used in the synthesis of the photocatalysts. The increase in the photodeposition time up to 120 min had a positive effect on the TiO_2 photoactivity for 0.5 wt% of Pd loading. Glucose conversion values obtained with the catalysts with 2 wt% of Pd content were lower than those obtained with the catalysts with 0.5 wt% of Pd loading, for both deposition times investigated. In particular, the best catalyst for the conversion of glucose was 0.5Pd120, for which the glucose conversion was as high as 26%.

This result is in agreement with the production of hydrogen and methane (Fig. 10); the highest production of H_2 and CH_4 was observed for the catalyst 0.5Pd120. In particular, on this catalyst it is possible to obtain 0.65 mmol/g_{cat} of H_2 and 0.18 mmol/g_{cat} of CH_4.

It is very important to note that the hydrogen production is very similar to that obtained by using a UV lamp with a nominal power of 125 W [29].

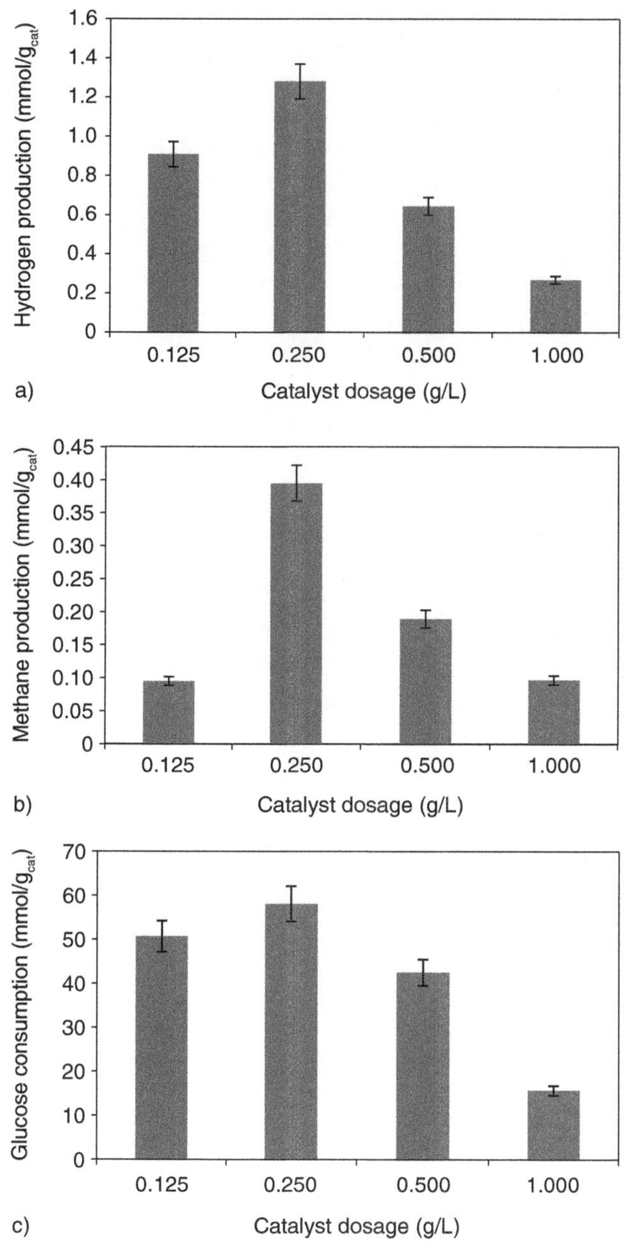

Figure 11

Optimization of catalyst loading according to hydrogen a) and methane b) production, and glucose consumption c).

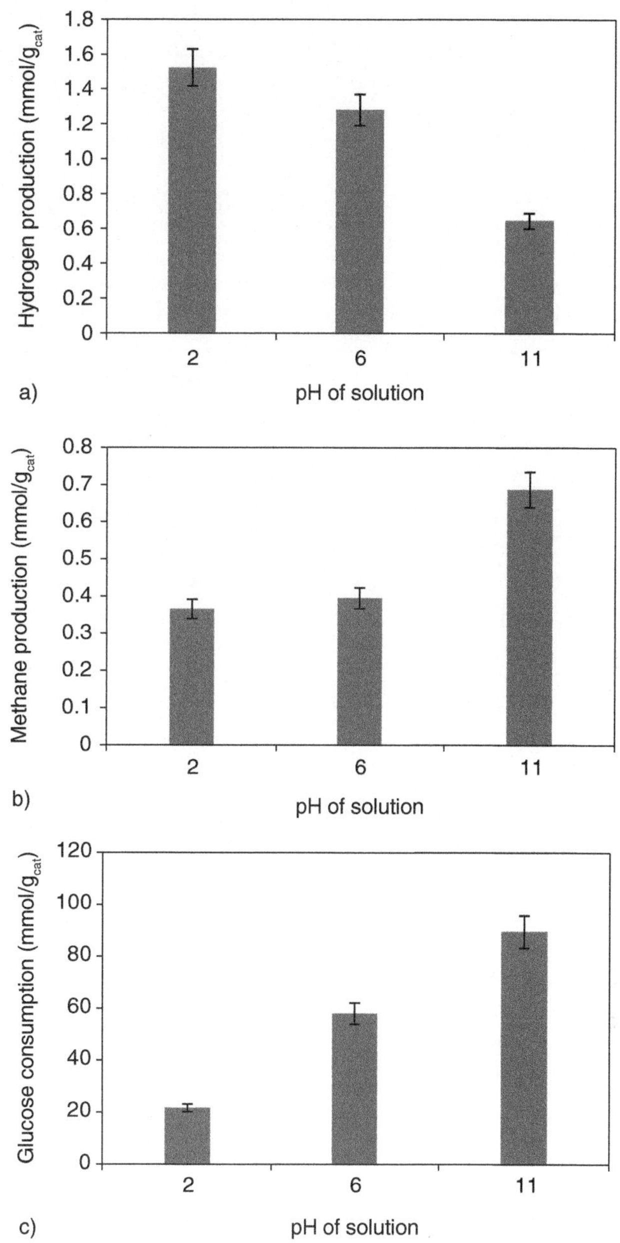

Figure 12

Effect of the initial pH on hydrogen a) and methane b) production, and glucose consumption c).

The behavior observed could be explained taking into account that in the analyzed photocatalysts partially oxidized Pd species and metallic Pd were detected. These species could be responsible for the differences observed in the glucose conversion and hydrogen production in the glucose reforming over Pd-TiO_2 photocatalysts. The catalyst prepared with 0.5 wt% Pd loading and 120 min of photodeposition time resulted in the highest glucose conversion and H_2 production. The highest performances are possibly related to a homogeneous Pd nanoparticle distribution over the TiO_2 surface with sizes lower than 4 nm (as observed by TEM analysis) and Pd^0 oxidation state (according to XPS analysis, metal preferentially occurs in its metallic state (Pd^0)).

For the catalysts with a Pd content higher than 0.5 wt%, the presence of $Pd^{\delta+}$ species on the TiO_2 surface could induce a preferential Pd photodeposition on these species, not over TiO_2, leading to an increase in the Pd particle size. Moreover, the metal oxidation state can modify the adsorption of glucose on the photocatalyst surface, thus determining a different catalytic activity.

2.2.2 Optimization of Catalyst Dosage for Photocatalytic Tests

The optimization of the catalyst dosage was carried out under UV irradiation by testing different concentrations of 0.5Pd120 photocatalysts, in the range 0.125-1 g/L. Photocatalytic efficiency increased as catalyst loading was increased up to 0.25 g/L. As catalyst loading was further increased, the conversion of glucose and the simultaneous formation of H_2 and CH_4 decreased (Fig. 11).

Possibly, the increase in the catalyst dosage over the optimum value resulted in a decreased light penetration through the solution because of the increased opacity of the aqueous suspension [46].

2.2.3 Influence of Initial pH of Solution

Finally, the effect of the pH on the production of hydrogen (Fig. 12a) and methane (Fig.12b) as well as on the conversion of glucose (Fig. 12c) was evaluated. The specific conversion of glucose increased as the solution pH was increased. The effects of pH on hydrogen generation are considered to be quite complex, involving the changes of the chemical state of glucose, redox potential of H^+/H_2 and relative position of the band edges of the semiconductor [47]. At lower pH values (pH = 3), the redox potential of H^+/H_2 would become more positive, which is advantageous for efficient hydrogen generation [47]. However, the most favorable condition for methane production is alkaline solution. Accordingly, the highest methane production was observed at pH 11 (about twice that at pH 3). Possibly, when the redox potential of H^+/H_2 becomes more negative and the pH of the solution is similar to the pk_a of glucose, the methane production is enhanced.

CONCLUSIONS

Pd deposition onto the sulfated TiO_2 surface throughout the photodeposition time was favorable to the production of H_2 or CH_4 by photocatalysis. These compounds were produced efficiently by photocatalytic reforming of aqueous solutions of glucose. The experimental conditions during photodeposition, such as deposition time, were found to have a strong influence on the final properties of the materials, and consequently on their photocatalytic activity in glucose reforming. A short deposition time (15 min) led to a smaller average particle size and higher dispersion of the palladium on the sulfated TiO_2 surface, as well as a higher fraction of the metal in its metallic state. These features resulted in a good photocatalytic performance in terms of glucose conversion. The best photocatalytic behavior was observed in the sample prepared with 0.5 wt% of palladium and 120 min of photodeposition time. By changing the pH of the solution, it is possible to modulate the performance of the photocatalytic reaction, producing more methane in alkaline conditions or more hydrogen in acidic conditions.

The process appears to be a promising, cost-effective method (also considering the high energy-efficiency lighting by UVA-LED) for the treatment of sugar water from industrial food processes, with a parallel recovery of products (CH_4 and H_2) which could be successfully used for energy production.

ACKNOWLEDGMENTS

J.J. Murcia would like to thank the CSIC for the concession of a JAE grant. CITIUS (University of Seville) is acknowledged for XPS and XRF measurements.
Diana Sannino wishes to thank the University of Salerno for funding the project "ORSA111873; *Processi chimici catalitici per la produzione di energia sostenibile e l'ambiente*".

REFERENCES

1 Turner J.A. (2004) Sustainable Hydrogen Production, *Science* **305**, 5686, 972-974.

2 Fu X., Long J., Wang X., Leung D.Y.C., Ding Z., Wu L., Zhang Z., Li Z., Fu X. (2008) Photocatalytic reforming of biomass: A systematic study of hydrogen evolution from glucose solution, *International Journal of Hydrogen Energy* **33**, 22, 6484-6491.

3 Kondarides D.I., Patsoura A., Verykios X.E. (2010) Anaerobic Photocatalytic Oxidation of Carbohydrates in Aqueous Pt/TiO_2 Suspensions with Simultaneous Production of Hydrogen, *J. Adv. Oxid. Technol.* **13**, 1, 116-123.

4 Azadi P., Otomo J., Hatano H., Oshima Y., Farnood R. (2010) Hydrogen production by catalytic near-critical water gasification and steam reforming of glucose, *International Journal of Hydrogen Energy* **35**, 8, 3406-3414.

5 Bičáková O., Straka P. (2012) Production of hydrogen from renewable resources and its effectiveness, *International Journal of Hydrogen Energy* **37**, 16, 11563-11578.

6 Moreno T., Kouzaki G., Sasaki M., Goto M., Cocero M.J. (2012) Uncatalysed wet oxidation of d-glucose with hydrogen peroxide and its combination with hydrothermal electrolysis, *Carbohydrate Research* **349**, 33-38.

7 Prüße U., Herrmann M., Baatz C., Decker N. (2011) Gold-catalyzed selective glucose oxidation at high glucose concentrations and oxygen partial pressures, *Applied Catalysis A: General* **406**, 1-2, 89-93.

8 Foster A.J., Jae J., Cheng Y.-T., Huber G.W., Lobo R.F. (2012) Optimizing the aromatic yield and distribution from catalytic fast pyrolysis of biomass over ZSM-5, *Applied Catalysis A: General* **423-424**, 154-161.

9 Hallenbeck P.C., Abo-Hashesh M., Ghosh D. (2012) Strategies for improving biological hydrogen production, *Bioresource Technology* **110**, 1-9.

10 Ntaikou I., Gavala H.N., Kornaros M., Lyberatos G. (2008) Hydrogen production from sugars and sweet sorghum biomass using Ruminococcus albus, *International Journal of Hydrogen Energy* **33**, 4, 1153-1163.

11 Lee H.S., Salerno M.B., Rittmann B.E. (2008) Thermodynamic evaluation on H_2 production in glucose fermentation, *Environmental Science & Technology* **42**, 7, 2401-2407.

12 McGinley J., McHale F.N., Hughes P., Reid C.N., McHale A.P. (2004) Production of electrical energy from carbohydrates using a transition metal-catalysed liquid alkaline fuel cell, *Biotechnology Letters* **26**, 23, 1771-1776.

13 Kawai T., Sakata T. (1980) Conversion of carbohydrate into hydrogen fuel by a photocatalytic process, *Nature* **286**, 5772, 474-476.

14 Bahruji H., Bowker M., Davies P.R., Al-Mazroai L.S., Dickinson A., Greaves J., James D., Millard L., Pedrono F. (2010) Sustainable H_2 gas production by photocatalysis, *Journal of Photochemistry and Photobiology A: Chemistry* **216**, 2-3, 115-118.

15 Song C. (2002) Fuel processing for low-temperature and high-temperature fuel cells: Challenges, and opportunities for sustainable development in the 21st century, *Catal. Today* **77**, 1-2, 17-49.

16 Hamelinck C.N., Faaij A.P.C. (2002) Future prospects for production of methanol and hydrogen from biomass, *Journal of Power Sources* **111**, 1, 1-22.

17 Bridgwater A.V. (1994) Catalysis in thermal biomass conversion, *Applied Catalysis A: General* **116**, 1-2, 5-47.

18 Huber G.W., Shabaker J.W., Dumesic J.A. (2003) Raney Ni-Sn catalyst for H_2 production from biomass-derived hydrocarbons, *Science* **300**, 5628, 2075-2077.

19 Vaiano V., Sacco O., Sannino D., Ciambelli P., Longo S., Venditto V., Guerra G. (2014) N-doped TiO_2/s-PS aerogels for photocatalytic degradation of organic dyes in wastewater under visible light irradiation, *J. Chem. Technol. Biotechnol.* **89**, 8, 1175-1181.

20 Vaiano V., Sacco O., Stoller M., Chianese A., Ciambelli P., Sannino D. (2014) Influence of the photoreactor configuration and of different light sources in the photocatalytic treatment of highly polluted wastewater, *Int. J. Chem. React. Eng.* **12**, 1, 1-13.

21 Mohamed R.M., Aazam E.S. (2012) H_2 Production with Low CO Selectivity from Photocatalytic Reforming of Glucose on Ni/TiO_2-SiO_2, *Chinese Journal of Catalysis* **33**, 2-3, 247-253.

22 Gomathisankar P., Yamamoto D., Katsumata H., Suzuki T., Kaneco S. (2013) Photocatalytic hydrogen production with aid of simultaneous metal deposition using titanium dioxide from aqueous glucose solution, *International Journal of Hydrogen Energy* **38**, 14, 5517-5524.

23 Zhang L., Shi J., Liu M., Jing D., Guo L. (2014) Photocatalytic reforming of glucose under visible light over morphology controlled Cu_2O: Efficient charge separation by crystal facet engineering, *Chemical Communications* **50**, 2, 192-194.

24 Carraro G., Maccato C., Gasparotto A., Montini T., Turner S., Lebedev O.I., Gombac V., Adami G., Van Tendeloo G., Barreca D., Fornasiero P. (2014) Enhanced Hydrogen Production by Photoreforming of Renewable Oxygenates Through Nanostructured Fe_2O_3 Polymorphs, *Advanced Functional Materials* **24**, 3, 372-378.

25 Kondarides D., Daskalaki V., Patsoura A., Verykios X. (2008) Hydrogen Production by Photo-Induced Reforming of Biomass Components and Derivatives at Ambient Conditions, *Catal Lett.* **122**, 1-2, 26-32.

26 Linsebigler A.L., Lu G., Yates J.T. (1995) Photocatalysis on TiO_2 Surfaces: Principles, Mechanisms, and Selected Results, *Chemical Reviews* **95**, 3, 735-758.

27 Chong R., Li J., Ma Y., Zhang B., Han H., Li C. (2014) Selective conversion of aqueous glucose to value-added sugar aldose on TiO_2-based photocatalysts, *Journal of Catalysis* **314**, 101-108.

28 Wu G., Chen T., Zhou G., Zong X., Li C. (2008) H_2 production with low CO selectivity from photocatalytic reforming of glucose on metal/TiO_2 catalysts, *Sci. China Ser. B-Chem.* **51**, 2, 97-100.

29 Colmenares J.C., Magdziarz A., Aramendia M.A., Marinas A., Marinas J.M., Urbano F.J., Navio J.A. (2011) Influence of the strong metal support interaction effect (SMSI) of Pt/TiO_2 and Pd/TiO_2 systems in the photocatalytic biohydrogen production from glucose solution, *Catalysis Communications* **16**, 1, 1-6.

30 Passio L., Rizzoa L., Fuchs S. (2012) Two-phase anaerobic digestion of partially acidified sewage sludge: a pilot plant study for safe sludge disposal in developing countries, *Environmental technology* **33**, 16-18, 2089-2095.

31 Li Y., Park S.Y., Zhu J. (2011) Solid-state anaerobic digestion for methane production from organic waste, *Renewable and Sustainable Energy Reviews* **15**, 1, 821-826.

32 Sakata T., Kawai T. (1981) Heterogeneous photocatalytic production of hydrogen and methane from ethanol and water, *Chemical Physics Letters* **80**, 2, 341-344.

33 Colón G., Hidalgo M.C., Navío J.A. (2003) Photocatalytic behaviour of sulphated TiO_2 for phenol degradation, *Applied Catalysis B* **45**, 1, 39-50.

34 Maicu M., Hidalgo M.C., Colón G., Navío J.A. (2011) Comparative study of the photodeposition of Pt, Au and Pd on pre-sulphated TiO_2 for the photocatalytic decomposition of phenol, *Journal of Photochemistry and Photobiology A: Chemistry* **217**, 2-3, 275-283.

35 Hidalgo M.C., Murcia J.J., Navío J.A., Colón G. (2011) Photodeposition of gold on titanium dioxide for photocatalytic phenol oxidation, *Applied Catalysis A: General* **397**, 1-2, 112-120.

36 Murcia J.J., Hidalgo M.C., Navío J.A., Araña J., Doña-Rodríguez J.M. (2014) Correlation study between photodegradation and surface adsorption properties of phenol and methyl orange on TiO_2 Vs platinum-supported TiO_2, *Applied Catalysis B* **150-151**, 107-115.

37 DuBois M., Gilles K.A., Hamilton J.K., Rebers P.A., Smith F. (1956) Colorimetric Method for Determination of Sugars and Related Substances, *Analytical Chemistry* **28**, 3, 350-356.

38 Albalasmeh A.A., Berhe A.A., Ghezzehei T.A. (2013) A new method for rapid determination of carbohydrate and total carbon concentrations using UV spectrophotometry, *Carbohydrate Polymers* **97**, 2, 253-261.

39 Navaladian S., Viswanathan B., Varadarajan T.K., Viswanath R.P. (2008) A Rapid Synthesis of Oriented Palladium Nanoparticles by UV Irradiation, *Nanoscale Research Letters* **4**, 2, 181-186.

40 Brun M., Berthet A., Bertolini J.C. (1999) XPS, AES and Auger parameter of Pd and PdO, *Journal of Electron Spectroscopy and Related Phenomena* **104**, 1-3, 55-60.

41 Siemon U., Bahnemann D., Testa J.J., Rodríguez D.Litter M.I. Bruno N. (2002) Heterogeneous photocatalytic reactions comparing TiO_2 and Pt/TiO_2, *Journal of Photochemistry and Photobiology A: Chemistry* **148**, 1-3, 247-255.

42 Bamwenda G.R., Tsubota S., Nakamura T., Haruta M. (1995) Photoassisted hydrogen production from a water ethanol solution a comparison of activities of Au-TiO_2 and Pt-TiO_2, *Journal of Photochemistry and Photobiology A: Chemistry* **89**, 2, 177-189.

43 Liu Y., Guo L., Yan W., Liu H. (2006) A composite visible-light photocatalyst for hydrogen production, *Journal of Power Sources* **159**, 2, 1300-1304.

44 Kenney J.F., Kutcherov V.A., Bendeliani N.A., Alekseev V.A. (2002) The evolution of multicomponent systems at high pressures: VI. The thermodynamic stability of the hydrogen–carbon system: The genesis of hydrocarbons and the origin of petroleum, *Proceedings of the National Academy of Sciences* **99**, 17, 10976-10981.

45 Ramachandran S., Fontanille P., Pandey A., Larroche C. (2006) Gluconic acid: Properties, applications and microbial production, *Food Technology and Biotechnology* **44**, 2, 185-195.

46 Sannino D., Vaiano V., Sacco O., Ciambelli P. (2013) Mathematical modelling of photocatalytic degradation of methylene blue under visible light irradiation, *Journal of Environmental Chemical Engineering* **1**, 1-2, 56-60.

47 Jing D., Liu M., Shi J., Tang W., Guo L. (2010) Hydrogen production under visible light by photocatalytic reforming of glucose over an oxide solid solution photocatalyst, *Catalysis Communications* **12**, 4, 264-267.

13

Large-Eddy Simulation (LES) of Spray Transients: Start and End of Injection Phenomena

Michele Battistoni[1,2*], Qingluan Xue[1] and Sibendu Som[1]

[1] *Argonne National Laboratory, Chicago, IL - USA*
[2] *University of Perugia, 67 via G. Duranti, 06125 Perugia - Italy*
e-mail: michele.battistoni@unipg.it

* Corresponding author

Abstract — *This work reports investigations on Diesel spray transients, accounting for internal nozzle flow and needle motion, and demonstrates how seamless calculations of internal flow and external jet can be accomplished in a Large-Eddy Simulation (LES) framework using an Eulerian mixture model. Sub-grid stresses are modeled with the Dynamic Structure (DS) model, a non-viscosity based one-equation LES model. Two problems are studied with high level of spatial and temporal resolution. The first one concerns an End-Of-Injection (EOI) case where gas ingestion, cavitation, and dribble formation are resolved. The second case is a Start-Of-Injection (SOI) simulation that aims at analyzing the effect of residual gas trapped inside the injector sac on spray penetration and rate of fuel injection. Simulation results are compared against experiments carried out at Argonne National Laboratory (ANL) using synchrotron X-ray. A mesh sensitivity analysis is conducted to assess the quality of the LES approach by evaluating the resolved turbulent kinetic energy budget and comparing the outcomes with a length-scale resolution index. LES of both EOI and SOI processes have been carried out on a single hole Diesel injector, providing insights in to the physics of the processes, with internal and external flow details, and linking the phenomena at the end of an injection event to those at the start of a new injection. Concerning the EOI, the model predicts ligament formation and gas ingestion, as observed experimentally, and the amount of residual gas in the nozzle sac matches with the available data. The fast dynamics of the process is described in detail. The simulation provides unique insights into the physics at the EOI. Similarly, the SOI simulation shows how gas is ejected first, and liquid fuel starts being injected with a delay. The simulation starts from a very low needle lift and is able to predict the actual Rate-Of-Injection (ROI) and jet penetration, based only on the prescribed needle motion. Finally, guidelines and future improvements of the model are discussed concerning the simulation of the transient injection phases.*

Résumé — **Simulations aux grandes échelles (*Large-Eddy Simulation*, LES) de transitoires de jets liquides : phénomènes de début et fin d'injection** — Le présent article rend compte d'études sur les transitoires de jets Diesel, prenant en compte l'écoulement interne de l'injecteur et le mouvement de l'aiguille, et démontre comment des calculs d'écoulement interne et de jet peuvent être réalisés dans le cadre d'une approche de type simulation aux grandes échelles (*Large-Eddy Simulation*, LES) en utilisant un modèle eulérien de mélange. Les contraintes sous-mailles sont modélisées avec un modèle de Structure Dynamique (*Dynamic Structure*, DS), un modèle LES à une équation non basé sur une viscosité turbulente. Deux problèmes sont étudiés avec un niveau élevé de résolution spatiale et temporelle. Le premier concerne un cas de fin d'injection (*End-Of-Injection*, EOI), où l'ingestion de gaz, la cavitation, et la formation de goutte à goutte sont résolues. Le second cas est une simulation de départ d'injection (*Start-Of-Injection*, SOI) qui vise à analyser l'effet du gaz résiduel piégé à l'intérieur du sac d'injecteur sur la pénétration du jet et le taux d'introduction de carburant. Les résultats de simulation sont comparés à des expériences réalisées à *Argonne National Laboratory* (ANL) en utilisant des rayons X synchrotron. Une analyse de sensibilité au maillage est réalisée pour évaluer la qualité de l'approche LES en évaluant le budget de l'énergie cinétique résolue et en comparant les résultats à l'aide d'un critère de résolution d'échelle de turbulence. Des LES de processus de fin et de début d'injection ont été réalisées sur un injecteur monotrou Diesel, donnant ainsi un aperçu de la physique des phénomènes de ces processus, avec des détails sur l'écoulement interne et externe, et liant les phénomènes de fin d'un événement d'injection à ceux du début d'une nouvelle injection. Concernant la fin d'injection, le modèle prédit la formation de ligaments et l'ingestion de gaz, comme observé expérimentalement, et la quantité de gaz résiduel dans le sac de l'injecteur est conforme aux données disponibles. La dynamique rapide du processus est décrite en détails. La simulation donne un aperçu unique de la physique en fin d'injection. De même, la simulation du début d'injection montre comment le gaz est tout d'abord éjecté et le carburant liquide commence à être injecté après un délai. La simulation débute à une levée d'aiguille très faible, et est capable de prédire le taux d'introduction (*Rate-Of-Injection*, ROI) effectif et la pénétration du jet, à partir de la seule prescription du mouvement de l'aiguille. Enfin, des directives et améliorations futures du modèle concernant la simulation des phases transitoires d'injection sont discutées.

NOMENCLATURE

ANL	Argonne National Laboratory
DS	Dynamic Structure
ECN	Engine Combustion Network
EOI	End-Of-Injection
HRM	Homogeneous Relaxation Model
LES	Large-Eddy Simulation
LSR	Length Scale Resolution
ROI	Rate-Of-Injection
SGS	Sub-Grid Scale
SOI	Start-Of-Injection

INTRODUCTION

Research on fuel injection technology, sprays and combustion is continuously motivated by the increase in the global energy demand, especially for the transportation sector. Improvements in combustion engine efficiencies, the introduction of alternative fuel blends and the increase of energy demand for commercial transportation continue to propel research in the area of both light and heavy duty vehicles.

Direct injection engines, in the quest for higher efficiency and pollutant reduction, are making use of higher and higher injection pressures, up to 3 000 bar for compression ignition engines. At such elevated pressures, the flow develops high velocities within the injector that can lead to cavitation (Arcoumanis *et al.*, 2000). Fast transients due to short injection durations are also of utmost importance as both direct injected gasoline and Diesel engines often adopt multiple injections, and possibly injection rate shaping, to cope with emission limits and fuel efficiency targets. Under these conditions, the transient part of spray behavior is most of the time the dominant part of an injection event. In Diesel common rail systems, commanded pulse duration can easily be as short as 200 μs, for pilot or post injections, even targeting injected mass as low as 1 mg/shot. Furthermore, the minimum dwell time between two consecutive injections has been progressively reduced, approaching nowadays 200 μs for solenoid injectors. As near nozzle region is concerned, most of the research has been devoted to the steady state part of a fuel jet, focusing on the steady liquid break-up, air entrainment, and mixing. Little information is available on the transient behavior during both opening and closing phases.

The shape of injection ramps affects spray penetration and mixing process, especially with short multiple injection strategies (Pickett *et al.*, 2013). It has been shown that after an injection pulse fuel dribble can be produced, and a significant amount of ambient gas is also ingested in the nozzle sac depending on factors like ambient back pressure, hole sizes, etc. (Swantek *et al.*, 2013, 2014; Battistoni *et al.*, 2014b; Eagle and Musculus, 2014). The start of a new injection event is consequently affected by the presence of residual gas in the sac. During the first 100 μs after SOI liquid penetration is shorter and evaporation rate is altered (Pickett *et al.*, 2013). Additionally, droplets dibbled after the EOI create locally fuel-rich pockets within an overall fuel-lean region near the injector, the combination of which may contribute to soot and/or unburned hydrocarbon emissions (Musculus *et al.*, 2013).

In the area of nozzle flow and sprays, the large variability of scales makes the two-phase flow modeling particularly difficult and computationally demanding. Some numerical methods track the liquid-gas interface explicitly like Level-Set (LS) or some formulations of Volume Of Fluid (VOF) methods, other methods do not pursue interface surface reconstruction (Gorokhovski and Herrmann, 2008). The first group of models requires very fine grids and large computational resources since it aims at directly resolving the mechanics of the liquid backup and drop formation. These approaches generally assume incompressible phases, no mass transfer due to phase change, relatively low Reynolds numbers, and generally limited to only external jet domain (Arienti and Sussman, 2014; Bianchi *et al.*, 2007; Bode *et al.*, 2014). To solve all scales these methods rely on extreme fine grids as no turbulence models can be easily incorporated. In addition, downstream they can be coupled with Lagrangian particle-tracking models to further simulate the motion of the smallest drops in the dilute spray region. The inclusion of complex geometries or moving boundaries is generally quite challenging, therefore these simulations, based on first-principles with high-fidelity, are limited to the external nozzle region, and require an inflow boundary derived from other type of in-nozzle flow calculations.

The second group of methods uses a diffuse-interface modeling approach. Indeed the liquid breakup is not directly resolved but it is modeled as diffusion, still thorough the aid of a fine supporting grid. These approaches offer the advantages of easy inclusion of the energy equation, with complex equation of states for the fluids, mass transfer due to phase change and compressibility. They generally allow the modeling of complex and moving geometries at high Reynolds numbers. Turbulence is closed either with Reynolds Average Navier-Stokes (RANS) for relatively inexpensive computations or with LES. All these approaches are Eulerian in nature, and can be formulated as single-fluid (Oefelein *et al.*, 2014; Schmidt *et al.*, 2010; Demoulin *et al.*, 2007; Xue *et al.*, 2015) or multi-fluid (Alajbegovic, 1999; Wang *et al.*, 2014) two-phase models. Eulerian models can be coupled with Lagrangian particle-tracking models in the dilute spray regions as well, resulting in the method termed the "Eulerian Lagrangian Spray and Atomization (ELSA)" model (Demoulin *et al.*, 2007; Vallet *et al.*, 2001). The modeling of phase change is also particularly relevant within these approaches. First principles can never be applied to resolve locally the dynamics of single bubbles, therefore simplified models based on Rayleigh-Plesset equation or on the homogeneous assumption (either equilibrium or non-equilibrium) are generally employed (Wang and Greif, 2006; Habchi, 2014). Significant improvements have been demonstrated if non-condensable gases are accounted for in the multi-phase model (Battistoni *et al.*, 2014a).

In this paper, we present an Eulerian method to simulate continuously the internal nozzle flow and near exit jet region, accounting for the needle motion, liquid compressibility, phase change due to cavitation and presence of non-condensable gas. The high pressure fuel flows through the injector needle valve, then through the orifice and is injected in an initially quiescent ambient gas. In this method, both the liquid and gas phases are solved in the same Eulerian coordinate system using a single fluid representation. The interface between these two phases is not tracked explicitly, but it is resolved only using the grid. The turbulence effects are modeled performing LES with the Dynamic Structure (DS) sub-grid scale model (Pomraning and Rutland, 2002).

Due to the high Reynolds number involved in practical Diesel injections (*Re* is greater than 500 000 in the present cases) addressing these problems is extremely challenging even using High Performance Computing (HPC) resources. In addition, because of the large differences in the length and time scales involved in the internal flow compared to the outward spray, the concurrent simulation of the internal and external domains is extremely demanding. This article presents an analysis of injection transient phenomena by means of high resolution LES, and, recognizing these challenges, part of the work focuses on monitoring the quality of these scale-resolving simulations and on quantifying the amount of resolved scales for assessing the model applicability. Despite the modeling challenges, the rest of the article highlights the physics of the injection transients in detail, and simulation results are compared with experimental data.

The remainder of the article is organized as follows. First, SOI and EOI phenomena are briefly reviewed. Then the Eulerian multi-phase model and the LES framework are described. Next, EOI and SOI cases are presented as applications. Results are compared against available synchrotron X-ray data and mesh requirements are discussed. Finally, the last section concludes the paper.

1 REVIEW

1.1 End of Injection

Yu *et al.* (1983) conducted a pioneering study showing that EOI rate shape, sac volume and nozzle geometry significantly affect the production of unburned hydrocarbons in Diesel engines. They identified local over-mixing, poor end of injection and fuel emptying from the sac as important sources of hydrocarbon production. Optical measurements of Diesel injection showed that the mixing between fuel vapor and air is enhanced during the decelerating period after the end of injection, leading to low local fuel concentration (Bruneaux, 2005). Musculus (2009) showed that

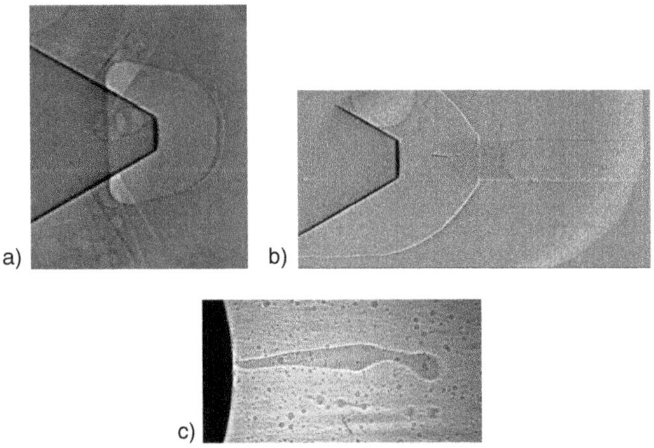

Figure 1

Synchrotron X-ray phase contrast imaging on EOI. a) Residual bubbles on a multi-hole nozzle; b) residual bubbles on a single hole nozzle; c) dribbles out of a single hole nozzle. (Images reproduced from Swantek *et al.*, 2013, 2014).

fuel-lean regions that form during the ignition dwell after the end of injection are likely a significant source of unburned hydrocarbons. A region of increased entrainment was identified. He also suggested that the deceleration rate could be shaped to achieve a desired mixing level after the end of a fuel pulse. It has also been shown that the EOI behavior affects soot formation and oxidation both for conventional Diesel combustion and Low Temperature Combustion (LTC) conditions (Musculus *et al.*, 2013).

To the best of our knowledge, few attempts have been made on simulating end of injection process within the injector and in the near nozzle region. Hu *et al.* (2010) conducted a LES of a transient air jet focusing on the EOI behavior in the external jet region. They suggested that the deceleration of the jet enhances flow instabilities, increasing small-scale turbulence and causes higher entrainment.

Only recently, in regard to the EOI behavior, the focus has shifted toward the near nozzle region and the internal nozzle flow. Research at ANL using different X-ray diagnostic techniques has revealed new features of the EOI behavior (Kastengren *et al.*, 2008). Figure 1 shows examples of ingested bubbles and dribble formation after an injection event for: a) a seven hole nozzle and, b), c) two single hole nozzles. Images are obtained using synchrotron X-ray phase contrast technique on metal Diesel injectors for the internal flow and the outside region. These results show that at the end of an injection pulse, rapid deceleration of the liquid jet eventually leads to its break up, forming poorly atomized spray and ligaments injected after the needle closes (dribbles). Correspondingly, a significant amount of ambient gas can be entrained in the orifice and in the sac of the injector nozzle, as the needle valve closes. Gas bubbles have been observed inside metal nozzles, and the volume of residual void after the injection process has been quantified for some injectors.

However, the details of the physics of this process still need to be fully understood. The dynamics of liquid ligament formation, and gas ingestion are very fast processes. Numerical simulations can help identifying the key parameters of the process, providing information in terms of pressure and void fraction evolutions inside the nozzle. These data should help answering questions like the following. Are internal bubbles due to local cavitation or due to gas ingested from the chamber? Is there a link between the extent of bubbles and the amount of external dribbles?

In this paper, numerical simulation were performed to characterize the EOI process in single hole injectors. Primarily, the focus is on the last part of the injection, encompassing the needle closure and extending for a large amount of time after it.

1.2 Start of Injection

Multiphase flow is generally expected in fuel injectors due to cavitation, but also due to ingestion of ambient gas in the nozzle sac after an injection event, as discussed in the previous section. The sac can be partially empty at the start of injection. The presence of hot ambient gas in the sac can be responsible for delayed liquid injection and higher liquid vaporization thereafter (Crua *et al.*, 2010; Pickett *et al.*, 2013).

Crua *et al.* (2010) and Stetsyuk *et al.* (2014) observed Diesel jets in hot conditions using ultra-high speed video and found that the initial stage of fuel injection is more complex than the accepted linear temporal evolution. A gaseous pre-jet, with spheroidal cap shape, was observed for a wide range of conditions and issues sub-sonically into hot and still compressed air. It may consist of cavitation pockets expanded after previous injection; ingested in-cylinder gases after previous injection; heated and vaporized fuel trapped in the injector holes between injections. The high speed liquid fuel jet initially trails the gaseous cloud but later its tip proceeds further ahead. This process is observed to occur within the first 1-2 mm of jet penetration.

Pickett *et al.* (2013) reports that in vaporizing conditions the liquid penetration of the Engine Combustion Network (ECN) spray A can be about 2 mm shorter in the first 200 μs than the steady state value. The behavior is attributed to the initial presence of residual gas or fuel vapor in the sac, which enhances vaporization rate of the first-injected fluid, as the gas trapped from the chamber in the sac is most likely warmer than the incoming liquid fuel. This also suggests that this issuing fuel can be ignited by the hot air mixed with remaining burnt products in the chamber.

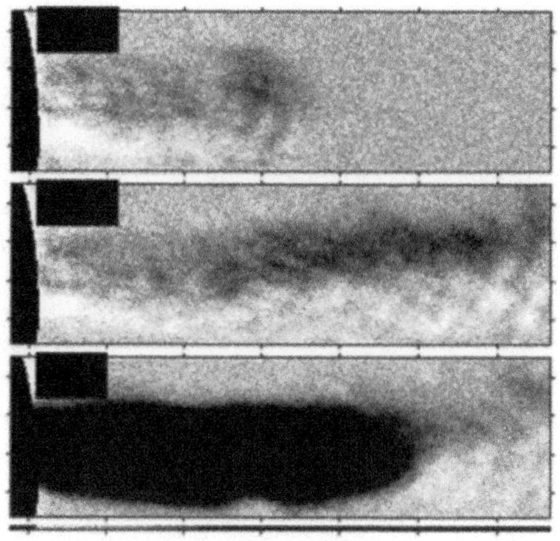

Figure 2

Microscopy high-speed imaging during the start of injection showing gas being injected before liquid SOI. (Image reproduced from Pickett *et al.*, 2013).

2 SIMULATION METHOD

2.1 Governing Equations

The compressible Navier-Stokes equations, which express the conservation of mass, momentum, energy and species, for a single-fluid mixture read as follows:

$$\frac{\partial \rho}{\partial t} + \frac{\partial \rho u_j}{\partial x_j} = 0 \tag{1}$$

$$\frac{\partial \rho u_i}{\partial t} + \frac{\partial \rho u_i u_j}{\partial x_j} = -\frac{\partial p}{\partial x_i} + \frac{\partial \sigma_{ij}}{\partial x_j} + F_{sfi} \tag{2}$$

$$\frac{\partial \rho e}{\partial t} + \frac{\partial \rho e u_j}{\partial x_j} = -p\frac{\partial u_j}{\partial x_j} + \sigma_{ij}\frac{\partial u_i}{\partial x_j} + \frac{\partial}{\partial x_j}\left(\lambda \frac{\partial T}{\partial x_j}\right) + \frac{\partial}{\partial x_j}\left(\rho D h_k \frac{\partial Y_k}{\partial x_j}\right) \tag{3}$$

$$\frac{\partial \rho Y_i}{\partial t} + \frac{\partial \rho Y_i u_j}{\partial x_j} = \frac{\partial}{\partial x_j}\left(\rho D \frac{\partial Y_i}{\partial x_j}\right) + \sum_{k,k\neq i} S_{ik} \tag{4}$$

where ρ is the mixture density (or pseudo-density), u is the velocity, p is the pressure, and σ_{ij} is the mixture shear stress tensor assumed to behave as Newtonian. F_{sf} is the surface tension, formulated as a volume force, and evaluated using the hypothesis of continuum surface force (Brackbill *et al.*, 1992). The momentum equation does not include any explicit term for sub-grid phase slip, therefore this contribution is neglected. However, as a consequence of the turbulent mass dispersion closure (cf. first right-hand side term in Eq. 4) a diffusion velocity difference between the two phases does exist (Demoulin *et al.*, 2007). Y_i denotes the mass fraction of species *i*. The energy Equation (3) is formulated in terms of specific internal energy *e* of the mixture. *T* is the mixture temperature, and h_k is the specific enthalpy of species *k*. In addition, λ is the thermal conductivity and *D* is the mass diffusion coefficient expressing the diffusion of a component in the mixture. *D* is assumed constant for each component. With this formulation, summation over *i* of Equation (4) yields global continuity (1). Continuity is solved to obtain the mixture density, and at the same time all the species transport equations are solved for the species mass fractions. A reconciliation procedure at the end of each loop ensures that any discrepancy due to round-off error is adjusted by renormalizing the species mass fractions to satisfy global continuity.

Lastly, S_{ik} are source terms related to species mass transfer due to phase change. Cavitation/condensation sources are calculated according to the Homogeneous Relaxation Model (HRM) of Bilicki and Kestin (1990) and Schmidt *et al.* (2010), described later in Section 2.2.

The multi-phase system is assumed to be a multi-component two-phase mixture comprising of:

- the liquid fuel (subscript 1),
- and multiple gaseous species, *i.e.*:
  - fuel vapor (subscript 2),
  - non-condensable gas mixed with the liquid (subscript 3),
  - other gaseous species, like ambient chamber species, etc. (subscripts 4, etc.).

The mixture has a variable density given by the weighted average of the single components densities. Gas phase densities are determined using the equation of state for an ideal gas:

$$\rho_i = p/R_i T \tag{5}$$

where R_i is the gas constant and *T* is the temperature. Liquid is considered compressible. The following barotropic ($\rho_{liq} = f(p)$) equation of state is used:

$$\rho_{liq} = \rho_{liq,0} \cdot e^{\frac{p-p_0}{K}} \tag{6}$$

where *K* is the liquid bulk modulus, and subscript 0 refers to an arbitrary reference condition.

The sum of all gas species will be referred to with the subscript *g*. The mixture density ρ is expressed through the following equation:

$$\rho = \sum_i (\alpha_i \rho_i) = \alpha_g \rho_g + (1 - \alpha_g)\rho_{liq} \tag{7}$$

where α_i is the volume fraction of each component. Volume fractions α_i and mass fractions Y_i are related through:

$$\rho_i \alpha_i = Y_i \rho \quad (8)$$

(without summation over repeated index). Using Equations (7) and (8), the cumulative void fraction α_g takes the form:

$$\alpha_g = \frac{Y_g/\rho_g}{\sum_i (Y_i/\rho_i)} = \frac{Y_g/\rho_g}{1/\rho} \quad (9)$$

This cumulative void fraction includes voids generated by phase change (cavitation), expansion of non-condensable gases, and mixing with ambient gas.

In this implementation, the void fraction α_g is not calculated with an explicit transport equation, as found in other VOF formulations, but the interface is resolved only using the grid. Individual species mass fraction Y_i of each fluid is tracked throughout the domain, by first solving the related species Equation (4). Then the void fraction is calculated using Equation (9).

This approach results in a diffused-interface modeling. As an example, the liquid break-up is modeled as diffusion and through the aid of a sufficiently fine grid. This approach, compared to geometric interface reconstruction methods usually employed in VOF, offers the advantages of easy coupling with energy and turbulence equations, inclusion of phase change, and modeling of complex and moving geometries for high Reynolds number flows.

2.2 Phase Change

Mass transfer due to phase change can occur at the liquid/gas interface. In general, in a multi-component mixture, it can occur between the liquid fuel and its vapor, or it can potentially occur when gas is released from or absorbed into the liquid solution. Different species are involved in the two cases and different mechanisms are involved.

2.2.1 Liquid-Vapor Mass Transfer Model

The model for mass exchange between liquid and vapor (cavitation) is based on the HRM model as developed by Bilicki and Kestin (1990). The model assumes a simple first order rate equation for the evolution of the instantaneous vapor quality x towards its equilibrium value $\bar{x}$ over a given time-scale Θ. This synthetic approach has been proved to be able to represent not only cavitation but also flash boiling (Schmidt *et al.*, 2010). These are phase change processes driven by a pressure drop. Usually, the process is referred to as 'cavitation' when a tiny amount of superheating is required to evaporate the liquid and the process is very fast, whereas 'flash boiling' is used when a higher degree of superheating is achieved, and the process takes a relatively long time. The HRM model expresses in a synthetic way the extremely complex process by which the two phases exchange heat and mass. The rate equation takes this form:

$$\frac{dx}{dt} = \frac{\bar{x} - x}{\Theta} \quad (10)$$

The instantaneous non-equilibrium vapor quality x (or dryness) is defined as:

$$x = \frac{Y_{vap}}{Y_{vap} + Y_{liq}} = \frac{m_{vap}}{m_{vap} + m_{liq}} \quad (11)$$

The equilibrium vapor quality $\bar{x}$ is a function of the thermodynamic properties at the local pressure, *i.e.*, $\bar{x}(p, h)$ is expressed by the following equation with bounds at zero and unity:

$$\bar{x} = \frac{h - h_{l,sat}}{h_{v,sat} - h_{l,sat}} \quad (12)$$

where h is the actual fluid enthalpy of liquid and vapor mixture, excluding air content, and $h_{l,sat}$ and $h_{v,sat}$ refer to saturated liquid and vapor, respectively. The time-scale Θ is evaluated using an empirical fit proposed by Downar-Zapolski *et al.* (1996):

$$\Theta = 3.84 \cdot 10^{-7} \alpha^{-0.54} \phi^{-1.76} \text{ (units of s)} \quad (13)$$

$$\phi = \left| \frac{p_{sat} - p}{p_{crit} - p_{sat}} \right| \quad (14)$$

where α is the gas phase volume fraction, and subscripts *sat* and *crit* denote saturation and critical pressures, respectively. Correlation provided by combining Equations (13) and (14) has been determined in experiments of flashing flows of water in pipes with upstream pressures of in excess of 10 bar. In the absence of specific values for other liquids and conditions, it has been applied in the present form in this study for representing Diesel fuel. The time integration of the rate Equation (10) allows to compute the source terms S_{12} and S_{21} $(= -S_{12})$ in Equations (4). Additional details on this source term evaluation are given in Battistoni *et al.* (2015).

2.2.2 Liquid-Air Mass Transfer Assumption

The model also account for the presence of non-condensable gas in the mixture (third component), however, the related species mass transfer source term is neglected, *i.e.*, $S_{13} = 0$. Within the time scale of the injector flow, events are assumed to be very rapid for significant mass transfer of contaminant gas to occur between a bubble and the surrounding liquid. The mass transfer of air from the liquid solution to the gaseous form (gaseous cavitation), and *vice versa*, is neglected. Indeed, mass transfer due to diffusion or dissolution is extremely slow compared to the time scales involved in the main flow (Brennen, 1995; Battistoni *et al.*, 2015), especially during SOI or EOI transients.

In the current implementation, the amount of air added to the liquid fuel is introduced directly in gaseous phase, homogeneously mixed with other components. In other words, we have not introduced a specific model for air adsorption and release from the solution, and in Equation (4), the related source term S_{13} is set equal to zero. As a result, as pressure in the fluid drops, air promptly expands, because it is already assumed to be in gaseous form. Conversely, if a cloud of expanded air exists and pressure rises back, gas will be compressed, but not reabsorbed in the liquid medium.

An effective constant amount of non-condensable air in gaseous form is assumed mixed with the liquid. In the current work, the value is set to $Y_3 = 2\times10^{-5}$ by mass, according to previous works (Battistoni *et al.*, 2014a, 2015), and the assumption is based on an analysis of the characteristic time-scale of air mass diffusion in the liquid solution.

2.3 Sub-Grid Models for Filtered Equations

LES have been performed covering the inner nozzle region and the outside near exit region. The code uses a modified cut-cell Cartesian method for generating the grid at run time (Senecal *et al.*, 2007). In this framework, the filter size Δ and shape are fixed by the computational cell size and shape, therefore the filter type is a box.

The set of Equations (1-4) is recast according to the LES decomposition into resolved fields and sub-grid fields, *i.e.*, for any flow variable $\phi = \bar{\phi} + \phi'$. For compressible flows Favre averaging is conveniently applied, yielding $\tilde{\phi} = \overline{\rho\phi}/\bar{\rho}$. The procedure results in additional Sub-Grid Scale (SGS) terms that need to be modeled.

As far as the momentum equation is concerned, following Pomraning and Rutland (2002), the DS model exploits the significant similarity between the SGS and the Leonard stresses that has been observed in real as well as numerical experiments. The sub-grid stress tensor, $\tau_{ij} = \bar{\rho}\left(\widetilde{u_i u_j} - \tilde{u}_i\tilde{u}_j\right)$, in the DS model is provided by:

$$\tau_{ij} = 2k_{sgs}\bar{\rho}\frac{L_{ij}}{L_{kk}}, \quad L_{ij} = \widehat{\tilde{u}_i\tilde{u}_j} - \hat{\tilde{u}}_i\hat{\tilde{u}}_j \tag{15}$$

where the Leonard stress tensor L_{ij} can be calculated from the resolved velocity field. The hat symbol $\widehat{(\cdot)}$ designates the test level filter, which is larger than the grid level filter, and in this implementation it is created from the center cell and the neighboring cells.

The DS model also involves an additional transport equation for the sub-grid turbulent kinetic energy, $k_{sgs} = (\widetilde{u_i u_i} - \tilde{u}_i\tilde{u}_i)/2$, which takes the form:

$$\frac{\partial\bar{\rho}k_{sgs}}{\partial t} + \frac{\partial\bar{\rho}k_{sgs}\tilde{u}_j}{\partial x_j} = -\tau_{ij}\bar{S}_{ij} - \bar{\rho}\varepsilon - \frac{\partial}{\partial x_j}\left(\frac{\mu}{Pr_{sgs}}\frac{\partial k_{sgs}}{\partial x_j}\right) \tag{16}$$

The sub-grid dissipation rate is given by:

$$\varepsilon = \frac{C_\varepsilon k_{sgs}^{1.5}}{\Delta} \tag{17}$$

where the model constant C_ε is set equal to 1. As a consequence of the non-eddy viscosity nature of the DS model, the SGS energy diffusion due to turbulent viscosity is not present in the transport Equation (16): *i.e.*, $Pr_{sgs} = 1$.

However, other sub-grid terms resulting in the energy and species conservation equations are modeled using an eddy viscosity concept. As an example, a turbulent diffusion concept is used and the turbulent diffusion is given by:

$$D_t = \frac{C_k \Delta k_{sgs}^{0.5}}{Sc} \tag{18}$$

Here, C_k is a model constant set to 0.5 and Sc is the turbulent Schmidt number set to 0.71. Further details about this implementation are provided in Senecal *et al.* (2013, 2014), Som *et al.*, 2012b and Xue *et al.* (2013).

2.4 Numerical Method

The overall set of equations is solved using the finite volume method. All variable values are co-located at the cell center and the Rhie-Chow algorithm (Rhie and Chow, 1983) is used to prevent pressure field decoupling (checker-boarding). First, the conservation equations of mass and momentum are solved using a pressure-velocity coupling iteration method, specifically the Pressure Implicit with Splitting of Operators (PISO) method of Issa (1986). Other transport equations, such as the energy and the species, are solved after the momentum predictor and the first corrector have been completed. Previous studies on nozzle flow simulations using this algorithm for a mixture model can be found in the work by Zhao *et al.* (2014) and Xue *et al.* (2014).

Spatial discretization is second-order accurate using central differencing schemes for all quantities wherever is possible. A minmod type of flux limiter is used to capture sharp gradients as well as to maintain high accuracy in the smoother regions (Quan *et al.*, 2014). Grid refinements are managed through time-dependent fixed embedding located in the key regions. Time integration is first order accurate running fully implicit Euler scheme, with fine time-stepping controlled by a velocity-based Courant–Friedrichs–Lewy (CFL) below 0.1 and a speed-of-sound-based CFL below 2.0. This multiphase flow model has been recently implemented in the Computation Fluid Dynamics (CFD) code Converge.

3 ASSESSMENT OF LES RESULTS QUALITY

It has to be appreciated that the practice of building a grid and applying LES using the grid itself as a turbulence resolution length scale makes the model incomplete (in the sense of Pope, 2004), *i.e.*, the specification is subjective and the grid becomes a parameter of the LES equations. Several efforts have been devoted to implement solution-adaptive gridding methodology (Pope, 2004; Celik *et al.*, 2005; Brusiani and Bianchi, 2008; Davidson, 2009; Piscaglia *et al.*, 2014; Xiao *et al.*, 2014), however, here we only undertook an assessment of the results quality after the grid was prescribed.

Many metrics have been proposed in the literature to evaluate the quality of LES results. Among them, two main types are reviewed here. One is based on the resolved fraction of turbulent kinetic energy and the other is based on the length-scale resolution. The first type of criteria, following Pope (2000, 2004), requires that the modeled turbulent kinetic energy be less than 20% of the total turbulent energy. It is expressed as:

$$M = \frac{k_{sgs}}{k_{sgs} + k_{res}} = \frac{k_{sgs}}{k_{sgs} + \frac{1}{2}\langle(\tilde{u}_i - \langle\tilde{u}_i\rangle)(\tilde{u}_i - \langle\tilde{u}_i\rangle)\rangle} < 0.2 \tag{19}$$

It is worth noting that the evaluation of the mean resolved turbulent kinetic energy, k_{res}, requires temporal averaging or ensemble averaging over multiple realizations of the experiment (the mean operator is denoted here by angle brackets$\langle\cdot\rangle$). In addition, to some extent the value of M is non-local, as it is affected by the grid resolution at other locations and at earlier times (Pope, 2004).

Another type of metric is based on the turbulence-resolution length scale. Following again Pope (2000, 2004) and Brusiani and Bianchi (2008), the criterion requires that the resolved scales be larger than the dissipation range scales. The demarcation length-scale, between the inertial and dissipative ranges, l_{DI}, is often estimated, according to Pope (2000), as $l_{DI} = 60\eta$, where η is the Kolmogorov length scale. The latter, $\eta = (\nu^3/\varepsilon)^{1/4}$, can be evaluated using the sub-grid energy dissipation rate, ε, which in this work is given by Equation (17). In synthesis, a Length Scale Resolution (*LSR*) parameter can be defined as:

$$LSR = \frac{\Delta}{l_{DI}} = \frac{\Delta^{3/4} C_\varepsilon^{1/4} k_{sgs}^{3/8}}{60\nu^{3/4}} \tag{20}$$

It is worth observing that expanding out its definition, *LSR* depends on k_{sgs}, therefore on the turbulence closure model. In addition, the evaluation of the demarcation length-scale l_{DI} is based on scaling arguments strictly valid for equilibrium, isotropic turbulence.

With that in mind, if *LSR* is below 1, all the scales in the inertial sub-range and above are resolved, while only the dissipative range is modeled. In practical applications, as suggested by Brusiani and Bianchi (2008), a value of *LSR* < 3-5 can still be accepted for a good LES calculation at a reasonable cost. One advantage of this criterion, compared to the previous one, is that it does not require statistical averaging of the resolved quantities.

Lastly, one should always keep in mind that regardless of the metrics being used (either Eq. 19 or 20, for example), the numerical discretization errors can play a significant role in the assessment. In fact, the numerical viscosity generates an additional contribution, which should be evaluated explicitly (Celik *et al.*, 2005). Therefore, the modeled SGS turbulent kinetic energy, which appears in both parameters (19) and (20), should be increased by the numerical viscosity counterpart. In both cases, the indexes would give a worse assessment. However, the estimation of the numerical error is beyond the scope of the present work and will not be included in the analysis.

4 APPLICATIONS TO TRANSIENT DIESEL JETS

A validation study of the proposed multiphase mixture model within the LES framework has been shown by Quan *et al.* (2014), for the case of Diesel sprays producing shock waves at relatively high injection velocities and low temperature ambient conditions.

In this work, the model is applied to study the transient behavior of jets issuing from Diesel injectors, with focus on the initial and final part of a typical injection event. The in-nozzle flow is part of the computation, since the physics mostly depends on what occurs inside the injector. Recently, these topics have taken on particular importance as short and multiple injections are often used in advanced engine combustion strategies as discussed in the Introduction section. However, a combination of complex physics (multi-phase, compressibility, phase change) and uncertainties in boundary conditions (accuracy of the internal geometry, moving boundaries, etc.) make the prediction of these flows a particularly challenging task.

4.1 End of Injection Simulations

4.1.1 Problem Description

First, the multi-phase model is applied to simulate a single-hole mini-sac Diesel injector nozzle, for which synchrotron X-ray imaging data, collected at the Advanced Photon

TABLE 1

Nozzle geometry features and operating conditions

Nozzle		
Orifice outlet diameter, D_{out}	0.180 mm	
Orifice length, L	1.0 mm	
Inlet edge rounding, R	0.050 mm	
k-factor, $(D_{in}-D_{out})/10$ (use μm)	2.4	
Sac volume, V_{sac}	0.15 mm^3	
Operating conditions		
Injection pressure, p_{inj}	1 500 bar (total)	
Chamber back-pressure, p_{out}	1 bar	~ 1.2 kg/m^3
Chamber temperature, T_{out}	300 K	
Wall temperature	300 K	
Needle lift	cf. Figure 3	
Fuel properties		
Type	Diesel	
Liq. density, $\rho_{liq,0}$ (Eq. 6)	848 kg/m^3 (300 K)	
Liq. bulk modulus (Eq. 6)	1.9×10^9 Pa	
Fuel inlet temperature	300 K	
Liquid viscosity, μ_{liq}	Variable with T (2.5×10^{-3} Ns/m^2 @ 300 K)	
Saturation pressure, p_{sat}	Variable with T (1 000 Pa @ 300 K)	
Non condensable gas mass fraction, Y_{air}	2×10^{-5}	

Source (APS) at ANL, are available (Swantek *et al.*, 2013, 2014). In the experimental campaign, different injection pressure (from 500 bar to 1 500 bar), back-pressure (from 1 bar to 20 bar) and orifice diameter (from 110 μm to 180 μm) have been investigated. Here, we present the simulation of the case with maximum expected ingested gas among the above mentioned conditions, *i.e.*, 1 500 bar injection pressure, 1 bar chamber pressure, and 180 μm orifice diameter. A summary of conditions and fuel properties, along with the main injector features is provided in Table 1. Also, the needle lift profile is shown in Figure 3.

Focusing on the simulation of the EOI process, we skipped the initial and central part of the injection event, and started the calculation at 1 900 μs where the lift is still high for this injector, being 150 μm. Results (shown later) demonstrate that starting from this lift the flow rate reaches a steady condition before the effective ramp-down starts.

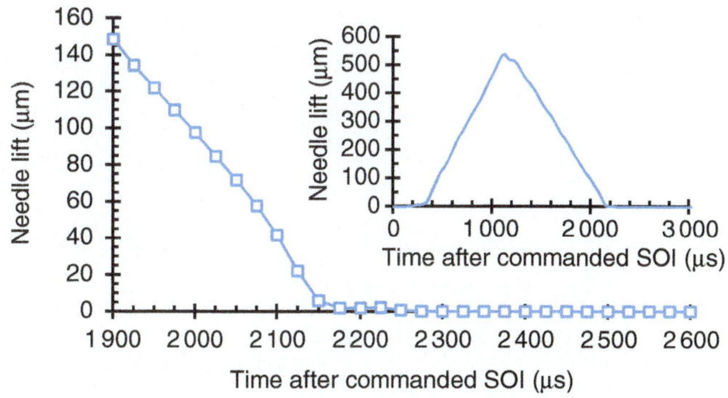

Figure 3

Needle lift profile *versus* time (from X-ray imaging experiments). The enlarged chart (on the left) focuses on the closure phase.

The objective is to reduce the computational time, yet provide a good estimate of the initial flow for the simulation of the EOI behavior. The needle closure occurs in between 2 150 μs and 2 175 μs, as visible in Figure 3 (experimental data have a temporal resolution of 25 μs). To deal with very low needle lifts, we assumed the needle valve closed below 10 μm. This occurs at 2 140 μs, therefore the flow is simulated as a single domain up to this point. After the closure we switch to two disconnected domains, and the simulation continues up to 2 500 μs, to ensure that all the relevant dynamics is covered. The minimum cell size is 5 μm, and it has been maintained throughout the simulation.

The main focus for the gridding strategy has been on the sac, the orifice, and the near nozzle regions. A view of the computational grid is shown in Figure 4. Grid is temporally refined and changed for better adapting to the relevant dynamics. The region around the orifice is discretized with the finest cell size throughout all the ramp-down phase and it is retained after the needle closure. To save computational time, during the start-up phase of the calculation (before 2 000 μs, at high lift) the finest cell size has been set to 10 μm.

Associated with the transient needle movement profiles, pressure and temperature are specified at the inlet boundary. The turbulent intensity is set to 0.1% at the inlet, in terms of k_{sgs} value. The Werner and Wengle (1991) wall model is employed at the wall boundaries. The outlet chamber is smaller than the real one, but its size is large enough to accommodate all the jet development in the time interval of interest. The length is 20 mm and the diameter is 16 mm. The bottom surface of the chamber is closed and the jet will eventually imping on it. The side walls are open and the static pressure is applied as boundary condition. The near nozzle region, below 10 mm, is unaffected by the

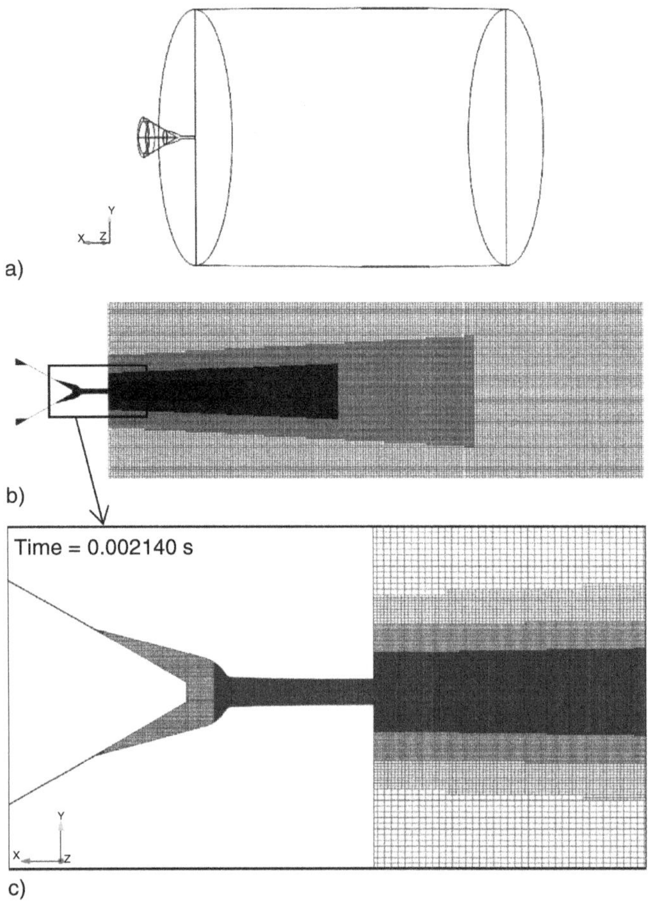

Figure 4

a) View of the computational domain. b) Global view of the mesh. c) Zoomed view of the nozzle showing the grid at the closing timing. Minimum cell size = 5 μm. Further information about the mesh is shown in Grid 2 in Table 2.

choice of the chamber size, yet the savings in terms of total number of cells are substantial. Overall, the mesh size for a fine grid (Tab. 2, Grid 2) ranges from 10 to 16 million cells over the simulated period. Sections 4.1.2 and 4.1.4 provide more details on the grid suitability and simulation cost.

4.1.2 Discussion of Results: Dynamics of the Flow

The main global parameters predicted by the LES are shown in Figure 5. The top chart displays the average velocity at the nozzle exit. The middle chart displays the average and minimum pressure in the sac and orifice regions. The bottom chart shows the global (integrated) void fractions in the same regions. Observing the behavior of these quantities, we have a synthetic view of the dynamics occurring during the EOI as predicted by the model.

TABLE 2

Grid characteristics for EOI simulations

	Grid 1	Grid 2
Base grid size	80 μm	80 μm
Orifice cell size	10 μm	5 μm
Sac cell size	10 μm	10 μm
Needle-seat gap cell size	5 μm	5 μm
Near exit region cell size	10 μm	5 μm
Total cell count (variable in time)	10.0-12.3 M	10.0-15.6 M

The first part of the simulation provides the initialization of the fuel jet while the needle lift is still high. The main flow features are basically steady for $t < 2\ 100$ μs, where the lift is above 50 μm, signifying that the flow losses are located in the orifice and not across the needle. After this point, the restriction imposed by the needle becomes relevant and ultimately causes the strong flow deceleration, which takes place in less than 40 μs, *i.e.*, from 2 100 μs to 2 140 μs (instant of closure). At 2 170 μs, the average orifice pressure reaches a minimum, dropping by four orders of magnitude in about 25-30 μs. The average sac pressure shows a slightly smoother drop, due to the averaging over a larger region. The minimum pressure drops very fast in both regions. Spots of extremely low local pressures are found in the sac and in the orifice, immediately after the needle closure. The minimum pressure level detected after the closure reaches 1-10 Pa (absolute), therefore locally a strong cavitation is initiated, since the fluid has a vapor pressure equal to 1 000 Pa at the reference temperature of 25°C. Vapor phase is formed and at the same time the dissolved gas in the fuel expands. The global result of this gases being formed or expanded is the global void fraction α_g shown in Figure 5c (integrated over each region). The effect of the abrupt deceleration of the liquid column at the needle closure is the formation of approximately 20% and 40% of void fraction in the sac and in the orifice, respectively. Later, from 2 250 μ to 2 350 μs, a back flow takes place, as displayed by negative velocities at the hole exit in Figure 5a. During this time interval, gas from the chamber is ingested, which eventually fills entirely the orifice and partly the sac (Fig. 5c after 2 400 μs). Residual gases are therefore predicted inside the nozzle after the EOI.

In order to have a clear view of the results, contour and iso-surface plots from the LES are shown in Figure 6, side by side with X-ray phase contrast imaging data. Contours are shown in a plane slicing the domain through the hole

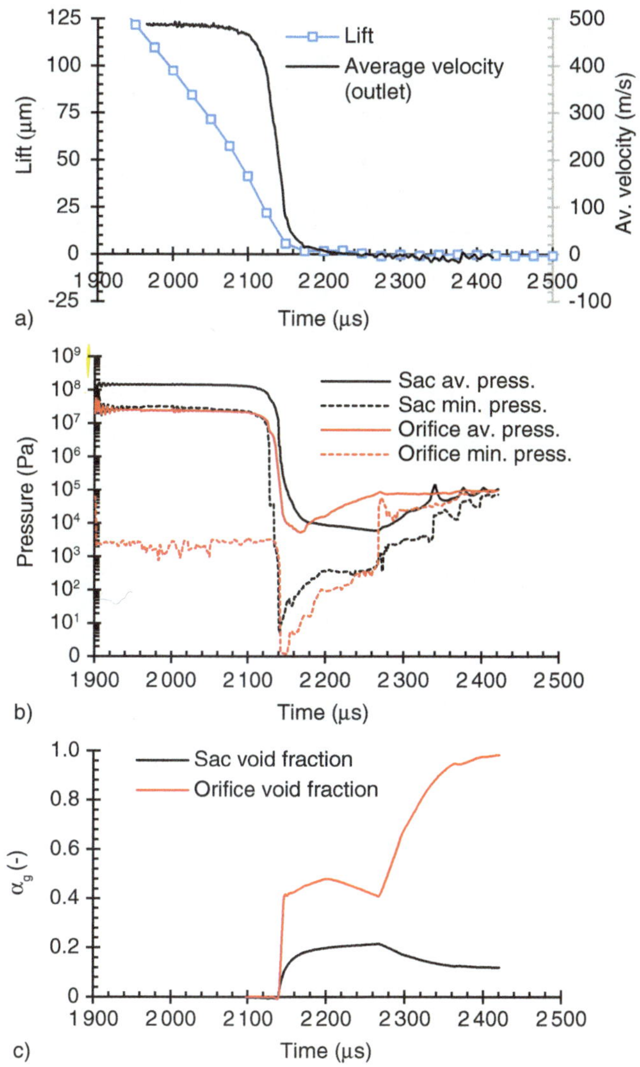

Figure 5

a) Results of the LES for the end of injection case: average velocity at the nozzle exit; b) average and minimum pressures in the sac and in the orifice regions; c) cumulative void fraction, α_g, in the sac and orifice regions. Results shown refer to Grid 2 (fine) in Table 2.

axis, and refer to liquid volume fraction, LVF $(= 1 - \alpha_g)$. The iso-surfaces are built at $LVF = 0.1$ and are colored by pressure. Different timings are presented, ranging from the high lift condition to the end of the process. The quasi-steady injection is shown at 2 000 μs, the ramp-down of the mass flow at 2 100 μs. In the model, the needle closure is at 2 140 μs, while the available experimental image is at 2 150 μs. Here, three calculated frames are shown, *i.e.*, at EOI and after 5 μs and 10 μs. Shortly after the EOI, other comparisons are provided at 2 160, 2 200 and 2 230 μs. Finally, results at 2 300 and 2 500 μs refer to a later stage.

In addition, Figure 7 shows turbulent flow structures visualized in terms of Q-criterion at 2 100 μs.

Observing Figure 6, a satisfactory qualitative agreement is found between the calculated LVF and the experimental imaging. At high lift, an intact liquid core exits from the orifice (cf. results at 2 000 μs). During the ramp down phase (2 100 μs) liquid decelerates and the liquid-gas interface shows some wrinkling. When the needle touches the seat, the fuel is subject to the highest deceleration. The positive tensile stresses on the liquid ultimately cause its breaks-up into ligaments (results from 2 140 to 2 200 μs). At 2 150 μs two cavitation spots are observed in the sac near the valve seat. Vapor in the sac then recondenses, as the local pressure rises back. Large areas of void fraction are also visible in the orifice as a result of the liquid jet rupture. Soon after, a pressure wave travels from the chamber towards the sac and brings back a liquid pocket. Behind the residual liquid, also some gas from the chamber flows toward the sac (cf. results from 2 200 μs to 2 300 μs). As a result of this phase, ambient gas will partially fill the sac.

The formation of ligaments in the near nozzle region is predicted to occur from the time of closure up to approximately 200 μs later. The shape and structure of dribbles are very similar to those observed in the experiments, however, the temporal persistence of these structures is not as long as in the experiments. This is likely due to the diffuse interface method used, which does not track the interface geometry, but rather models the mass dispersion as a diffusive mixing process. Mass content and distribution are therefore correctly reproduced, but the atomization is not resolved in details.

At the end of the process, a fraction of the sac is empty. An interesting comparison is given in Figure 8, where the last available simulation time-step is compared to experimental images taken before the start of an injection pulse. It is found that the predicted void fraction left inside the injector matches quite well the experimental evidence of a residual gas bubble lying at the tip of needle. The experimental image is taken long time after the EOI, which cannot be practically simulated, therefore the last available time-step is used for the comparison. The visual comparison with the bubble detected in the experiments is certainly encouraging. The location and the amount of void fraction predicted by the model are also in good agreement with the X-ray image.

Overall, the case analyzed here clearly shows that a large amount of ambient gas can be pulled into the sac after the EOI, as the liquid jets decelerates and eventually breaks up forming dribbles. The mechanism of liquid deceleration, breakup and gas ingestion is well captured. The model needs to be further improved with respect to the preservation of the liquid-gas interface, and this is highlighted by the shorter

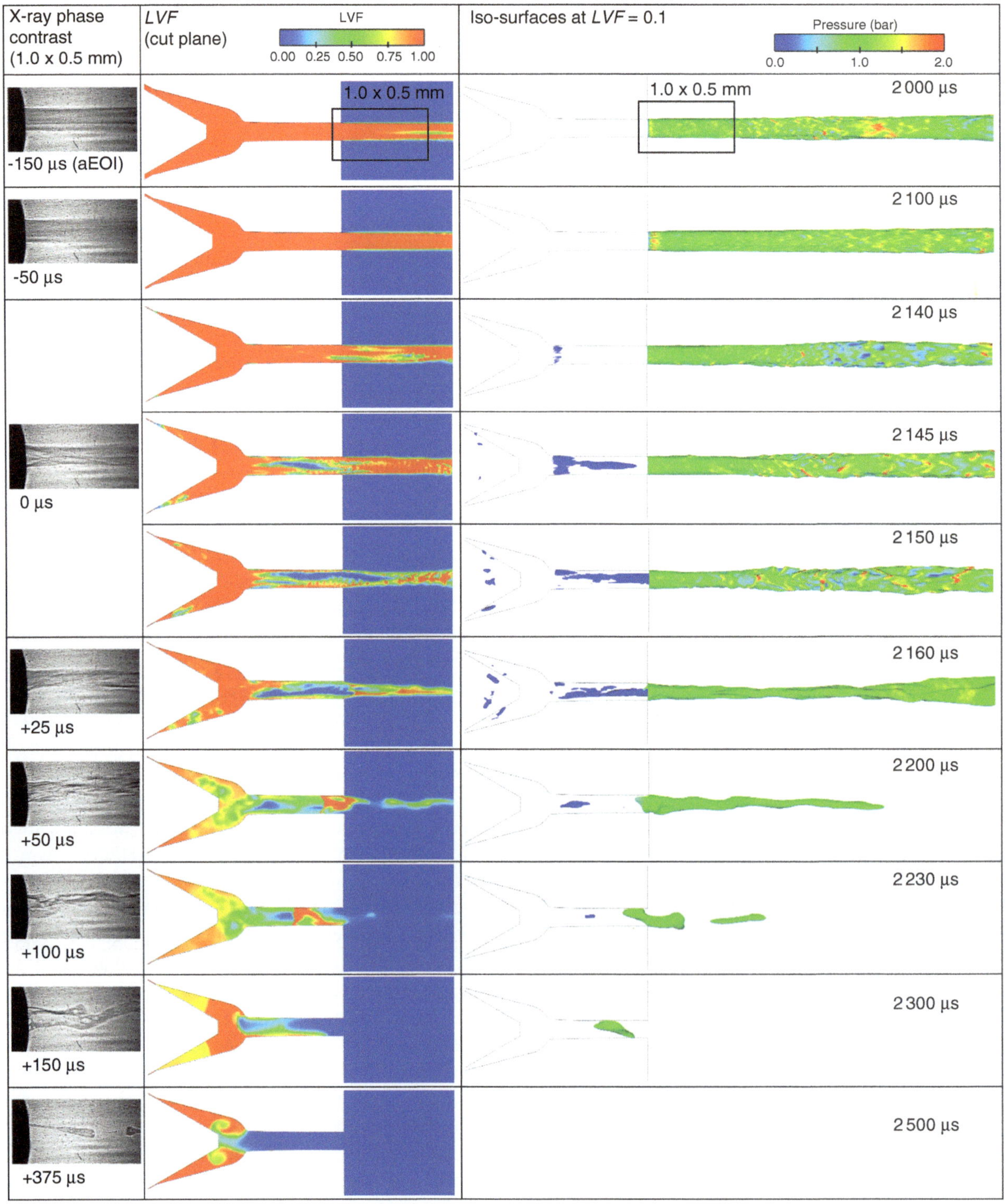

Figure 6

EOI behavior at different timings: X-ray imaging on the left, simulation results on the right. Each row shows X-ray phase contrast imaging of the external near nozzle jet with exposure time of 5 µs (Swantek *et al.*, 2013), computed LVF on a cut plane, and iso-surfaces at *LVF* = 0.1 colored by pressure. Results shown refer to Grid 2 (fine) in Table 2.

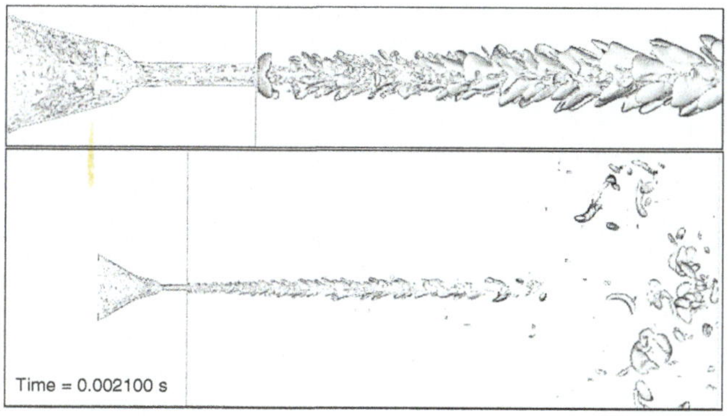

Figure 7

Vortices visualized as Q-criterion iso-surface, marking a threshold that localizes the most coherent turbulent structures, at t = 2 100 μs. Top view shows details in the nozzle region; bottom view shows the whole field.

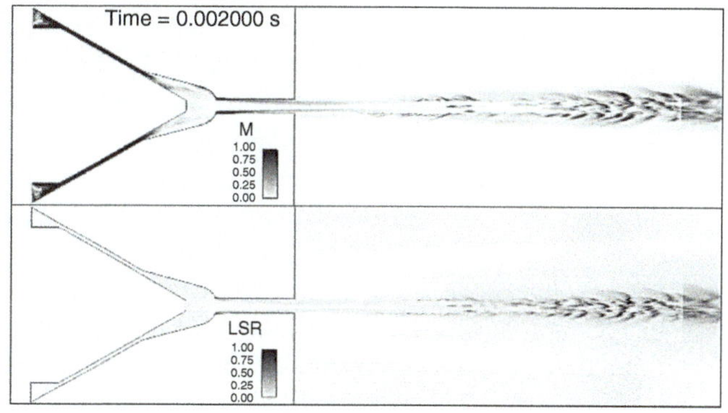

Figure 9

Measures of LES quality for Grid 2 (fine) at high needle lift (t = 2 000 μs). Top plot displays Pope's criterion, M, (Eq. 19) – threshold is 0.2; bottom plot shows contours of *LSR* criterion (Eq. 20), – threshold is 1.

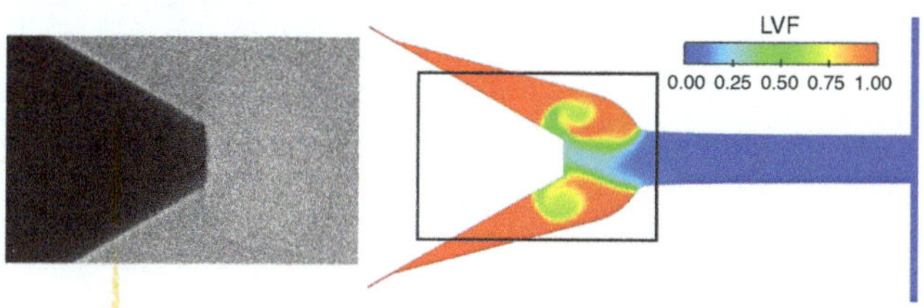

Figure 8

Residual gases ingested at the end of the process. X-ray phase contrast image on the left, simulated *LVF* on the right (simulation time t = 2 500 μs; grid 2 – fine – in Table 2).

duration of the ligament phase compared to the experiments. An High Resolution Interface Capturing (HRIC) scheme has been used to prevent interface smearing (Zhao *et al.*, 2014), however, this test case appears particularly challenging in respect to this, due to the very long physical time that needs to be simulated.

4.1.3 Grid Sensitivity and Evaluation of LES Quality

We investigate here the quality of the LES results, after the flow features have been already presented in the previous section. Two grids with different level of refinement have been built (Tab. 2). The evaluation of the simulation accuracy is done comparing the energy budget resolution parameter (Eq. 19) and the *LSR* parameter (Eq. 20).

Results in terms of M and *LSR* indices are shown in Figure 9 for the fine grid (Grid 2) at high needle lift. The evaluation of M requires the calculation of turbulent statistics. Due to the computational cost of a single realization, ensemble averaging is not practical for this kind of analysis on injector closing transients. Therefore, the only viable option is temporal averaging: this has been accomplished by averaging the resolved field for 20 μs, from 1 980 μs to 2 000 μs, where the flow rate is substantially steady, as demonstrated in Figure 5. Temporal averaging is done at run time from the specified start-timing by tracking non-transport scalars.

Values for M are meaningful in the sac, orifice and near exit regions, while they may not be accurate close to the needle moving surface and at the spray tip for the intrinsic unsteadiness of the problem.

Focusing on the LES quality assessment, the contours of the turbulent energy resolution parameter, M, show that the grid is fine enough in the sac, in the core of the orifice and downstream up to 3-4 mm from the exit, where M is well below 0.2. On the contrary, observing the results in terms of *LSR* parameter, globally a better picture is perceived. The needle gap region seems well resolved (the lift here is above 100 μm), as *LSR* is well below 1. In addition, the mesh resolution at the orifice inlet edge is satisfactory based on the *LSR* metrics. We can comment on the comparison between the two quality criteria, by noting that the temporal averaging required for M makes this parameter not local (Pope, 2004) and the interpretation of its value requires special care:

- at the domain inlet and upstream of the needle-seat area, M is very large because of the turbulence introduced at the inlet boundary. This suggests the use of a larger reservoir upstream of the needle seat to minimize the effect of unphysical turbulence introduced at the inlet;

- the flow through the gap region is highly dependent on the grid resolution. Also, the moving boundary may affect the results in terms of M, which appears to be very high, *i.e.*, close to 1. On the other hand, the *LSR* value in the needle-seat gap is very small. Based on this opposed answers, we might conclude that the use of these parameters in this narrow area needs to be further checked and tested;
- at the hole inlet edge, there is some recirculation and low velocities and at the same time high SGS turbulence is generated. This leads to very high M values. Additional cell refinement would likely be needed in this area;
- streaks of high M values are observed extending from the inlet edge down to the hole exit: this is probably due to the advection effect intrinsic in the definition of M. Conversely, the *LSR* parameter, which does not suffer from this non-locality problem, does show good grid quality in the orifice.

After this assessment on the comparison of the energy budget and the length-scale resolution parameters, the grid suitability at different timings during the closing transient can be conveniently appreciated by using the *LSR* index, which is deemed a proper tool for dynamic resolution evaluations, especially in the presence of boundary movement and temporally changing flows.

Figure 10 displays the contours at 2 100 μs, 2 140 μs and 2 170 μs. For each timing, a close-up and a global view are shown. Far from the nozzle (*i.e.*, at distances greater than 5 mm) the *LSR* index becomes greater than 1, but these spots are located outside the region of interest for the current analysis, where the mesh is made coarse to reduce the total cost of the simulation. The focus is therefore inside the injector and close to the nozzle. The first contour plot (2 100 μs) refers to the jet deceleration phase, where velocities are still quite high (around 400 m/s). Here, the *LSR* results suggest again that near the hole inlet edge, the grid can be further refined. Elsewhere, the grid spacing is adequate. The second timing refers to the instant of needle closure. Here, the grid is satisfactory in the whole region of interest. At most, some refinements could be applied in the tail of the detaching jet and at needle seat exit. The last image refers to a timing when ligaments are being formed. Here, the turbulent flow is well resolved due to the low liquid speeds (few tens of m/s) and the large flow structures.

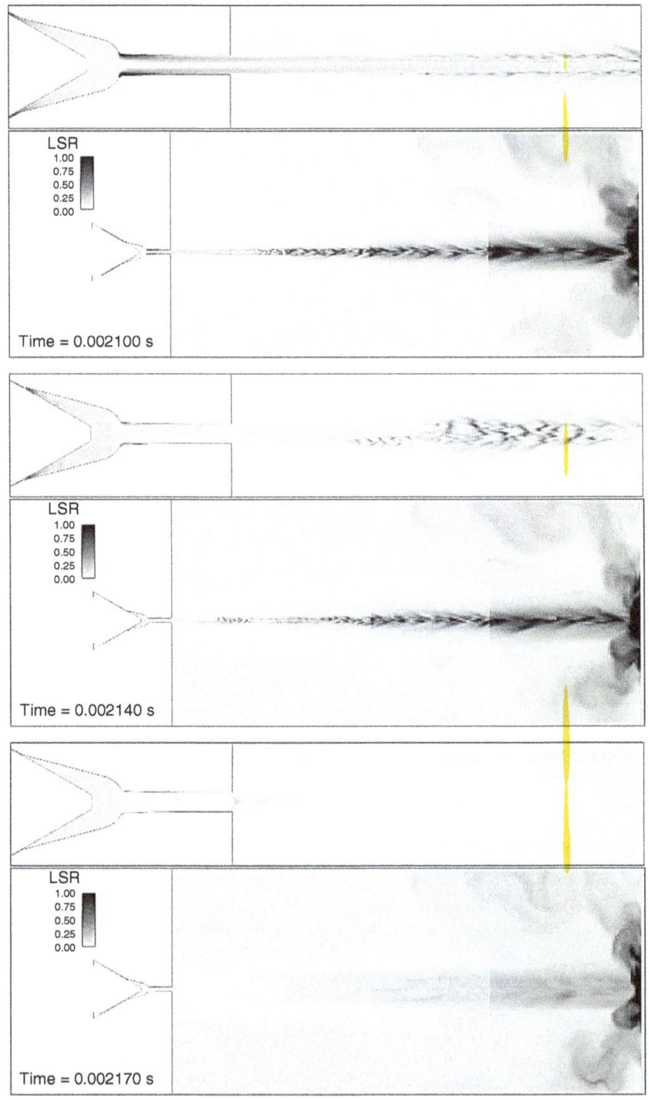

Figure 10

Dynamic resolution evaluation: *LSR* contours during the jet deceleration (top, 2 100 μs), at the closure (middle, 2 140 μs) and after the end of injection (bottom, 2 170 μs).

Computations with different grid resolutions have also been considered. A 10 μm grid is compared to a 5 μm grid (min cell size in the orifice and near nozzle region, Tab. 2). Results in terms of average pressure in the sac are shown in Figure 11. The prediction using the fine resolution shows that the pressure rises back earlier, therefore the back-flow after the EOI occurs in advance. In addition, Figure 12 shows contours of *LVF* at three timings during ligament formation (2 160, 2 180 and 2 220 μs). The fine grid predicts an appreciably longer and sharper liquid dribble compared to the lower resolution mesh. This is probably due to a lack of small turbulence structures in the jet using the 10 μm grid, which are dissipated faster. As a matter of fact, the Eulerian mixture model is based on a diffuse interface assumption. Alternatively, a front tracking method could provide a better reconstruction of the phase interface for capturing ligament formation, however generally at the cost of increased complexity due to the absence of SGS turbulence or with additional limitations like phase incompressibility.

Overall, this mesh resolution analysis suggests that the prediction of ligaments being formed after the needle closure requires a very fine grid in the nozzle region, and an even smaller cell size could be adopted. Instead, the global parameters describing the dynamics of the process do not change drastically with the resolution, therefore, trends are essentially well predicted within the accuracy of the model assumptions.

4.1.4 RANS-LES Comparison

The assessment of LES results can be complemented by the comparison with RANS results on the same grid and with similar numerical framework. Figure 13 shows the results of a RANS calculation with the standard k-ε turbulence model on the low-resolution mesh (grid 1). The *LVF* contours can be directly compared to LES results shown in Figure 12 – left column. All the time-steps shown have a common feature: the RANS turbulence model makes the flow extremely axisymmetric, as expected, but the consequence is that the liquid column rupture occurs with a very different mechanism compared to the LES prediction. The RANS model tries to capture the ensemble average behavior, which might be fully correct, but it cannot show the instantaneous features of a single realization as obtained with LES. Focusing inside the nozzle and the hole, the RANS model predicts a dynamics that has a one-dimensional character, therefore mass transfer, pressure, or other quantities propagate through quasi-planar wave fronts. In the RANS predictions, the residual liquid back flow occurs without fragmentation of the bulk mass, while in LES liquid breaks up generating complex three dimensional structures.

Another feature of interest is observed slightly after the needle closure (2 160 μs). In the RANS approach, an annular void region stems from the needle seat, and then re-condenses as the pressure rises back. This is not observed in LES, where spots of low pressure regions are produced even in areas detached from the needle seat.

The near exit region is also different. No ligaments are being formed using RANS, but rather a diffusive process takes place, which disperses the liquid in the chamber. With the same level of grid refinement, LES captures some low speed jet waving and eventually liquid islands. However, in the multi-phase mixture model the geometric reconstruction of the liquid-gas interface is not performed, therefore, even using LES neat pockets are not detected.

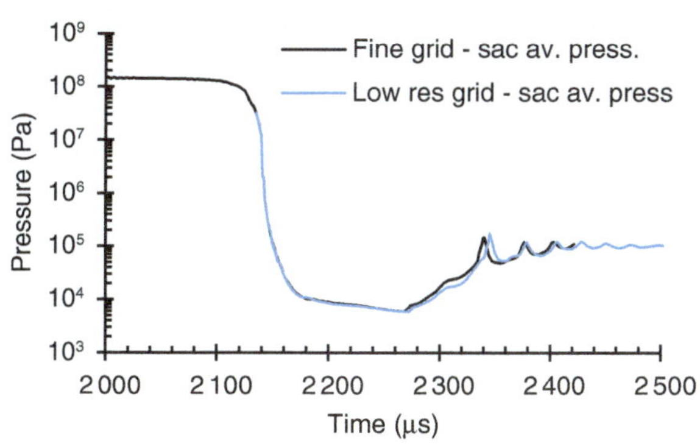

Figure 11

Effect of mesh refinements (Tab. 2) on a global quantity: average sac pressure *versus* time.

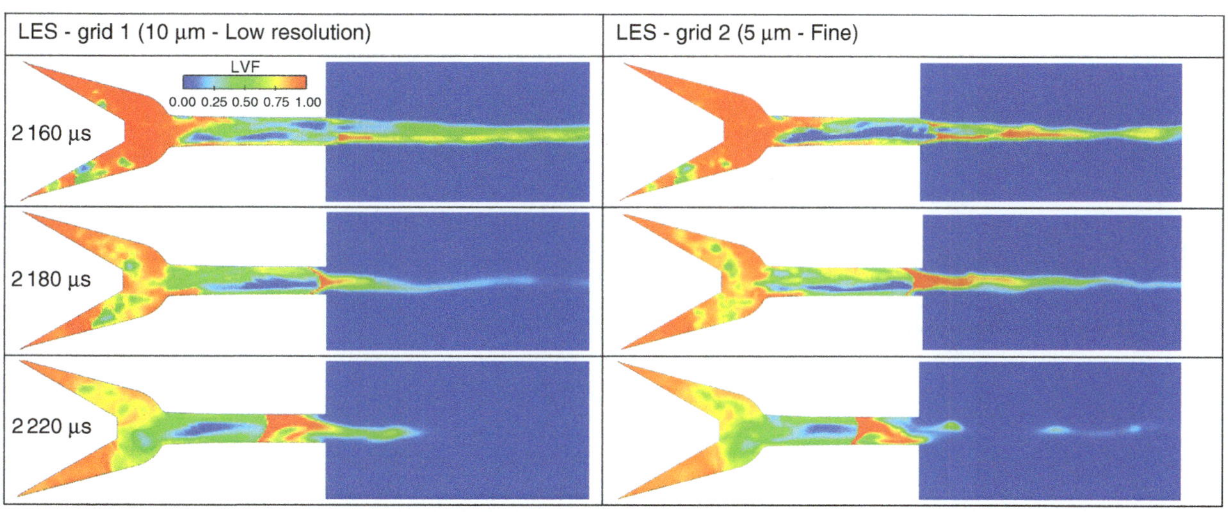

Figure 12

Effect of mesh refinements (Tab. 2). Contours of *LVF* on a cut-plane.

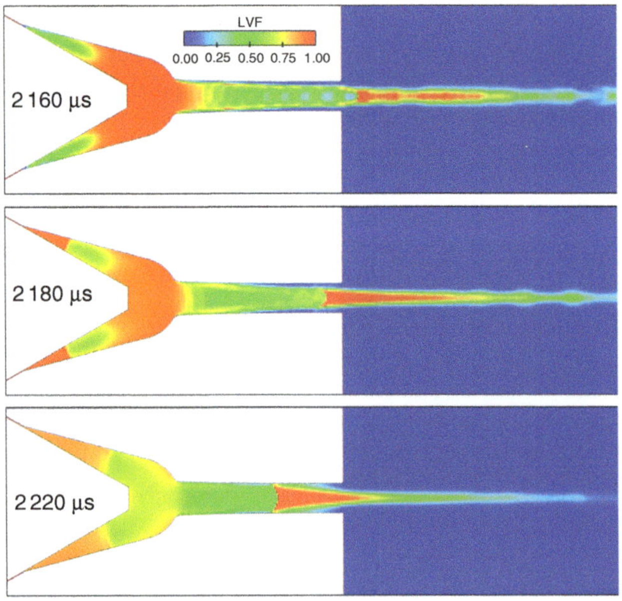

Figure 13

RANS results on grid 1 (10 μm – low-resolution grid, Tab. 2). Contours of *LVF* on a cut-plane.

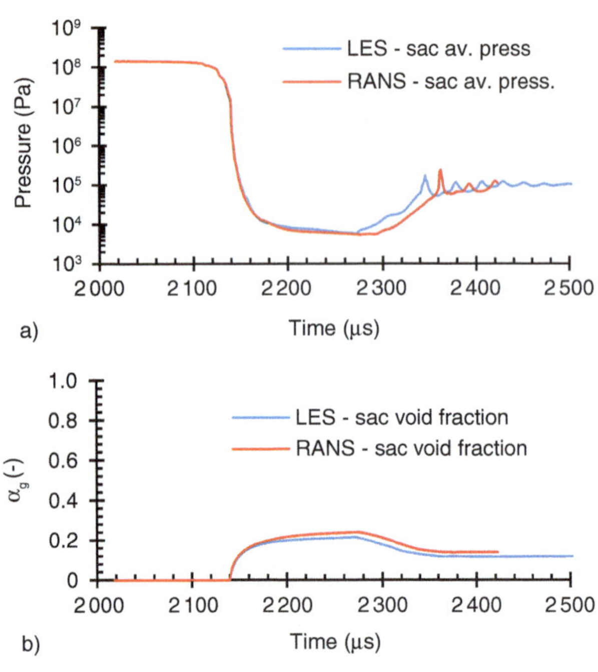

Figure 14

LES *versus* RANS comparison. Predictions of average pressure and total void fraction in the sac (Grid 1, low resolution, Tab. 2).

Lastly, Figure 14 shows the comparison of RANS and LES in terms of global quantities like pressure and void fraction averaged/integrated over the sac region. The pressure recovery in the sac takes more time with the RANS model. This is likely due to the "one-dimensional" character of the flow that has been observed in the RANS results. This point certainly requires further investigation, but it is our opinion that a fully three-dimensional pattern as observed with LES might be more realistic for a single realization. It is very likely that if several LES realizations had been run and averaged, results would be close to those obtained with RANS.

4.1.5 Computational Cost

Simulations were run using a time step controlled by a velocity-based CFL set to 0.1. With reference to the fine grid with 15.6 million cells (grid 2 in Tab. 2), the resulting time-step varied from 0.8×10^{-9} to 3.0×10^{-9} s during the simulated time. The total number of time steps was in the order of 300 000. The LES required about 33 days of wall clock time on 256 cores. Using the coarse grid the wall clock time was about 21 days on the same number of cores.

4.2 Start of Injection Simulations

4.2.1 Problem Description

In this section, we present an analysis of the start of injection of the "spray A" single-hole mini-sac Diesel injector, which is widely characterized within the ECN. The multi-phase model is applied to simulate the needle opening phase starting from a very small needle lift (2.5 μm). The simulation aims at capturing the initial ROI transient, the internal nozzle flow features occurring at very low needle lift, the sac filling process with liquid fuel, and also the initial spray features. In particular, one of the objectives of this demonstration study is to assess the initialization of the internal nozzle regions and its effect on the initial jet behavior.

As discussed in the Introduction and shown in the previous EOI test case, there are several evidences that confirm presence of gas in the sac at the start of an injection event. Here, as initial conditions we have assumed that the residual gas completely fills the sac and the hole volume at the beginning of the simulation, while liquid fuel only fills the region upstream of the needle-seat contact area.

A summary of nozzle geometry, operating conditions, and fuel properties is provided in Table 3, Figure 15, and Figure 16 (Tab. 4). Taking advantage of current efforts by the ECN to deeply characterize automotive injectors, the spray A internal geometry is thoroughly investigated. A high-resolution geometry reconstructed based on X-ray tomography is available, with surface features in the order of 1 μm (Fig. 16). The nozzle is tapered and has an equivalent outlet diameter of 89.4 μm. The hole

TABLE 3
Spray A nozzle data and operating conditions

Nozzle details (SN 210675)		
Outlet hole diameter, D_{out}	0.0894 mm	
Nominal hole length, L	0.75 mm	
Inlet edge radius, R	~ 0.025 mm	
Nominal k-factor, $(D_{in}-D_{out})/10$ (use μm)	1.5	
Sac volume, V_{sac}	0.2 mm^3	
Surface CAD model	High resolution STL based on X-ray tomography	
Operating conditions ("cold" ANL conditions)		
Injection pressure, p_{inj}	1 500 bar (total)	
Chamber back-pressure, p_{out}	20 bar	~ 22.8 kg/m^3
Chamber temperature, T_{out}	303 K	
Wall temperature	338 K	
Needle lift	cf. Figure 15	
Fuel properties		
Type	n-dodecane	
Liq. density, $\rho_{liq,0}$ (cf. Eq. (6))	750 kg/m^3 (303 K)	
Liq. bulk modulus (cf. Eq. (6))	1.9×10^9 Pa	
Fuel inlet temperature	338 K	
Liquid viscosity, μ_{liq}	variable with T (1.3×10^{-3} Ns/m^2 @ 303 K)	
Saturation pressure, p_{sat}	variable with T (33 Pa @ 303 K)	
Non condensable gas mass fraction, Y_{air}	2×10^{-5}	

axis has a significant offset from the nominal centerline. Needle motion, with three-dimensional trajectory, is available from high speed X-ray imaging through the metal injector operated in "cold" conditions, as tested at ANL (Kastengren *et al.*, 2014). The lift curve is shown in Figure 15, and focuses on the initial phase. The needle opening starts at 150 μs after the electrical command. The first part of the rising ramp has a small velocity and after 150 μs the needle has only reached 9 μm lift. After this point, the main fast rise begins. In this study, we considered only the vertical lift motion and neglected the off axis components. In addition, the needle lift prescribed in the model is slightly different from the experimental data and it is shown by the linearized ramp in Figure 15. As a matter of fact, due to the uncertainty on the lift measurements

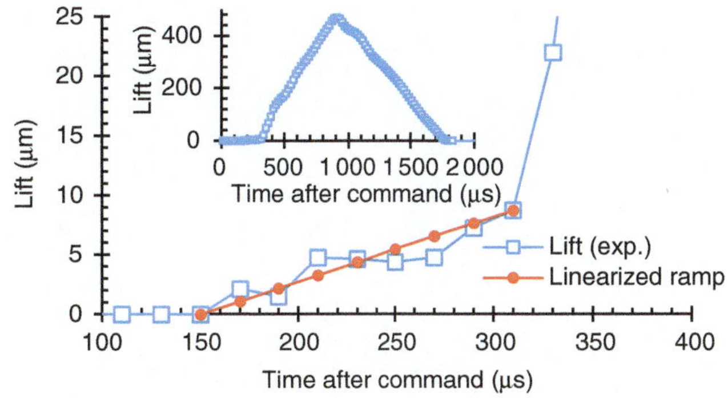

Figure 15

Spray A needle lift profile *versus* time, from X-ray imaging experiments (Kastengren *et al.*, 2014). The main chart focuses on the starting phase. The insert shows the whole injection duration.

(estimated to be around 2 μm) and to the noisy shape of the initial ramp, a smoothed linear rise from 150 μs to 310 μs is preferred. In addition, since a zero gap cannot realistically be modeled, the simulation starts at $t = 200$ μs, thus at point in time where the lift is about 2.5 μm. The simulated period extends up to $t = 350$ μs. *i.e.*, encompassing only the initial injection stage.

The fuel is n-dodecane introduced at 338 K. The injector body is at 303 K. Ambient conditions are non-evaporating (different from standard spray A specifications), but the chamber density matches the nominal density value of 22.8 kg/m^3 (pressure and temperature are correspondingly reduced). These conditions are those used for X-ray imaging and radiography at Argonne.

The high resolution geometry (Fig. 16) encompasses the orifice and the bottom part of the sac, while the needle-seat contact is not covered. Therefore, this part has been reconstructed in the CAD using two coaxial conical surfaces, with the same 30 degree nominal inclination with respect to the injector axis.

The computational grid (Fig. 16) has a base cell size of 40 μm and uses 4 levels of embedding to obtain a minimum cell size of 2.5 μm. The finest resolution is employed in the needle-seat gap and for 3 layers on the hole walls. Refinements are permanent and Adaptive Mesh Refinement (AMR) is not used. The outlet chamber is 16 mm × 8 mm (length × diameter) and allows an adequate simulation of the jet up to 350 μs, as the tip penetration is around 10 mm at that time. Overall, the mesh size is 15.9 million cells. An additional "coarse" grid has been tested, which has 80 μm base size and 5 μm minimum cell size. Details are provided in Table 4.

The available experimental pressure trace has been applied at the inlet surface as total pressure. The chamber is large enough to be modeled as a closed vessel.

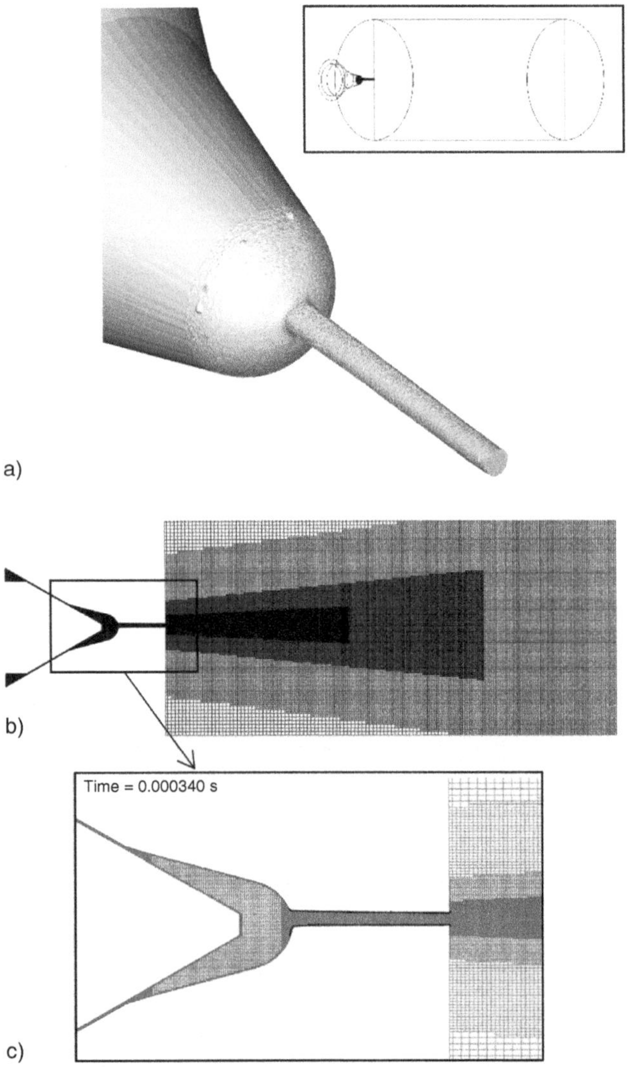

Figure 16

a) View of spray A high-resolution surface with insert showing the whole domain. b) Global view of the mesh. c) Close up of the nozzle showing grid details. Shown is the fine grid, cf. Table 4.

TABLE 4

Grid characteristics for SOI simulations

	Grid 1 (coarse)	Grid 2 (fine)
Base grid size	80 μm	40 μm
Orifice cell size	5-10 μm	2.5-5 μm
Sac cell size	10-20 μm	5-10 μm
Needle-seat gap cell size	5 μm	2.5 μm
Near exit region cell size	10 μm	5 μm
Total cell count	7.1 M	15.9 M

4.2.2 Discussion of Results

The representation of the moving needle and the use of the inlet pressure boundary condition enable a realistic calculation of the mass flux variation in time through the orifice. To show this, the global features of the transient simulation are shown in Figure 17. The top chart displays the liquid and gas mass flow rates exiting from the orifice. The bottom chart reports the liquid tip penetration, compared against available experimental data.

The liquid outlet flow rate is negligible until the SOI is approached. Instead, the gas phase flow rate is basically always positive from the beginning of the simulation (200 μs), denoting the ejection of the residual gas from the sac. Evidently, the gas phase mass flux at the orifice exit is quite small because of the density difference compared to the liquid. Based on the liquid mass flow rate at the hole exit, the predicted fuel injection starts around $t = 300$ μs, when the curve has a steep increase. Experimental ROI curve is not available with sufficient time resolution and accuracy at the SOI, as needed for this validation.

A complementary assessment method is based on the analysis of the tip penetration, which is presented in the bottom chart of Figure 17. Experiments based on X-ray radiography (Argonne data, square symbols, by Kastengren *et al.*, 2014) indicate that the SOI is approximately at 305 μs. Sandia penetration data, based on diffuse back-illumination imaging, are also considered, but the available time information is only relative to the apparent SOI, which we assumed to be the same as in the X-ray dataset. Numerical penetration is calculated based on a radiographic projection of density distribution (Battistoni *et al.*, 2014a), in analogy with the experimental methodology. The numerical predictions and the experimental data match quite well in terms of both start timing and slope of the liquid penetration curve.

The main result of these analyses is that if the sac is initially filled with gas, the liquid SOI, which occurs several tens of μs after the start of needle movement, is in good agreement with the experimental evidence. This delay is in the order of 100 μs, and it is compatible with the duration of the first slow-rising part of the needle movement. It is evident that any other types of sac initialization, *e.g.* partially filled or fully filled with liquid, would cause a premature SOI, which would be severely off compared to the experimental data.

We now switch the focus to the analysis of the flow features. The evolution of the sac filling process is displayed in Figure 18, with cut planes at different timings showing contours of Liquid Volume Fraction (LVF) and of residual gas fraction (calculated as transported passive scalar). The dynamics depicted in the sequence of images shows that at $t = 210$-220 μs, that is 10-20 μs after the start of the simulation, the liquid coming from the upstream high pressure

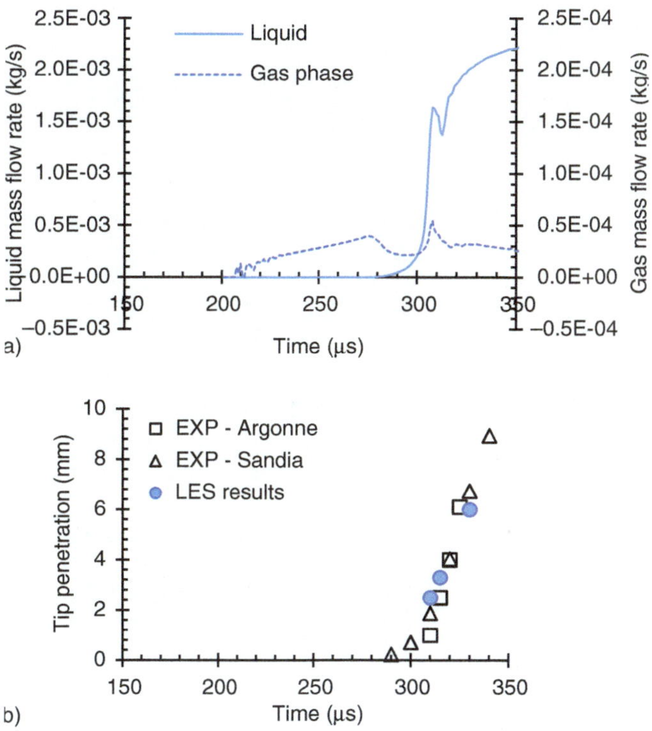

Figure 17

a) Predicted mass flow rate (liquid and gas phase) at the nozzle exit. b) Liquid tip penetration based on projected mass density. Shown are results obtained with grid 1, cf. Table 4.

region begins to fill the sac. The liquid exiting from the gap area diffuses and occupies the space available. The predicted outlet velocities from the needle-seat gap are in the order of 150 m/s (in this initial phase). At the nozzle outlet, the residual gas jet emerges in a sort of 'mushroom' shape.

Moving forward in time, at t = 250 μs a substantial amount of residual gas is still present in the sac. A small fraction of void is created at the needle-seat exit, after the expansion. Because of the off-axis location of the orifice, the liquid front does not proceed symmetrically. At this point in time, only gas phase is detected at the orifice outlet. Fine turbulent structures are observed in the gaseous jet.

At t = 270-290 μs, the residual gas concentration in the outlet chamber decreases, signifying that some high speed liquid has approached the hole exit and is passing through the residuals cloud. The near exit region, where fuel arrives, shows less turbulent structures because of the high density of the liquid phase.

At the time of t = 310 μs the liquid injection has newly started, and the jet is in an early developing stage. The features of an highly under-expanded jet are visible at the orifice outlet, with the formation of the Mach disk.

In addition, at this time liquid tip reaches the residual gases tip. Later in time, at t = 330 μs the fuel jet penetrates well beyond the gas plume and residuals are pushed away and diffused into the ambient air.

From a physical perspective, it is evident that the fuel mixes with the residual ejected gases first and with the ambient air later. It is expected that this mixing process can change the vaporization rate of the first part of the injection. In particular, since residual gases might be hotter than the ambient gases in a real engine combustion chamber, the initial liquid tip is expected to vaporize faster and produce a shorter liquid length. Further investigations are needed to support this interpretation, even if literature data do show a reduced liquid length for the first 100 μs after SOI for spray A in hot chamber conditions (Pickett *et al.*, 2013). Even other injectors reported in the literature do behave in the same way under hot conditions: *e.g.*, spray H experiments with *n*-heptane show that initial liquid length is shorter than the steady value (Som *et al.*, 2012a). The simulations shown here are restricted to a cold environment, therefore the observations based on the model can only refer to the mixing process and not to the vaporization rate. Simulations of vaporizing jets are planned for the near future, to further explore these physical processes.

4.2.3 Grid Sensitivity

Figure 19 displays the evolution of turbulent structures both inside and outside the nozzle. Two grids are compared, the main features have already been shown in Table 4. Iso-surfaces mark a suitable Q-criterion threshold that identifies vortices and maximizes their rendering. Iso-surfaces are also colored by liquid mass fraction, Y_{liq}, that gives an immediate perception of where the fuel is actually located at each time.

Looking at the outlet region, we notice that generally the two grids give a similar representation of the vortical structures, but the fine grid is able to provide more details (the residual gas jet tip at t = 250 μs, and the very near exit region at t = 340 μs).

At the bottom of Figure 19 a close-up view of the orifice is also provided. This comparison shows quite clearly that the coarse grid, using 5-10 μm resolution, is not adequate for capturing the smallest turbulent structures in the holes. On the contrary, the fine grid, that uses 2.5 μm cells near the wall and 5 μm cells in the core of the hole, is able to capture small turbulent structures in the orifice itself.

An interesting additional feature that the simulations capture is the formation of a shock wave at the tip of the liquid jet during the early stage of jet penetration, as the fuel penetrates in to the chamber. A snapshot is shown in Figure 20 for both grids. Density contours are plotted in the restricted range 20-30 kg/m^3 for better visualization. The prediction of

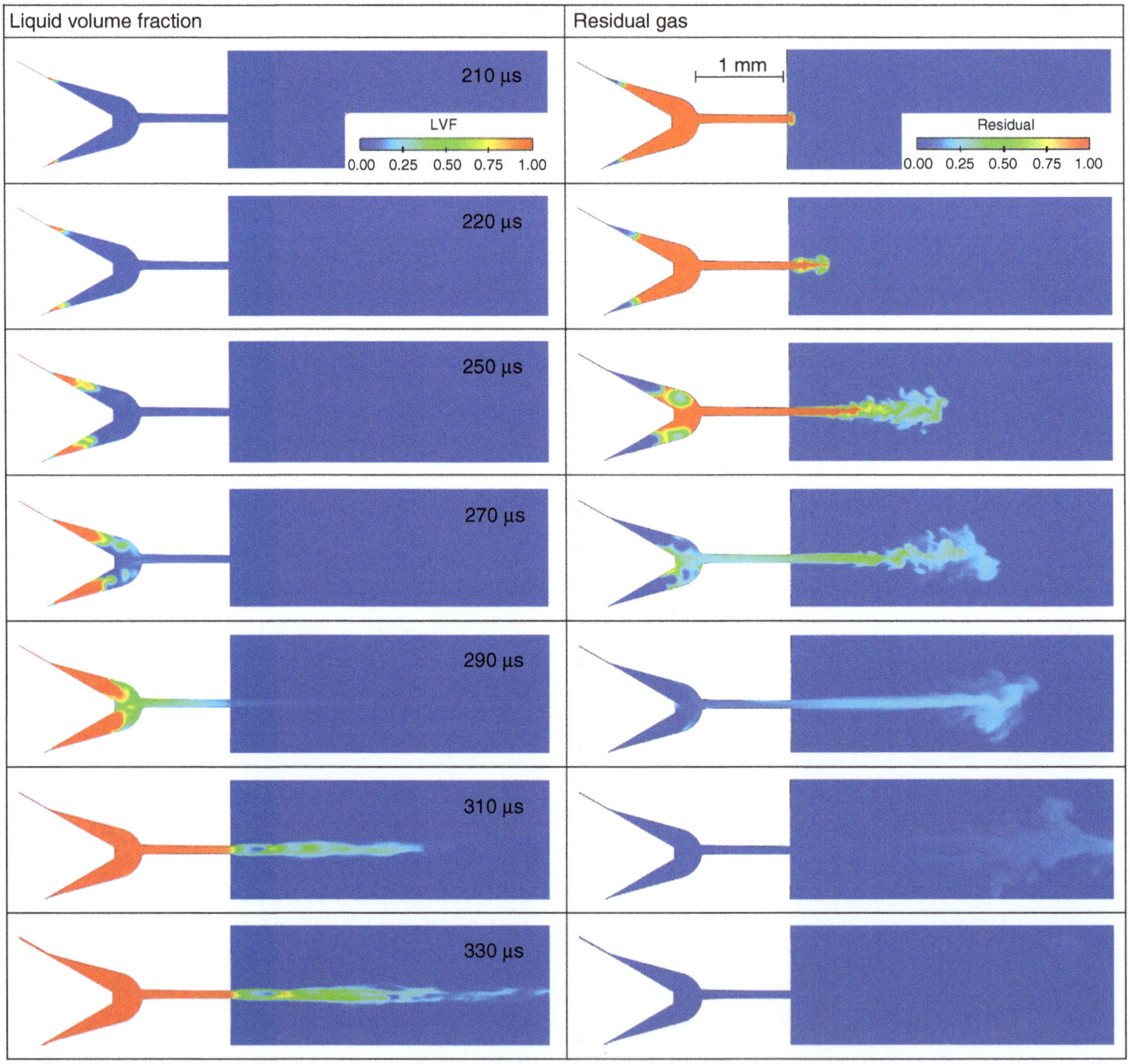

Figure 18

SOI evolution at different timings. Liquid volume fraction, left column; residual gas fraction, right column. Results refer to the fine grid (grid 2 in Tab. 4).

shock wave occurrence matches quite well with the experimental X-ray radiographic projections discussed by Kastengren *et al.* (2014), as they found shock waves at the early stage of the jet penetration, around 20 μs after the SOI, as it is seen in the simulations. LES predictions of shock location and occurrence timing are therefore in very good agreement with experiments. The formation of shock waves at the jet tip are due to a relatively cold environment and to the high liquid tip velocity which slightly exceeds the speed of sound of the ambient gas. Concerning the grid effect, we observe a slightly more intense shock wave predicted by the coarse grid compared to the fine grid at $t = 320$ μs. At $t = 325$ μs the situation appears to be reversed. Some minor differences in the liquid tip velocities with the two grids cause this behavior. It is also interesting to observe that the fine grid allows capturing several secondary waves located behind the main shock wave.

The mesh resolution analysis shown here is not complete, as grid converged results are not clearly achieved yet (cf. turbulent vortices inside the orifice in Fig. 19).

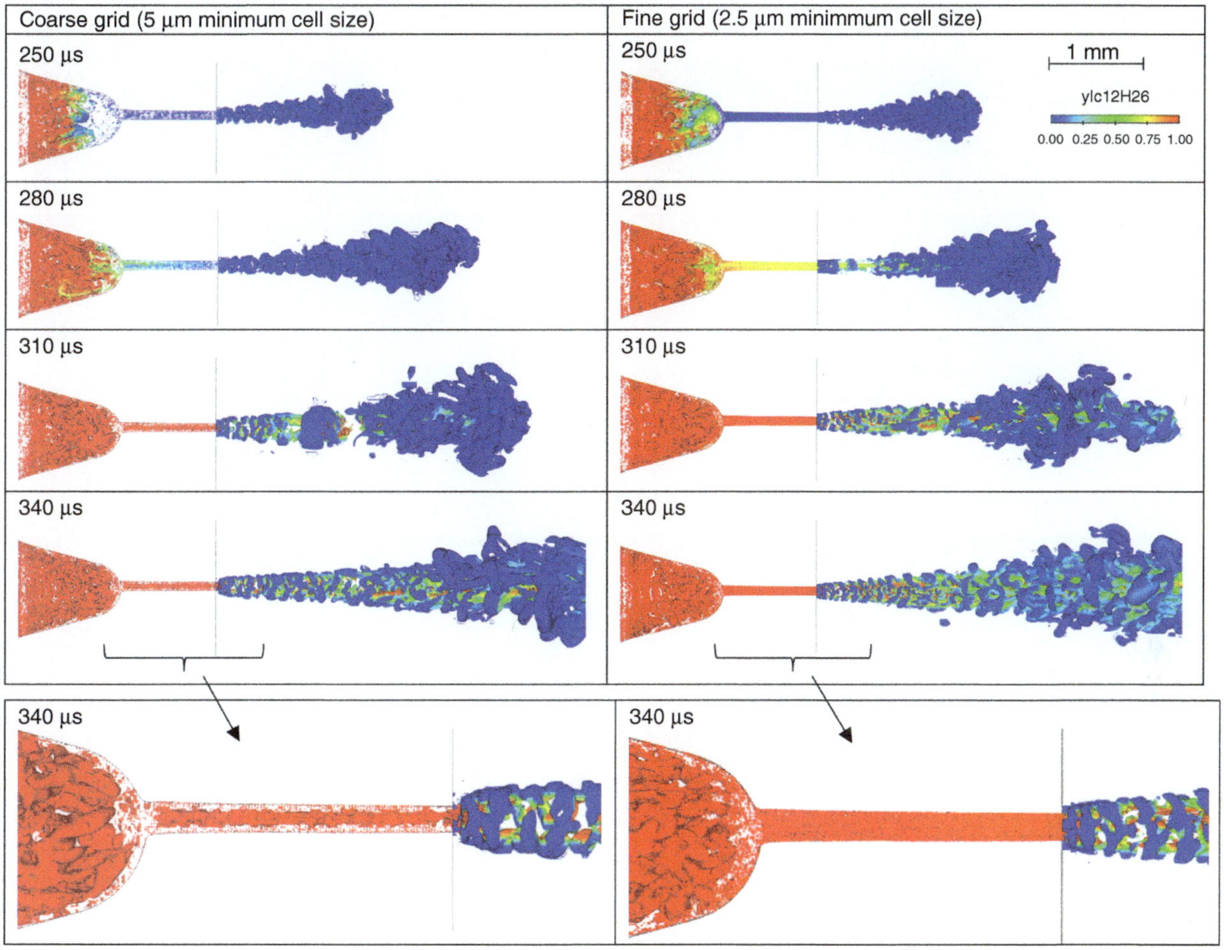

Figure 19

Turbulent structures inside and outside of the nozzle. Iso-surfaces mark a suitable Q-criterion threshold that maximizes vortex visualization. Colors refer to liquid mass fraction, Y_{liq}. Coarse grid results are displayed on the left column, fine grid on the right column.

However, from these results we believe it is possible to get some useful indications for future directions in terms of mesh requirements. In a seamless simulation of the internal nozzle flow and outward jet, the main challenges remain the strong variation in the relevant length scales involved in the various regions. If 5 μm grid resolution seems adequate to provide a reasonable description of the turbulent flow in the free jet region, the wall bounded flow that takes place inside the orifice certainly requires a much finer grid spacing, as a 2.5 μm grid size may still not be sufficient for a proper flow representation. In addition, the flow through the needle-seat passage at very low lifts (below 5 μm) is also critical and constitutes an area for future investigation. It is also acknowledged that wall-models are essential for enabling LES of realistic problems at high Reynolds numbers, as in this work. An effective wall model should be able to give reasonably accurate predictions of the wall shear stress irrespective of the grid resolution near the walls. The Warner and Wengle model here used has been proven to work quite well in various applications (Ameen *et al.*, 2015), as with different near wall resolutions the predicted wall shear stress is substantially the same. Nevertheless, inclusion of more SGS physics, like usage of non-equilibrium wall models, will be considered for future work on this subject.

The total computational cost of the simulation covering the SOI with the 15.9 million cells grid (cf. fine grid in Tab. 4) is 12 days of wall clock time on 256 cores. Simulations were run using a time step controlled by a velocity-based CFL set equal to 0.2. The resulting time-step is in the order of 1.0×10^{-9} s to 2.0×10^{-9} s. The total number of time steps required to cover 150 μs of physical time was ~80 000.

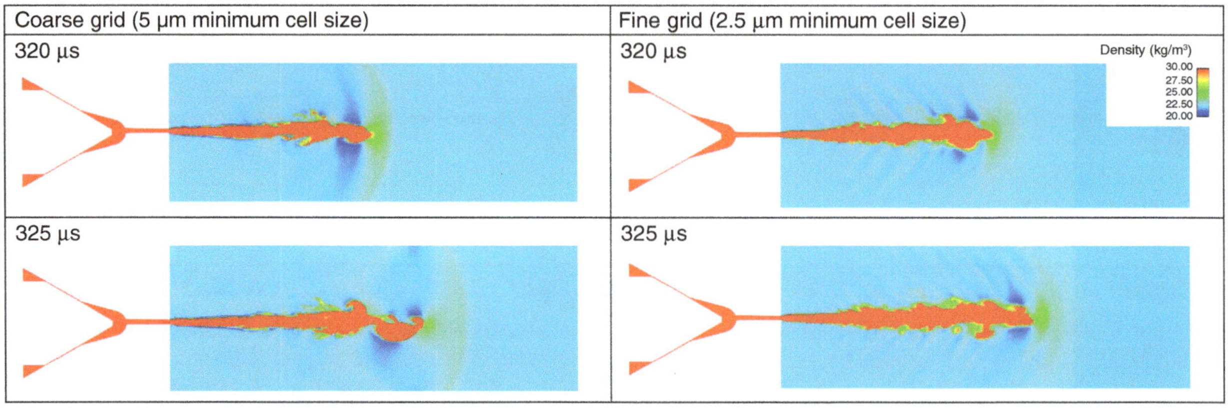

Figure 20

Shock waves at the liquid jet tip, during spray A SOI. Coarse grid, left. Fine grid, right.

CONCLUSION

The objective of this study is to evaluate the feasibility and the possible benefits of multiphase LES applied to the simultaneous modeling of in-nozzle and near exit flow in Diesel injectors. Transient part of injections are simulated in single-hole injectors using a two-phase three-components Eulerian mixture model. High-resolution (2.5 μm) LES are devised and carried out to capture phenomena at small needle valve lifts and on small orifices. The multiphase model accounts for compressibility of both liquid and gas phase, and it also includes phase change due to cavitation. Transient jet features at the SOI and EOI are simulated and results compare well with synchrotron X-ray imaging and radiography data. Simulation results demonstrate the potential to comprehensively understand the physics of SOI and EOI transients complementing experimental data.

Part of the article focuses on the description of the fast dynamics of these processes, which is resolved in great detail. Concerning the EOI behavior, liquid jet breakup, gas ingestion, in-nozzle bubble formation and dribbles are predicted. For the SOI behavior it is reported the initial ejection of residual gases trapped in the sac. Prescribing the needle motion from a very low lift, the timing of the first injected fuel is also captured. In addition, some potential effects on the initial evaporation rate and liquid length of the spray are highlighted.

Part of the article also focuses on the mesh assessment and suitability for LES of nozzle-flow and jet region. Two criteria are considered, one based on the percentage of resolved turbulent kinetic energy, and one based on the ratio between the SGS turbulent length-scales and grid spacing. Due to the high computational cost of the simulations, the collection of statistical turbulence quantities is performed in quasi-steady conditions *via* temporal averaging, and not as ensemble averaging over multiple realizations. This poses some specific challenges that will be subject of new work. Instead, the length-scale resolution index allows a local dynamic evaluation of the mesh resolution that can be used either for *a posteriori* grid assessment or for adaptive LES in the future. The grid resolutions employed, of about 2.5-5 μm results adequate for the resolution of the external spray, but may still not suffice for the extremely challenging requirements of the internal wall-bounded injector flow.

Based on these findings, future work will be devoted to the prediction of these phenomena at higher temperatures, more relevant to engine conditions, by developing a more comprehensive phase change model which is able to account for fuel evaporation. Improved wall models, particularly in wall-regions where cavitation/roughness is present and heat transfer is relevant, will be considered as well. Lastly, an equation of state for fuel density with simultaneous pressure and temperature dependency will be implemented.

ACKNOWLEDGMENTS

The submitted manuscript has been created by UChicago Argonne, LLC, Operator of Argonne National Laboratory (Argonne). Argonne, a US Department of Energy Office of Science laboratory, is operated under Contract No. DE-AC02- 06CH11357. The US Government retains for itself, and others acting on its behalf, a paid-up nonexclusive, irrevocable worldwide license in said article to reproduce, prepare derivative works, distribute copies to the public, and perform publicly and display publicly, by or on behalf of the Government. This research was funded by DOE's Office of Vehicle Technologies, Office of Energy Efficiency and

Renewable Energy, under Contract No. DE-AC02-06CH11357. The authors wish to thank Gurpreet Singh and Leo Breton, program managers at DOE, for their support.

We gratefully acknowledge the computing resources provided on Fusion and Blues, computing clusters operated by the Laboratory Computing Resource Center at Argonne National Laboratory.

REFERENCES

Alajbegovic A. (1999) Three-Dimensional Cavitation Calculations in Nozzles, *Second Annual Meeting of the Institute for Multifluid Science and Technology*, Santa Barbara, CA, March 14-18, pp. 97-103.

Ameen M.M., Yang X., Kuo T.W., Xue Q., Som S. (2015) LES for Simulating the Gas Exchange Process in a Spark Ignition Engine, *ASME paper number* ICEF2015-1002, submitted to *ASME ICEF2015*, Houston, TX, USA, Nov.

Arcoumanis C., Flora H., Gavaises M., Badami M. (2000) Cavitation in Real-Size Multi-Hole Diesel Injector Nozzles, *SAE Technical Paper* 2000-01-1249.

Arienti M., Sussman M. (2014) An embedded level set method for sharp-interface multiphase simulations of Diesel injectors, *International Journal of Multiphase Flow* **59**, 1-14.

Battistoni M., Som S., Longman D.E. (2014a) Comparison of Mixture and Multifluid Models for In-Nozzle Cavitation Prediction, *J. Eng. Gas Turbines Power* **136**, 6, 061506-061506-12, GTP-13-1436.

Battistoni M., Kastengren A.L., Powell C.F., Som S. (2014b) Fluid Dynamics Modeling of End-of-Injection Process, *Proceedings of ILASS Americas*, Portland OR.

Battistoni M., Duke D.J., Swantek A.B., Tilocco F.Z., Powell C.F., Som S. (2015) Effects of Noncondensable Gas on Cavitating Nozzles, *Atomization and Sprays* **25**, 6, 453-483.

Bianchi G., Minelli F., Scardovelli R., Zaleski S. (2007) 3D Large Scale Simulation of the High-Speed Liquid Jet Atomization, *SAE Technical Paper* 2007-01-0244.

Bilicki Z., Kestin J. (1990) Physical aspects of the relaxation model in two-phase flow, *Proc. Roy. Soc. Lond. A.* **428**, 379-397.

Bode M., Diewald F., Broll D.O., Heyse J.F., Le Chenadec V., Pitsch H. (2014) Influence of the Injector Geometry on Primary Breakup in Diesel Injector Systems, *SAE Technical Paper* 2014-01-1427.

Brackbill J.U., Kothe D.B., Zemach C. (1992) A Continuum Method for Modeling Surface Tension, *Journal of Computational Physics* **100**, 335-354.

Brennen C.E. (1995) *Cavitation and Bubble Dynamics*, Oxford University Press.

Bruneaux G. (2005) Mixing Process in High Pressure Diesel Jets by Normalized Laser Induced Exciplex Fluorescence Part I: Free Jet, *SAE Technical Paper* 2005-01-2100.

Brusiani F., Bianchi G. (2008) LES Simulation of ICE Non-Reactive Flows in Fixed Grids, *SAE Technical Paper* 2008-01-0959.

Celik I.B., Cehreli Z.N., Yavuz I. (2005) Index of Resolution Quality for Large Eddy Simulations, *ASME Journal of Fluids Engineering* **127**, 5, 949-958.

Crua C., Shoba T., Heikal M., Gold M., Higham C. (2010) High-Speed Microscopic Imaging of the Initial Stage of Diesel Spray Formation and Primary Breakup, *SAE Technical Paper* 2010-01-2247.

Davidson L. (2009) Large eddy simulations: How to evaluate resolution, *Int. J. Heat Fluid Flow* **30**, 1016-1025.

Demoulin F.X., Beau P.A., Blokkeel G., Mura A., Borghi R. (2007) A new model for turbulent flows with large density fluctuations: Application to liquid atomization, *Atomization and Sprays* **17**, 4, 315-345.

Downar-Zapolski P., Bilicki Z., Bolle L., Franco J. (1996) The non-equilibrium relaxation model for one-dimensional flashing liquid flow, *Int. J. of Multiphase Flow* **22**, 3, 473-483.

Eagle W.E., Musculus M.P. (2014) Cinema-stereo imaging of fuel dribble after the end of injection in an optical heavy-duty Diesel engine, *Thiesel Conference Proceedings*, Valencia.

Gorokhovski M., Herrmann M. (2008) Primary atomizing modeling, *Annu. Rev. Fluid Mech.* **40**, 343-366.

Habchi C. (2014) A Gibbs Energy Relaxation Model (GERM) for Cavitation Simulation, *Atomization and Sprays* **25**, 4, 317-334.

Hu B., Musculus M., Oefelein J. (2010) Large Eddy Simulation of a Transient Air Jet with Emphasis on Entrainment during Deceleration, *SAE Technical Paper* 2010-01-1133.

Issa R.I. (1986) Solution of the implicitly discretised fluid flow equations by operator-splitting, *J. Comput. Phys.* **62**, 40-65.

Kastengren A.L., Powell C.F., Wang Y.-J., Wang J. (2008) Study of Diesel Jet Variability Using Single-Shot X-Ray Radiography, *ASME J. Eng. Gas Turb. Power* 032811-1.

Kastengren A.L., Tilocco F.Z., Duke D., Powell C.F., Zhang X., Moon S. (2014) Time-resolved X-ray radiography of sprays from Engine Combustion Network spray A Diesel injectors, *Atomization and Sprays* **24**, 3, 251-272.

Musculus M. (2009) Entrainment waves in decelerating transient turbulent jets, *J. Fluid Mech.* **638**, 117-140.

Musculus M.P.B., Miles P.C., Pickett L.M. (2013) Conceptual Models for Partially Premixed Low-Temperature Diesel Combustion, *Progress in Energy and Combustion Science* **39**, 2-3, 246-283.

Oefelein J., Lacaze G., Dahms R., Ruiz A., Misdariis A. (2014) Effects of Real-Fluid Thermodynamics on High-Pressure Fuel Injection Processes, *SAE Int. J. Engines* **7**, 3, 1125-1136.

Pickett L., Manin J., Payri R., Bardi M., Gimeno J. (2013) Transient Rate of Injection Effects on Spray Development, *SAE Technical Paper* 2013-24-0001.

Piscaglia F., Montorfano A., Onorati A., Brusiani F. (2014) Boundary Conditions and SGS Models for LES of Wall-Bounded Separated Flows: An Application to Engine-Like Geometries, *Oil & Gas Science and Technology – Rev. IFP Energies nouvelles* **69**, 1, 11-27.

Pomraning E., Rutland C.J. (2002) A Dynamic One-Equation Non-Viscosity LES Model, *AIAA Journal* **40**, 4, 659-701.

Pope S.B. (2000) *Turbulent Flows*, Cambridge University Press.

Pope S.B. (2004) Ten questions concerning the large-eddy simulation of turbulent flows, *New Journal of Physic* **6**, 35.

Quan S., Dai M., Pomraning E., Senecal P.K., Richards K., Som S., Skeen S., Manin J., Pickett L.M. (2014) Numerical Simulations of Supersonic Diesel Spray Injection and the Induced Shock Waves, *SAE Int. J. Engines* **7**, 2, 1054-1060.

Rhie C.M., Chow W.L. (1983) A numerical study of the turbulent flow past an isolated airfoil with trailing edge separation, *AIAA* **21**, 11, 1525-1532.

Schmidt D.P., Gopalakrishnan S., Jasak H. (2010) Multidimensional simulation of thermal non-equilibrium channel flow, *Int. J. of Multiphase Flow* **36**, 284-292.

Senecal P.K., Richards K., Pomraning E., Yang T. (2007) A New Parallel Cut-Cell Cartesian CFD Code for Rapid Grid Generation Applied to In-Cylinder Diesel Engine Simulations, *SAE Technical Paper* 2007-01-0159.

Senecal P.K., Pomraning E., Richards K., Som S. (2013) An Investigation of Grid Convergence for Spray Simulations using an LES Turbulence Model, *SAE Technical Paper* 2013-01-1083.

Senecal P.K., Pomraning E., Xue Q., Som S., Banerjee S., Hu B., Liu K., Deur J. (2014) Large Eddy simulation of vaporizing sprays considering multi-injection averaging and grid-convergent mesh resolution, *ASME Journal of Engineering for Gas Turbine and Power* **136**, 11, 111504: 1-13.

Som S., D'Errico G., Longman D., Lucchini T. (2012a) Comparison and Standardization of Numerical Approaches for the Prediction of Non-reacting and Reacting Diesel Sprays, *SAE Technical Paper* 2012-01-1263.

Som S., Longman D.E., Luo Z., Plomer M., Lu T., Senecal P.K., Pomraning E. (2012b) Simulating flame lift-off characteristics of Diesel and biodiesel fuels using detailed chemical-kinetic mechanisms, *ASME Journal of Energy Resources Technology* **134**, 3, 032204.

Stetsyuk V., Crua C., Pearson R., Gold M. (2014) Direct imaging of primary atomization in the near-nozzle region of Diesel sprays, *Proceedings of ILASS Europe*, Bremen, Germany.

Swantek A.B., Duke J.D., Sovis N., Tilocco F.Z., Powell C.F., Kastengren A.L. (2013) End of Injection Behavior in Single Hole Diesel Fuel Injectors, *Advanced Engine Combustion (AEC) Meeting*, Southfield, MI.

Swantek A.B., Duke J.D., Tilocco F.Z., Sovis N., Powell C.F., Kastengren A.L. (2014) End of Injection, Mass Expulsion Behaviors in Single Hole Diesel Fuel Injectors, *Proceedings of ILASS Americas*, Portland OR.

Vallet A., Burluka A.A., Borghi R. (2001) Development of a Eulerian Model for the 'Atomization' of a Liquid Jet, *Atomization and Sprays* **11**, 619-642.

Wang D.M., Greif D. (2006) Progress in Modeling Injector Cavitating Flows with a Multi-Fluid Method, *ASME Paper* No. FEDSM2006-98501.

Wang Y., Qiu L., Reitz R.D., Diwakar R. (2014) Simulating cavitating liquid jets using a compressible and equilibrium two-phase flow solver, *International Journal of Multiphase Flow* **63**, 52-67.

Werner H., Wengle H. (1991) Large Eddy Simulation of Turbulent Flow Over and Around a Cube in a Plane Channel, *Proceedings of Eighth Symposium Turbulent Shear Flows*.

Xiao H., Wang J., Jenny P. (2014) Dynamic Evaluation of Mesh Resolution and Its Application in Hybrid LES/RANS Methods, *Flow Turbulence Combust* **93**, 141-170.

Xue Q., Som S., Senecal P.K., Pomraning E. (2013) Large Eddy Simulation of Fuel Spray Under Non-Reacting IC Engine Conditions, *Atomization and Sprays* **23**, 10, 925-955.

Xue Q., Battistoni M., Som S., Quan S., Senecal P.K., Pomraning E., Schmidt D. (2014) Eulerian CFD Modeling of Coupled Nozzle Flow and Spray with Validation Against X-Ray Radiography Data, *SAE Int. J. Engines* **7**, 2, 1061-1072.

Xue Q., Battistoni M., Powell C.F., Longman D.E., Quan S., Pomraning E., Senecal P.K., Schmidt D.P., Som S. (2015) An Eulerian CFD Model and X-ray Radiography for Coupled Nozzle Flow and Spray in Internal Combustion Engines, *International Journal of Multiphase Flow* **70**, 77-88.

Yu R., Kuo T., Shahed S., Chang T. (1983) The Effect of Mixing Rate, End of Injection, and Sac Volume on Hydrocarbon Emissions from a D.I, Diesel Engine, *SAE Technical Paper* 831294.

Zhao H., Quan S., Dai M., Pomraning E., Senecal P.K., Xue Q., Battistoni M., Som S. (2014) Validation of Three-Dimensional Internal Nozzle Flow Including Automatic Mesh Generation and Cavitation Effects, *ASME J. Eng. Gas Turbines Power* **136**, 9, 092603-092603-10, GTP-14-1111.

Permissions

All chapters in this book were first published in OGST, by EDP Sciences; hereby published with permission under the Creative Commons Attribution License or equivalent. Every chapter published in this book has been scrutinized by our experts. Their significance has been extensively debated. The topics covered herein carry significant findings which will fuel the growth of the discipline. They may even be implemented as practical applications or may be referred to as a beginning point for another development.

The contributors of this book come from diverse backgrounds, making this book a truly international effort. This book will bring forth new frontiers with its revolutionizing research information and detailed analysis of the nascent developments around the world.

We would like to thank all the contributing authors for lending their expertise to make the book truly unique. They have played a crucial role in the development of this book. Without their invaluable contributions this book wouldn't have been possible. They have made vital efforts to compile up to date information on the varied aspects of this subject to make this book a valuable addition to the collection of many professionals and students.

This book was conceptualized with the vision of imparting up-to-date information and advanced data in this field. To ensure the same, a matchless editorial board was set up. Every individual on the board went through rigorous rounds of assessment to prove their worth. After which they invested a large part of their time researching and compiling the most relevant data for our readers.

The editorial board has been involved in producing this book since its inception. They have spent rigorous hours researching and exploring the diverse topics which have resulted in the successful publishing of this book. They have passed on their knowledge of decades through this book. To expedite this challenging task, the publisher supported the team at every step. A small team of assistant editors was also appointed to further simplify the editing procedure and attain best results for the readers.

Apart from the editorial board, the designing team has also invested a significant amount of their time in understanding the subject and creating the most relevant covers. They scrutinized every image to scout for the most suitable representation of the subject and create an appropriate cover for the book.

The publishing team has been an ardent support to the editorial, designing and production team. Their endless efforts to recruit the best for this project, has resulted in the accomplishment of this book. They are a veteran in the field of academics and their pool of knowledge is as vast as their experience in printing. Their expertise and guidance has proved useful at every step. Their uncompromising quality standards have made this book an exceptional effort. Their encouragement from time to time has been an inspiration for everyone.

The publisher and the editorial board hope that this book will prove to be a valuable piece of knowledge for researchers, students, practitioners and scholars across the globe.

List of Contributors

C.C. Ezeuko, J. Wang, M.S. Kallos and I.D. Gates
Department of Chemical and Petroleum Engineering, Schulich School of Engineering, University of Calgary - Canada

M. Adelinet and M. Le Ravalec
IFP Energies nouvelles, 1-4 avenue de Bois-Préau, 92852 Rueil-Malmaison Cedex - France

C. Dorbath, M. Calò and L. Dorbath
École et Observatoire des Sciences de la Terre, EOST, Université de Strasbourg, Strasbourg Cedex – France

Yuta Ogura, Mao Yoshiba and Yasuo Izumi
Department of Chemistry, Graduate School of Science, Chiba University, Yayoi 1-33, Inage-ku, Chiba 263-8522 – Japan

Mohammad Sadeq Shahamat and Louis Mattar
IHS Canada, Sun Life Plaza, East Tower Suite 800, 112 -4th Avenue S.W., Calgary, AB, T2P 0H3, Calgary - Canada

Roberto Aguilera and Hamidreza Hamdi
University of Calgary, 2500 University Dr. NW, Calgary, AB, T2N 1N4, Calgary – Canada

Yacine Haroun and Ludovic Raynal
IFP Energies nouvelles, Rond-point de l'échangeur de Solaize, BP 3, 69360 Solaize – France

Zhiping Li, Dazhen Tang and Hao Xu
School of Energy Resources, China University of Geoscience, Beijing 100083 - China

Yumin Lv
School of Energy Resources, China University of Geoscience, Beijing 100083 - China
New Energy Research Center, CNOOC Research Institute, Beijing 100027 – China

Xiaozhi Chen
New Energy Research Center, CNOOC Research Institute, Beijing 100027 – China

L. Raynal, F. Augier, F. Bazer-Bachi, Y. Haroun and C. Pereira da Fonte
IFP Energies nouvelles, Rond-point de l'échangeur de Solaize, BP 3, 69360 Solaize – France

Frank Hartmann, Stefan Buhl, Florian Gleiss and Christian Hasse
Chair of Numerical Thermo-Fluid Dynamics, TU Bergakademie Freiberg, Fuchsmühlenweg 9, 09599 Freiberg - Germany

Philipp Barth, Martin Schild and Sebastian A. Kaiser
Institute for Combustion and Gas Dynamics, University of Duisburg-Essen, Lotharstraße 1, 47057 Duisburg – Germany

J.X. Zhou
College of Mechanical and Electrical Engineering, Wenzhou University, Wenzhou 325035 - PR China

B. Moreau, C. Mounaïm-Rousselle and F. Foucher
Université d'Orléans, laboratoire PRISME, 8 Rue Léonard de Vinci, 45072 Orléans – France

Stefan Buhl, Frank Hartmann and Christian Hasse
Chair of Numerical Thermo-Fluid Dynamics, TU Bergakademie Freiberg, Fuchsmühlenweg 9, 09599 Freiberg – Germany

Maxine Akhurst, Sarah D. Hannis, Martyn F. Quinn and David Long
British Geological Survey, West Mains Road, Edinburgh, EH9 3LA - UK

Ji-Quan Shi
Imperial College London, South Kensington Campus, London SW7 2AZ - UK

Marielle Koenen and Maarten Pluymaekers
TNO, Princetonlaan 6, 3584 CB Utrecht - The Netherlands

Florence Delprat-Jannaud, Jean-Claude Lecomte and Daniel Bossie-Codreanu
IFP Energies nouvelles, 1-4 avenue de Bois-Préau, 92852 Rueil-Malmaison Cedex - France

Stanislaw Nagy and Łukas Klimkowski
AGH University of Science and Technology, 30 Mickiewicza Avenue, 30-059 Krakow - Poland

Davide Gei
Istituto Nazionale Di Oceanografia E Geofisica Sperimentale, Borgo Grotta Gigante, 42/C - 34010 Sgonico (TS) – Italy

Vincenzo Vaiano, Giuseppina Iervolino, Giuseppe Sarno and Diana Sannino
Department of Industrial Engineering, University of Salerno, via Giovanni Paolo II, 132, 84084 Fisciano (SA) - Italy

Luigi Rizzo
Department of Civil Engineering, University of Salerno, via Giovanni Paolo II, 132, 84084 Fisciano (SA) - Italy

Julie J. Murcia Mesa
Universidad Pedagógica y Tecnológica de Colombia, Avenida Central del Norte, Tunja, Boyacá - Colombia
Instituto de Ciencia de Materiales de Sevilla (ICMS), Consejo Superior de Investigaciones Científicas (CSIC), Universidad de Sevilla, Américo Vespucio 49, 41092 Sevilla - Spain

Maria C. Hidalgo and Jose A. Navío
Instituto de Ciencia de Materiales de Sevilla (ICMS), Consejo Superior de Investigaciones Científicas (CSIC), Universidad de Sevilla, Américo Vespucio 49, 41092 Sevilla - Spain

Qingluan Xue and Sibendu Som
Argonne National Laboratory, Chicago, IL - USA

Michele Battistoni
Argonne National Laboratory, Chicago, IL - USA
University of Perugia, 67 via G. Duranti, 06125 Perugia - Italy

Index